下沙村，张云摄于 2023
Xiasha Village, photograph by Zhang Yun, 2023

and the other is an urban village which formed spontaneously, growing wildly

白石洲，2021 © 都市实践
Baishizhou, 2021 ©URBANUS

南头古城，张超摄于 2022
Nantou Old Town, photograph by Zhang Chao, 2022

has brought city and village together: one is a modern new city

认购
88
日丰管

岗厦村，孟岩摄于 2006
Gangxia Village, photograph by Meng Yan, 2006

白石洲，2011 © 左氏文化
Baishizhou, 2011 ©Zeus Culture

shadows creating the strange phenomenon of two heterogeneous urban states

信豪家具
批发零售
各类坚果
瓜子杏仁
各地特产
潮汕甜汤
电话：26600069
新旧买卖
成长快
旅馆
湘菜馆
高价回收 家私家电
各类火锅
新
旧
货

vibrant intensity, urban villages comprise a stock of spaces that can still inspire

白石洲，孟岩、曲韦世摄于 2010、2013 © 都市实践
Baishizhou, photograph by Meng Yan, Travis Bunt, 2010, 2013 ©URBANUS

安信电脑
维修中心
电脑出售 回收旧电脑
宽带上网 50元包月
电 13530017223
话 15323717535
便利店转让
单房招租
新房
新楼
本楼
13714622247
免手续费
天天1.5角
精洗8元/桶
干洗
XINGX
宽带上网
88833557
中医推拿

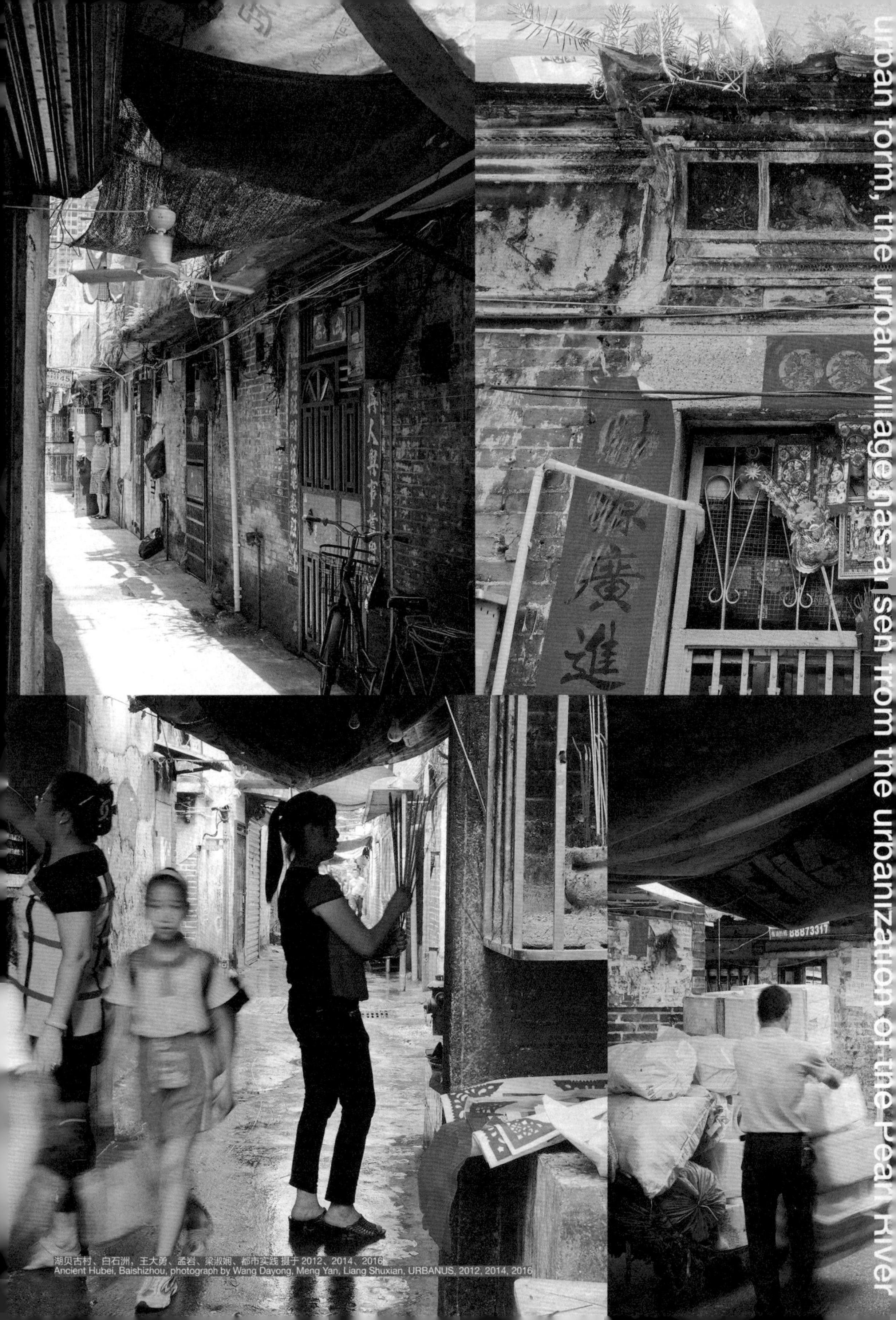

湖贝古村、白石洲，王大勇、孟岩、梁淑娴、都市实践 摄于 2012、2014、2016
Ancient Hubei, Baishizhou, photograph by Wang Dayong, Meng Yan, Liang Shuxian, URBANUS, 2012, 2014, 2016

老川菜馆

大芬油画村，孟岩、都市实践摄于 2007、2009、2020、2021
Dafen Oil Painting Village, photograph by Meng Yan, URBANUS, 2007, 2009, 2020, 2021

but also a dense spatial texture that weaves together a complex, large and

深圳市
承接订单
现货清仓
100
元

南头古城、白石洲，张超、黄扬、孟岩、都市实践摄于 2010、2016—2018
Nantou Old Town, Baishizhou, photograph by Zhang Chao, Huang Yang, Meng Yan, URBANUS, 2010, 2016-2018

cultures and diverse populations. Neither village, nor city, it is both village and

小偷
摄像头
旺旺
全新家私
批发
零售
专业租房 搬家 网线
水电维修 疏通下水道
旧货回收 家电维修
旧货买卖
回收傢俬家电
I MISS YOU!
业主写字楼厂房
13650874121
13650874121
业主写字楼厂
1365087412
共你
CITY GROW TOGETHER
IKEA
新房

沙井村，孟岩摄于 2021
Shajing Village, photograph by Meng Yan, 2021

白石洲，2014 © 都市实践
Baishizhou, 2014 ©URBANUS

drop of dramatic change, power and capital have joined forces, making a large

大芬村，2021 © 都市实践
Dafen Village, 2021 ©URBANUS

undergo a one-time complete eradication and re-formatting. Con[illegible]ted wi

白石洲，2021 © 都市实践
Baishizhou, 2021 ©URBANUS

heterogeneity have wrought, people wonder if an alternative exists.

南头古城，张超摄于 2022
Nantou Old Town, photograph by Zhang Chao, 2022

城村共生 URBAN VILLAGE COEXISTENCE

都市实践编著
EDITED BY URBANUS

上 海 文 化 出 版 社
SHANGHAI CULTURE PUBLISHING HOUSE

城村共生：
中国城市的另一种未来图景

孟岩

深圳一直被描述为一座奇迹般速生之城，它的面貌如此年轻，似乎未曾承载任何厚重的过往。其实，在它变身为“城”之前，有着悠久历史的“村”一直存在。深圳的高速发展让“城”与“村”相遇：一个是自上而下，依照经济理性和社会理想规划出来的现代新城；另一个则是自下而上，借力于前者的发展自发形成并野蛮生长的城中村。二者如影随形，铸就了两种异质的城市状态并行不悖、相互缠绕、共同生长的奇特现象。城中村聚合了超乎想象的城市密度和活力强度，是平庸的当代城市中仍然能激发无限想象的空间存量。

城中村是珠三角地区城市化进程中产生的特殊聚落形式，不仅承载着源自农耕时代悠久的人文历史，其致密的空间肌理中更是编织了一个复杂、庞大和自我组织的社会关系网络。城中村包容了多元的人口和杂糅的文化，它非村非城，亦村亦城。它自身不断演进，极具活力和创造力。可以说城中村是在快速城市化进程中演化、变异而成的、独特的当代城市遗产。尽管如此，在过去的数十年间，对现代性的执念和对单一未来线性发展观的信奉，使得人们选择对这座城市的另一部分视而不见；在权力与资本的合力下，大量“非正规”的城市形态正在经历一场对历史层积和多元现实的一次性清除与“格式化”。面对这幅不断消除异质、快速迈向同质化的未来城市图景，人们不禁要问，是否存在另一种具有替代性的可能？

深圳已然步入一个城市发展的新阶段，亟须寻找新的理论和新的模式。以“城市共生”为主题的 2017 年深港城市 \ 建筑双城双年展（深圳）（简称 2017 深双）便是一次描绘不同于主流想象的中国城市未来图景的尝试，策展理念“城市共生：从城中村开始⋯⋯”，宣告一个“城村共生”时代的来临。这场基于南头古城的观念实验与介入实践不止步于一次对城中村——这一中国城市化特殊样本的全方位解析，更着眼于当今世界和中国城市化的现实，并以城中村为参照进行批判性解读。对城中村的研究和介入实践并不局限于城中村本身，更不是一个怀旧的话题，城中村提供了一个不同的视角，让人们重新审视当今城市的处境与明日都市的构想，城中村的未来映射着城市的未来。这场双年展汇集了鲜活立体的全球智慧和在地经验，为期三个月的展览将城中村推到城市舞台的最前沿。这与城中村话题在首届深双的亮相时隔 12 年，期间许多城中村已被高楼取代。

深圳作为中国城市化最为剧烈的现场，其特殊的历史背景、别样路径和未来模型能否为当代城市提供更加多元且更具想象力的样本？“城市共生”以城市策展为媒介，在南头这个千年古城和当代城中村的共生体中，成就了一次行动上的动员和观念上的碰撞。在后 2017 深双的当下，如何借助研究者和实践者的多元视角，在宏大叙事与个人叙述的交织中，建构一个承载了历史、现实和未来的“城村共生”愿景，并进一步推进其融入当代城市的蓝图就显得愈发迫切。

《城村共生》一书既是一部回顾性的文献，也是对城市未来的思考。它是对第七届深双（南头）“城市共生”实验一次必要的学术梳理和延展，旨在于当下中国城市发展的拐点，秉持“城村共生”理念，修复城中村与地方历史的链接，重新定义和拓展已经固化的城、村二元分化的空间与社会场域的边界。本书试图构建一个关于城中村和城市未来的话语空间。这里汇聚了错综交叠的不同观点：不仅包括规划师和建筑师以及地理学、社会学、人类学、经济学、管理学等领域的学者，也包括租户、原村民与村集体、各级政府、地产、运营商的代表，还包括以城中村为背景进行创作的艺术家。本书以“城市档案”“非常经验”“特别实验”和“都市实践”四个章节为叙事线索，从文献研究的宏观叙事、人文观察的集体和个人记忆到多元的实践经验， 展开各个学科视角下的城中村研究，力图兼顾各利益主体的代表性言论以及专业团队的介入实验和方法。

《城村共生》一书不仅是一场展览的档案记录，更试图为城中村与城市发展研究提供具有多重视角的知识库和工具包。在城中村有机更新的新阶段，以地产开发为主导的发展模式已经捉襟见肘，而从地产思维转向城市思维则是必然选择。在中国城市化进程的当下，“城村共生”下的“城”与“村”正在逐渐走出二元对峙，走向相生与相融的第三类型——“城村”。“城村共生“是一个新的起点、新的宣言和新的栖居类型，力图重建一种杂糅与共生的城市多元主义体系。愿借此书的出版感召更多的学者和实践者，加入这一关于城市发展模式的讨论和实践，并由此出发共同谋划、想象和描绘中国城市未来的另类图景。

Urban Village Coexistence: An Alternative Vision of the Future Chinese Cities

Meng Yan

Shenzhen has always been described as a miracle city of rapid growth, so young that it seems to have never experienced a heavy past. In fact, before it became a "city," a "village" with a long history existed. Shenzhen's rapid development has brought "city" and "village" together: one is a modern new city planned from the top down according to economic rationality and social ideals, and the other is an urban village which formed spontaneously, growing wildly from the bottom up in interaction with the former. The two follow each other like shadows, creating the strange phenomenon of two heterogeneous urban states growing in parallel and intertwined. Combining unimaginable urban density and vibrant intensity, urban villages comprise a stock of spaces that can still inspire infinite imagination despite the banality of the contemporary city.

As a special urban form, the urban village has arisen from the urbanization of the Pearl River Delta region, which not only has a long human history that dates to the agrarian era, but also a dense spatial texture that weaves together a complex, large and self-organized network of social relations. The urban village embraces a mix of cultures and diverse populations. Neither village, nor city, it is both village and city, evolving itself, full of vitality and creativity. Urban villages comprise a unique contemporary urban heritage that has evolved and mutated during the process of rapid urbanization. Nevertheless, a long-standing obsession with modernity and a belief in the linear development of future have led people to choose to ignore this other part of the city. Over the past few decades, against the backdrop of dramatic change, power and capital have joined forces, making a large number of "informal" urban forms despite historical layers and multiple realities undergo a one-time complete eradication and re-formatting. Confronted with the gradual homogenization of the future urban landscape that the elimination of heterogeneity have wrought, people wonder if an alternative exists.

Shenzhen has entered a new stage of urban development and urgently needs a new theory and model. Through the theme of "Cities, Grow in Difference," the 2017 UABB (Shenzhen Hong Kong Bi-City Biennale of Urbanism\Architecture (Shenzhen)) offered a picture of China's urban future that differs from the mainstream imagination, declaring the arrival of the era of "urban village coexistence." In addition to an all-encompassing analysis of the urban village as a particular typology of Chinese urbanization, this conceptual experiment and intervention practice started in Nantou Old Town looked at the reality of urbanization worldwide and in China today, critically interpreting these processes with respect to urban villages. The research on urban villages presented in the exhibition did not limit itself to the urban villages themselves, much less interpret them as objects of nostalgia. Urban villages provide a different perspective for re-examining the situation of today's cities and the vision of tomorrow's cities—the future of the urban village maps the future of the city. Bringing together vibrant global wisdom and local experiences, the three-month exhibition brought urban villages to the forefront of the urban scene. This happened twelve years after the issues of urban villages debuted at the first Shenzhen Biennale. In the time between those two editions of the Biennale, many of those urban villages were replaced by high-rise buildings.

As the site of the most dramatic urbanization in China, Shenzhen's special historical background begs the question: do its different paths and future models provide a more diverse and imaginative sample for contemporary cities? Through the medium of urban curation, "Cities, Grow in Difference" mobilized action and conceptual collision in the symbiosis of Nantou, a thousand-year-old city and a contemporary urban village. In the post-2017 UABB era, it has become ever more urgent to construct a vision of "Urban Village Coexistence" that accommodates history, reality and future, interweaves grand narratives and personal accounts with the multiple perspectives of researchers and practitioners, and promotes their integration into the contemporary urban blueprint.

The book, *Urban Village Coexistence* is both a retrospective document and a consideration of the future of the city. It is an essential academic review and extension of the 7th UABB "Cities, Grow in Difference" experiment in Nantou. Written at the pivotal moment in China's contemporary urban development, the book foregrounds the ideology of "urban-village co-existence," aiming to restore the link between urban villages and local history, and to redefine and expand the boundaries of the spatial and social field that have been solidified as the urban-village dichotomy.

Urban Village Coexistence attempts to construct a discourse about the urban villages and the future of cities. Different and interrelated viewpoints are brought together: including not only scholars from the fields of architecture, urban planning, geography, sociology, anthropology, economics and management but also tenants, original villagers, village collectives, governments at all levels, relevant real estate and property managers as well as artists inspired by urban villages. *Urban Village Coexistence* is organized into four sections: "City Files," "Tool Kits," "Test Ground," and "URBANUS." These narrative threads weave through macro narratives of documentary research, humanistic observations of collective and personal memory, pluralistic practical experiences, and studies of urban villages from various disciplinary perspectives, representative discourses of interested parties, and intervention experiments and methods of professional teams.

Urban Village Coexistence is not only an archival record of a variety of voices from the exhibition, but also an attempt to provide a multi-perspective knowledge base and toolkit for the study of urban villages and urban development. Given that the property development-led renewal model has been stretched to its limit, we have entered a new phase of exploring the organic regeneration of urban villages, making the transformation of urban development from real estate thinking toward urban thinking inevitable. In the current urbanization process in China, instead of settling into a binary confrontation between city and village, *Urban Village Coexistence* looks toward a different typology in which the two are integrated as Urban-village. *Urban Village Coexistence* is a new start, a new manifesto, and a new dwelling typology. It attempts to rebuild a hybridized and symbiotic urban system that celebrates diversity and inclusiveness. Therefore, it is hoped that the publication of this book will call more scholars and practitioners to join the discussions and practices of the "urban-village co-existence" development model, and begin to jointly plan, imagine and depict an alternative vision of the future of Chinese cities.

目录

CONTENTS

城市档案

CITY FILES

捕捉历史的多重叙述

Catching the Diverse Narratives of History

不同于绝大多数的城市，深圳在塑今的同时溯源，它既建立在一种没有历史的幻象中，又试图基于这种幻象构建一种历史观。然而，在广泛传播的官方文本以外，关于它的记录繁复、游移、含糊，甚至相互矛盾与冲突。特区的超高速发展产生了两种并置的城市化现实，相斥又相生，反映在自上而下规划生成的城市和自下而上自发形成的城中村这两种不同肌理的城市形态的对峙中。这幅城市图景已然拒绝了单一视角、单一线索、单一声道对其化约，“城市档案”的写作也必然无法绕开其发展历程中的城中村坐标。我们试图在赞誉与争议中解读城中村的真正意义，由此反观并重新思考今日城市生活的本质。

《城村大事纪》（见折页）呈现深圳城市和城中村在多种因素影响下的生成和演化脉络，以及它们在数个转折时期的空间形态。《续自发中国——新常态下城乡关系的韧性转型》深度剖析城、村的“自发”与“自觉”现象，论述二者交锋融合关系的演变和产生的影响。《对城中村进行未来想象的考古学》从城中村的改造中解析关于深圳的多种想象以及它所错过的另一种城市未来。《城市规划与城中村：谁来改造谁？》讲述从以城市规划方法看城中村到以城中村现象看城市规划的转变，进而提出借鉴城中村的经验以实现城市理想和解决城市问题。《关于城中村的一种文献综述》从多个尺度梳理城中村的定义，从而点明城中村研究课题中个案与区域、局部与整体、地方与全球的具体关系。

Unlike most cities, Shenzhen shapes the present by building upon the illusion of a city without history, even as it creates an origin story based on this illusion. However, beyond the widely known official story, records of the city are complicated, uncertain, vague, and even contradictory. The rapid development of the Special Economic Zone (SEZ) has produced two juxtaposed urban realities, which simultaneously resist and reinforce each other. These two realities are manifested in the confrontation of the top-down planned city and the bottom-up informal urban villages. This city refuses to be simplified into a single perspective or a single story. Hence the articles of City Files need to include urban villages as coordinates within the city's urbanization process. We attempt to understand the nature of urban villages through praise and critique, speculating and rethinking the spirit of today's urban life.

"Chronology of Shenzhen City and Urban Villages" (see pamphlet) illustrates the formation and evolution of Shenzhen's formal urban areas and its urban villages with respect to various factors as well as portrays their spatial forms at several turning points. "Informal China II: Resilient Transformation of Urban-Rural Relations in the New Normal" breaks down the urban phenomenon of formal city and informal village, and discusses the development and impact of their conflicting yet integrated relationship. "Excavating the Future as Imagined in an Urban Village" analyzes how various futures have been implemented and overlooked in the transformation of urban villages. "Urban Planning Vs. Urban Villages, Who Should Transform Whom?" describes a shift in perspective from defining urban villages by the means of urban planning to rethinking urban planning through the lens of urban villages. It also further proposes to resolve urban issues and achieve urban ideals by learning from urban villages. "A Literature Review on the Urban Village" studies the definition of the urban village on multiple scales, thereby identifying specific associations in urban village studies between individual cases and the larger area, part to whole, and local practices to global discourse.

续自发中国
——新常态下城乡关系的韧性转型

姜珺
研究型建筑师、文献编辑、独立策展人，现任中国美术学院教授、社会与策略研究所所长，曾任《城市中国》杂志创刊主编（2005—2010），莫斯科史翠卡建筑设计与媒体学院课题导师（2010—2011），牛津大学 ESRC 中心访问学者（2011—2012），威尼斯国际建筑双年展中国馆总策展人（2014），“蛇口议事：2025”策展人（2015）。
本文由莫思飞中译英。

2006 年，名为“中国当代”的建筑大展在荷兰建筑学院开幕，是中国当代实验建筑师在海外第一次成规模的集体亮相（图 1）。在此半年前，策展人琳达 · 弗拉斯乌曾邀请包括笔者在内的一个专家团队前往鹿特丹提供策展咨询。我的建议是在展示中国当代建筑之外，考虑与之相关的三个关系：“内容与语境”，作为“内容”的“自觉与自发”，以及作为“语境”的“现实与理想”。建议被采纳，并在之后的展览中呈现为由 OMA 设计的一个空间结构：悬空和落地的两组矩阵，上下分别对应“语境”与“内容”。展览内容中，我通过《城市中国》杂志的同名特刊（图 2），为落地的“内容”提供了一组名为“自发中国”的民间设计，以对应实验建筑师们的自觉设计，并为悬空的语境部分提供来自数百个城市的上千幅场景，对应一组由水晶石公司提供的同样数目的建筑效果图，作为城市中国的“现实”与“理想”（图 3）。

展览开幕同期，荷兰的 *Volume* 杂志发布题为《条条大路通中国》的中文特刊以作策应（图 4）。我撰写了以《自发中国》为题的同名文章，但不同于参展作品的副标题“物品 / 建筑 / 城市在中国的潜意识发明”，文章副标题名为“一部控制与失控的历史”，以一组纵贯 2000 年中国历史的横向制度对比，包括战国的中央集权与古代资本主义、唐宋的宵禁与解禁、元朝初末的游牧与农耕、明清的海禁与特区、民国的租界与黑社会、共和国前三十年的计划与自发，以及后三十年的军队与特区等，粗线条勾勒潜藏于中国文明演进中穿朝越代的治理困境。虽以“自发”为名，文章的实质却是正式与非正式、组织与自组织的间性与张力。结合参展作品对“物品 / 建筑 / 城市在中国的潜意识发明”的谱系化，我只是希望表达一个容易忽略的常识：一切显见范畴都是潜在秩序的（有意识或无意识的）呈现。因此，如果治理可以定义为一种集权与分权之间取得平衡的技术，“物品 / 建筑 / 城市”等显见范畴则是可以为之提供“考现”的公共领域，这也成为当时我创办《城市中国》的理念。

《自发中国》的末章为“全球化与共产主义”，将时间定格在中国加入以世界贸易组织为代表的全球体系的 2001 年，既是传统天下体系的历史延续，也是国家与资本持续博弈的大政之始。文末将之形容为“将全球语境可能带来的变量作为‘中国特色’纳入‘社会主义’恒量的历史进程”，并对此持观望态度。此后，中国历经 2007 年股灾、2008 年华尔街金融危机、2015 年股灾、2018 年中美贸易战等一系列外源性危机。时至今日，判断国家与资本博弈胜负还为时尚早，从国家在国际变局下对“国内大循环”[1] 的频繁表态中我们却不难预见：始于 2001 年的“全球化 1.0”已是危若累卵，在被更可

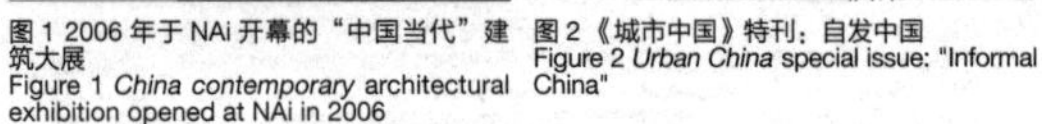

图 1 2006 年于 NAi 开幕的“中国当代”建筑大展
Figure 1 *China contemporary* architectural exhibition opened at NAi in 2006

图 2《城市中国》特刊：自发中国
Figure 2 *Urban China* special issue: "Informal China"

1　国内大循环并非“闭关锁国”，而是“国际国内双循环”之主体。中国因其特殊的自然条件，可在基本建设领域将过剩的生产力通过财政负债、金融投资等方式进行激活，而当下海量的生产性资产能否激活恰是“大众创业、万众创新”的重要基础，也是该决策的前提。

持续的“全球化2.0”取代之前，一个充斥着新冷战气氛的“去全球化”过渡期或许在所难免。因此在“革故”与“鼎新”的扑朔迷离之间，通过《自发中国》的空间视野，尤其是通过2006年之后加速的城乡关系演进，分析“城市-乡村”关系背后的“规划-市场”“实业-金融”“国家-资本”“公有-私有”等方面渐趋韧性化的制度设计，续写“全球化与共产主义”这一未尽的章节或许正当其时。更重要的是，是否可以从中国入世20年的治理轨迹中，一瞥“全球化2.0”的雏形。

1 土地：国家与集体

《城市中国》杂志始于2003年，创刊号主题为《中国式造城》（图5），“中国式”即“中国特色”。这一年的两件事对中国的城乡现代化进程产生了深远影响：7月，央行启动“721”汇改，天量外资在人民币升值预期下大举进入房地产，中国城市化加速；10月，《中共中央关于制定国民经济和社会发展第十一个五年规划的建议》提出“建设社会主义新农村”。次年元旦，延续千年的农业税被历史性地宣告废除，中国全域性的乡村建设由此启动，《城市中国》随后发布《社会主义新农村》专辑（图6）。中国的城市化，意味着乡村与城市的此消彼长，而与乡村中国相关的主题，几乎构成这辑杂志主题的半壁江山，包括中国制造、移民、土地、气候变化、危机管理等，事实上是将“中国式造城”和“新农村”作为这一进程的一体两面。“中国式造城”历经1994年分税制、1998年房改、2003年土地招拍挂[2]，逐渐将楼市构建为“以房地产信贷对标货币增发”的资金池，将天量流动性涌入中国所造成的超级通胀锁定其中，同时将“人民币升值预期”转化为“房地产升值预期”，进而通过“地方政府土地财政”和“中央财政转移支付”，将土地收益“溢价归公”并投入于基础设施建设和乡村建设，从而形成数十万亿的“新农村”投资。今天看来，2005年的里程碑意义，正在于将“新农村”建设平行于大规模的“中国式造城”，以缓和在依附性原始积累中高度紧张的城乡关系；而全域性的“城乡再平衡”（不只是局域性的“城村再平衡”）则为之后数次缓解外源型危机的冲击提供了系统化的韧性。

真正的中国特色，在于这种城乡发展得以平行推进背后的土地制度。中国的土地制度筑基于20世纪50年代的建政之初，历经土地改革和社会主义改造，形成城市国有土地和农村集体土地的二元所有制结构，进而通过城乡二元户籍制度将人地挂钩，形成土地基础上的“大国家”和“强集体”，是民主主义革命和国家权力主导下的制度变迁结果。作为一个有着数千年农耕历史和海量人口、周期性陷于内卷的后发现代化国家，通过城乡二元的“剪刀差”[3]汲取农业剩余，成为实现社会主义原始积累的唯一选择。中华人民共和国前后历经两次“剪刀差”：一次是前期通过“工农业产品剪刀差”提取农业剩余以支持工业化，一次是后期通过“城乡土地剪刀差”以支持城市化。前期的主体是中央政府，后期的主体是地方政府。1994年分税制之后，在中央政府增大税收比例的同时，地方政府可以通过垄断土地一级市场获取土地收益。通过“土地财政”主导土地供给与城市开

图3 分别作为城市中国“现实”与“理想”的“民间设计”（自发）与“大师设计”（自觉）
Figure 3 Informal design as "Reality" and formal design as "Ideality" of urban China

发、放大级差地租，地方政府得以成为对地方经济深度介入的准市场化组织，一方面将来自公地出让的收益投入于平台性、重资产的基础设施建设和公共服务，提高城市在招商引资方面的竞争力；另一方面可以通过地方国企介入城市规划和开发，将国有资产作为级差地租溢价的收益主体，进而降低地方政府财政对税收的依赖，既体现了“中国特色”的“地方竞争”和“市场收益”，又体现了“社会主义”的“溢价归公”。二者在推进“中国式造城”的同时，对“中国制造”形成价格优势提供了正式或非正式的补贴。地方政府土地财政越自主，中央政府财政转移支付就越能面向更广泛的欠发达地区，相对于前者直接的“溢价归小公”，后者是间接的“溢价归大公”。“溢价归公”最早来自孙中山的民生主义，其思想源头则可追溯到19世纪末的美国经济学家亨利·乔治[4]，他的“单一土地税”理论中所内涵的“地租社会化”思想，在一个世纪后“将全域化的新农村建设平行于大规模的中国式造城”的中国实践中得到参照。

2 韧性：依附与自主

尽管土地财政极大地促成了中国的地方原始积累，但是它促成的“高地价-高房价”的马太效应也广受诟病。2017年雄安新区横空出世并定位为“千年大计”，表面上看是“疏解北京非首都功能”，内里则是中国决心以实业立国并开启土地财政转型的号角，是经济发展模式由“依附”转向“自主”的昭示。回顾在1994年、2005年、2015年二十多年三次汇改期间步步做大的土地财政，本质是将外向型的“中国制造”在持续顺差下生成和放大的天量流动性，锁定在楼市的“权宜之计”，以避免超

2 “分税制”“房改”和“土地招拍挂”及其相关内容详见于《国务院关于实行分税制财政管理体制的决定》《国务院关于进一步深化城镇住房制度改革加快住房建设的通知》和《招标拍卖挂牌出让国有土地使用权规定》众政策。

3 “剪刀差”概念产生于20世纪20年代的苏联，并于30年代介绍至中国。针对较为特殊的中国国情及其需求，此概念已被发展和广义化。国内学者普遍认为，“剪刀差”是指在工农产品交换的过程中，因工业产品价格高于其价值、农业产品价格低于其价值导致非等价交换，从而形成的“剪刀状差距”。参见：孔祥智，何安华. 新中国成立60年来农民对国家建设的贡献分析 [J]. 教学与研究，2009(9):5.

4 亨利·乔治，美国经济学家和土地改革家。在1879年发表的《进步与贫困》一书中，他提出了一套以“土地”为核心的经济学，建议国家针对所有因使用土地而获得的收入征税，并主张废除其他税种。乔治认为，政府将从该“单一税”（single tax）中获得巨量的财政收入，其盈余亦可用于公共建设（地租社会化）。

图 4 *Volume* 中文特刊：条条大路通中国
Figure 4 *Volume* special issue: "All Roads Lead to China"

级流动性涌入对通胀更敏感的日用品市场。“中国制造”始于与“1994 年税改”同年的“1994 年汇改”，人民币与美元挂钩并大幅贬值，再通过将城市的基础设施、公共服务与乡村的廉价劳动力相结合，对全球产业资本形成虹吸效应，通过贸易顺差和出口结汇，对冲增发为具有高度依附性的基础货币；“2005 年汇改”之后的“中国式造城”则通过“土地金融”的内部机制，创造出天量内生性的“土地货币”，以在美元周期性波动的潮汐效应中保持相对自主。同时通过与“中国式造城”相平行的“2006 年新农村”和“2013 年美丽乡村”建设，将“2015 年汇改”之后的乡村打造为下一个堪比“中国式造城”体量的资本市场，一个建立在中国广袤生态人文资源基础上的货币内生机制。时至今日，在《中国制造 2025》[5] 的“近未来”预期下，中国已成为全球产业门类整全度最高的头号工业大国，以“中国制造”为主体的国际大循环为“进”，以“新农村”为主体的国内大循环为“退”。“中国式造城”的历史作用在于以其货币内生机制沉淀货币主权，为建构城乡二级资本市场、在两个大循环之间的“可进可退”中积累实力并赢得时间。由此，“二元一体”的土地制度基础上的中国制造、中国式造城与新农村三大战场（对标出口、消费、投资三驾马车），构成中国模式可进可退的韧性机制。

“韧性”概念最早由加拿大生态学家霍林[6] 提出，指系统对外界风险变化的抵御、适应与复原能力，之后这一概念的内涵从生态韧性发展到工程韧性、经济韧性和社会韧性。2020 年，“韧性城市”理念纳入中国的国家战略规划，但偏向于生态与工程的“技术韧性”。相比之下，“土地公有 + 城乡二元”偏向经济与社会的“制度韧性”，使社会在外源性经济风险的冲击下，能够快速恢复到稳定状态。“2001 年中国入世”加剧了全球产业链的“极化分工”，同时也加速了西方资本从产业资本向金融资本的“脱实向虚”。这意味着中国的“入世风险”也随“入世收益”的加大而倍增，尤其是美元通过周期性波动对外转嫁经济危机的种种前车之鉴。在缓冲外源性危机的冲击方面，“土地公有 + 城乡二元”体现出双重韧性：一方面，作为“资本池”的城市，以非贸易品的不动产与国际货币体系保持相对独立，结合个人外汇业务管控等种种“闸门”，为防范大规模资金外逃起到“堰塞湖”效应；另一方面，作为“劳动力池”的乡村，通过与农业户口挂钩的宅基地与承包地，结合地方财政维持的最低保障，为大规模劳动力返乡安置提供缓冲。“2006 年新农村”建设启动三年之后，中国便遭遇“2008 年华尔街金融海啸”，海外市场急遽萎缩，大规模打工者因下岗集体返乡，集体土地成为最低保障。

相较于“2006 年新农村”的全域性一级土地开发和“2013 年美丽乡村”的乡村内需建设，“2017 年乡村振兴”着力于乡土生态资源的自主货币化，即通过集体经营性建设用地入市，实现集体土地和国有土地同价同权。“城市中国”开始向“乡村中国”进行经济分权，意味着从“2006 年新农村”到“2013 年美丽乡村”，高达数十万亿的投资[7] 形成的沉淀资产被“2017 年乡村振兴”激活。时至今日，城乡人口双向流动，乡建模式百花齐放，乡村中国逐渐具备利于返乡创业的平台条件，结合“全国统一大市场”的启动，以“自主内循环”重构“新型外循环”，这是中国得以从容直面“去全球化”的系统韧性。

3 城中村：公与私

作为“土地公有 + 城乡二元”的产物，城中村不仅与生俱来就反映着城乡关系的张力，同时也内涵制度背景的韧性。近年城中村的经济社会价值被广泛认可，尤其是作为廉价的落脚性聚落，城中村为海量非户籍就业人口提供了自发社会住宅的保障作用，而且也衍生出成规模的草根经济生态，通过降低廉价劳动力的生活成本，对“中国制造”的价格竞争力形成“非正式补贴”。

城中村之所以能形成“低保障、广覆盖”[8]，是承继中国以精耕细作、有机循环的小农经济自主养活高密度人口的千年红利，中国主要城市的城中村均以百计，其数量与分布反映了农耕中国留给城市中国的历史遗产。以沿海城中村为“进”，内地空心村为“退”，是中国城市化没有形成大规模贫民窟的韧性，根源在于“土地公有 + 城乡二元”中集体土地的保障作用。但值得注意的是，无论是土地财政还是城中村对实体经济形成的“非正式补贴”，本质上都来自城乡二元的土地价格剪刀差，也即土地财政中地方政府对土地一级市场的垄断，折射为城中村“小产权房”的不可入市交易。城乡二元的土

5　2015 年，国务院印发《中国制造 2025》，提出：“力争通过三个十年的努力，到新中国成立一百年时，把我国建设成为引领世界制造业发展的制造强国，为实现中华民族伟大复兴的中国梦打下坚实基础。”

6　加拿大生态学家克劳福德·斯坦利·霍林曾于 1973 年在《生态学与系统学年鉴》上发表《生态系统的韧性与稳定性》一文。这篇极具影响力的文章对工程韧性和生态韧性进行了区分，促进韧性内涵的发展，并将韧性的思想应用至系统生态学。

7　中国三农问题专家温铁军曾于 2017 年在受访节目中坦言，“十年下来（2006—2017 年），中国农村得到的国家基本建设和社会建设投入，高达十万亿之巨，乃历史上前所未有……世界上没有任何国家能向农村做这么大规模的投入。”此外，在中华人民共和国财政部的历年报告中显示，中央财政用于“三农”的合计支出已于 2011 年突破一万亿元，而其递增速度保持在 10% 以上。参见：https://v.qq.com/x/page/a03884q5mdz.html.

8　“低保障，广覆盖”原指有关部门在建立和践行社会保障服务时所遵循的原则，对于保障性住房而言亦如此。然而，当保障性住房因建设数量和分配制度的限制无法保障外来务工人员时，他们一般会选择在集体宿舍或相对便宜的“城中村”租房暂住。陶然等人的文章表明，“有估计显示，全国 1.2 亿进城民工中，半数住在 5 万个城中村。”参见：陶然，汪晖．中国尚未完成之转型中的土地制度改革：挑战与出路 [J]. 国际经济评论，2010(2): 107.

地价格剪刀差反映为国有土地的“正式商品房”和集体土地“非正式廉租房”两个市场，前者提高了高端产业的公共服务，后者则降低了低端产业的制造成本。但由于二者在区位上的毗邻，由商品房市场对廉租房市场形成水涨船高的价格带动，客观上造成土地财政“溢价归私”的代价，城中村农民成为坐享城市土地增值收益的不劳而获者，他们与来自内地的打工者们同为农民，命运却因所处区位的差异而有天壤之别。这一与社会主义理念背道而驰的治理安排，本质上是在城市保障房建设未能及时对海量流动性人口实现“广覆盖”的条件下，商品房作为超发货币池的“金融保障”与城中村廉租房作为自发社会住宅[9]的“社会保障”两大韧性机制之间不可调和的结果，因为二者都直接或间接地指向同一个目标：遏制通胀。

4 新常态：平台与动能

“2015 年股灾”之后，时值“去库存”周期的中国房地产市场，再次作为对内遏制通胀、对外防范外逃的资金池，“暂托”股市出逃资金。同期，最高层定位中国经济已进入“新常态”，其深层含义，是经过长达十年与美元挂钩的信用宽松周期后，被中国房地产市场锚定的广义货币发行已经超过美国，这意味着从“中国制造”开始的依附性货币政策已经通过“中国式造城”完成货币的“自主化”，并且在进一步面向乡村资源的“再锚定”过程中完成“内生化”，中美货币终于到了可以分庭抗礼的拐点（只不过其载体分别选在“中国楼市”和“美国股市”）。这是“2015 年汇改”开启人民币对美元脱钩的背景，也预示“2017 年雄安”新型城镇化模式的横空出世，以及“2017 年乡村振兴”的全面推进。以雄安为拐点，中国楼市价格被锁定在历史最高位，而其锁定流动性的“暂托”，也将成为中国应对输入性通胀的最后一次“权宜之计”。

作为人民币国际化、汇率利率市场等中国货币主权建构的一部分，深圳蛇口工业区于2015年升级为自贸区。蛇口曾因其改革的先锋作用而被称为“特区中的特区”，特区政策的升级，意味着这一分形特征将在以自贸区为龙头、以“一带一路”为躯干、以人民币为本位的国家战略中放大。工业区过去在制造、物流和金融方面所积累的生产力，将转换为自贸区在国际经济事务中的话语权，与之伴随的是蛇口区位价值的提升及其土地开发策略的转变。年底，2015 深双在蛇口大成面粉厂开幕[10]，针对这一“地缘 - 币缘”变局，笔者以 2015 深双为平台策划名为“蛇口议事：2025”（简称“蛇口议事”）的特展（图 7）。此前，蛇口浮法玻璃厂通过 2013 深双转型为“价值工厂”的示范（图 8），导致招商局在大成面粉厂片区的开发策略从整体拆除到存量改造的转向。深双也借此形成独特的模式，通过展览的定位、选址和策划，全面策应城市发展战略和总体规划，同时将产业布局、空间规划、城市设计等专业领域，以及决策、分配、施工、管理等内部过程通俗化为展览议题对社会开放，形成一种开放而开源的造城机制。展览策划作为“文化规划”的一部分，因此有机会与（过去偏重于产业和空间的）城市规划以“自发 + 自觉”的方式相结合。作为深双“以活动策划带动城市更新”模式的延续，“蛇口议事”一方面以“文化总体规划”和“事业 - 产业集群”介入过去以产业主导的土地规划与经济模式，以内容作为蛇口可持续开发中的核心附加值；另一方面以“公共参与”将这一内容置于国家与社会的互动互补之间，呈现包括蛇口国际城市学中心、大路社、艺术特区、国际设计教育联盟、工业遗产、城中村等 13 个文化集群议题（图 9），展览场地也因此设计为议事进程的样板间，成为以精细化的内容策划（包括业态运营与节庆活动）前置于规划环节带动社会协同的原型，对在 2015—2017 年“楼市暂托股市”的过渡期中尚未成形、却又呼之欲出的“城市中国 2.0”之“新常态”进行“打样”。

图 5《城市中国》创刊号：中国式造城
Figure 5 *Urban China* first issue: "We Make Cities" (Chinese-style City Making)

图 6《城市中国》专辑：社会主义新农村
Figure 6 *Urban China*: "Socialist New Village"

“房住不炒”政策下的房地产去金融化，是中国建成的住宅存量已经全面过剩[11]，以及与之挂钩的广义货币发行量超过美国，金融层面启动自主国内大循环的准备已经就绪。在此背景下的“新常态”，即在政策自主的货币主权支撑下，将金融与实业“脱实向虚”的关系重建为“脱虚向实”，转向“2017 年雄安”内涵的“金融支持实业”的经济模式，以供给侧的结构优化应对“去全球化”的外需下降、人口红利窗口期关闭、地方债风险等新常态危机。这意味着城市化模式也进入“新常态”，地方政府将从一次性收入的土地财政和土地金融，转型为可持续获取税收的产业财政；城市化的主导方向，也将由“增量开发”的大规模扩张转型为“存量更新”的内涵式发展。“蛇口议事”正是在这一土地财政向产业财政转型的趋势下，对即将到来的“平台 - 动能”关系所作的系统化设想。

过去，政府主导的土地供给及其基础设施，可视为旧常态下城市化的“平台 1.0”，在“政府搭台”基础上的“招商引资”则构成其“动能 1.0”，因此，旧常

9 有学者认为，在西方发达资本主义国家的语境下，“社会住宅”与“市场住宅”是一对二元概念：“市场住宅”既是“社会住宅”的反义词，也是其存在的前提。然而，在中国应将其定义为：1949—1998 年的“公有住房”和 1998 年至今的“保障性住房”之总和。究其根本特征，当指切实满足人民基本居住需求的房屋。参见：王韬 . 保障性住房相关概念 [J]. 住区，2013(4): 14-17.

10 笔者曾为深双撰文《策展城市》。参见：深圳城市 \ 建筑双年展组织委员会 . 看（不）见的城市：下 [M]. 北京：中国建筑工业出版社，2017: 324-349.

11 根据国家统计局近期公布的《中国人口普查年鉴 2020》显示，我国人均住房建筑面积为 41.76 平方米；平均每户住房间数为 3.2 间；平均每户住房建筑面积为 111.18 平方米。其中，城市家庭人均住房建筑面积为 36.52 平方米。

图 7 “蛇口议事：2025”特展
Figure 7 *Shekou Roundtable: 2025* Special Exhibition

图 8 “浮法玻璃厂”到“价值工厂”的示范性转型
Figure 8 The transformation of Shekou Float Glass Factory into the "Value Factory" became a paradigm

态下的城市规划更多体现出“亲资本”的粗放型特征；“2017 年雄安”之后，泡沫化的房地产市场在信用紧缩政策下定向冻结，从而事实上终结了过去以高杠杆撬动高周转和高增长的“房地产金融”模式。无法在融资枯竭的极限条件下以现金流维持生存的房企，将在新旧常态的交替中以债务爆雷、贱卖资产的方式出局，而肩负“逆周期调节”[12] 功能的央企国企，则通过国家信用背书接盘兜底。这意味着部分过剩的住房资产将在“房住不炒”的去金融化过程中以国有化的方式与土地融为一体，从而作为城市公共服务的一部分成为“平台 2.0”；而那些能够成功将存量转变为持续现金流的房企，将从“开发商”转型为“运营商”（以万科为代表[13]），从旧常态下资本密集型的固定资产投资模式，转向新常态下内容密集型的运营模式，成为“动能 2.0”。新常态下的城市规划也因此更多体现出“亲内容”的精细化特征，旧常态下土地财政一次性收益，将被新常态下动能生态的持续现金流运营机制所替代。

平台是动能赖以发展的生态，而平台升级则是旧动能在新动能的推陈出新中沉淀为平台公共服务的过程，因此，“平台 - 动能”天然具有“公 - 私”的相对性，因而也具有“差序化”的公私关系。平台在更高层级可能作为“私动能”（比如竞争性的地方政府），动能在低层级则可能作为“公平台”（比如城中村的集体股份制企业）。平台公共服务的价格，既取决于平台相对于其他平台的“竞争性”，也取决于平台相对其内在动能的“公共性”，平台在形成垄断之后，也因其“公 - 私”相对性，在公共服务定价上呈现出“溢价归公”和“溢价归私”的制度性差异。中国作为社会主义国家在“总平台”层面的公有性，决定了平台系统性地降低公共服务价格以降低小微动能创业成本的倾向，相对于动能通过竞争追求“经济收益”，公有制下的平台则倾向于通过垄断追求“社会价值”。

问题在于，基于垄断平台的公共事业往往表现出公共性有余、竞争性不足的惰性，而动能性的民营产业则竞争性有余、公共性不足。“蛇口议事”的核心，就是在城市规划由“内容亲资本”向“资本亲内容”精细化转向中，提出“公共性平台事业”和“竞争性动能产业”相整合的模式（图 10）。动能的创新力和多元化决定了平台总体价值的增长，平台的公共性则保障了小微动能的创业环境和公共服务的普惠性，本质上，依然延续了土地财政的“溢价归公”。由于平台通过沉淀旧动能的资本与技术“增密”，新常态下“城市中国 2.0”将更能够胜任重资产的平台打造，使得竞争性动能更有条件选择轻资产的内容生产。重资产平台与轻资产动能的互补共生，就是“国进民进”，这是“社会主义”的原本含义，是“公有制 + 社会化大生产”的马克思政治经济学原理对“国营 - 民营”二分法的解构，也是“自发 - 自觉”阴阳互根的中国式韧性。

5 遗产：城市与乡村

尽管从城市化率的定量来看，城市中国与乡土中国是一个此长彼消的历史进程，但从生活方式的定性来看，二者关系并非泾渭分明，而是可以“阴阳互根”，即城中有村、村中有城。“城乡二元”不是永远如两次剪刀差时期的“城乡二分”，而内涵了一种对传统文化和基层社会的保护机制，即在高度组织化的国家中内嵌一套相对自组织的网络，从而也预留城乡之间“自觉 - 自发”的余地。因此，由于乡村中国为城市中国提供了一个系统性的历史遗产，城中村、村中城、城乡接合部等“城乡不分明”的领域，将成为物质与非物质、正式与非正式的传统遗产与现代性交锋融合的现场。

作为千年农耕文明的载体，传统中国乡村不是现代意义上“社会化分工”的一个农业生产部门，而是生态、生产、生活有机关联的整体，一个“麻雀虽小、五脏俱全”的“小乾坤”，其核心特征在于围绕着以土地为核心的生态资源而展开的有机循环小农生产，以及在此基础上展开的“小农 + 小工 + 小商”的混合业态：“桑基鱼塘、稻鸭共养”的农副混态，“男耕女织、农工相辅”的农工混态，“半耕半读、耕读传家”的士农混态，乃至风水思想、宗族谱系、宗教信仰等系统化的文化景观，其核心模式可归纳为“精细化可持续的小微动能平台”，并由此通过京杭运河、茶马古道、丝绸之路等自发与自觉的基建系统，联网成为货通天下的国内外双循环大平台。

12 “逆周期调节”也称“反周期调节”，旨在熨平经济运行中过度的周期性波动，并降低由此积累的系统性风险，包括我国在内的全球主要经济体均采用：通过在经济上行时“及时刹车”避免经济过热催生泡沫，在经济下行时“强力助推”阻止经济衰退陷入萧条。“逆周期调节”减少了由经济剧烈波动带来的潜在损失，有利于经济稳健运行。参见：乔海曙 . 为什么要逆周期调节？[J]. 求是 ,2014(6):61.

13 深圳万科企业股份有限公司成立于 1984 年，于 2014 年将其“三好住宅供应商”的定位延展为“城市配套服务商”，又于 2018 年将这一定位迭代升级为“城乡建设与生活服务商”。

“城市中国 1.0”曾以廉价土地和劳动力换出口完成原始积累，但也在旧常态粗放式不可持续的发展模式下，付出牺牲生态资源与割裂文明传统的代价，中国城市在“亲资本”的城市化中被系统性地扁平化和平庸化；与此相对，“城市中国 2.0”在“精细化可持续”方面与乡村社会的千年传统相吻合，在城市中早已失忆的传统，很多在乡村得以留存，作为农耕中国留给城市中国最重要的历史遗产，“精细化可持续的小微动能平台”的生态文明模式将在新常态下被系统性地激活。与“2017年雄安”同步的“2017 年乡村振兴”，一方面体现为与“实业兴国”相平行的“百业兴乡”，从生态农法、饮食习惯到材料工艺、聚落营造等方面系统性地“礼失求诸野”；另一方面则是“智能 +”大基建平台加持下的城市文创，通过整合城乡供应链赋能与活化乡土知识，尤其是“士”面向“小农 - 小工 - 小商”的“轻资产的内容精细化再生产”的“万众创新”，由此构成新常态下城乡关系的“平台 + 动能”。

“轻资产的内容精细化再生产”意味着动能主体的多元化，在“三权分置”[14]的制度框架下，多元动能抱团联合而成的“合作性社会企业”将成为主角，将资源性的集体经济和赋能性的社会经济相融合，兼顾平台的规模效应与小微动能的差异性，本质是建立在新型城乡关系上的“合作社 2.0”：一个政府、企业、机构、社团、院校、媒体、村民和市民共同参与构建的产事业共同体，其资产升值的过程，也是公共品投资社会化的过程，即乡村在地化的生态资源与人文遗产，通过社会企业追加投资形成增量收益。这一合作社与共同体的“再组织化”进程也将发生在城市，即“蛇口议事”中描绘的差异化内容集群（图 11），在城市中产阶层兴起的趋势下，通过社会协同的内容规划，重构产事业关系和优化内容精细度，以多主体业态集群的模式对本地资源进行深耕与升级，进而形成增量收益的社会化分配。二者都接近亨利 · 乔治所设想的“地租社会化”。在这一意义上，新常态下的城市社区和乡村也将“城乡不二分”。

内容增密，不仅将填补和重构空心村的闲置存量，也将令城中村的城市区位和空间密度更有价值，但前提是对地方资源与历史遗产的充分认识，将旧常态下被扁平化和低估的本地资源精细化（从“山水林田”的生态和“士农工商”的生产到“食衣住行”的生活和“体娱游艺”的消费），以激活在平台建设中长期沉淀下来的资金。2017 深双将“定向激活”的主场选在南头古城这一城中村（图 12），在深双的空间更新与内容赋能下，正是这一历程让作为正式遗产的“古城”和作为非正式遗产的“城中村”成为新常态下“城乡不二分”的样本，而将“非正式遗产活化”纳入正式的城乡更新，既是“城市中国 2.0”包容性的体现，也是农耕中国活力与韧性的延续。

生态文明是华夏文明的内涵，也是唯一可持续的真文明，几千年间虽屡遭劫毁，其天人一体的整全思维却生生不息延续至今，以“百姓日用而不知”的方式遍布中国乡村，并藏匿于生产生活的种种自发或自觉的传统之中，包括生态维度的“全域”，产业维度的“全链”，社会维度的“全民”，并进一步延展为政治维度的“全局”，以及文化维度的“全息”。新常态下，城市与乡村的重心将从“物”的投资转向“人”的建设，而只有在“全域 - 全链 - 全民 - 全局 - 全息”的整全性之中，在社会化大生产的劳动分工下被“局部化”乃至“异化”的人，才能够重新成为一个完整的人。不可持续的“城市中国 1.0”将通过对农耕中国遗产的系统学习与实践，成为“城市中国 2.0”，借助生态文明的内在韧性缓冲城市化风险，在城乡关系的韧性转型中吸收现代性冲击。不可持续的“全球化 1.0”只有在“全域化 - 全链化 - 全民化 - 全局化 - 全息化”的世界大势中才能成为可持续的“全球化 2.0”。从这个视角看，“一带一路”所指向的是“全域化和全链化”，基础则是“天下为公”的“全局化和全息化”，核心是“人类命运共同体”的“全民化”。所谓“不谋全局者，不足谋一域”，是指“谋划”而言，具体到“行动”，则是“千里之行，始于足下”，在即将到来的“统一大市场”和“自主内循环”的“一域”之中，我们也许可以得见“全球化 2.0”的“全局”。

2022 年 8 月 10 日，杭州

14 承包地“三权分置”指的是农村土地集体所有权、农户承包权、土地经营权“三权”分置并行。改革开放之初实行家庭联产承包责任制，所有权归集体，承包经营权归农户，称之为“两权分离”。2016 年，国务院颁布《关于农村土地所有权承包权经营权分置办法的意见》，将土地承包经营权再分为承包权和经营权，形成所有权、承包权、经营权“三权分置”格局。

Informal China II: Resilient Transformation of Urban-Rural Relations in the New Normal

Jiang Jun
Jiang Jun is a research architect, archive editor and independent curator. Jiang is also a Professor at the China Academy of Fine Arts (CAFA) and director of the Research Institute of Society & Strategy of CAFA. Prior to joining CAFA, he was the founding editor-in-chief of the magazine *Urban China* (2005-2010). He taught at Strelka Institute for Media, Architecture and Design in Moscow (2010-2011) and at Oxford ESRC Center as visiting scholar (2011-2012). He curated the China Pavilion of the 2014 Venice Biennale of Architecture and *Shekou Roundtable 2025* during the 2015 UABB.
The article was translated from Chinese to English by Mo Sifei.

In 2006, the architectural exhibition, *China Contemporary* opened at the Netherlands Institute of Architecture (NAi). This was the first overseas appearance of a collective of Chinese contemporary experimental architects (Figure 1). Six months before the opening, curator Linda Vlassenhood invited a team of experts including me to Rotterdam for advice. My idea was to consider three related relationships in addition to showcasing Chinese architecture: "Content and Context," with "Formal and Informal" as the aforementioned content, and "Reality and Ideality" as its context. OMA realized the idea by designing one grounded and one suspended spatial structure, which corresponded to content and context respectively. For the exhibition content, I offered a group of informal designs titled "Informal China," eponymous with a previous issue of *Urban China* (Figure 2). These designs engaged the formal designs of the experimental architects. In addition, I provided thousands of scenes from hundreds of cities corresponding to a set of architectural renderings by Crystal Digital Technology Co., Ltd to differentiate between the "Reality" and "Ideation" of urban China (Figure 3).

The Dutch magazine, *Volume* published a companion issue for the exhibition, "All Roads Lead to China" (Figure 4). I contributed to the issue, an article also entitled "Informal China," but with a different subtitle, "A History of Control & Out-of-Control." The article parallels the chronology of 2,000 years of Chinese history, producing a comparative history of the present, including topics such as centralization and capitalism in ancient times, curfew and the lift of curfew during the Tang and Song dynasties, nomadic pastoralism and farming during the early and late Yuan Dynasty," the ban on maritime trade and the 'special zones of the Ming and Qing Dynasties, foreign concessions and triad organizations during the Republic of China, and planning and informal generation during the first 30 years of the People's Republic of China (PRC), as well as the Armed Forces and Special Zones over the next 30 years. The article outlined the governance dilemmas that have been intrinsic to the evolution of Chinese civilization. Although entitled "Informal," nevertheless the article focuses on the co-evolution of and tension between the formal and informal and organized and spontaneous order. Through a genealogy of "Subliminal Invention on Objects/Architectures/Urbanism," I hoped to present an implicit but often overlooked common sense: all perceptible categorizations are (conscious or unconscious) presentations of an underlying order. Therefore, if governance can be defined as a means of striking a balance between centralization and decentralization, the perceptible genealogy of "Objects/Architecture/Urbanism" can be understood as a sphere through which to study modern Chinese social phenomena. This insight was also the concept behind the founding of *Urban China*.

The last chapter of "Informal China" is

"Globalization and Communism," which takes 2001, the year when China joined the World Trade Organization (WTO) as a representation of the country's entry into the global system. China's entry into the WTO marked a historical continuation of the traditional global system and the beginning of constant interaction between the nation and global capital. The article takes a wait-and-see attitude to what these changes mean, concluding that "the variables that are likely to be brought by globalization can all be tackled as elements with 'Chinese characteristics' and be incorporated into the ongoing development of 'socialism.'" Since 2001, China has experienced a series of exogenous crises such as the 2007 stock market crash, the 2008 Wall Street financial crisis, the 2015 stock market crash, and the 2018 China-United States trade war. Today, it is still too early to determine the outcome of the interaction between the nation and global capital. Yet in frequent statements on "domestic circulation," a term that refers to China's internal cycle of production, distribution, and consumption,1 it is not difficult to see that the form of globalization that characterized the post-2001 world is in danger. I refer to this order as "Globalization 1.0." Before "Globalization 2.0" can emerge, de-globalization as a transition period immersed in the atmosphere of the new Cold War seems inevitable. Therefore, the blurred edges between the "revolution" of the old order and the establishment of the "new order" have created an opportune moment to intervene in the historic evolution of the nation from the spatial perspective of "Informal China." This has become even more true since 2006, when the evolution of urban-rural relations accelerated. This perspective would facilitate analysis of the increasingly resilient institutional design behind urban-rural relations in terms of co-evolutionary pairings, including planning and the market, industry and finance, the nation and global capital, and public and private spheres. More importantly, it is about the possible opportunity to glimpse the embryonic form of globalization from the governance trajectory of China's 20-year participation in the WTO.

1 Land: Nation and Collective

The magazine, *Urban China* was founded in 2005. The title of the first issue was "The Taxonomy of Contemporary Chinese Cities" (in Chinese) and "We Make Cities" (in English) (Figure 5), referring to a Chinese-style of city-making. That year, two events profoundly impacted modernization in China's urban and rural areas. In July, the central bank (People's Bank of China) launched the reform of China's exchange rate regime (moving away from a fixed exchange rate). In anticipation of RMB appreciation, this move attracted massive foreign capital investment in real estate, further accelerating domestic urbanization, and; in October, the *11th Five-Year Plan Outline Proposal* put forward the idea of "building socialist new villages." On New Year's Day, 2006 the millennia-old agricultural tax was abolished, launching China's nationwide rural construction. In turn, *Urban China* released the dedicated issue, "Build Socialist New Villages: Experiments on Modernity in a Century" (Figure 6). In China, urbanization refers to shifting relations between cities and villages. Almost half the articles published in *Urban China* have been related to rural China, including topics such as domestic manufacturing, migration, land, climate change, and crisis management. Chinese-style city-making and the country's new villages are two aspects of this co-constitutive process. The ongoing evolution of Chinese-style city-making has included tax-sharing reform (1994), housing reform (1998), and the tender, auction and listing policy (2003).[2] During this process, the real estate market was gradually established as a capital pool of real estate debt benchmarking through money issuance, preventing the hyperinflation that would otherwise result from a large capital inflow. Meanwhile, the expectation of RMB appreciation was transformed into the expectation of real estate appreciation. Then through local government land finance and central government transfer payments, the premium of land transfer revenue could be returned to the public through investment in infrastructure construction as well as rural construction. The result was tens of trillions of yuan invested in new village projects. Currently, the year 2005 is a milestone in the

1 "Domestic circulation" is not a closed-door policy, but rather the intended object of "international and domestic dual circulation." The special natural condition of China allows the activation of excess productivity in infrastructure construction through "fiscal liabilities" and "financial investment." Whether the existing massive productive assets can be activated is precisely the foundation of the "mass entrepreneurship and innovation initiative" as well as the premise of this strategy.

2 The tax-sharing system, housing reform, the tender, auction and listing policy, and their related content are detailed in the *Decision of the State Council on Implementing the Financial Management System of the Tax-sharing System, Notice of the State Council on Promoting the Reform of Urban Housing System and Accelerating Housing Construction* and *Provisions on Transfer of the Right to Use State-owned Construction Land through Tender, Auction and Listing.*

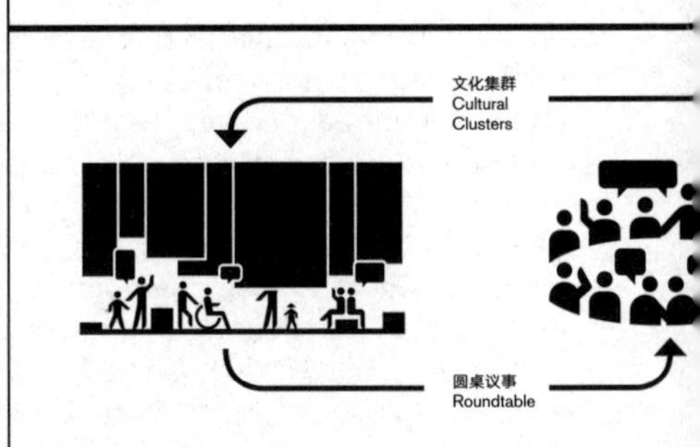

图 9 "蛇口议事"在"公众参与"方面的 13 个文化集群议题 Figure 9 The *Shekou Roundtable* addressed thirteen cultural issues with respect to public participation

construction of new villages. This parallel manifestation of large-scale Chinese-style city-making has eased urban-rural tensions through a dependency on primitive accumulation. At the national level, urban-rural re-balancing (rather than localized urban-rural re-balancing) has mitigated shocks brought on by subsequent exogenous crises.

The essence of Chinese-style city-making is the land system that supports this parallel advancement of urban and rural development. China's current land system was founded in the early years of the PRC, during the early 1950s. After the Land Reform Movement and socialist transformation, two types of land ownership—state-owned and collectively-held-were instituted. City land was state-owned and rural land was collectively held. People were integrated into this system via the household registration system (*hukou*). The formation of a "great state" in urban areas and "powerful collectives" in rural areas is the result of institutional changes led by violent revolution and state power. China is an ancient state that has seen cyclical involution due to millennia of agricultural production and a large population. In turn, as a late-developing modernized country, the PRC's only option for primitive socialist accumulation was through the "price scissors" that transferred agricultural surpluses to cities.[3] In the 70 years since the establishment of the PRC, two "price scissors" have been generated: one that extracted agricultural surpluses to support early-stage industrialization in cities, and the other that has funded late-stage urbanization by redeploying rural land. The central government administered the first price scissors, while local governments have used the latter price scissors. After the 1994 tax-sharing reform, the central government increased its proportion of tax revenue by allowing local governments to obtain fiscal revenue through their monopoly over the primary land market. In turn, this generated income through land use rights transfers. Consequently, local governments have become quasi-market-oriented organizations deeply involved in the local economy by means of land finance, regulating land supply and urban development while increasing differential rents. The second price scissors has two components. On the one hand, revenue generated through the transfer of land use rights has been invested in asset-heavy infrastructure and public services to improve the competitiveness of cities in attracting investment. On the other hand, local state-owned enterprises have intervened in urban planning and development, earning revenue that was generated by differential rents. This reduced local governments fiscal reliance on taxation, embodying how local competition and market benefits are defined with respect to so-called "Chinese characteristics," as well as the premium return to the public as defined under socialism.

Through the promotion of Chinese-style city-making, these two price scissors have in fact provided formal and informal subsidies to "Made in China," resulting in price advantages. The more autonomous a local government's land

3 The "price scissors" concept originated in the Soviet Union in the 1920s and was introduced to China in the 1930s. This concept has been developed to a broader understanding of the Chinese domestic conditions and needs. Most Chinese scholars believe that the price scissors refers to the scissors-like difference created by the non-equivalent exchange due to overpriced industrial products and underpriced agricultural products. KONG X., HE A. Analysis of the contribution of farmers to national construction since the founding of New China 60 years ago[J]. *Teaching and Research*. 2009(9): 5.

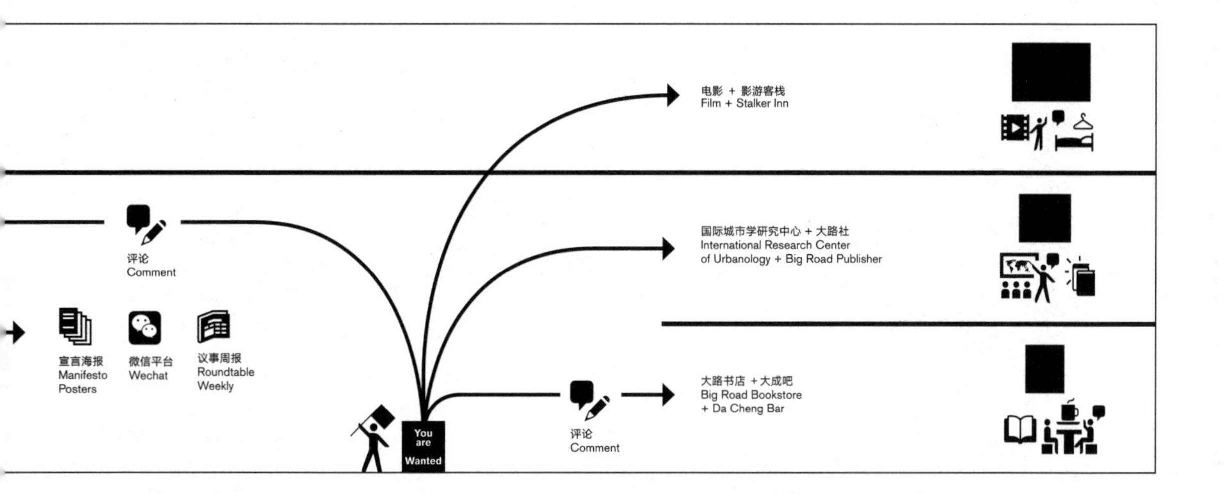

finances are, the more central government transfer payments can be directed to a wider range of underdeveloped areas. Land transfer revenues have gone directly from the central government to the public on a small scale while moving indirectly from local governments to the public through mega-scale urbanization. The concept that land transfer profits should be returned to the public originated in Sun Yat-Sen's principle of the people's livelihood. It can also be traced to the late nineteenth-century American economist Henry George.[4] Indeed, a century after George theorized the socialization of land rent through a single tax, China realized this goal through its practice of the parallel construction of new villages through nationwide, large-scale city-making.

2 Resilience: Dependency and Autonomy

Although land finance has greatly contributed to local primitive accumulation in China, the Matthew effect of "high land price - high housing price" has also been widely criticized. In 2017, the Xiong'an New District planning was positioned as "a strategy of millennial significance." It appears to be "helping phase out functions unrelated to capital from Beijing." However, it has been China's determination to build on industry and begin the transformation of land finance that signaled a shift in the economic development model from being dependent to being autonomous. Over the past twenty years, exchange rate reforms occurred in 1994, 2005 and 2015. Generated and amplified by the continuous surplus generated by China's export-oriented economy, a large withholding of capital inflow into the real estate market has facilitated land finance development, in turn preventing a huge capital influx into the leading indicator of inflation, the commodity market.

As a national economic model, Made-in-China began in 1994 with exchange rate reform. That same year, the RMB was pegged to the US dollar, depreciating sharply. Through the integration of urban China's infrastructure, public services and cheap rural market, Made-in-China siphoned global industrial capital. Simultaneously, trade surplus and the settlement of foreign exchange through exports hedged the base currency against strong dollar dependency due to money issuance. Since the 2005 exchange rate reform, Chinese-style city-making has created an endogenous land-backed currency. This currency operates through an internal regime of land finance, maintaining relative fiscal autonomy from the fluctuations of the US dollar cycle. The construction of New Villages (2016) and Beautiful Villages (2013) paralleled the Chinese-style city-making model of development, with an endogenous currency regime based on China's vast ecological and cultural resources emerging in 2015.

In 2015, the state council issued *Made in China 2025*, proposing to "strive to build the country into a world's leading manufacturing power by the time of 100th anniversary of New China through three decades of hard work, laying

4 The American economist, Henry George sparked reform movements. In *Progress and Poverty* (1879), he proposed economic theories focused on land, advising the state to tax all income derived from the use of land and arguing for the abolishment of other taxes. George believes that the government would receive high fiscal revenue from this single tax, which in turn could be used for public construction. This was known as the socialization of land rent.

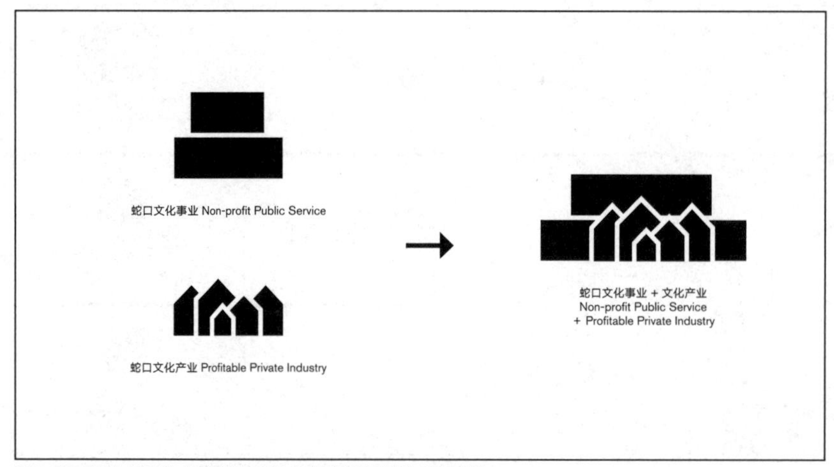

图 10 “蛇口议事”中提出的“公共性平台事业”和“竞争性动能产业”相整合的模式
Figure 10 A model to integrate public platform service and competitive dynamic industry was proposed in the *Shekou Roundtable*

a solid foundation for the Chinese Dream of the great rejuvenation of the Chinese nation." Today, through the 2025 Made-in-China policy[5], China has become the world's largest manufacturing power, boasting the most complete array of industry sectors. The historical role of Chinese-style city-making has been threefold: to accumulate monetary sovereignty through its endogenous money regimes, to develop and gain time for the establishment of urban and rural secondary capital markets, and to "advance and retreat" within this dual circulation. Together, Made-in-China, Chinese-style city-making and New Village construction based on the "dual-in-one" land system (benchmarking export, consumption and investment) constitute the resilience regime of the Chinese model that can either advance or retreat (in the military sense of expanding into or withdraw from territory). Under this model, international circulation which depends upon Made-in-China is an advance, while domestic circulation through New Village development is a retreat.

The Canadian ecologist Crawford Stanley Holling first introduced the concept of ecological resilience, which refers to the system's ability to resist, adapt and recover from external disturbances.[6] Later, this concept was extended to include ideas about engineering, economic and social resilience. In 2020, the idea of a "resilient city" was incorporated into China's national strategic plan, leaning toward technological resilience via ecological and engineering interventions. In contrast, the state-owned land and urban-rural dual system promotes the institutional resilience of the social economy, through social absorption of and recovery from external shocks. When China joined the WTO in 2001, for example, the polarized division of labor in the global industrial chain intensified, accelerating the transition of western capital from industry to finance, and from real to virtual industries. In turn, China's benefits from joining the WTO have also increased economic risks, especially through fluctuation cycles of the US dollar. In terms of buffering the impact of exogenous crises, the integration of the State-owned land and urban-rural dual systems has provided two sites of domestic resilience. First, cities have functioned as capital pools, maintaining relative independence from the international monetary system through non-tradable real estate, preventing large-scale capital flight much as a dam prevents the outflow of water. Second, rural areas have functioned as labor pools, buffering the large-scale resettlement of laborers. The outflow of labor has been prevented through homesteads and contracted land linked to rural hukou, as well as through local governments providing guaranteed minimum income. For example, just three years after the start of New Village construction in 2006, the 2008 world financial crisis caused overseas markets to shrink, leading to domestic layoffs. China was able to absorb this shock because migrants returned to their hometowns and villages. Subsequently, collectively held land became the source of guaranteed basic income.

The 2006 New Village movement can be understood as a global development of state-owned as well as collective land, while the 2013 Beautiful Village movement represents the expansion of local demand. In contrast, the 2017 Rural Revitalization Strategy has focused on

5 In 2015, the State Council of the People's Republic of China issued *Made in China 2025*. It proposed to develop the manufacturing power that leads the world's manufacturing industry through three decades of hard work to lay a solid foundation for the Chinese dream of the Great rejuvenation of the nation at the 100th anniversary of the People's Republic of China.

6 Canadian ecologist, Crawford Stanley Holling published "Resilience and Stability of Ecological System" in the *Annual Review of Ecology and Systematics* in 1973. This influential article distinguishes between engineering resilience and ecological resilience promoting the development of the concept of resilience while applying the ideas to systems ecology.

self-initiated monetization of local ecological resources, allowing collectives to commercialize land for construction, and entering the market with an eye toward equal pricing and rights for collective and state-owned lands. This process has allowed urban prosperity to be shared with rural areas through trillions in subsidized assets. Specifically, the 2017 Rural Revitalization Strategy has activated investments made through the 2006 New Village and 2013 Beautiful Village projects.[7] Today, through the two-way flow of urban and rural populations and diverse rural development models, rural China has gradually acquired a platform for and conditions to benefit returning start-ups. Along with the launch of a unified national market to reconfigure international circulation with autonomous domestic circulation, China has emplaced conditions for system resilience in the face of deglobalization.

3 Urban Village: Public and Private

A product of the state-owned land and urban-rural dual systems, the urban village not only indicates tensions between city and village, but also implies the resilience of a system that allows the existence of urban villages. In recent years, the socioeconomic value of urban villages has been widely recognized, especially as affordable settlements. Urban villages serve as informal social housing for many non-registered working populations while maintaining a large-scale grass-root economic ecology that reduces the living cost for this population, lowering the cost of labor. In fact, this ecology provides an informal subsidy to Made-in-China, increasing the country's comparative advantage. Urban villages have provided cities with their "guaranteed basic need and wide coverage." Urban villages are an inheritance from China's age-old division into a peasant economy of extensive farming and organic recycling that has fed the high-density urban populations.[8] The quantity and distribution of hundreds of urban villages throughout China's major cities reflects this historical legacy. The spatial form of resilience in China's urbanization takes urban villages in coastal cities as territorial advances and inland hollow villages as sites for territorial retreat, thereby preventing the formation of large-scale slums.[9]

It is worth noting that both land finance and informal subsidies provided by urban villages are essentially the results of the aforementioned price scissors, which is a consequence of the disparity between urban and rural land prices. This disparity is caused by local government monopolies over the primary urban land market and is manifested in land finance as the restriction on collective properties being formally exchanged on the market. Hence, the urban and rural land price scissors are characterized by the formal commercialized housing and informal affordable housing markets of collective land. In this scissors, formal commercialized housing improves public services for the high-end industry, while the informal affordable housing market reduces the manufacturing costs of low-end industries.

Due to the spatial proximity of these two markets, rising costs of formal housing have driven concomitant increases in affordable housing rents. Land transfer premiums have been pocketed by urban village members, who have benefitted from urban land appreciation and obtained their wealth for nothing. Thus, the respective hometowns of migrant rural workers and urban village members have determined their different fates, despite their shared status as farmers. This administrative arrangement runs counter to the basic concept of socialism. It is, however, essentially the result of two irreconcilable resilience mechanisms that aim to curb inflation directly or indirectly. On the one hand, commercialized housing products have served as the financial security for an over-issued currency, and on the other hand, informal affordable housing in urban villages has served as the informal safety net for the country's floating population.[10]

7 In 2017, Wen Tiejun, an expert on China's "Three Rural Issues" said, "After ten years (2006-2017), the investment in national infrastructure and social development in rural china has reached an unprecedented amount in history, 10 trillion yuan. No country in the world can make such a large-scale investment in rural area." In addition, it is stated in the annual reports of the Ministry of Finance of the People's Republic of China that the total expenditure of the central government on "Three Rural Issues" exceeded one trillion yuan in 2011 with a growth rate more than 10% per year. Details see https://v.qq.com/x/page/a03884q5mdz.html.

8 "Guaranteed basic need and wide coverage" was a principle followed by relevant departments in the practice of establishing and implementing social security services including affordable housing. Yet when the limitation of quantity and allocation system excluded migrant workers, most of them choose dormitories or relatively cheap "urban villages" as accommodation. In many articles like Tao Ran's, it is stated that "an estimate of half of the 120 million migrant workers of the country live in 50,000 urban villages."

TAO R, WANG H. China's Unfinished Transformation of Land System Reform: Challenges and Solutions[J]. *International Economic Review*. 2010(2): 107.

9 A micro geographic phenomenon, hollow villages display two essential features: ① a significant loss of land resources in rural areas due to housing developments annexed massive amounts of land in recent years; ② a decrease in the number of residents in the village due to urbanization.

10 Some scholars believe that in the context of western developed capitalist countries, social housing and commercialized housing are paired concepts. In this formulation, commercialized housing is the opposite of social housing, while being the premise of the existence of social housing. However, in China, social housing was defined as public housing from 1949-1998, and subsequently as affordable housing from 1998 to the present. The fundamental characteristic is housing that effectively meets the basic living demands of people.

WANG T. Concepts related to affordable housing[J]. *Community Design*. 2013(4): 14-17.

4 The New Normal: Platform and Momentum

After the stock market crash of 2015, China's real estate market was in a destocking cycle. However, it once again served as a capital pool to curb domestic inflation and prevent the flight of capital that had been pulled out of the stock market. Meanwhile, the highest authority asserted that China's economy had entered a "new normal." This meant that after a decade of credit easing by pegging the RMB to the US dollar, the broad money supply anchored by China's real estate market has surpassed that of the United States. It also means that domestic monetary policy has become autonomous through Chinese-style city-making, as well as achieving "endogenization" by further anchoring rural resources. Moreover, the currencies of China and the US have finally hit an inflection point where they can compete against each other (except that one is associated with the Chinese real estate market and the other is associated with the U.S. stock market). The background of the 2015 exchange rate reform and gradual decoupling of the US dollar and PRC yuan foreshadowed the emergence of a new urbanization model in Xiong'an and the advancement of rural revitalization in 2017. Taking Xiong'an as a turning point, China's soaring real estate prices are locked at an all-time high and the concomitant locked inflows are one last stopgap ameliorating China's reaction to imported inflation.

In 2015, as part of gaining monetary sovereignty through mechanisms such as RMB internationalization and the establishment of exchange rates and interest markets, the Shekou Industrial Zone was upgraded to a Free Trade Zone (FTZ). Shekou was once known as a "special zone within a special zone" because of its avant-garde role in early reform. Through the upgraded policy of the special zone, the fractal feature was to be amplified in a national strategy led by the FTZ, backboned by the Belt and Road Initiative, and based on the RMB. In Shekou, manufacturing, logistics and finance allowed for accumulation that is now being deployed as a form of discourse power in international economic affairs. There has also been a shift of Shekou's location value and a change to its land development strategy.

The opening of the 2015 UABB was held at the Dacheng flour factory in Shekou.[11] UABB has established its unique model responding to the urban development strategy and overall planning by the means of positioning, site selection, and curation of the exhibition. The Shekou Float Glass Factory had been transformed into the "Value Factory" (Figure 8), a venue for the 2013 UABB. That exhibition became a paradigm, leading China Merchants to shift its development strategy for the Dacheng Flour Factory district from complete renewal to brownfield revitalization. For the 2015 exhibition, I curated *Shekou Roundtable: 2025* (Figure 7), which responded to the "Geopolitical to currency" transformation. Through the exhibition, professional fields such as industrial strategy, planning, and urban design are introduced to the public, as well as internal processes such as decision-making, distribution, construction, and management. The popularization of this knowledge as exhibition discourses has established an open-source city-making mechanism. As part of cultural planning, exhibition curation could be integrated with urban planning (focusing on industry and space in the past) through the approach of "informal + formal."

Shekou Roundtable extended a curation model that uses event curation to drive urban renewal. On the one hand, the traditional urban planning and economic model is led by industry. The *Roundtable* intervened in this model through the "Cultural Master Planning" and "Public Service-Industry Cluster" sections. This intervention made the target of renewal the core content, adding Shekou's sustainable development to the value of the project. On the other hand, the *Roundtable* introduced public participation into the complementary interactions between the nation and the society, presenting 13 cultural cluster issues (Figure 9), including "International Research Center of Urbanology," "Big Road Publisher," "Art Special Zone," "International Design Education Alliance," "Industrial Heritage," and "Urban Village." The venue was turned into a demonstration of the dialogue process, becoming a prototype for refined content curation that can motivate social synergy before the urban planning process takes place. It was also a paradigm for a new normal which could have emerged as urban China transitioned from the real estate - stock market model of urbanization (described above) circa 2015-2017.

Under the policy of "houses are for living in, not for speculation," the de-financialization of

11 I authored "City as Tactics: Curation & Reaction" for UABB. Shenzhen Biennale of Urbanism\Architecture Organizing Committee Office. *(IN)VISIBLE CITIES: 10 Years of UABB Research*, Vol.2 [M]. Beijing: China Architecture & Building Press, 2017: 324-349.

real estate suggests that China's residential unit overstock and its broad money have exceeded that of the U.S., allowing the start of an autonomous financial circulation domestically.[12] In the context of this new normal, that is, with the support of monetary sovereignty through autonomous policies, the relationship between finance and industry has aimed to reverse a shift from the substantial economy to the virtual economy, instead shifting from the virtual to the substantial economy. At Xiong'an in 2017, a turn toward an economic model in which finance supports industry was implied. This model optimizes supply-site structural reform to cope with crises of the new normal such as the decline in external demand due to de-globalization, the expiration window of the demographic dividend, and the risks associated with local debt. Another aspect of the new normal in Chinese urbanization is that local governments are transitioning from land finance and one-time revenue to sustainable tax revenue as their fiscal model. The focus of urbanization has also shifted from extensive vacant land development to intensive brownfield redevelopment. The *Roundtable* offered a systematic hypothesis of the upcoming relationship of "platform-momentum" as a shift from land finance to industrial finance.

The government-led land supply and its infrastructure could be considered the "Platform 1.0" of urbanization in the old normal. In turn, attracting investment on the platform established by the government would constitute its "Momentum 1.0." Urban planning in the old normal favored capital. However, since 2017 Xiong'an, real estate in the housing bubble has been directionally frozen through a credit tightening policy. In fact, this move ended the model of highly leveraged real estate finance to lever the high turnover rate and rapid growth. Real estate developers in extreme conditions of financing exhaustion cannot survive on cash flow because it will be out of the market via debt explosion and underpriced asset sales in the alternation between old and new normals. Central and State-owned enterprises with responsibility for counter-cyclical adjustment take over the assets through national credit endorsement.[13] Therefore, part of the surplus housing assets will be nationalized and integrated with the land in the process of de-financialized housing advocated by the "houses are for living in, not for speculation" policy. In turn, this housing could become part of public services as "Platform 2.0." Real estate developers with the ability to convert assets into cash flow could transform from developers to property managers (represented by Vanke),[14] shifting from capital-intensive asset investment of the old normal to content-intensive operational model of the new normal, hence Momentum 2.0. Urban planning under this new normal would therefore express a refined characteristic of being "pro-content." The one-time revenue from land finance of the old normal will be replaced by a sustainable operating mechanism of continuous cash flow.

The platform is an ecology for the development of momentum, while the upgrade of the platform is the process in which the old momentum becomes a public service platform during the emergence of new momentum. Therefore, depending on social organization, the platform-momentum relationship can correspond to different orders of relationship: a higher order platform is considered to express private momentum (such as competitive local government), while momentum at lower level can be a public platform (such as an urban village stock-holding corporation). The price of a given public service on the platform would not only depend on the competitiveness of the platform relative to other platforms, but also on the public nature of the platform relative to its inner momentum. After the monopoly of the platform and its concomitant public-private ordering, policy differences between the premium of land transfer revenue return to the public and private would emerge. As a socialist country, the public nature of China's central platform would determine the platform's tendency to systematically lower the price of public service in order to reduce the cost of small entrepreneurship. Instead of aiming for economic benefits through competition, a common ownership platform tends to pursue "social value" through monopoly.

12 According to the *China Census Yearbook* (2020), which was released by the National Bureau of Statistics, the average floor area per person of Chinese housing is 41.76 square meters; the average number of bedrooms per household is 3.2; the average floor area per household is 11.18 square meters, and; the average floor area per person of Chinese urban housing is 36.52 square meters.

13 Counter-cyclical adjustments are tools used by the international economy including China to neutralize possible negative effects and accumulated systematic risks of economic cycles. Adjustments aim to slow down rapid economic expansion to avoid an overheated market that is at risk of a bubble because economic contracting prevents the economy from falling into recession. Counter-cyclical adjustment minimizes the potential losses due to severe economic fluctuations which allow for a stable economic operation.
QIAO H. Why Counter-cyclical adjustment?[J] *Qiushi*. 2014: 61.

14 China Vanke Co. Ltd. was established in Shenzhen in 1984. In 2014, it extended its positioning from "Good Real Estate Developer" to "urban service provider." In 2018, this positioning was upgraded to "urban and rural developer and service provider."

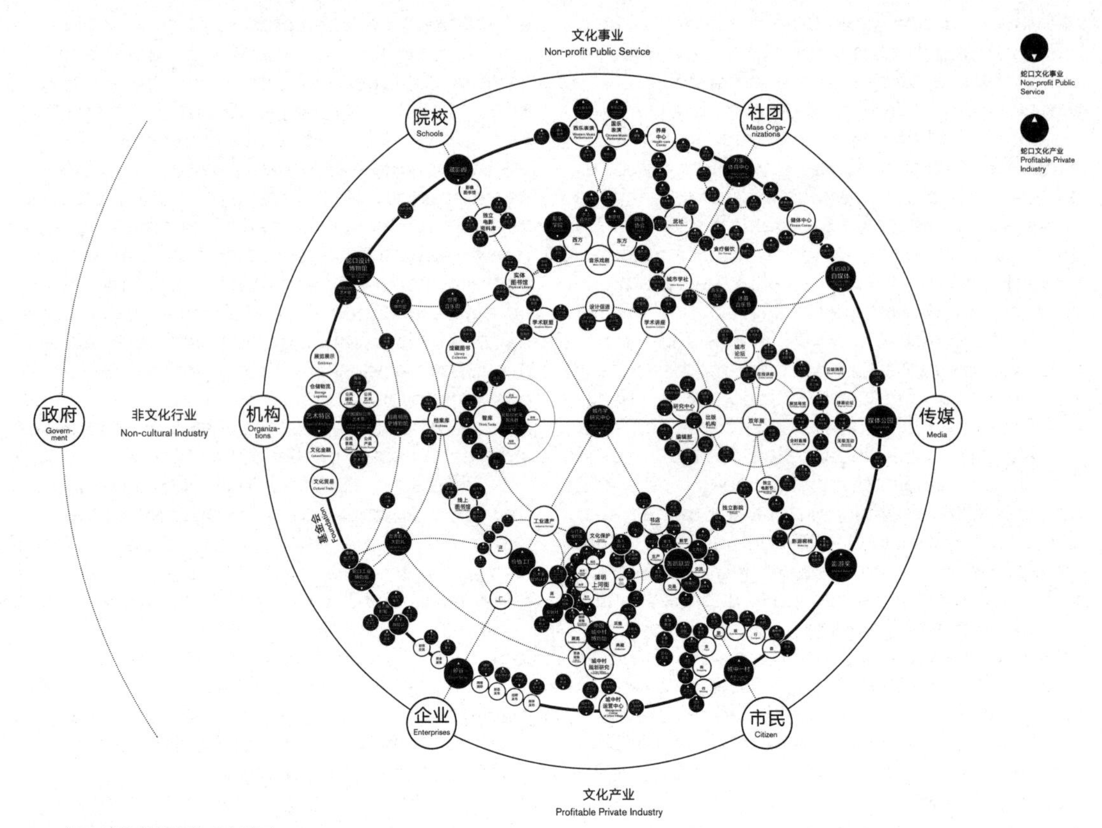

图 11 “蛇口议事”中描绘的差异化内容集群
Figure 11 Differentiated content clusters depicted in the *Shekou Roundtable*

Public services based on a monopoly platform are often inefficient because although they are public, nevertheless they are insufficiently competitive. In contrast, the private sector based on momentum is competitive enough, but it is not public. The *Roundtable* aimed to propose a model that integrated a platform of public service and the momentum of competitive industry (Figure 10) during the refined transformation of urban planning from content serving the needs of capital to capital serving the needs of content. The innovation and diversity of momentum would decide the value growth of the platform, while the public nature of the platform would protect the universality of the entrepreneurial environment and public services. This model continues the practice of the premium of land transfer revenue being returned to the public. As the platform accumulates capital and densifies technology of the old momentum, the new normal of Urban China 2.0 would be more capable of asset-heavy platform building, allowing the competitive momentum to choose asset-light content production. The complementary symbiosis of the asset-heavy platform and the asset-light momentum would be the manifestation of people advancing with the nation, which is the original meaning of socialism. It would destroy the dichotomy between state- and privately-owned property through the Marxist political economy principle of public ownership and socialized large-scale production. This model also manifests a Chinese-style, yin-yang resilience of the integration of the formal and informal.

5 Legacy: City and Village

In quantitative terms, urban China has been growing and rural China has been shrinking through the historical process of urbanization. Yet, in the qualitative terms of lifestyle, they are complimentary like yin and yang, taking forms such as a village-in-the-city and a city-in-the-village. The current urban-rural duality is not always comparable to the urban-rural dichotomy of the price-scissors period. Instead, the current duality implies a protection mechanism for traditional culture and grassroots society, that is, a set of relatively self-organized systems embedded in a highly organized country. It has also left room for the informal-formal connections that integrate city and village. Therefore, rural China has provided urban China with a systematic legacy in which the urban village, rural city, and urban-rural junction share blurred boundaries, which have become sites of collision between and integration of tradition and modernity as well as the formal and informal.

As the carrier of China's 1,000-year-old agrarian civilization, the traditional village is not agricultural in the modern sense of a social division of labor into different sites of production, but rather an organically integrated ecological, productive and living—a small, but complete universe. Its essence lies in the organic circulation that arises when small-scale farming comprises land as its fundamental ecological resource and mixed programs, including small-scale farming, manufacturing and small business. The traditional village further extended this model by integrating agriculture and sideline productions, such as the mulberry fish pond and rice-duck co-culture farming system, as well as a cultural geography that integrates fengshui, family lineage, and religious beliefs. Traditional proverbs such as "men farm, women weave; farmers and workers supplement each other's work" and "plowing half the time, studying the other" express this mix of farming and manufacturing, or farming and scholarship. In our terms, the fundamental model can be summarized as a "refined and sustainable small platform of momentum" which became the platform of domestic and international dual circulation through informal and formal infrastructure systems and networks, such as the Beijing-Hangzhou Canal, the Ancient Tea-Horse Road, and the Silk Road.

Since 1979, Urban China 1.0 has achieved primitive accumulation by exchanging underpriced land and labor for exports, paying for this achievement by sacrificing ecological resources, and fragmenting civilization and tradition. This was the extensive and unsustainable development model of the old normal. Chinese cities have been systematically flattened and mediocre in their pro-capital urbanization. In contrast, Urban China 2.0 is in line with the millennial tradition of rural society in terms of refined sustainability (as defined above). Many long-lost traditions have survived in rural areas and are the most important legacy of agrarian China for urban China. The ecological civilization model of a refined and sustainable small platform of momentum will be systematically activated in the new normal. Synchronized with "Xiong'an 2017," there has been the revitalization of rural areas, or the "rural revitalization 2017." On the one hand, in parallel to the "rejuvenating the nation by industry," this would be manifest as the "rejuvenating the rural by many industries," which would take the form of a search for lost traditions, such as ecological farming techniques, culinary habits, materials, and craftsmanship, and settlement making. On the other hand, urban cultural and creative industries supported by the "smart+" large infrastructure platform will shape the "platform+momentum" of urban and rural relationships in the new normal. The integration of urban and rural supply chains will empower and activate local knowledge especially through "mass innovation" of asset-light refined content reproduction by scholars addressing the small scale farming, manufacturing and business."

"Asset-light refined content reproduction" refers to the diversity of the subjects of momentum. Within the institutional framework of "three rights separation," a cooperative social enterprise formed by grouping diverse momentums will become the protagonist.[15] This model integrates the resource-based collective economy and the empowering social economy and also considers the differences between the large-scale platform and small momentums. Its nature is "Cooperative 2.0," which is based on the new urban-rural relationship: a combination of industry and public service jointly constructed by the government, enterprises, institutions, associations, colleges, media, villagers, and citizens. The process of asset appreciation is also the process of socialization of public goods investment in which localized ecological resources and cultural heritage in rural areas generate incremental income through additional investment from social enterprises.

图 12 2017 深双将"定向激活"的主场选在南头古城这一城中村
Figure 12 2017 UABB selected an urban village, Nantou Old Town as the site of "directional activation"

This reorganization of cooperative and combination of industry and public services will also take place in cities as was depicted as differentiated content clusters in the *Roundtable* (Figure 11). With the emergence of an urban middle class, the relationship between industry and public service is reshaped and the finesse of content is optimized through social collaboration in content planning. Local resources are developed and improved through the model of multi-agent business clusters, which establishes the social distribution of incremental revenue. Both rural and urban models are like the socialization of land rent envisioned by Henry George (above), and in this sense, urban communities and villages will be undifferentiated in the new normal.

Content densification will not only refill and reconstruct the brownfield of hollow villages, but also make the urban location and spatial density of urban villages more valuable. Yet the premise is to fully understand local resources and historical heritage, refining local resources that were flattened and overlooked in the old normal (from an ecology of mountains, water, forests and fields and scholar, farmer, worker and merchant producers to a living essence of food, clothing, housing and transportation and consumption via sports, entertainment, games and the arts) to activate the funds accumulated in the construction of the platform. The 2017 UABB selected the urban village, Nantou Old Town (Figure 12) as the site of directional activation which would be the beginning of this process. With the spatial renovation and content empowerment of the UABB, the old town as formal heritage and the urban village as informal heritage stand for the paradigm of undifferentiated urban-rural areas in the new normal. Incorporating informal heritage regeneration into formal urban and rural renewal is not only a manifestation of the inclusiveness of "Urban China 2.0," but also a continuation of the vitality and resilience of agrarian China.

Ecological civilization is the essence of Chinese civilization, and it is also the only sustainable civilization. Although it has been repeatedly destroyed over the past thousand years, holistic

15 The "three rights separation" is the separation of collectively owned property rights, contractual rights (right to contract for the use of rural land), and rural land management rights. At the beginning of the reform and opening up, the "Household Contract Responsibility System" policy granted collective property rights and individual household contractual and management rights, which are known as "two rights separation." In 2016, the state council promulgated the "Opinions on the Separation of Rural Land Ownership, Contract, and Management Rights." Under this policy, land contractual and management rights are subdivided into contractual and management right forming the "three rights separation."

thinking about the integration of nature and humankind has continued to the present because it is embedded and unrecognized in the daily life of Chinese villages and hidden in traditions of informal and formal making and living. It includes all areas of ecology, all chains of industry, all people of society, and is further elaborated as all circumstances of politics and all dimensions of culture. In the new normal, the focus of cities and villages will shift from the investment in things to the establishment of people. Only in the integration of all area-all chain-all people-all circumstances-all dimension can people who have been localized or even alienated in the division of labor in socialized large-scale production become full again. The unsustainable Urban China 1.0 will become Urban China 2.0 through a systematic study and implementation of the legacy of agrarian China, buffering the risk of urbanization with the inherent resilience of ecological civilization and absorbing the impact of modernity in the resilient transformation of urban-rural relations. The unsustainable Globalization 1.0 can only become a sustainable Globalization 2.0 in a worldwide trend of "globalization-full-chain-nationalization-globalization-holographic." From this perspective, the Belt and Road Initiative refers to globalization and full chain founded on the concept of all aspects and all dimensions of the world is for everyone with the core of all people of a community with a shared future. "Those who do not plan for all circumstances are not able to plan for one area" is an expression that refers to planning, while "a thousand miles begins with a single step" refers to action. In the upcoming unified national market and autonomous domestic circulation, we may be able to see the holistic view of Globalization 2.0.

对城中村进行未来想象的考古学

马立安

握手 302 艺术空间合作创始人，人类学博士。已出版研究成果《向深圳学习》，并且在虔贞女校艺展馆促进文化与教育计划。2017 年深港城市\建筑双城双年展龙华（大浪）分展场总策划人，2018 年深圳商报年度创意影响力人物之一。

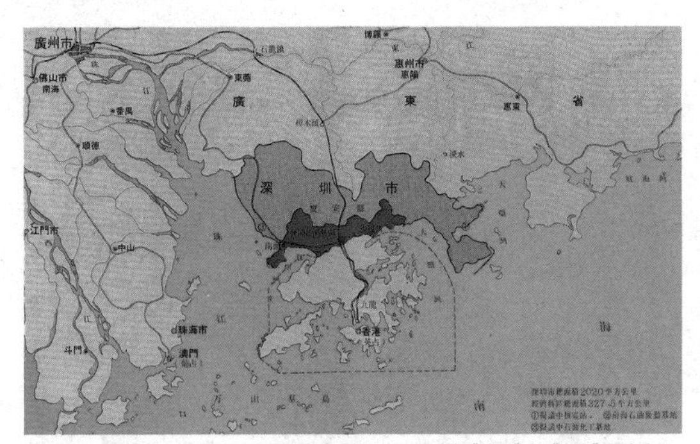

图 1 1981 年深圳市由经济特区与宝安县组成的，以二线分为“特区”与“宝安”两个大区。当时，“特区”的未来想象是成为制造业城市，“宝安”的未来思象则是将农亚产品供给“特区”与香港两座城市。

Figure 1 In 1981, Shenzhen was comprised the SEZ and Bao'an County, which was divided by the second line. At that time, Shenzhen's future was imagined as comprising manufacturing in the "special zone" and agricultural production in "Bao'an" for urban Shenzhen and Hong Kong.

自 1980 年成立以来，深圳经济特区对中国未来的设想和追求起到至关重要的作用。20 世纪 80 年代，“特区”象征着计划经济的未来，而目前，深圳作为中国第四个一线城市和世界硬件创意中心，巩固了其在中国和世界其他发展中城市的典范作用。因此深圳与创造未来城市的项目有两个关联。一方面，因为深圳的规划和定位一直向前看，所以其诞生、发展和存在的价值一直与未来相关；另一方面，因为深圳成功的现代化，所以它已经成为一种典范。无论是在中国还是在整个工业化世界，“深圳”已经成为一个用来想象不发达国家的未来的例证。

苏珊·巴克·莫尔斯关于瓦尔特·本杰明对 19 世纪巴黎物质文化的解读启发了我对田面进行未来想象的考古学探讨，并且尝试将这种未曾实现的未来想象，融入当今城市的建筑环境的可能性。本杰明的《拱廊计划》旨在创造一种唯物主义的教育方法，挖掘历史留下来的潜力。[1] 这种潜力是内在的，因为它包含一些符号和形式，例如现在的建筑师对阿戈拉的建筑引文，其中提到更具包容性的社会乌托邦想象。在这篇文章中，我特别感兴趣的是城市形式的能力，如何适应不同的未来想象。在本文中，作为对未来想象起点的“现在”确定为 2004 年，当时深圳第一个城中村旧改项目已经完成，深圳正处于从制造业城市向后工业城市转型的风口浪尖。如果说那年是深圳城市转型的机遇，那么与此同时我们也看到一种被错过的城市化模式。今天重提这些被错过的城市模式只能以对另一个未来城市想象的虚拟语境来表达。即便如此，这样一种对想象世界的重构也许还能对当下的深圳城市想象，甚至对现实的城市更新模式的演进与方向提供参考和启发。

除了对理论的兴趣，我写这篇文章的动机还源于我在深圳居住和生活了近三十年的个人经历。在这段时间里我就近观察了深圳城市发展的进程，目睹了几次大的转变时机。这些变迁形塑今天深圳的格局，也决定了这个城市未来很长一段时间的发展趋势。当我今天再次回忆那些转变时刻出现的许多微妙的事实时，才能体会到其中蕴含着不同的历史可能性，以及被错失的机会中显现的价值与启发。

我于 1995 年第一次来到深圳，做文化人类学博士论文的田野调查。开始住在靠近深圳大学的粤海门村，后来搬到南新路的墩头村。与南新路平行的还有一条路，被本地人称作“南头老街”。事实上，这是明清两代新安县的政府所在地——今天的南头古城通往赤湾唯一的一条官道。沿着这条老街有一连串传统的自然村，从南头古城外一直排列到南头半岛顶端的大南山，翻过这座山就是蛇口，那里还有一些被安置上岸居住的渔民村落。我刚来的时候，这些古老的村庄刚经历了一轮自发的城市化过程，许多往昔的农民在新分配的宅基地上盖起了两层高的小别墅。这些别墅是独立的，户与户之间的距离不远，但因为楼层不高，没有后来城中村握手楼之间那种不见天日的恐怖。这些新分配的宅基地一般都是成片安置在离旧村不远的地方，所以被称作“新村”。

当我修完博士并且在芝加哥大学人类学系做了一年的访问教授之后再次回到深圳，那是在2003年。此时虽距我离开不过短短数年，深圳的城市景观已经大变，原来新村中两三层的小别墅常常被加高到9层以上，而且有了一个新的称呼“城中村”，而沿深南大道带玻璃幕墙的摩天楼也纷纷拔地而起。南头半岛因为填海造地足足加宽了一倍。原来我从深圳大学教师宿舍就能望到海岸，现在需要行走40分钟才能走到深圳湾海边。我也回到过南头半岛另一侧，那里在我离开的时候曾经有一个火车站叫深圳西站，每天都有发往湖南等内地的火车，有许多打工的外地人从这里进出深圳。但我回来后不久这个火车站就废弃了。因为那时深圳正处在从一个出口导向的工业城市向金融和高科技城市转型的关键时期，深圳不再需要那么多的打工者，而是需要资本和受过教育的专业白领。

这也是深圳的房地产业暴涨的开始。我以翻译兼学者的身份参与观察田面新村旧改的全过程，并且在旧改结束后获聘成为旧改配套学校——当时称“北京景山学校深圳分校”，后来更名为“深圳城市绿洲学校”的首任外籍校长。作为某种努力的结果，我也观察到正如地产业的繁荣造就了无数新的中产阶级一样，新一波的私立学校和国际学校的出现也让中产阶级的子女具有了不同于工薪阶层和普通市民教育背景和身份认同。他们中的大多数能够赴欧美留学，之后或定居国外或回国进入一些高薪职业。深圳在进入21世纪的第二个十年以后，我离开了教育界，开始从事公共艺术，因为这个机遇，我进入白石洲这个深南大道沿线最大的城中村。2013年我和我的伙伴们成立了一个艺术工作室“握手302”，直到2019年夏天因为白石洲拆迁而离开。在白石洲的日子里给了我从另一个角度理解深圳的机会。尤其是一个非正式经济、非规划区域、非户籍人口聚集的城市空间能够带给深圳何种惊喜、遗憾和想象。在这篇文章中我聚焦的重点是2004年田面村的旧改，那是一个门刚刚被打开的时刻，是一个万事皆有可能的起点，一个历史的十字路口。

1 构成深圳基底的历史文化地理

1978年前，宝安县的文化地理至少由三个整合的集镇和村庄网络组成，这些集镇和村庄通过河、海岸的网络，以及稻田和荔枝园的路径连接起来。在深圳西部，珠江河岸的码头方便牡蛎养殖户和渔民在广州和香港之间航行。宝安村民讲东莞广府话，在沙井、固戍、西乡等地赶集；中部村民讲客家话，在观澜、布吉等地赶集，而在梧桐山以东的客家村民在沙头角赶集；在连接罗湖福田和香港新界的边界地带，村民们讲围头话，在深圳墟购买日常用品，也会过境在香港新界耕种历代属于他们村庄的土地。

1979年，广东省将宝安县升格为深圳市。1980年，中央政府确定了深圳、珠海、汕头、厦门四个经济特区，表明改革开放正式启动了，也开启了如何想象和创造未来的机遇。在1981年，深圳市被重新划分为深圳经济特区和宝安县，一条叫“二线”的边界隔开了这两大区域。深圳市管理着特区，而新的宝安县城建立在新安。然而，宝安县现存的粤语、客家、围头话三方文化地理对深圳未来的建设起到决定性的作用。简单而言，新建立的深圳市，其空间格局相当于按照一种文化地理的边界进行了某种划分。首先，讲围头话的文化地理部分被划分出来，由“二线”与深港边界包围，成为“深圳1.0版”（图1）。在二线关外原来的宝安县，讲东莞广府话的文化地理部分相当于本地的老宝安市区，而其东边是讲客家话的文化地理部分相当于老龙岗市区。（如果我们注意到龙岗是1958年从惠州划给今天属于深圳的宝安县，则能更清晰地了解今日在深圳范围内讲客家话的部分原本就是传统客家文化区的惠州的一部分。）

2 1980年代：起点上的城市空间与想象的开始

经济特区的建设始于罗湖—上步，起点靠近历史悠久的深圳墟，也就是今天的东门地区，向西延伸至今天的上海宾馆。位于围头地峡的早期特区通过九广铁路和文锦渡检查站进入香港，使八卦岭、上步等工业区实现和世界经济连接的功能。因此，罗湖—上步的发展不仅采用香港的基础设施标准、商业模式和语言（港式粤语，其实就是围头话的一个变种），而且还推进消费主义、建筑美学和各种娱乐（曾被毛时代的中国称作“资产阶级生活方式或资本主义的腐朽现象”）。日常生活中，住在深圳其他地方的居民（包括住在蛇口工业区的新移民）习惯把从东门到上海宾馆——一个新兴的城市中心区——称作“深圳”，而在罗湖的热闹街道和丰富市场里，一种新的未来想象诞生了（图2）。

要强调的是，经济特区的早期建设不仅集中在有正式规划的罗湖—上步，同时更有大量的不在规划内的本地人自发的建设。本地的集体企业向香港出售农产品并建立基于村一级集体所有的工业区，和香港亲戚朋友合作进行加工出口。这些实验也都展现出对未来城市的一种新想象，这种想象是农民自发进行“农村城市化”乡镇企业的先兆。更重要的是，这种早期的集体投资，成为早期工业发展的原型和动力之一。在没有政府直接投资和监督的情况下，宝安县粤、客家语地区与罗湖地区同步发展，但发展方式明显不同。在连接松岗和文锦渡检查站的G107国道和珠江海岸线之间，展现出来一条延伸到邻近东莞的工业走廊。当年，位于珠江沿岸的集体企业依然向香港出口牡蛎、禽类等农业产品。通过现有的河岸运输网络，这里也方便从佛山和广州进口建筑材料。相比之下，除了布吉之外，位于深圳山区的客家村落很少有机会与香港发展关系。就连位于深港边界的沙头角因为地貌的原因，很难像西部那样进行工业化。因此，这里的未来想象体现为一种特殊的市内移民。许多来自龙岗的村民移居到罗湖—上步寻找工作机会，有些在工厂打工，有些在东门商业区摆摊（图3）。

1980年代的深圳文化地理植根于政治改革、经济实验和地方村镇之间的互动，受到香港市场乃至世界经济的影响。新成立的深圳行政区对宝安原有的文化地理重新划分，更是催化了深圳市未来的文化地理。在罗湖—上步，深圳最早的城市建设展现了当年对未来的一种想象：一座以加工业和商业为经济基础的边陲城市。更重要的是，这里早已实现的小康生活以及实现它的具体条件，对规划外地区的居民开始产生巨大的影响和示范效应。因此，罗湖—上步的文化地理也带动城市其他地方对未来的想象和自发城市化。在深圳西部，宝安的集体企业正将传统农业资产重组为本地制造业基地。在深圳东部，集体企业很难直接进入工业化，因此这一片客家老区将为深圳的繁荣提供食物和劳动力，而近年来，它的不发达又使绿色规划和郊区住宅区的试验成为可能。

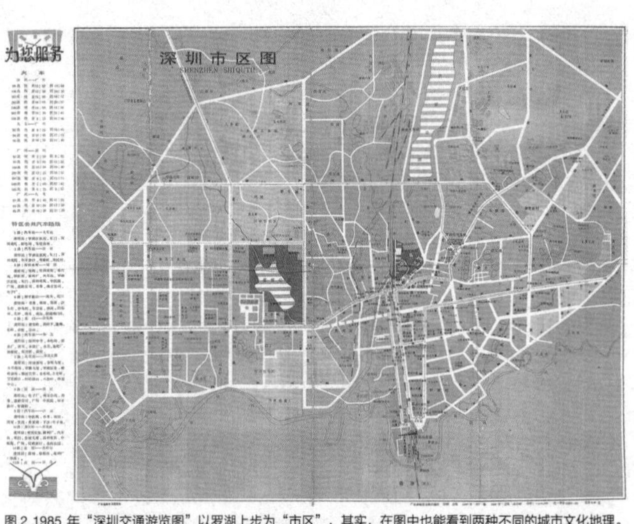

图2 1985年"深圳交通游览图"以罗湖上步为"市区"，其实，在图中也能看到两种不同的城市文化地理。在东门，城市化的地理架构是历史留下来的地貌与交通网，而在上步，城市化的地理架构是推平的地貌和理性的方块交通网。两种文化地理酝酿着不同的未来想象：罗湖以自发贸易为导向，上步以城市规划为导向。重点是两种模式的互补关系决定了1980年代"特区"的城市化。

Figure 2 In 1985, the Shenzhen Traffic Tour Map took Luohu-Shangbu as the city's "urban area." In fact, two different cultural geographies can be seen on this map. The eastern section near Dongmen, the geographical framework of urbanization was inherited from the historical landform and transportation network. In the western section, the geographical framework of urbanization was the landform and rational square transportation network of modern new town planning. These cultural geographies manifest two different future imaginaries. Luohu was guided by spontaneous trade and urban planning, while Shangbu was influenced by rational urban planning. At the same time, the complementary relationship between these two spatial modes determined urbanization of the SEZ in the 1980s.

3 1990年代：调整未来想象的时代

1990年前，关于深圳未来的想象还是把关内外的关系理解为城市和郊区的关系。以罗湖—上步为实现未来想象的平台。不过，1990年后，"特区"显然已经突破了最初的地理范围和随之而来的未来想象。因此，深圳对二线内外的区域进行了调整，建立了目前由市、区、街道办事处和社区组成的四级体系。1992年，深圳征用关内农村的土地，村民也得到非农户口。[2] 这个过程被理解为"农村城市化"改造，因而重建后的"新村"酝酿着一种自发的未来想象。[3] 这种未来想象基于罗湖—上步、香港城区、旧村建筑物、集体工业区等元素，其结果就是目前城中村的雏形。

新村主要的经济主体是股份公司，其责任包括组织建设以及在新村范围内提供水、电、卫生和道路维护等"自发城市基础设施"的工作。城市化在这些新村具有双重重心——空间的非正式城市化和官方规划城市体系的收编。1996年，深圳颁布了第二次总体规划，对城市未来提出新的想象，将关内外的分散城区组成一个整体。随后，统一城市的理念对深圳提出新的挑战——将新村改造成建设城市未来的重要场所（图4）。

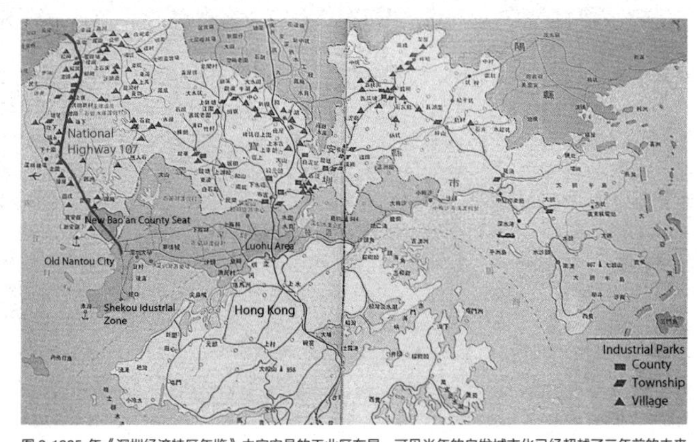

图3 1985年《深圳经济特区年鉴》中宝安县的工业区布局，可见当年的自发城市化已经超越了三年前的未来想象。

Figure 3 The image included in the 1985 *Shenzhen Special Economic Zone Yearbook* shows the layout of early industrialization in Bao'an County. This map reflects the extent of informal urbanization in Shenzhen as opposed to simply within the SEZ or the Luohu-Shangbu area.

4 城中村未来想象的"东南西北"——以田面为例

到了2000年，对深圳未来的想象已经变得很具象——玻璃塔与国际生活方式。在这个时代，当地新村开始显得"过期"了，"农村城市化"所产生的城市环境被鄙视为"脏乱差"的"城中村"。正在此时，福田区实现了第一个城中村改造项目，也就是田面村的旧村更新项目。改造后的田面（2004年左右）让我们看到25年城市化历程中曾有的各种未来想象。为了挖掘城中村所酝酿的各种未来想象，我从田面村的文化地理变迁说起，画一个解析网络，把"田面"当作城市的"中心"。我认为，这样"细读"一片具象的文化地理不但能让我们进一步了解城中村的诞生，也能让我们挖掘已经被"放弃"的未来想象。

田面位于深南大道北侧，在罗湖区和福田区的交界地带。因此，这个解析网络的东西轴穿越华强北和市中心区，而南北轴线连接内地和香港（图5）。20世纪80年代，田面利用区位优势，兴建了16家工厂。从这里，电子产品和玩具被装在托盘上，装入集装箱，通过皇岗口岸经由香港国际海港运往国外。这种经济模式，也带来对未来的想象，就是通过制造业积累资本，在村里投资现代化建设，经营商店，随后建设现代化生活。

1996年的总体规划将未来的市民中心放到田面西侧，田面成为城市更新的早期目标。2004年，田面"村"已经是深圳市机关内的一个小区，与城市水电系统相连，通过公交地铁网络到达全城。股份公司、富荔花园、花园格兰云天酒店、城市绿洲住宅区及绿洲学校等不同主体分享了村里的城市空间。在接下来的几年里，田面工业区被重新定位为"设计之都创意产业园"。虽然改造后的田面空间布局还保留着八九十年代的街道面貌，但是这一环境的主要功能已经不再是生产，而是提供方便的住宅，大多数居民要采用汽车、公交、地铁通勤往来。也就是说，虽然田面是一座展现了1980年代对非正式聚落未来想象的新村，但是2004年以后， 其功能已被整合至1996年总体规划首次提出的统一城市想象中。例如，从田面向东走，在经过华强北到东门的路径上，至少会遇到两个过去的未来想象。第一个未来是1980年代“市区与郊区的想象”，其痕迹包括中心公园里的荔枝树和已经成为世界最大电子市场和山寨手机生产“零地带”的华强北，还包括用斯大林式混凝土建造的商业建筑、1990年在东门开营的中国第一家麦当劳。第二个未来是1990年代的统一城市规划想象的城市，就在田面以东、东门以西的深南大道上的第一批高大上的摩天楼，以地王商厦（1997年）、罗湖书城（2001年）和中国第一家豪华购物中心万象城(2004年)为代表(图6)。

从田面向南走到香港边境，我们就碰上一个突然开放的未来，带有潜藏的诱惑、私欲和危险。紧邻田面南部的皇岗口岸于1989年12月29日起允许货物运输，1991年8月8日开始启用个人过境。香港商人和日间游客激增，导致这一带发廊、SPA、按摩院、卡拉OK酒吧和“二奶楼”的发展。与此同时，香港的商品让深圳人体会到购物的乐趣。2006年，深圳从两个方面打断了这个未来。首先，深圳在沙嘴进行扫黄运动；第二，福田区拆除了渔农村在口岸旁违法修建的17栋高层公寓，为官方批准的旧村改造模式开路（图7）。

从田面往西，我们到达2004年5月31日起成为深圳城市中心的市民中心。当年，市民中心的建成预示着关于深圳未来的想象：玻璃幕墙高楼、环保建筑，以及沿8车道高速公路延伸的、景观优美的绿地。从田面步行到市民中心时，就会路过深圳最典型的未来想象：在这里，居民可以进入建筑形式独特且开放的政府机构办事、在写字楼的办公室工作，在高档住宅居住、在干净的公园里玩耍、在购物中心放松。2006年，深圳宣布“来了就是深圳人”这样的未来想象，而福田中心区就是吸引这些还没来的深圳人的文化地理（图8）。

从深圳向北的路相对更长，经过梅林“二线关”，穿越深圳东部的山脉，直至进入东莞。一路延伸的空间即后来所谓的“城中村”构成的大片区域。此处的城中村文化地理体现农村城市化的两种不同模式。这种文化地理包含两类不同的农民，一种是深圳当地的村民，也是早期深圳农村城市化的受益者；另一种是外地农民，在城中村租房的打工者。越往内地走，未来的想象越有“错位”的感受，一个人要离开内地移民到深圳，在城中村落脚，才能启动对自己的“农村城市化”改造，毕竟从城中村出来后，自己真能成为深圳人（图9）。

2004年，田面的文化地理展现了深圳的新村转向城中村过程中的各种未来想象。这种想象主要有两条可能的发展路径。第一条路径是中国式的，深圳的农村城市化使计划经济得以试验集体所有资源的市场化道路；另一条路径则是新村的城市化形式通过物流、贸易、不断变化的文化规范和美学将中国与资本主义世界联系起来。田面集体所有权图解表明了其布局和建筑是如何协调这两条发展路径。当我们把田面作为十字路口，放在东门（东）、香港（南）、市民中心（西）和内地（北）的基本关系上时，我们才看到一个城中村能包容多少不同的未来想象——东门代表的自由市场、香港所代表的高端消费主义、市民中心的理性规划、内地的无奈和外流（图10）。

虽然2004年田面所展现的城市化模式并不能代表每一个深圳城中村所有的未来想象，但这种多元想象环境的基本要素却存在于其他城中村。例如，在1980年代罗湖一上步的新村，我们发现国际化新生活方式是当地很重要的未来想象导向，这种导向也带动新体制、新机构、国际化审美等未来想象。又例如，1990年代宝安、龙岗两个区，对市场经济体系的未来想象是自发城市化的主要导向，这也意味着多样化的获取利益的方式。2004年田面依然维护城中村原有的包容和多元化，凝练了过去25年里在“村里”扎根的未来想象。换言之，在2004年左右，城中村及城中村旧改都包含着多元化的未来想象。然而，自2005年去工业化以来，深圳越来越追求通过集中化的未来想象，而这种未来想象体现于整体连片“拆迁”和“替换”的城市化模式。

5 2005年后：错过的未来想象

几乎在2004年田面的改造完成后，深圳立即完成了对关外集体土地的征用，并通过了一系列立法，使城中村旧改成为城市的统一未来想象的关键环节。突然间，这座城市的未来想象转变为一种宣传运动，用于促进“纠正”城市在空间历史演进中自然达成的结果与局面。而在这样的未来想象中，城市空间“多元化”经常被视为“脏乱差”。2009年，田面旧改完成5年后，深圳颁布了《深圳城市更新办法》，鼓励开发商制定改造计划，使城中村与深圳作为世界城市的形象相一致。当年，全市共批准了93个村庄更新项目，使改造投资增加了3倍，从2009年的6.723亿元人民币跃升到2010年的25.1亿元人民币。[4]

与早期版本的未来想象不同，因为制造业变少了，所以城市的未来想象很少包容外来工。对比田面与原来跟它同属于一个大队的岗厦就能看到“拆迁”的未来想象。

2004年田面的发展模式至少包括三类住房和个体户商业活动场所，由高端商业、高端住宅、中档商住、低端商住四大板块组成，也就是说田面城市

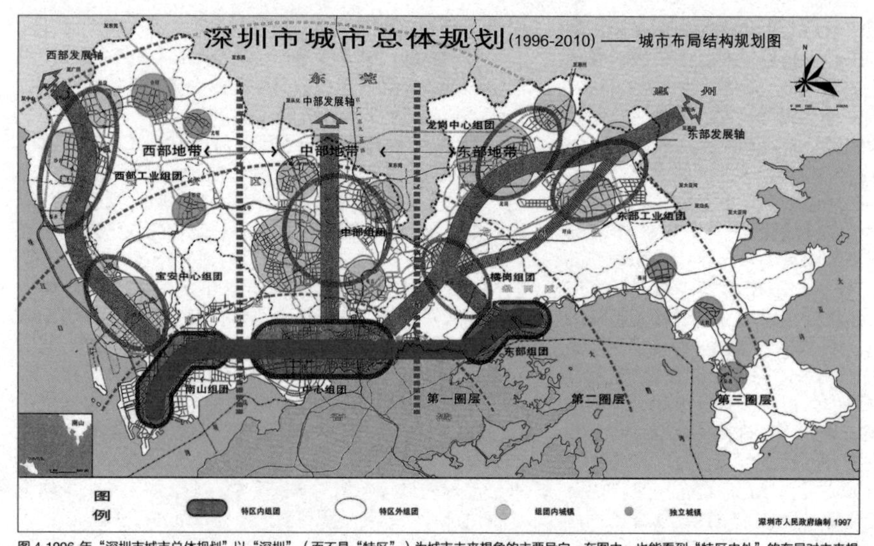

图 4 1996 年“深圳市城市总体规划”以“深圳”（而不是“特区”）为城市未来想象的主要导向。在图中，也能看到“特区内外”的布局对未来想像依然起作用，“关外”的功能也依然是支持“关内”的发展，尤其是福田的发展。不过，与 10 年前的未来想象不一样的是，1990 年代“特区外”的未来想象已经是城市化导向（而不是农业导向）。就这样，1990 年代关于深圳未来的想象以“市区 + 郊区”为导向。
Figure 4 In 1996, the *Comprehensive Urban Plan of Shenzhen* took "Shenzhen" (not the SEZ) as the main direction for the city's imagined future. In the 1996 map, we observe that the layout between "inside and outside the SEZ" still played a role in how the future was imagined, while inside the second line continued to be the focus of development. However, in contrast to how the future city had been imagined ten years earlier, by the mid-1990s, the entire area was imagined as being urbanized without any rural production. In this way, the future imagination of Shenzhen in the 1990s was guided by a model of "urban area + suburb."

更新的未来想象非常包容。这种建筑布局使城中村的居民和生计多样化，也让更多不同的未来想象在这里扎根。与 2004 年的发展模式不同，2009 年岗厦的发展模式将密集、多用途的城中村环境替换为高档、封闭式社区、办公楼和购物中心，也同时让岗厦的未来想象变得非常单一。相比之下，岗厦的旧改先拆平了原有的城中村，迁走了 10 万人左右，随后建设了新的城市环境。问题就来了，除了获得“回迁房”的本村民外，搬进新岗厦的是公司、连锁店和高层白领等主体。[5]

随着岗厦和其他高端楼盘取代了深圳市中心的城中村，我们看到，转型期深圳的解放潜力不仅源于其多用途的建筑环境，也源于城中村容纳非正式、半正规和正规经济的未来想象；在这样的环境里才能想象“来了就是深圳人”能成真。2004 年田面的未来想象容纳工人家庭和个体户。相比之下，2009 年岗厦的未来想象被金融和商业精英占据，着眼于吸引跨国资本。事实上，将岗厦的未来想象进行考古分析的话，就会发现一个与田面布局截然不同的未来想象。岗厦以东是中心公园，以南是福田口岸，以西是购物中心，以北是市民中心。可见，岗厦的文化地理与周围的环境已经成为一新城市的典型片区，其中少有空间实现非主流的未来想象。

白石洲是关内深南大道沿线最后将被改造的城中村。拆迁开始于 2019 年前，白石洲曾是一个容纳多元化的未来想象的、绝无仅有的城市空间。该村与城市供水和电网相连，国家通过邮局、两个诊所和三所小学改善了这个城中村的基础设施和社会服务。在商业上虽有大型连锁店和购物中心在村口附近，然而，大多数村内的生鲜市场、商店、餐馆、微型生产与加工企业以及物流中心是由个体或非大企业经营的。初创企业与物流公司、汽车设计工作室、按规格组装小批量电路板的车间以及青年文化的时尚中心——剑道工作室、黑客空间、两个微型酿酒厂和一个有现场音乐的酒吧——共享这些原来属于群集体的工厂空间。对许多深圳人来说，这一大批创业企业以其便利、廉价的住房，代表着曾经让深圳成为一个可以想象和实现个人未来梦想的机会之地。对于在白石洲和整个深圳不同的城中村中，2010 年以后出现的，这种小规模的士绅化、青年亚文化和具有本土和国际混合情调的社区，似乎是比大规模重建更具包容性的另一种城市未来。不过如今，我们只能从“错过的未来”这种虚拟语境中谈论这个新的城市想象。

Excavating the Future as Imagined in an Urban Village

Mary Ann O'Donnell

An anthropologist by training, O'Donnell is a co-founder of Handshake 302 Art Space and co-editor of *Learning from Shenzhen: China's Post-Mao Experiment from Special Zone to Model City*. She participated in the promotion and educational outreach of the Longheu Girl's School Gallery, was the executive curator of the Longhua (Dalang) Sub-venue of the 2017 Bi-City Biennale of Urbanism\Architecture (Shenzhen), and in 2018 received a Shenzhen Economic Daily influential creative award.

Since its establishment in 1980, the Shenzhen Special Economic Zone (SEZ) has played a vital role in China's vision and pursuit of the future. In the 1980s, the "special zone" symbolized China's post-planned economic future. At present, Shenzhen is both China's fourth first tier city and the world's hardware innovation center, consolidating its exemplary role in China and developing cities worldwide. Thus, Shenzhen has two connections with the project to create a future city. On the one hand, because Shenzhen's planning and positioning have been looking forward, the value of its birth, development and existence has always been related to the future. On the other hand, Shenzhen has become a model because of its successful modernization. Whether in China or in the entire industrialized world, "Shenzhen" has become a symbol used to imagine the future of underdeveloped countries.

Susan Buck Morss's reading of the *Arcades Project*, Walter Benjamin's interpretation of 19th century Parisian material culture inspired my exploration of the archaeology of the future as imagined from Tianmian. I attempt to integrate this unrealized future imagination into the architectural environment of today's city. Benjamin's *Arcades Project* aimed to create a materialistic pedagogy by tapping the potential of historical ruins.[1] This potential is inherent to urban forms because the city itself cites shared symbols. For example, present-day architects cite the architecture of the agora, using Greek history to evoke a more inclusive social Utopian imagination. I am particularly interested in the capacity of urban forms to adapt diverse future imaginaries. In this paper, the ethnographic present of the future imaginary is 2004. That year not only saw the completion of Shenzhen's first urban village renewal project, but also witnessed the cusp of its transformation from a manufacturing to a post industrial city. If that year was an opportunity for Shenzhen's urban transformation, it was also a missed opportunity. Today, we can only access these missed urban models in the virtual context of imagining another future city. Even so, such a reconstruction of how past urban residents imagined the future may still provide reference and inspiration for how Shenzhen currently imagines its future. This imaginative work may also inspire the evolution and direction of the real urban renewal model.

In addition to theoretical interests, my personal experience of living and working in Shenzhen for almost thirty years motivates this essay. During this time, I have observed the process of urban development in Shenzhen and witnessed several major changes. These changes have shaped contemporary Shenzhen's urban form and will determine the shape of the city in the future. When I recall small events in light of the present, I have realized that they contain different historical possibilities and missed opportunities which offer fresh perspectives on the present.

I came to Shenzhen in 1995 to conduct

图 5 在 1986 年《深圳经济特区的现状图》能看到当年田面位于“市里”和“郊区”的边界，其位置让它在 1990 年代以后的发展中获得地理优势，也是不同未来想象扎根的“圃田”。
Figure 5 On the 1986 Current Status Map of the Shenzhen SEZ, Tianmian's position at the boundary between the "city" and its "suburbs." This location not only gave the village geographical advantages during development in the 1990s, but also made it a "nursery" for future urban imaginaries.

fieldwork for my doctoral dissertation in cultural anthropology. My first apartment was in Yuehaimen Village near Shenzhen University, and then I moved to Duntou village on Nanxin Road. "Nantou Old Street" ran parallel to Nanxin Road. In fact, during the Ming and Qing Dynasties (1368-1912), this was the only official road from Nantou Ancient City to Chiwan Tianhou Temple. Along this road was a string of historic villages that ended at Nanshan Mountain. Shekou, which had been settled by fishermen villages was on the other side of the mountain. When I first arrived, these ancient villages had just experienced a round of spontaneous urbanization. Many former farmers had built two-story villas on their allocated homestead land. Although these houses were freestanding and located near each other, the original layout was relatively open because the buildings were not tall. As these homesteads were generally located near the old village, they were called "new villages."

After finishing my Ph.D., I worked as a visiting adjunct in the Department of Anthropology at the University of Chicago for one year, before returning to Shenzhen in 2003. From my arrival in 1995 to 2003, the urban landscape of Shenzhen had changed greatly. The two or three story houses that had defined the original new villages had been demolished and replaced with tenement buildings that were at least six stories tall, but were often higher. Moreover, these "new villages" were now called "urban villages." Outside the villages, the Nantou peninsula coastline had been reclaimed to make room for new development, while glass skyscrapers had redefined Shennan Avenue. In 1995, I could see the coast from the teachers' dormitory of Shenzhen University. However, by the early 2000s, it was a 40 minute walk to the Shenzhen Bay seaside. During the 1990s, trains to and from Hunan and other provinces came into Shenzhen west railway station. However, by the 2000s, that railway station had been abandoned. Indeed, by 2003-2004, Shenzhen was undergoing a critical period of transformation from an export-oriented industrial city to a financial and high-tech city; the city no longer needed workers, but capital and educated professional white-collar workers.

The early 2000s were also the beginning of the emergence of Shenzhen's real estate boom. As a translator and scholar, I observed the renewal of Tianmian New Village, including working in a local school. At that time, the school was called Beijing Jingshan School Shenzhen Branch, and later renamed Shenzhen Green Oasis School. Through these efforts, I also observed that just as the prosperity of the real estate industry created countless new middle classes, the emergence of a new wave of private schools and international schools has also given middle-class children a different educational background and identity from the working class and ordinary citizens. Most of them can study in Europe and America, and then either settle abroad or return home to enter high paying occupations. After Shenzhen entered the second decade of the 21st century, I left the education sector and began to engage in public art. Because of this opportunity, I entered Baishizhou, the largest urban village along Shennan Avenue. In 2013, my partners and I set up an art studio "Handshake 302," which we operated until the demolition of Baishizhou in the summer of 2019. My days in Baishizhou gave me the opportunity to understand Shenzhen from another angle. In particular, the my experiences in Baishizhou gave me fresh perspective on my experience in Tianmian. In this article, I focus on Tianmian village in 2004, when Shenzhen was facing an historic crossroads.

1 Restructuing Shenzhen's Historic Cultural Geography

Before 1978, the cultural geography of Bao'an county comprised at least three integrated cultural regions, which were connected through market towns and village networks, riparian and coastal transportation, rice fields and litchi gardens. In the west of Shenzhen, along the banks of the Pearl River, oyster farmers and fishermen sailed between Guangzhou and Hong Kong. Bao'an villagers spoke Dongguan Cantonese, marketing in Shajing, Gushu, and Xixiang. Villagers in central Bao'an spoke Hakka, gathered in Guanlan and Buji Markets, while Hakka villagers who lived east of Wutong Mountain shopped at Sha Tau Kok. In the border zone connecting Luohu, Futian and the Hong Kong New Territories, villagers spoke Weitou dialect, buying daily necessities at Shenzhen Market and crossing the border to cultivate historical village land.

In 1979, Guangdong Province upgraded Bao'an County to Shenzhen City. In 1980, the central government identified four special economic zones: Shenzhen, Zhuhai, Shantou and Xiamen, indicating that reform and opening up was not only officially launched, but also a new opportunity for imagining and creating the future. In 1981, Shenzhen was redistricted into the SEZ and Bao'an County. A border called the "second line" separated these two areas. The city government managed the SEZ, while the new Bao'an county seat was established at Xin'an. Nevertheless, the existing Cantonese, Hakka and Weitou cultural areas have played decisive roles in the construction of Shenzhen and its diverse future imaginaries. Generally speaking, the boundaries of the SEZ reflected Bao'an's historical cultural geography. First, the Weitou area was divided and surrounded by the "second line" and Shenzhen Hong Kong boundary, becoming "Shenzhen version 1.0" (Figure 1). In the original Bao'an County outside the second line, the western Cantonese section became Bao'an District, while the Hakka area became Longing District. (Longgang was transferred from Huizhou to Bao'an County in 1958. Thus, eastern Shenzhen was originally a part of Huizhou, a traditional Hakka cultural area.)

2 The 1980s: Imagining the Future City

The construction of the SEZ began in Luohu-Shangbu, starting from the historic Shenzhen market, that is, today's Dongmen area, and extending westward to today's Shanghai Hotel. The earliest iteration of the SEZ was located on the Weitou isthmus, entering Hong Kong via the Kowloon Canton Railway and Man Kam checkpoint, connecting Bagualing, Shangbu and other local industrial parks to the world economy. Therefore, the development of Luohu-Shangbu not only adopted many of Hong Kong's infrastructure standards, but also Hong Kong business model and language (Hong Kong Cantonese, in fact, is a variant of Weitou dialect). The cultural geography of the early SEZ also promoted consumerism, modern architecture and commercial entertainment, (which was known as "bourgeois lifestyle or capitalist corruption" in Mao era China). In daily life, residents living in other parts of Shenzhen (including new immigrants living in Shekou Industrial Zone) referred to this area as "Shenzhen." In the busy streets and rich markets of Luohu, a new vision of the future was born (Figure 2).

It should also be emphasized that the early construction of the SEZ was not limited to formal planning in Luohu-Shangbu, but included informal development by local collectives. Local collectives sold agricultural products to Hong Kong and established industrial parks. These examples of local rural urbanization also included a new imagination of the future city. More importantly, this early collective investment became one of the prototypes and driving forces of early industrial development through the city more generally. In the absence of government direct investment and supervision, the Guangdong and Hakka speaking areas of Bao'an county nevertheless developed at the same time as Luohu-Shangbu, albeit with a different model. Between the G107 national highway, which connects Songgang to Wenjindu checkpoint and the Pearl

River coastline, an industrial corridor developed. In addition, collective enterprises along the Pearl River continued to export oysters, poultry and other agricultural products to "Shenzhen" and Hong Kong. The extent river transportation network also facilitated the importation of construction materials from Foshan and Guangzhou. This larger context promoted the development of western Bao'an (the Cantonese section of the county). In contrast, apart from Buji, Hakka villages in the mountainous areas of eastern Shenzhen rarely had opportunities to develop relations with Hong Kong. Even Shatoujiao, located at the boundary between Shenzhen and Hong Kong, did not industrialize as completely as the eastern section of Bao'an. Therefore, the future imagination here is displayed in a special kind of urban immigrants. Many villagers from Longgang moved to Luohu-Shangbu to look for job opportunities. Some worked in factories and some set up stalls in Dongmen business district (Figure 3).

The cultural geography of Shenzhen in the 1980s was rooted in political reform, economic experiments and the interaction between local villages and towns. It was also affected by the Hong Kong market and world economy. The establishment of the SEZ restructured the original cultural geography of Bao'an, which nevertheless influenced the city's future cultural geography. In Luohu-Shangbu, the earliest urban construction in Shenzhen imagined the future city as a border city based on processing industry and commerce. More importantly, the realization of a good life here and the specific conditions for its realization impacted residents in areas outside the formal plan. Therefore, the cultural geography of Luohu-Shangbu also drove the imagination and spontaneous urbanization of the future in other parts of the city. In the west, Bao'an's collective enterprises redeployed traditional agricultural assets into local manufacturing bases. In the east, it was more difficult for collective enterprises to directly participate in industrialization. Therefore, this old Hakka area became an important source of food and labor for the rest of Shenzhen. Significantly, in recent years, the underdevelopment of the east has facilitated experiments with green planning and suburban residential areas—an altogether different future from that imagined in Luohu-Shangbu during the 1980s.

3 The 1990s: Adjusting the Future Imaginary

Before 1990, Shenzhen's future vision was to understand the relationship between the downtown area (Luohu-Shangbu) and its suburbs (Bao'an County). However, after 1990, the city had obviously broken out of its original geographical area and concomitant future imaginary. This included administrative restructuring both inside and outside, establishing a four-level system of the city, district, street offices, and communities. In 1992, Shenzhen expropriated the land of inner district villages, with local villagers receiving urban hukou.[2] This process was understood as formal "rural urbanization," which catalyzed new future imaginaries.[3] This new imaginary was based on elements such as Luohu-Shangbu, the Hong Kong urban area, old village buildings and collective industrial parks. The result was the prototype of present-day urban villages. The main economic body of the new village was the joint-stock company, whose responsibilities included organization and construction, as well as providing informal urban management of water supply, electricity, sanitation and road maintenance within the new village footprint. In these villages, urbanization had a double focus—informal urbanization of the space and entering the officially planned city. In 1996, Shenzhen promulgated its second master plan, which put forward a new imagination for the future city, integrating the SEZ and its suburbs. Subsequently, the idea of the integrated city presented Shenzhen with a new challenge—transforming new villages into important sites for building the future (Figure 4).

4 Tianmian North South East and West: Envisioning the Future from an Urban Village

By 2000, Shenzhen's future imaginary had become very explicit—glass towers and international lifestyles. In this era, new villages seemed to have "expired," and the urban environment generated by informal rural urbanization was despised as a "dirty, chaotic and substandard," producing "urban villages." At this time, Futian District completed Shenzhen's first urban village renewal project in Tianmian circa 2004. The reconstructed village manifested 25 years of accumulated (but changing) visions of the future. In order to explore the diverse future imaginaries that fermented in new/urban villages, I start with the cultural and geographical changes in Tianmian village, proposing an analytical grid with "Tianmian" as the city's "center." This close reading of a concrete cultural geography can not only let us further understand the birth of villages in the city, but also let us explore the future imagination that has been "abandoned."

Located in the north of Shennan Avenue, at the junction of Luohu District and Futian District, Tianmian used to be situated at the border between Luohu-Shangbu and its suburbs. The east-west axis of the analytic grid runs between Huaqiangbei and the downtown area, while the north-south axis connects the mainland and Hong Kong (Figure 5). In the 1980s, Tianmian took advantage of its location and built 16 factories. From here, electronic products and toys were loaded on pallets, loaded into containers, and then transported to the port of Hong Kong international via the Huanggang checkpoint. This economic model also included a future imaginary: accumulating capital through manufacturing, investing in modernization in the village, operating stores, and building a modern life.

On the 1996 *Shenzhen Comprehensive Plan*, Tianmian was located east of the future Civic Center, which became the early goal of urban renewal. In 2004, Tianmian "Village" was already a community within the Shenzhen municipal apparatus, connected to the urban water and power grids, as well as to the bus and subway networks. Within the village itself were the offices of the village stock company, Fuli Garden Estates, the Garden Grand Sky Hotel, Green Oasis residential estate, and Green Oasis School. Over the next few years, Tianmian industrial park would be repositioned as the "City of Design Tianmian Creative Industrial Estate." Although the urban form of renewed Tianmian inherited the street life of the 1980s and 1990s, nevertheless the primary function of the village was no longer production, but rather to provide convenient housing and most residents went to work by car, bus or subway. In other words, although Tianmian appeared to be a new village, sharing a future imaginary with informal developments from the 1980s, nevertheless, its function after 2004 was incorporated into the unified city imaginary that had first appeared in the 1996 *Comprehensive Plan*. For example, when going east from Tianmian one encounters two residual future imaginaries. The first (discarded) future is the imagined "city and suburbs" model of the 1980s. Traces of this imaginary include the Central Park litchi trees, Huaqiangbei, which has become the world's largest electronic market and "ground zero" for the production of fake mobile phones, and Stalin-style concrete buildings and China's first McDonald's in Dongmen. The second latent future is the unified city as it was imagined by urban planning in the 1990s.

图 6 2006 年田面工业区将 20 世纪 80、90 年代“制造业城市”的未来想象展现出来。
Figure 6 This picture was taken in 2006, when the repurposed buildings in the Tianmian Industrial Park manifested the future imaginary of Shenzhen as a "manufacturing city," which was popular in the 1980s and 1990s.

Shenzhen's first generation of tall skyscrapers appeared east of Tianmian and west of Dongmen along Shennan Avenue, including Diwang Commercial Building (1997), Luohu Book City (2001) and the MixC, China's first luxury shopping center (2004) (Figure 6).

Walking south from Tianmian toward the Hong Kong border, we encounter a suddenly open future with potential temptations, selfish desires and dangers. Located south of Tianmian, Huanggang checkpoint opened for cargo shipping on December 29, 1989 and for personal transit on August 8, 1991. The surge of Hong Kong businessmen and day tourists into the SEZ led to the development of hair salons, spas, massage parlors, karaoke bars and "mistress houses" in this area. At the same time, Hong Kong products stimulated consumerism in Shenzhen. In 2006, Shenzhen interrupted this future in two ways. First, Shenzhen carried out the anti pornography campaign in Shazui. Second, Futian District demolished 17 high-rise apartments that had been illegally built by Yunong Village next to the Huanggang checkpoing, opening the way for the officially approved old village reconstruction model (Figure 7).

Heading west from Tianmian, we arrive at Citizens' Center, which opened on May 31, 2004, anticipating Shenzhen's future imaginary: tall glass buildings, sustainable architecture, and beautiful green spaces that extend along an eight-lane expressway. When you walk from Tianmian to the Civic Center, you traverse Shenzhen's most iconic future imaginary: here, residents can enter the unique and open government institutions, work in modern office buildings, live in high-end houses, play in clean parks, and relax in the shopping center. In 2006, Shenzhen announced the future imaginary of "come and

图 7 2003 年的田面还留下来 20 世纪 80、90 年代的另一种未来想象，私欲的满足意味着危险与死亡。
Figure 7 This picture was taken in Tianmian in 2003, suggesting a dystopian future as imagined in the 1980s and 1990s, when satisfying selfish desires was understood as a path to danger and death.

图 8 2007 年的田面展现了玻璃幕墙建筑和机构，其对巩固“统一城市”的未来想象非常重要。
Figure 8 Taken in 2007, this photograph of Tianmian includes the glass buildings and institutions that have been important to consolidating the "unified city" future imaginary.

you are a Shenzhener." The Futian central district embodies the cultural geography to attract these future Shenzheners (Figure 8).

The road leading north from Shenzhen is relatively long. After passing the second line in Meilin, one traverses the mountains and valleys of eastern Shenzhen before entering Dongguan. The spatial spread between Meilin and Dongguan comprises "urban villages." Here, this urban village cultural geography expresses two different forms of rural urbanization. On the one hand, local Shenzhen villagers were the beneficiaries of the early rural urbanization in Shenzhen. On the other hand, migrant farmers from Shenzhen's hinterland rent houses in these urban villages. The further inland one goes, the more "misplaced" Shenzhen's future imaginary seems because these urban villages are the gateway to personal urbanization. After all, migrant farmers become Shenzheners only after moving out of a village into high-end Shenzhen (Figure 9).

In 2004, Tianmian's cultural geography included diverse future imaginaries that suggested different ways for the transition from new village to urban village in Shenzhen. These imaginaries relied on two possible development paths. The first path was domestic. The rural urbanization of Shenzhen enabled the planned economy to test a market-oriented road for collectively held resources. A second path, the urban form of new villages also connected China with the capitalist world through logistics, trade, changing cultural norms and aesthetics. In fact, mapping Tianmian suggests how its layout and architecture mediated these two paths to the future. When we regard Tianmian as the nexus of these two paths—one moving east-west from the Civic Center to Dongmen and one moving north-south from the hinterland to Hong Kong—we can see how many different future imaginaries an urban village can actually sustain: a free market future (represented by Dongmen); high-end consumerism (represented by Hong Kong); the rational urban planning (represented by the Civic Center), and; rural helplessness and outmigration (represented by Shenzhen's hinterlands) (Figure 10).

Although Tianmian's urbanization model in 2004 can not represent the future imaginaries that took root in every Shenzhen urban village, nevertheless the basic elements of this diverse imaginary did exist in other urban villages. For example, in the new villages of Luohu during the 1980s, international lifestyle was an important feature of future imaginaries, which also drove how people imagined future systems, institutions and aesthetics. Or again, in Bao'an and Longgang Districts during the 1990s, market economy institutions informed the future imaginary that guided informal urbanization, which in turn suggested diverse ways for making money. Circa 2004, Tianmian still maintained the original inclusiveness and diversity of these earlier iterations of the urban village, refining the future imaginaries that had taken root in the village for roughly 25 years. In other words, circa 2004, urban villages and imagined urban renewal thereof still contained diversified future imagination. However, once Shenzhen began to deindustrialize in 2005, the city increasingly pursued the "unified city" imaginary, which takes its most concrete form in the ongoing process of urban village "demolition" and "replacement."

5 After 2005: Missed Futures

At roughly the same time that the renewal of Tianmian was completed, Shenzhen finished expropriating collectively held land outside the second line, making the renewal of urban villages a key part of the city's "unified" future imaginary. Suddenly, the city's future imaginary was transformed into a publicity campaign to promote the "correction" of the results and situations that had been achieved through informal urbanization in the city. According to this future imaginary, the "diversity" of urban space has often been regarded as "dirty, chaotic and substandard." In 2009, five years after land expropriation was completed, Shenzhen promulgated measures for urban renewal, encouraging developers to formulate plans to remake urban villages in ways that were consistent with its imagined future as a world city. That year, the city approved 93 village renewal projects, which tripled investment in urban village renewal from 672.3 million yuan in 2009 to 2.51 billion yuan in 2010.[4]

Shenzhen's current future imaginary differs from earlier versions because as manufacturing has decreased, so too has the place of migrant workers in the city's imagined future. When we compare Tianmian with Gangxia, we see how "demolition" includes a specific vision of the future. In 2004, the development model of Tianmian comprised at least three types of housing and individual businesses: high-end commerce, high-end housing, medium-sized commercial housing and low-end commercial housing. In turn, the city as imagined from Tianmian was very inclusive. This architectural layout not only diversified the residents and livelihoods of the urban villages, but also accommodated diverse futures. In contrast, the development model of Gangxia in 2009 replaced the dense and multi-purpose urban village environment with high-grade, closed communities, office buildings and shopping centers. This urban space also included a vision of the future, but when imagined from new Gangxia this future is singular and exclusive. To create new Gangxia, the old buildings were demolished, about 100,000 people were displaced, and a new urban form was constructed. Suddenly, the problems of this future imaginary became apparent. With the exception of local villagers who received compensation housing, residents were displaced by companies, chain stores, and managerial class workers.[5]

As Gangxia and other high-end renewal projects have replaced urban villages in the center of Shenzhen, we can how the city's potential during the 2004 transition stemmed not only from its multi-purpose building environment, but also from a future imaginary that accommodated the informal, semi formal and formal economies. In such an urban environment, we can imagine that "come and you are a Shenzhener" is a true possibility. In 2004, Tianmian's future vision accommodated working families and self-employed households. In contrast, in 2009, Gangxia's future vision was occupied by financial and business elites, with a focus on attracting transnational capital. When we perform an archaeology of the imagined future as built in Gangxia, we discover just how different its future was from the futures that we have tracked in Tianmian. To the east of Gangxia is Central Park, to its south is Futian port, to its west is the Central Axis shopping mall, and to its north is Shenzhen Civic Center. Clearly, the cultural geography and surrounding environment of Gangxia represent a unified vision of the future, in which there is little space to imagine and realize alternative futures.

In 2019, the demolition and evictions of Baishizhou began the renewal of the last urban village to be transformed along Shennan Avenue. Baishizhou was once a unique urban space that accommodated a diversified future imaginary. The village was connected to the urban water supply and power grid. The state has improved the infrastructure and social services of the urban village, opening a post office, two clinics and three primary schools there. Although there were large chain stores and a shopping mall near the entrance of the village, most of the fresh food markets, shops, restaurants, micro production and processing enterprises and logistics centers in the village were operated by individuals or small enterprises. Startups shared repurposed factory spaces that with logistics companies, automobile design studios, workshops for assembling small batches of circuit boards according to specifications, and fashion centers of youth culture—a Kendo studio, hacker spaces, two micro breweries and a bar with live music. For many

图 9 2004 年的田面还经常见到内地很普及的小个体户经济模式。
Figure 9 This photograph was taken in 2004, suggesting that the entrepreneurial model of hinterland rural areas was also an important component of Tianmian's economic environment.

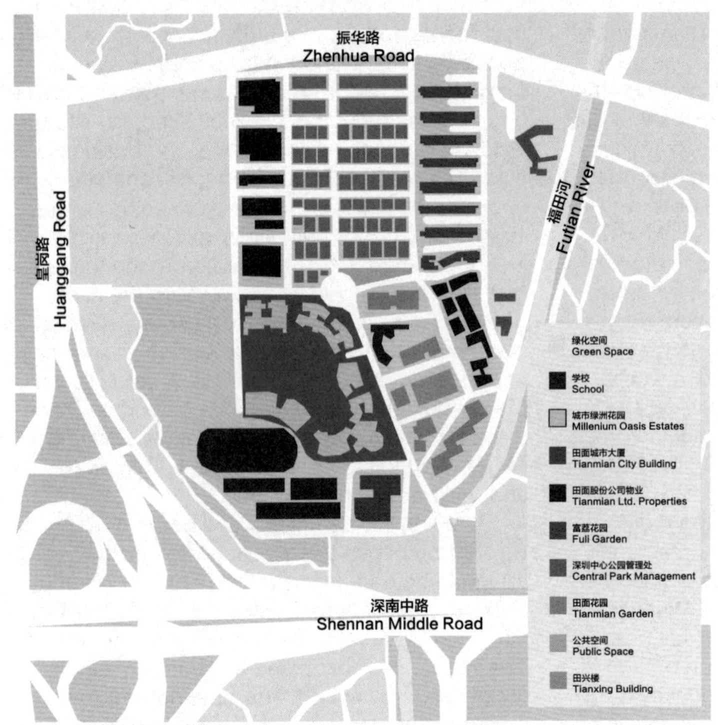

图 10 在 2004 年田面未来想象的类型分析图，可见田面这种文化地理包含对新的管理体制、市场制度、国际化审美、消费主义生活方式等四种未来想象的导向。在这样丰富的文化地理．不同身份的居民都有机会在田面落脚，成为深圳人。
Figure 10 This 2004 map of Tianmian's suggests the diversity of imaginary futures that were built into the landscape. Tianmian's cultural geography comprises sites for new management systems, market institutions, international aesthetics and commercial lifestyles. In such a rich cultural geography, residents of different identities had diverse opportunities to settle in Tianmian and, by extension, become Shenzheners.

people in Shenzhen, this large number of start-ups, with their convenient and cheap housing, represented a place that once made Shenzhen a site to imagine and realize personal futures. After 2010, Baishizhou and other urban villages supported youth subcultures and alternative futures that were more inclusive than large-scale reconstruction. However, since 2019, we can only talk about these alternative imaginaries in the virtual context of "missed futures."

[1] BUCK-MORSS S. The Dialectics of Seeing: Walter Benjamin and the Arcades Project [M]. Cambridge, Mass.: The MIT Press, 1989.
[2] 社会治理演变中的深圳社区居委会 [EB/OL]. http://sz.people.com.cn/n2/2016/0420/c202846-28181193.html.
[3] HUANG W. The Tripartite Origins of Shenzhen: Beijing, Hong Kong, and Bao'an [M]// O'DONNELL M A, WONG W, BACH J, eds. Learning from Shenzhen: China's Post-Mao Experiment from Experiment to Model City. Chicago: University of Chicago Press, 2017.
[4] 刘颖，林艳柳．城中村社会分层与城市权利：基于深圳市岗厦村改造的实证研究 [R]. 岗厦罗生门．深圳：土木再生，2015: 1-14.
[5] 时娜，彭超．深圳岗厦村拆迁集体暴富，造就十个亿万富豪 [N]. 上海证券报，2009-12-25. http://news.qq.com/a/20091225/001471.htm.

城市规划与城中村：谁来改造谁？

黄伟文

未来+学院联合创始人，城市研究者和城市设计工具开发者。曾任深圳市城市规划和国土资源委员会副总规划师（2011—2012）、深圳公共艺术中心和深圳城市设计促进中心主任（2011—2017）、深圳市规划局城市与建筑设计处副、正处长（2003—2009）。哈佛大学设计研究生院 Loeb 访问学者（2009—2010）、深港城市\建筑双城双年展的主要创建、组织与推动者（2005—2017）、土木再生城乡研究所联合发起者与推动者（2008—）。

本文首次发表于《住区》, 2011(5): 102-105；作者为本书补充，2021 年；由罗祎倩中译英。

作为一名城市规划师，我与深圳城中村的接触与研究已有 16 年。我个人对城中村的认识也在不断转变中，在这里和大家分享交流。

1　规划改造城中村？

1995 年，我在深圳市城市规划设计研究院，参与了新洲村与大冲村的改造规划（图 1）。那时候，城中村这个名字还没叫开，意味着城市还没有将农村完全包围，两个村子仅仅是以个案进入我的视野。借助研究生时参与导师朱自煊教授承担的北京什刹海、黄山屯溪老街等历史风貌保护更新规划的经验，我提出了城中村的人文观察与调研、城市肌理研究和拆差建新、新旧共生的规划，其中也有获得院里好评和广东省优秀规划成果奖的。那时我相信，好的城市规划与设计，能够扬长避短地改进已有的城中村空间形态。至于这种物质形态规划会如何被实施，就不是当时从学院走出、成为生产型规划师没多久的我要考虑的。

1996 年我开始参与深圳市中心区规划管理机构，成为管理型规划师。面对中心区范围内岗厦村河园片区不断的自发扩张，我曾经忧心忡忡，觉得这些越来越密集的村民楼，严重影响了经过多轮国际咨询、不断改进的中心区城市设计水平；同时也不愿看到这种反差不断加剧的后果，有可能激发更强的铲除这些违章建筑的舆论与决心。

我一方面主张规划部门要与村集体接触、谈判，另一方面努力寻找政府和村民都能接受的规划。那是 2000 年左右，上海新天地还没开张，但从资料中看到这种大胆利用旧建筑来创造开放、时尚和适于步行的城市空间的做法，令我耳目一新。我向领导、规划师、区政府和岗厦村的人士推荐这样的改造方式，得到当时福田区领导的初步支持，邀请参与新天地项目设计的日建公司建筑师宋照青来为岗厦村做改造规划。

宋照青的规划选择性拆除了一半（17 万平方米）村民住宅，将剩下一半村民住宅通过改建并与新建（30 万平方米）住宅一道，组成能顺应原有街巷走向与结构的 10 个院落街坊，并延续和升级了当时远近闻名的岗厦食街（图 2）。这个方案得到岗厦村的赞同，但奇怪的是当时的技术委员会专家因为担心村民之间可能无法在拆谁家留谁家的问题上达成共识，以缺乏可行性为由将方案搁置。

2002 年国际建筑师协会主席、巴西库里蒂巴市前市长、建筑师詹姆·莱纳先生提出“城市庆典”概念设计国际竞赛，号召专业人员关注城市问题。我将多年城市规划管理中对城中村的认识和思考作了一个整理，与同事及都市实践合作，于 2002 年底提交了“有机整改深圳城村”的方案，获中国建筑学会“城市庆典”概念设计竞赛国内评选二等奖。这个方案被都市实践整理出版的《村·城 城·村》收录（图 3），还参加了 2005 年深圳双年展、2008 年中法“街道是我们大家的！”巡回展等。

这个阶段，我只是把城中村看作一种既有的城市现状和空间文脉，觉得可以找到一种理想的规划方法，将其结构肌理进行梳理和延续，并进行空间整理和新的建设控制。这总比那些要么推倒重来当一张白纸规划，要么拿现状束手无策的两种极端做法来得有意思吧！我也不希望自己的主张被误解为一种历史文化保护观点。我对别人更多的解释是，这种新旧共生、不把已有房屋全部变为垃圾的改造方式，是一种更加具备经济可行性和可持续发展的选择。为了劝阻少拆房子，我甚至向那些厌恶城中村建筑光秃形象的人描绘，只要对城中村立面形象加以改造，深圳城中村也能像上海新天地，甚至威尼斯与阿姆斯特丹那样的欧洲老城老镇那样风情万种。没想到财富积累起来的城中村及地方政府，这几年真的

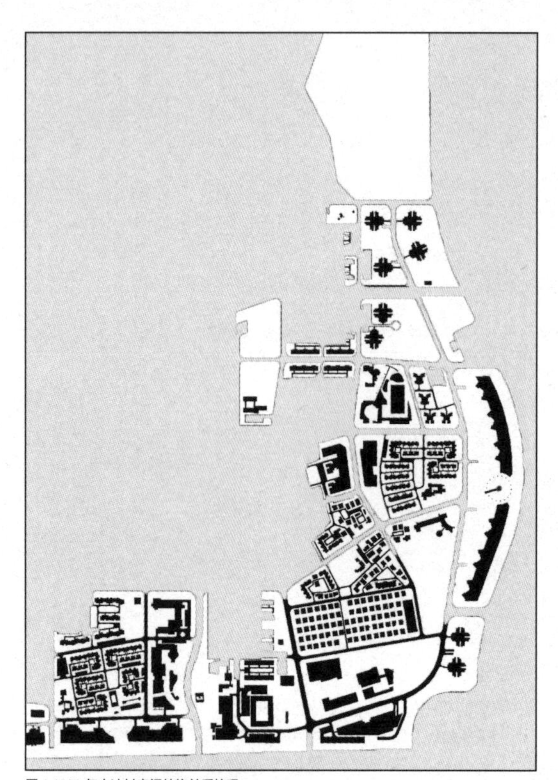

图 1 1996 年大冲村空间结构关系梳理
Figure 1 Relationship of the spatial structure of Dachong Village in 1996

将面向城市大干道的村民房子都“穿衣戴帽”成了欧洲风情。尽管这种伪装有助于延长城中村在城市中的生存时间，也与我七八年前的提议没有什么关联，但我还是为自己曾经有过这样的设想感到吃惊和惭愧。

而我孜孜以求通过规划来改造城中村的那些主张都未能实现。现实里能够实现的案例，是通过城市规划来彻底铲除，而不是改造城中村，如渔民村、渔农村、岗厦村、大冲村。而这种将现状痕迹铲除得一干二净的城市规划，从绿色可持续、社会结构和城市活力延续性、城市空间与建筑多样性的角度看，我个人认为是非常不道德的城市规划。

2 如何评价城中村?

深圳市 2004 年发起的城中村改造动员，当时的决策者、规划师、主流媒体基本上是将城中村当作城市负面形象来看待。比较极端的城中村负面评价主要有“城市毒瘤”“城市包袱”“城市疮疤”“城市癌症”。普遍认为的城中村问题有：形象丑、密度高、环境差、卫生糟、治安乱、公共空间不足、社区配套和市政设施缺乏、消防不符合规范，等等。

我在 2006 年《村·城 城·村》中的文章 [1] 认为：“城中村实际上自发地对城市规划与管理的不足进行了补充，有助于改善城市土地功能和效益、健全廉价住宅体系、丰富城市服务内容、降低城市服务和创业成本。”城中村住房的补充调剂，使得深圳避免了在快速城市化进程中因大量人口涌入城市而形成棚户区贫民窟的可能。“城中村主观上是原村民突破各种政策法律限制追求自身利益的结果，客观上是城市土地按照其经济规律以扭曲形式释放其应有价值的必然”“是对将城市土地限制为福利用品、无视其经济价值的计划经济式土地政策的一种纠正，也是原村民贡献其土地支持城市建设之后，以所剩土地分享城市发展成果的一种权利”。

我在 2006 年博客文章《城市的下半身》[2] 中更指出，城中村犹如城市不可缺少的下半身，它支持了城市的功能正常运作，即使“上下半身之间的要素可以流动转化，但其上下半身的相互依存和机能分工却是永恒存在的。换句话说，如果要把大部分城中村推倒重建成高档楼盘，不管城市乐意与否，那些原城中村的租户会转移到新的区域、新的犄角旮旯中，塑造新的下半身形态和机能，因为他们和城市相互需要”。

2011 年我在 *Domus* 中文版的文章《深圳：如何快速塑造一座城市》[3] 中又有了进一步的认识：“城中村成为城市飞地和自治社区，也是能自我调适、有序支持城市运作的自组织系统。城中村承担起深圳多年保障房的作用，不仅因为村民自建房成本低（没有地价及报建登记等手续）、设计经济实用、紧凑开放而且租金便宜，更因为全市村落均匀散落在市域范围，基本上能以步行的距离，覆盖和服务与其相邻的中心区、商业区、工业区。这种农业时代以‘日出而作’的步行距离控制的村落分布，奇妙地转化为位置分布均匀的深圳‘中低收入阶层居住布局’，弥补了众多版本城市规划对中低收入人群居住用地安排的长期忽略，平抑了政府保障房建设不足时期的廉租房需求，同时也减缓了城市交通的恶化程度。”

2005 年深圳双年展的城中村专题展及论坛后，专业界及公共领域对城中村有了更多全面客观的研究和讨论，其中对我有启发性的有：深圳综合开发研究院李津逵很早从事的城中村社会系统调研；深圳首届双年展助理策展人杜鹃在组织城中村展览研讨后，进一步做的白石洲调研和皇岗村改造研究，将城中村视作中低收入居住地的同时，也是维持城市交通运作的双保障系统；曾经的哈佛大学设计研究生院的马可·森扎蒂教授认为深圳城中村比深圳城市化区域更具城市性；张永和提出的按城中村方式改造市民广场的建议；规划前辈约翰·弗里德曼考察深圳的观点——“与华侨城相比，下沙更能代表深圳文化，只有后者的社会风貌和社会氛围才是深圳的特色。”

3 城中村改造城市规划！

既然现有城市规划不能真正改造而只是铲除城中村，而城中村又是一种积极和正确的存在和真正的深圳城市特色（图 4），那么可以判断：出问题的不是城中村，而是当下的城市规划；需要改造的，不是城中村，而是当下的城市规划。可以启发和改造城市规划的，也恰恰是被城市规划威胁的城中村。

当下中国城市规划所遵循的现代主义理论基础，其实早在 20 世纪 60 年代的美国已被新闻记者及社会活动家简·雅各布斯成功改造，甚至说是颠覆清除。当时美国现代主义城市规划也是大行其道，到处推倒重来改造城市老区。忍无可忍的小妇人简奋起与纽约最大开发商

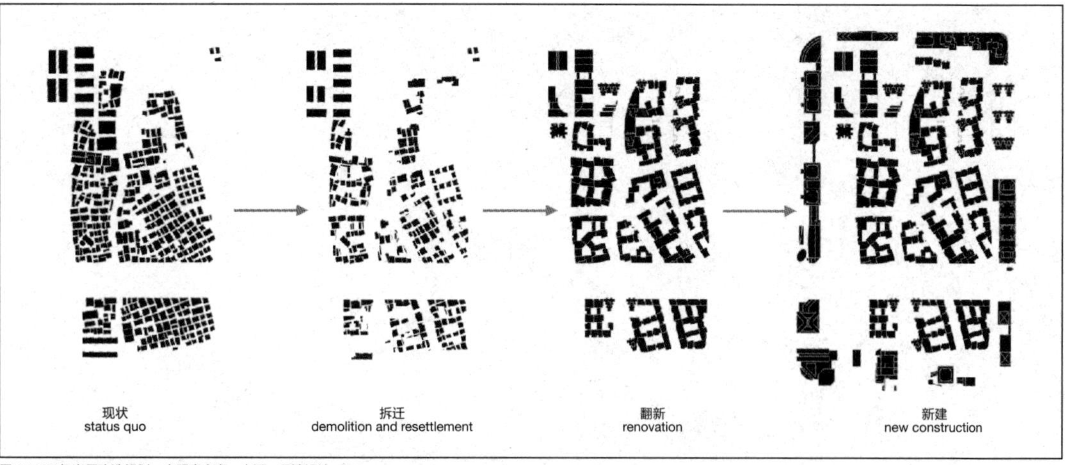

图 2 2000 年岗厦改造规划，宋照青方案，来源：日清设计
Figure 2 Song Zhaoqing's scheme of Gangxia Renewal Planning in 2000, source: Lacime Architects

角力，通过发动社区和写文章，挽救了要被规划改造颠覆、被高速公路穿越的格林威治区，其所著《美国大城市的死与生》也改造了甚至颠覆了现代主义城市规划。

虽然迄今为止的历史观察发现，日益扁平的地球一端发生的事情，也总会出现在另一端。没有曼哈顿格林威治区浪漫优雅的城中村，凭什么能改造中国城市规划呢？十多年来我曾经以为自己的规划知识和设计技巧可以改进城中村，但随着对城中村和新规划城区的深入观察和比较，我越来越对没有规划师的自我建设所呈现出的个体智慧的群体效果充满尊重和深受启发，而对有规划师控制建设的新城区的精英阶层短视盲区充满警惕和反思。过去我认为城中村需要改进的划地形状、市政和消防问题，相比较其巨大和珍贵的自治自主自足的低碳生活以及和谐社会建设的标本意义来说，已经是微不足道的问题。2003 年的一次座谈发言中我曾偏激地认为：城中村需要改造是个伪命题，通过这么多年的城中村观察思考，终于明白，通过城中村来改造城市规划才是真命题。迄今为止，日益通过城市规划被铲除的深圳城中村，可以给当下城市规划教育进行补习甚至改造的课程有：

（1）城市规划如何可以有自下而上真正的公众参与？

当下城市规划的做法是公示征集意见。

城中村答案：让公众拥有土地。

（2）如何能让公众而不仅仅是大开发商成为城市建设主体？

当下城市规划偏爱大地块，规定小于 3000 平方米土地不能开发，而且正在流行大型综合体。

城中村答案：土地细分到 100~2000 平方米 / 块。

（3）如何创造步行友好的城市？

当下城市规划在布置完高架快速路或拓宽道路后，再建议设置人行桥或地下通道。

城中村答案：没有围墙，街巷纵横密布，道路间距 10~40 米，路幅宽度 2~8 米。

（4）如何创造活力城市空间？

喜欢功能清晰甚至单一的当下城市规划通常把这个问题留给商业策划。

城中村答案：凡没有绝对影响到安全健康的功能都可以自由混合和灵活调整。

（5）如何创造社区认同感？

当下城市规划会依赖图形化标志性形态设计来避免所谓“千城一面”。

城中村答案：自我建设基础设施并自治管理，保留或更新牌坊、祠堂、健身广场（通常还有一小块荔枝林风水林）作为社区共享空间设施。

（6）如何提供低成本住宅及其生活环境？

当下城市规划几乎失语。

城中村答案：个人或集资合作建房，免除地产开发利润和税费，或保留小产权房状态，允许日照时间不足的住房存在。配置菜市场，允许种养、小作坊、摆摊和临时搭建。

（7）如何创造低碳环保城市？

当下城市规划会侧重在绿地和绿色指标上。

城中村答案：参见以上（3）（4）（5）（6）条。

（8）如何建设卫生、治安、消防模范城市并使不同阶层和谐共处？

当下城市规划有苛刻的日照、消防、绿化规范支持建设高档社区，保障人群则越来越往远郊地区集中。

城中村答案：社区自治、民兵组织、缩小消防车尺寸、通过积累和投入不断更新基础设施。合作建设少数高档楼盘散布在中低收入社区边上，分享社区资源。

（9）如何争取居民权益最大化？

当下城市规划需要考虑这个问题吗？

城中村答案：股份合作，全民选举，并有代表本社区利益的代表。

综上，城中村不仅仅为规划建设公众参与、宜居宜业、和谐低碳城市的样板，还是城市不可缺少的基础框架。如果任由当下的城市规划以改造之名行消灭城中村之实，伤害的不仅仅是点状散落的样板区域，而是在损害这个城市的支持系统。是任其继续瓦解城市支持系统直至瘫痪，导致真正患上城市 / 社会生态失衡的“癌症”？还是从这一支持系统中获取灵感和知识资源来改造当下的城市规划？当下的城市规划师和城市更新决策者都面临着选择和挑战。

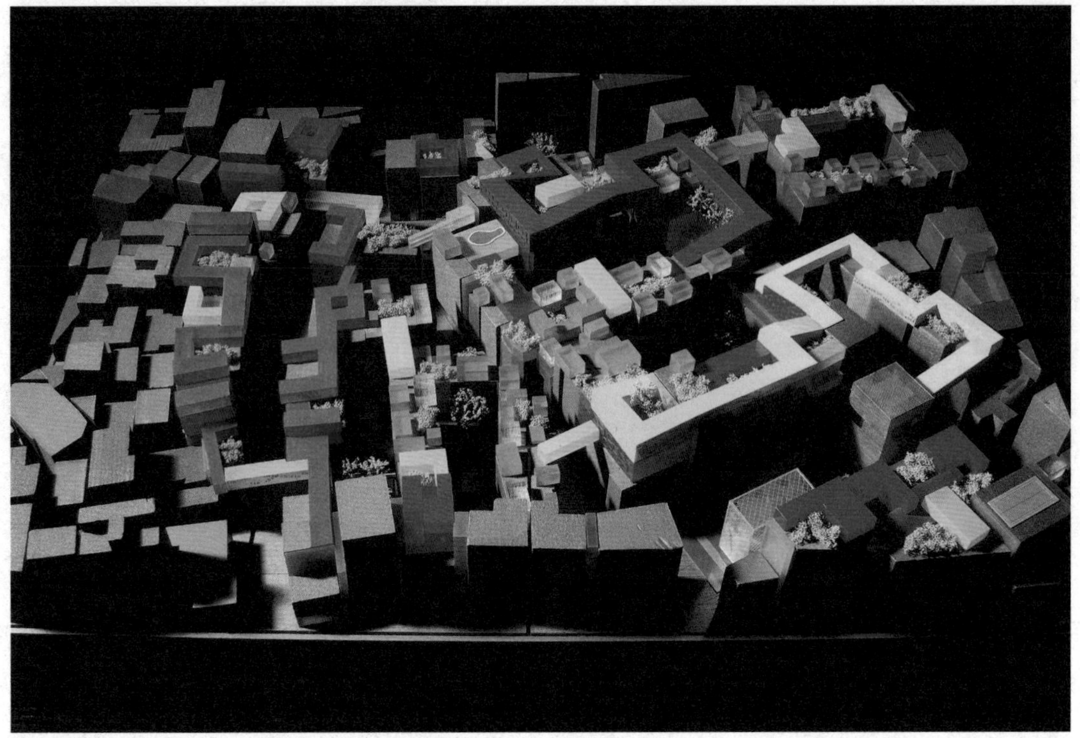

图 3 发表在 2006 年《村·城 城·村》中，岗厦村改造设计概念方案，都市实践与黄伟文、张建辉团队合作
Figure 3 Published in 2006 *Village/City City/Village*, Gangxia Village renovation design proposal by URBANUS and Huang Weiwen and Zhang Jianhui team

4 后记：六维评估体系

2011 年我曾撰文讨论城中村和城市规划各自的问题，提出城中村可以启发城市规划的优点。有鉴于每个人对事物认识的角度、立场和侧重不同，对城中村以及其他规划社区的优缺点也会有较大的评价差异。这些差异能否放到一个各方都能认同的系统框架上来比较和讨论呢?

结合我一直以来对当下城市规划设计理论和实践的反思、批判，以及未来 + 学院的产学研成果，我逐步探索出一套六维评估方法，强调从可持续发展（社会、经济、生态）及其延展（空间、时间、治理）维度来全面描述和评价城中村的形成机制及现有干预方式（图 5）。未来 + 还将六维评估开发成手机应用小程序，方便每个人都能按照各个维度系统要素的提示，凭个人经验针对一个城中村社区以及改造方案进行各维度的判断。

这些简易的比较肯定是主观和难以量化的，但面对相同讨论对象的人如果都能使用这一系统框架进行评估（具体做法是自己可应用小程序打分），就具备了更好的交流与对话的频道和基础，甚至大家参与的结果就能够进行统计，从而得到某种量化结果，能更全面反映不同人对城中村及现有干预措施的优缺点的判断。

比如拿城中村与规划建成区及拆旧建新的片区相比，我的评估结论是：城中村自发建设的市政基础、房屋间距、物业管理及资产价值上肯定有很多不足，但优势体现在社会维度的效益上，则以低成本提供了政府本该通过正规建设提供，但远远不能满足外来移民落脚乃至就业需求的廉租保障房。另外的优势还体现在通过自治管理、自发分散建设带来的功能混合丰富、空间渐进多样和生态冲击相对较小等优点上。要提升城中村品质，关键是发挥其社会、历史维度的长项，同时将其他维度的短板通过各种创新方式予以补齐，而不是陷入追求经济效益的推倒重来或空间景观效益的穿衣戴帽这些简单模式中。任何的改造更新政策、具体方案以及措施，最终需要评估其六个维度影响的优劣和均衡度。

Urban Planning Vs. Urban Villages, Who Should Transform Whom?

Huang Weiwen
A researcher and strategist on urbanism, Huang Weiwen is the co-founder of FuturePlus. He served as the Deputy Chief Planner of the Urban Planning, Land and Resources Commission of Shenzhen Municipality (2011-2012), Director of Shenzhen Public Art Center and Shenzhen Center for Design (2011-2017), and Director of the Design Department of Bureau of Planning of Shenzhen Municipality (2003-2009). He was a visiting Loeb Fellow at the Graduate School of Design at Harvard University (2009-2010), the main founder, organizer and promoter of the Bi-City Biennale of Urbanism\Architecture (2005-2017), and the co-founder and promoter of Retumu Urban/Rural Institute.
The article was first published in *Design Community*, 2011(5): 102-105; amended by the author for this publication in 2021; translated from Chinese to English by Luo Yiqian.

As an urban planner, I have engaged and researched Shenzhen urban villages for 16 years, during which time my views on urban villages have constantly changed. In this article, I share my experiences with an eye toward deepening the conversation.

1 Can we transform urban villages by urban planning?

Under the supervision of my graduate advisor, Zhu Zixuan, I participated in the protection and regeneration planning of Shichahai, Beijing and Tunxi Old Street, Huangshan and other historical areas. Based on those experiences, I proposed that planning schemes should consider humanities observations and investigations, research the urban fabric, reconstruct condemned buildings, and accommodate the co-existence of new and old buildings. In 1995, while working at the Urban Planning & Design Institute of Shenzhen, I participated in the renewal planning for Xinzhou Village and Dachong Village (Figure 1), which came to my attention as individual cases. At that time, the name "urban village (village in the city)" was not commonly used, which indicates that the villages were not yet enclosed by the city. Some of these planning schemes earned praise from the Urban Planning & Design Institute of Shenzhen as well as won a Guangdong Excellent Urban Planning Achievement Award. At the time, I believed that good urban planning and design could critically develop the spatial form of urban villages. However, as a fresh graduate just beginning my career as a practitioner of urban planning, I did not fully consider how urban plans would be physically implemented.

In 1996, I became a management planner in the planning and management organization of Shenzhen Central District. Almost immediately, the continuous informal expansion of the Heyuan block of Gangxia Village in the city's Central District concerned me. The densifying concentration of villagers' buildings seriously affected the urban design of the Central District which had been improved through several rounds of international consultation. I also worried that the increasing contrast between the two areas would stir up public opinion and strengthen the determination to demolish illegal buildings.

My effort was two-pronged. I suggested that the planning department should contact and negotiate with the village collective; I also sought a plan that would be acceptable to both the government and villagers. This occurred around 2000 before the opening of Xintiandi Shanghai. Through archival research, the bold idea of creating open, stylish, and pedestrian-friendly urban space by re-purposing old architecture suddenly inspired me. I recommended this transformation model to officials, planners, the district government, and Gangxia Village. With preliminary support from Futian District officials, we invited Song Zhaoqing, the architect of Nikken Sekkei who participated in Xintiandi design, to conduct renewal planning for Gangxia Village.

Song's plan selectively demolished half of the villager-built housing (1,700,000 square meters), transformed the other half, and built new housing (300,000 square meters) to form ten courtyard neighborhoods, which followed the original street orientation and structure as well as retained and upgraded the renowned Gangxia Food Street (Figure 2). The scheme was approved by Gangxia Village, however the technical committee experts inexplicably rejected the plan as lacking feasibility because they reckoned it would be difficult to secure villagers' consensus on whose buildings would be demolished and whose would be retained.

In 2002, Jaime Lerner who was the president of the International Union of Architects, former mayor of Curitiba, Brazil, and an architect, sponsored an international competition for "Celebration of Cities" conceptual design, calling for professional attention to urban issues. I reflected on my experience with urban villages during my years in urban planning management, collaborating with my colleagues and URBANUS to submit the proposal "Organic Transformation of Shenzhen Urban Villages" at the end of the year. The Architectural Society of China awarded the proposal second place in the domestic selection. Later, the proposal was collected in the publication *Village/City City/Village* by URBANUS (Figure 3), exhibited in the 2005 Shenzhen Biennale of Urbanism\Architecture, and included in the 2008 China-France touring exhibition *The Street Belongs to All of Us!*

At this stage, I regarded urban villages as a given urban condition and spatial context, believing that there existed an ideal planning that could sort out and extend urban village structure and fabric, organize its space, and establish new construction controls. At the least, this planning would be more interesting than extreme cases, either demolishing everything through *tabula rasa* planning or doing nothing and maintaining the status quo. Also, I hoped that my proposal would not be misconstrued as historical culture protection. I have always maintained that instead of turning existing buildings into waste, a transformation model that allows for the co-existence of the new and the old would be an economically feasible and sustainable choice. To dissuade those who hated the urban village buildings from pursuing demolition, I created illustrations that showed how by changing building facades, Shenzhen urban villages could be as charming as Xintiandi Shanghai and even older European old cities and towns, such as Venice and Amsterdam. What I did not expect was that in recent years, urban villages and local governments would use accumulated wealth to implement beautification initiatives for the village buildings that abutted main roads, imitating European styles (locally known as *chuanyi daimao*, meaning to dress up). Although such disguises extended the lifespan for urban villages and shared little in common with my earlier suggestions, nevertheless, today I feel shocked and ashamed by my earlier formulations.

None of the proposals to use urban planning to renew urban villages were realized. Instead, urban villages such as Yumin, Yunong, Gangxia, and Dachong were eradicated, rather than transformed, by urban planning. From the perspectives of green sustainability, continuity of social structure, urban vitality, and the diversity of urban spaces and architecture, I personally believe that the urban planning that wiped out existing traces has been immoral.

2 How do we evaluate urban villages?

In 2004, Shenzhen began mobilizing to transform its urban villages. At that time, decision-makers, planners, and mainstream media generally regarded urban villages negatively. Extremely negative evaluations included calling urban villages "urban tumors, burdens, scars, and cancer." Problems widely ascribed to urban villages included having an ugly image, high density, poor environment and sanitation, disorderly public security, insufficient public space, community facility and municipal infrastructure, and substandard fire control.

As I wrote in *Village/City City/Village* in 2006,[1] "Urban villages, in fact, make up for the deficiencies of urban planning and management, contributing to improving land functions and efficiency, maturing affordable housing system, enriching urban services, and reducing urban service and start-up company cost." The housing supplementation and adjustment provided by urban villages prevented slums from forming in Shenzhen during rapid urbanization when there was a large population influx. "Subjectively speaking, urban villages are the result of villagers breaking policies and regulations in pursuit of their personal interests, and objectively speaking, they were the inevitable result of releasing urban land from a distorted form to seek adjustment according to economic laws. It was a correction of the land policy under the planned economy, whereby urban land was treated as a welfare benefit and its economic value was ignored. It was also the villagers' right to develop the remaining land after they had contributed to the city's achievements by giving their land to support urban development."

As I further pointed out in my 2006 blog article "The City's Lower Body,"[2] urban villages have played an indispensable role in the city, supporting its functioning like the lower half of the body. Although "the upper and lower body have flowable and convertible elements, they are interdependent and functionally different. In other words, if most urban villages are demolished to make room for building high-end real estate, no matter whether the city likes it or not, the former tenants of urban villages will move to new corners in other areas. The city's lower body will acquire a new form and function, as they are interdependent with the city."

I elaborated further in my article "Shenzhen: The Rapid City-making," published in *Domus* in 2011.[3] "Urban villages have become enclaves and autonomous communities in the city, and a self-organizing system that supports the city's operation through internal adjustment. Urban villages have undertaken the function of affordable housing for Shenzhen for years, not only because the villager-built housing has lower capital investment (without land and construction registration costs), cheap rents, and practical, economic, compact and open design, but also because urban villages are well-distributed across the city, providing services for adjacent central, commercial, and industrial districts within walking distance. During the agricultural era, the walking distance from home to fields determined the distribution of villages, which has marvelously transformed into the 'habitation layout for Shenzhen's lower middle class.' This has made up for the long-term neglect of residential land for the lower middle class in several urban planning schemes, relieved the demand for affordable housing when the government supply has been insufficient, and slowed down the deterioration of urban transportation."

图 4 深圳岗厦村宜居宜业的活力和谐景象，2023 © 都市实践
Figure 4 A vibrant and harmonious scene of livability and business vitality in Gangxia Village, 2023 ©URBANUS

Since 2005, when the first Shenzhen Biennale included a dedicated exhibition and forum on urban villages, more comprehensive and objective studies and debates in professional and public spheres have emerged. I have been inspired by the following results. Li Jinkui at China Development Institute conducted long-term research on the social system of urban villages. Juan Du, assistant curator for the first Shenzhen Biennale, studied the transformation of Baishizhou and Huanggang Villages, after organizing the dedicated exhibition and debate. She regards urban villages as a dual protection system that both accommodated the lower middle class and maintained the functioning of urban transportation. Marco Cenzatti, a professor at Harvard Graduate School of Design, believes that Shenzhen urban villages have developed more urbanity than the city proper. Yung Ho Chang proposed transforming the Civic Square according to the way of urban villages. Finally, after visiting Shenzhen, John Friedman, an early pioneer in urban planning, recognized that compared with Overseas Chinese Town (OCT), the social features and atmosphere of Xiasha Village better represent Shenzhen traits.

3 Urban planning should be transformed by urban villages!

As current urban planning cannot transform but only eradicate urban villages, while urban villages have a positive and rational existence that reflects Shenzhen's urban characteristics (Figure 4), it can be asserted that the problem does not lie in the urban villages, but rather in urban planning.

Indeed, it is not urban villages, but current urban planning that needs to be transformed.

The modernist theories adopted by current urban planning in China were transformed and even overturned in the United States by journalist and social activist Jane Jacobs as early as the 1960s. At that time, the modernist planning model was also widely practiced in the United States, leading to wholesale demolitions and reconstructions of old districts. Pushed beyond forbearance, Jacobs fought with the largest developer in New York. She saved Greenwich Village from being cut by the highway and subverted by urban planning through mobilizing the community and writing articles. Her book *The Death and Life of Great American Cities* transformed and even overturned modernist urban planning.

When we study history, we discover that things which occur on one side of the world will happen on the other. Nevertheless, how can urban villages that are not as beautiful and elegant as Greenwich Village, Manhattan transform urban planning in China? For more than a decade, I tried to improve urban villages with planning knowledge and design skills. However, through observation of and comparison between urban villages and planned districts, I have come to respect more and been more inspired by the collective outcome of individual wisdom reflected in the self-built districts without planners and also been alerted to and reflected on the blind spots and short-sightedness of new districts constructed under elite planners' control. The problems of plot shape, municipal facilities, and fire protection of urban villages, which I had once thought needed to be improved, became minor concerns when compared to the great and valuable significance of urban villages as a model for constructing an autonomous, independent, self-sufficient, and low-carbon life as well as a harmonious society. In a symposium in 2003, I put forward a radical view: it is a false proposition that urban villages need to be transformed. Through years of observation and study of urban villages, I have finally realized that the real proposition is to use urban villages to transform urban planning. To this day, despite being eradicated through urban planning, Shenzhen urban villages can supplement and even transform current urban planning education, providing answers to the following questions:

(1) How can we practically involve bottom-up public participation in urban planning?

Current urban planning uses information publicity.

Urban villages' answer: allow the public to own land.

(2) How can we make the public become a major actor in urban construction, not limiting agency to large developers?

Current urban planning prefers large plots and stipulates developing no less than 3,000 square meters of land, and large complexes become popular.

Urban villages' answer: divide the land into 100-2,000 square-meter plots.

(3) How can we build a pedestrian-friendly city?

Current urban planning sets highways and expressways or widens roads before suggesting where footbridges and underground pedestrian walkways should be located.

Urban villages' answer: lay crisscross streets and alleys densely, with 10-40 meters in between, and 2-8 meters in width and without fences.

(4) How can we create dynamic urban space?

Current urban planning that prefers defined and even unitary functions usually leaves the question to commercial planning.

Urban villages' answer: allow programs that cause no harm to public safety and health to mix freely and adjust flexibly.

(5) How can we form community identity?

Current urban planning relies on design forms, using graphical and symbolical means to avoid the so-called "one expression for a thousand cities (*qiancheng yimian*)."

Urban villages' answer: self-construct and self-manage infrastructure, retain or transform village gates (*paifang*), ancestral halls, and exercise squares (usually a small plot of Litchi land serving as a fengshui woodland) within spaces and facilities shared by the community.

(6) How can we provide low-cost housing and living environments?

Current urban planning is almost aphasic.

Urban villages' answer: allow individuals or the collective to build housing, eliminate profits and taxes from initial real estate development, or retain limited property buildings, and allow housing with insufficient sunshine to exist. Set up food markets, and allow for the construction of urban vegetable plots and small-scale poultry coops, workshops, stalls, and other temporary constructions.

(7) How can we build a low-carbon and environmentally sustainable city?

Current urban planning focuses on green land and greening indexes.

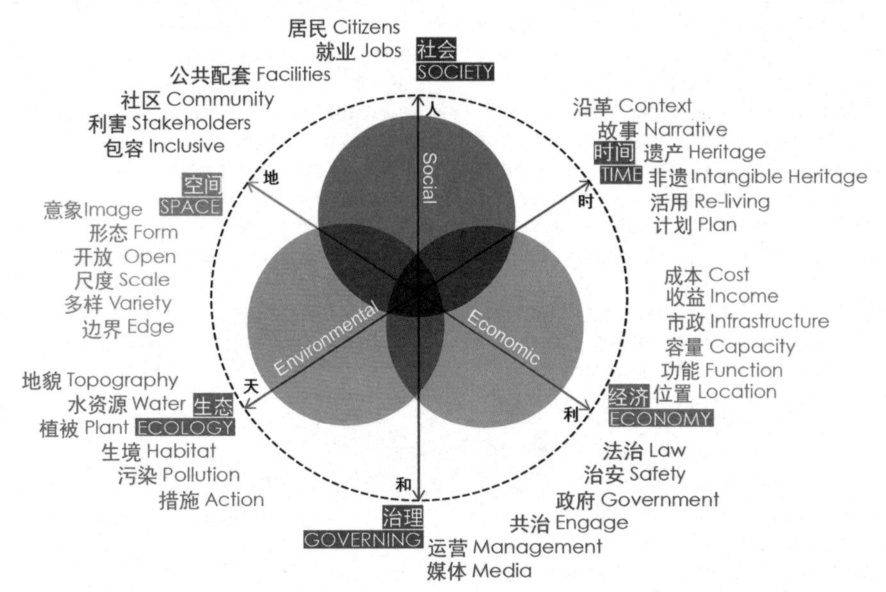

图 5 资源与影响六维评估体系
Figure 5 Six Dimensions Assessment of the resources and influences

Urban villages' answer: refer to points (3) (4) (5) (6) above.

(8) How can we build a sanitary, safe, and fire-protected city as well as allow various classes of residents to live in harmony?

Current urban planning has rigid standards in sunshine duration, fire protection, and greening to support high-end neighborhood construction, while affordable housing tends to gather in urban outskirts.

Urban villages' answer: allow community autonomy and self-policing, reduce fire engine size, keep upgrading infrastructure by accumulation and investment; cooperatively build a few high-end neighborhoods scattered around lower-middle-class neighborhoods, sharing resources.

(9) How can we maximize inhabitants' rights and interests?

Does current urban planning need to consider this question?

Urban villages' answer: have stock cooperation, democratic elections, and people's representatives to represent the local community.

In conclusion, urban villages are not only a model for planning and constructing a harmonious city that accommodates public participation, provides livable and workable environments, and achieves low carbon emissions, but also an indispensable framework for the city itself. If we allow current urban planning to eradicate urban villages in the name of transformation, it will not only harm the scattered model districts, but also the city's support system. Shall we watch the city disintegrate until it is paralyzed, suffering the real "cancer" of urban/social-ecological imbalance? Or shall we retrieve inspiration and knowledge from the supporting system to transform current urban planning? It is a choice and a challenge faced by urban planners and decision-makers in urban regeneration.

4 Postscript: Six Dimensions Assessment

In 2011, I wrote the above article to discuss the respective issues facing urban villages and urban planning, suggesting that some advantages of the former could inspire the latter. Each individual's perspective, standpoint, and focus are different, and we also have great differences in the assessment of the pros and cons of urban villages and planned neighborhoods. Is it possible to have a widely accepted systematic framework to compare and discuss these differences?

I have gradually developed the Six Dimensions Assessment through long-term reflection and criticism of urban planning and design theory and practice as well as production, teaching, and research results at FuturePlus. The Six Dimensions Assessment focuses on the holistic description and evaluation of the formation of and intervention in urban villages from the perspective of sustainable development (society, economy, ecology) and its extended dimensions (space, time, governing) (Figure 5). To facilitate individual assessment of an urban village and its concomitant transformation scheme, FuturePlus has developed a mobile application of the Six Dimensions Assessment, which guides users through tips to systematically explore the elements of each transformation dimension based on their perception.

Indeed, simple comparisons are subjective and difficult to quantify. However, when people use the same systematic framework to evaluate the same object (by giving scores on the application), we have a better channel of and basis for communication. Furthermore, with the statistics of participation, we can have certain quantified results which reflect different users' assessments of the pros and cons of urban villages and current intervention measures.

Through comparison of urban villages to both established planned areas and renewal areas of demolition and reconstruction, I have reached the following conclusions. First, although the self-constructed infrastructure, proximity to each other, property management, and asset value of buildings in urban villages are not ideal, nevertheless they perform much better along social dimensions than do formal developments. Specifically, urban villages have done what government construction should have, but failed to do: provide affordable housing to meet the living and employment demands of migrants during Shenzhen's boom times. Second, the autonomous management and scattered, informal construction of urban villages are also superior with respect to integrated and extensive programs, growing diversified space, and minimizing ecological impact. To improve the quality of urban villages, it is crucial to maximize their social and historical advantages, while using innovative measures to strengthen their weaknesses, instead of being trapped in crude models such as demolition and renewal for economic profit or beautification initiatives that only fix landscape visuals. The Six Dimensions Assessment program facilitates the necessary evaluation of the impacts of any regeneration policy, scheme, and measure, tracking pros and cons as well as balance.

谁来
Transform
改造谁

[1] 都市实践 . 村・城 城・村 [M]. 北京 : 中国电力出版社 , 2006: 2-15.
[2] 黄伟文 . 城市的下半身 [EB/OL]. 黄伟文的新浪博客 . http://blog.sina.com.cn/s/blog_7275adaa0100oelk.html.
[3] 黄伟文 . 深圳：如何快速塑造一座城市? [J]. Domus 中文版 , 2011: 102-105.

关于城中村的一种文献综述

安太然

普林斯顿大学建筑历史理论专业博士候选人，2017 年深港城市\建筑双城双年展（深圳）文献研究员。

本文由陈一苇中译英。

1 写在前面：一次关于尺度的写作试验

2017 年，在深圳南山区南头主展场的南楼，深双以“世界 | 南方”板块为整个双年展铺陈了视野和立场。若干“南”的重叠或许不仅仅是地名、方位和研讨对象的巧遇。诸南相汇之处，暗含了一个关于尺度的问题：岭南在大陆之南，深圳在岭南之南，而北纬 22° 的南头城仅以咫尺，将将踏入地理学意义上的热带。在“世界 | 南方”的概念之下，或者说，在作为热门学术研究领域的“全球南方”之中，北靠羊台山、南临伶仃洋的这座弹丸之城诚然构成一个“案例”。但尚未言明的，是“城中村”这一城市研究课题中个案与区域、局部与整体、地方与全球的具体关系。这些关系在 2017 深双得到些微的透露，在这个意义上，2017 深双尚未完成（而且，也不必完成——如总策展人之一孟岩所说——“‘城市共生’是一个‘进行时’的展览”）。[1] 本文通过对城中村研究的综述，尝试对这些关系进行最初步的、同时也是迟来的厘清，将观察的尺度由细微拉向深远，目的是将深圳提供的案例和中国城中村议题带入“全球”和“南方”的视野之下，并借助城中村，反问“全球”和“南方”究竟为何，城中村又因何关系重大。城中村正在批量消失，在它们彻底成为历史和记忆之前，本文想要调查，这种特定的城市形态和生活图景为城市研究提出了哪些新的问题？在城中村研究的学术史上，具体来说，它们启发了哪些思考和关切、诱导了哪些调查和研究方法？城市化的轴心在根本上从世界北方转向南方、从西方转向东方的当下，一个跨越地理和时间尺度的研究框架可能必须以“文献综述”的方式走出第一步。从这个意义上讲，本文和“世界 | 南方”一样，是一次跨越观察尺度的试验。

因本文观察尺度上的较大跨度——从微小琐碎的具体生活到全球南方研究的学术前沿——此处不妨提出三点假设，有待后续写作的检验。第一，对跨尺度的观察有文献价值的材料应该也是“跨体裁”的，换言之，“文献”的概念需要重新定义。除了惯常意义上的各类学术著作、期刊论文、书评展评等学术写作体裁之外，一些非学术性写作和其他类型的实践活动也应该纳入“文献”的范畴——包括但不限于新闻报道、调查访谈等媒体写作，以及来自学者、独立撰稿人和艺术家的独立调查、独立写作、口述史写作和艺术创作，等等；第二，观察角度和尺度的多样必然意味着调查方法、学科传统、研究策略的差异，也就是说，跨尺度的文献调查取向必然会导致跨学科的调查结果，而不同学科之间究竟以何种方式产生对话或互相启示则有待进一步的检视；第三，如果说，与任何其他档案工作一样，文献综述不应该、也不可能以全面和中立为目标，那么本文不妨诚恳面对出发点的局限、当下的偏见和局部的特异。作为一次迟到的——或者说，为 2017 深双补写的——文献再调查，本文从这场展会（2017 年 12 月 15 日—2018 年 3 月 15 日）至今的这个时间缺口写起：之后，深圳已有若干与城中村直接相关的城市更新政策出台，这些政策及其相应的举措已经并且正在发生巨大的实质影响。这些短期内的巨变意味着，对文献调查的补写必然同时也是对参考文献的重写：从 2017 年—2021 年这一时间段、深圳这个地理片段出发，倒叙一个巨大而缓慢的非正规城市变化过程，从中探测出梳理文献的线索。也就是说，不是先有文献目录，再有文献综述，而是反过来：一个全新而不全面的文献目录将在文献调查的过程中一点一点生长出来。

2 序言：废城白石洲与时代寓言

2021 年 3 月 23 日，在个人博客“深圳笔记”上，久居深圳的美国人类学家马立安以《深圳的城中村时代是不是就此结束了？》为题，进行了简短更新。她做出一个大胆的推断：“我认为，深圳的城中村时代已经结束了。我把城中村时代的发端放在 1992 年，即关内农村开始城镇化之际；而城中村时代的终结大约就在此时此刻的当下，以白石洲全面拆迁为标志。”[2] 在马立安看来，作为深圳最大的城中村，同时也是深南大道两侧最后一个主要城中村，白石洲 2019 年至今的集中清退和拆改标志着一个时代的落幕。[1]

废城白石洲被高楼大厦彻底取代只是时间问题。早在 2005 年，白石洲就被南山区政府列入旧村改造研究计划之中；2014 年 7 月，白石洲项目更名为“沙河五村城市更新单元”，并列入深圳市城市更新单元计划；2018 年底，白石洲改造专项规划获得深圳市规土委正式批复，正式启动拆改清租。[2] 随后的几年，房东们陆续与开发商绿景中国地产签约，租户逐步清退，所有握手楼也已经或即将被悉数拆除。旧改后的新城计划将于 2029 年落成，项目最终打造成 31 栋 49~65 层住宅、21 栋公寓及 4 栋写字楼，总面积约 0.6 平方公里，共有出租屋 2500 多栋、5 万多间（套）、商铺约 2310 家、片区总人口超过 14 万（其中外来居民超过 12.7 万）的白石洲终于全面展开了早已写进城市规划的拆改进程。[3] 但应该指出，白石洲不仅行将化作物理意义上的残垣断壁，更摇摇欲坠于一片话语的废墟之上。倘若我们暂且不去细究马立安所谓“深圳的城中村时代”的具体断代，而是相信这个关于“结束”的推断并非姑妄之言，或具洞见之明，那我们不妨对这一断言的内涵稍作追问：从终结之处回看，我们可以从近二十年来围绕白石洲所展开的大量研究、实践和公众讨论中翻捡出哪些不该被忘掉的议题？之所以以白石洲的近况为引，逆时序展开本文的写作，也因为这一重要的个案与马立安的激进猜想暗合本文对于尺度问题的关切：白石洲诚然仅仅是深圳成百上千座城中村中的一例，但对其进行聚焦的文献调查或可以启迪对当下和历史的整体判断。[4] 此处从白石洲出发，回溯 2005 年以来（同时也是首届深双以来）一系列以城中村为对象的学术研究、公众讨论和独立观察，不仅意欲显露某种时代的剪影，更想要从语言和文字的瓦砾中察觉一些从未调和的矛盾、标记一些从未找到答案的问题。脱胎于沙河农场和周边五座旧村的白石洲即将彻底成为历史，它正加速向前走进一个后城中村新时代，到那时，所有的矛盾和问题都将隐去、灭失、遗忘。[5]

2019 年以来白石洲的集中旧改引发了大量来自主流媒体和独立个人的记录、调查和批判性写作。[6] 其中，独立撰稿人袁艾家发表于《野人》杂志微信公众号的《“值钱”的白石洲：城中村的发展主义》一文（2020 年 8 月 17 日）给出一种分析白石洲的思路。该文指出，白石洲的历史显示，这座巨大的城中村不啻两种城市观念的角力场：一边是一味向前、服膺房地产逻辑的线性发展主义；另一边是对线性发展主义的“知识分子式”的反动。然而令人遗憾的是，后者从未获得一个面对面地与前者对垒的机会。“白石洲之类的社区去与留，本来能引发一场‘城市到底因何存在’的社会观念大辩论。”袁文给出“知识分子式”努力的两个具体示例：其一是本节开篇提及的马立安，其二是“白石洲小组”。前者及其发起的“握手 302”公共艺术和社区实践项目（创立于 2013 年）驻扎在白石洲多年，致力于培育一种城中村经验的集体自觉——“不是我们帮助白石洲，是白石洲帮助了我们”。[3] 她素来反对把白石洲视为等待被更新方案去“拯救”的贫民窟，认为白石洲能够反过来提供另一种“深圳生存经验”。后者（创立于 2015 年 6 月）同样力图培育白石洲经验的自觉表达。[7] 在袁艾家看来，在奉行房地产逻辑的城中村发展主义的要求下，社区层面的、内向的居住和商业生态必然被城市层面的、外向的经济要求无情迭代。[8]

该文的悲观论调必然地奠基于这样一个事实之上：2005 年以来围绕白石洲的全部“知识分子式”努力——

1　在这篇简短的博客中，马立安并未给出作出这一断言的详细依据——如她开篇指出，这是一篇“猜想性的帖子”。

2　关于白石洲旧改的时间线，可参看来自白石洲项目的开发商绿景中国的报道：http://www.lvgem-china.com/report/2020/05/27/6110.html。

3　深圳南山区政府的调查数据显示白石洲有租户 11 万人，但其他数据来源显示，租户总数可能接近 15 万。 和深圳其他城中村的人口数字一样，因各数据来源互不相同，白石洲的真实人口数字将无从考证。此处列出的数字来源为白石洲兴趣小组 2016 年的调研，该数字被广泛采纳。

4　在方法论层面，本文受惠于意大利历史学家卡洛 · 金斯伯格的一系列有关（案例的）特异性与（推论的）一般化之间的关系的重要论断。金斯伯格指出，对异常个案的深入研究能够启示对常规的认识，而反过来则不成立。参见 GINZBURG C. Microhistory: Two or Three things That I know about it [J]. TEDESCHI J, TEDESCHI A C, trans. Critical Inquiry, 1993, 20(1): 10-35.

5　关于白石洲自其前身沙河农场以来的历史已有众多文献记录，此处不再赘述。例如 O'DONNELL M A. Laying Siege to the Villages: The Vernacular Geography of Shenzhen [M]// O'DONNELL M A, WONG W, BACH J, eds. Learning from Shenzhen: China's Post-Mao Experiment from Experiment to Model City. Chicago: University of Chicago Press, 2017: 120; O'DONNELL M A. Of Shahe and OCT [EB/OL]. Blog "Shenzhen Noted," 2010-1-31. http://maryannodonnell.wordpress.com/2010/01/31/tangtou-baishizhou. 关于对深双的历史回顾，可参看深圳城市 \ 建筑双年展组织委员会 . 看（不）见的城市：“深双”十年研究（上卷）[M]. 北京：中国建筑工业出版社，2017; 黄伟文 . 看不见的城市：“深双”十年九面 [J]. 时代建筑 . 2014(4): 48-56.

6　2019 年 7 月，白石洲的租户陆续收到通知，被要求 8 月底或 9 月中清空搬离，因清租通知突然而转插学位紧张，数千流动学童面临难以就近上学的困难，引发《南方都市报》《界面新闻》《澎湃新闻》《财新周刊》等国内官方媒体、《南华早报》等境外媒体以及许多自媒体平台的热切关注。如梁宙发表于《界面新闻》的文章《深圳最大城中村“白石洲”清租困局：被断档的学位》（https://m.jiemian.com/article/3282575.html?from=timeline&isappinstalled=0）；蔡宇晴发表于《南方都市报》的报道《众“深漂”第一站将旧改，房东突然清租，租户们遇子女上学等难题》（https://m.mp.oeeee.com/a/BAAFRD000020190703178296.html?&layer=6&share=chat&isndappinstalled=0）；Sisyphus 发表于 NGOCN 的文章《深圳最大城中村白石洲清租，近千学生读书受阻》（https://chinadigitaltimes.net/chinese/614043.html）；Phoebe Zhang 于 2019 年 7 月 22 日发表于《南华早报》的报道“Migrant workers forced out as one of Shenzhen' s last 'urban villages' faces wrecking ball”等。长期关心白石洲和城市更新议题的策展人郑宏彬、建筑师张星和艺术家坚果兄弟等人联合发起“深圳娃娃”计划。他们向白石洲的孩子们征集了 400 个玩偶娃娃，并于 8 月 4 日在深圳与惠州边界用一台 29 吨的巨型抓机夹着娃娃一个个从深圳界内抓到界外。这次艺术行动使许多人第一次关注到这些因城市政策变动而面临上学困境的儿童，让白石洲在 2019 年 7—8 月短暂成为舆论的焦点。有关“深圳娃娃”计划的详细方案描述，可参看于 2019 年 7 月 24 日见于“ARThing 艺术活”的《“深圳娃娃”计划正在收集》一文（https://www.arthing.org/archives/2019/07/post2410857.html）。另外一些回响则未见于主流官方媒体，而散布于各类自媒体平台，这些声音碎片被广泛传播，激发大量网络讨论，其文献价值不应被忽略。如长期关注公民社会议题的自媒体 NGOCN 发表 2019 年 8 月的《被拆迁的“深圳娃娃”》一文、青年左翼平台“马各庄青年”于 2020 年 6 月发表于豆瓣网的《非暴力拆迁作为一种治理技术——迟到的白石洲札记》一文、“白石洲小组”连载于微信公众平台的“看不见的深圳人”系列等。这些文章从各自的视角梳理了白石洲家长们就流动儿童教育问题开展的一系列艰难协商。值得提及的相关独立写作还包括由署名“文山湖”的作者发表在澎湃新闻“湃客”创作者平台的白石洲非虚构写作系列连载，包括《从清租下的白石洲退场，一场深漂人的集体别离》《深漂老租户，在最大城中村建起自己的江湖》和《消逝中的白石洲，那走了多少深漂人的避难所》等。

7　“白石洲小组”由建筑师段鹏等人于 2015 年 6 月成立，开辟专栏“看不见的深圳人”，邀请白石洲的居民，举着填上姓名、性别、职业、年龄和故乡的白色纸板，站在白石洲的街道上拍下一张照片。

8　根据《野人》杂志编辑来福的介绍词，该文最早写于 2018 年，本来是为一本“记录深圳‘多元’的城市空间和历史脉络”的出版物而写。“知识分子式”这一说法，与马立安的一个论断形成照应：后者将一批深圳建筑师、艺术家、学者和各行各业的建言人士归纳为“第二代深圳知识分子”。不同于第一代建设深圳的知识分子，在情感和身份上仍然认同童年故乡，第二代深圳知识分子认同自己的“深圳人”身份，并且“致力于拓宽深圳城市形态的公共讨论，而不是阿附政府和开发商的声音。”见 O'DONNELL M A. Heart of Shenzhen: The Movement to Preserve "Ancient" Hubei Village [M]// BANERJEE T. LOUKAITOU-SIDERIS A. eds. The New Companion to Urban Design. London: Routledge, 2019: 480-493.

来自一大批建筑师、规划师、艺术家、作家、学者、社会工作者和其他独立观察者——都加起来，也未能阻止发展主义逻辑最终将白石洲原地铲平。2016 年“湖贝古村 120 城市公共计划”（简称湖贝 120）的小范围成果（最终的湖贝旧改方案中，古村得到部分认可和保护），以及 2017 深双展场南头古城的“城市共生”理念，这些“成功”先例都没有机会在白石洲旧改上得到复制和实践。就南头和湖贝而言，恰恰是历史古城、古村与当代城中村这两个命题的相互掣肘，帮助它们免遭连根拔除的命运（而代价是，南头——如都市实践研究员罗祎倩所说——“背上了创意文化的指标，外来案例疾速填充，南头自身生长的本土文化反而被忽略”[4]）；没有古城古村这张牌可打的白石洲则在劫难逃。对于“为什么湖贝 120 可以取得一定的成功，而白石洲不行？”建筑师段鹏的回应是，“可能根本上，湖贝 120 保卫的是古迹，白石洲保卫的是居民。”[9] 的的确确，局部变成文物的湖贝和重新塑造古城的南头在事实上均未避免原有租户的流失；对“人口结构”和“产业结构”进行“升级”的潜台词，就是用一种居住和商业生态来淘汰另一种居住和商业生态，用一部分人来淘汰另一部分人。就保卫白石洲的城中村居民而言，大获全败的不仅仅是传统意义上的激进主义者，甚至也包括力图“让城中村的重建和保育不再是一对矛盾”的中间派实干家[10]——2017 年策展南头的都市实践团队对白石洲抱持多年关注（都市实践团队甚至一度考虑将 2017 深双的主展场放在白石洲，但因各方阻力未能实现）。他们一直致力于“在现实的经济因素与城市设计和保护的理想目标之间寻找平衡点”，从而把“难以解决的社会学问题”变成一个“有解决的可能的设计问题”[5-6]，但在都市现代化的神话和冰冷无情的事实面前，哪怕是这些带有实用主义色彩的尝试也显得理想主义。

倘若我们以此回看马立安“深圳的城中村时代已经结束了”的断言，走向终结的或许远不止城中村这一非正规城市形态及其所庇护的非正规城市生活，还有深圳的“知识分子们”与地产思维主导的单一城市开发模式博弈的一块重要“战场”。这场围绕城中村的对弈已成残局，我们必须看一看，除了一座语言、图像和记忆的废墟，城中村还给我们留下了什么。

3　微观：田野视角、设计研究与深圳案例

2005 年，供职于北京非常建筑事务所的建筑师杜鹃南下深圳，与一座烟火气十足的夜市不期而遇，这是她第一次造访一座城中村。事后她了解到，这座城中村正是白石洲。[7]4-5,[8] 杜鹃彼时担纲首届深双助理策展人，该届深双首次对城中村进行专题观察。自那时起，她多年来无数次探访白石洲和其他深圳城中村，并于 2020 年出版英文专著《深圳试验：中国速生城市的故事》。这本为深圳神话祛魅的著作以她 15 年前与白石洲的偶遇开篇，以白石洲的清退和拆改作结。她在正文结尾处写道：

> *在提升经济活力和社会韧性方面，城中村能够为城市提供来自过去的知识和面向未来的策略，而不是仅仅呈现为一些需要解决的问题。不应该把深圳的成功简单归于政策或核心规划。这座城市某些最重要的创新恰恰源自于局部对于自上而下的总体规划的应对，这些应对方式有时甚至与城市规划截然对立。*[7]304

几乎可以认为，“与城市规划截然对立”不仅事关深圳城中村的创造性生存经验，同时也映照了该书的写作目的。杜鹃试图在书中揭示，通过增长神话、速生神话、处女地神话、渔村神话等一系列神话建构过程，深圳城市史的叙事服务于改革开放政治议程。换言之，主政者们讲了一个基于城市规划的神话“故事”，“故事”里南海边画圈并发表南方讲话的老人将这座南方的弹丸渔村勾画为中国走向现代、走向国际的“南大门”。因而对这一“故事”的祛魅，就是对自上而下、自北向南的城市规划造梦机器的祛魅。这一“故事”较少触及村和城数百年来的复杂层积，以及几十年来几百万外来务工人员的适应和奋斗、他们惊人的创造力以及作出的巨大贡献。在城市规划这部造梦机器——同时也是一部造史机器中——见微知著，从微观尺度出发、从南方出发，凭借人类学家般的眼睛进行观察、以非虚构写作者的方式展开在地的写作，提供另一种叙事方式。作者证明，深圳自有其绵延跌宕的生命历程，非任何规划蓝图能框定。作为一种深埋在历史遗留问题泥淖中的非正规城市空间，深圳城中村不是深圳历史的他者，它的空间史、政治史和文化史就是深圳历史本身；研究城中村就是研究城市，城中村研究是对城市研究的重要补充。

二十余年来，已有一批人类学家、社会学家和城市研究者着手进行为深圳城中村“去他者化”的工作，力图将城中村纳入“深圳经验”的范围之内。由马立安、乔纳森·巴赫和黄韵然主编的《向深圳学习：中国改革开放时期从经济特区到模范城市的试验》（英文版 2017 年，中文版 2020 年）是英文世界深圳研究的重要论文集，其中马立安、乔纳森·巴赫、黄韵然等作者在各自的章节中对城中村及其荫蔽的外来务工人员作出细致的批判性观察。[11] 书名不仅暗指“文革”前夕“工业学大庆，农业学大寨”的工农生产口号，同时是对罗伯特·文丘里、丹尼斯·布朗和史蒂文·艾泽努尔出版于 1972 年的著作《向拉斯维加斯学习》的援引。这一点在英文语境下不言自明，但在中文书名翻译中需要澄清：正如拉斯维加斯的消费主义美学扰动了现代主义建筑的纯粹主义叙事，深圳的非正规城市化也启发了一种上与

9　这句话引自段鹏的复述，见沈于渊．深圳人生样本 01｜建筑师建造了他们的城市，也被异化为工具；又如“握手 302”成员刘赫所说：“我们和‘居民们’并没有密切的联系，因为我与‘居民’并没有区别。和所有白石洲的人一样，我们在这或工作或生活，也有熟悉的邻居和陌生的路人。”“握手 302”公共艺术项目于 2013 年由马立安、张凯琴、雷胜、吴丹、刘赫等人组成的“城中村特工队”创立，成立于深圳市白石洲上白石村二坊 49 栋 302 号。她们驻扎在城中村，进行多种空间实验和行为艺术，鼓励握手楼的居民表达观点。关于“握手 302”项目，可参见马立安的访谈：https://soundcloud.com/uncuttalks/mary-ann302；另有马立安发表于 2020 年的文章 O'DONNELL M A. The Handshake 302 Village Hack Residency: Chicago, Shenzhen, and the Experience of Assimilation [M]// FORREST R, REN J, eds. The City in China: New Perspectives on Contemporary Urbanism. Bristol: Bristol University Press, 2020: 125-140.

10　参见都市实践官网对“白石洲五村城市更新 2013”的项目描述：http://www.urbanus.com.cn/projects/baishizhou。

11　就中文世界之外的深圳研究而言，比该书更早的学术贡献还包括 Linda Vlassenrood 编著的 *Shenzhen: From Factory of the World to World City* (Rotterdam: nai010 publishers, 2016) 一书，该书编自荷兰 The International New Town Institute (INTI) 于 2014 年 12 月 12 日举行的同名研讨会。

下的颠倒，本土的、民间的经验同样值得具体的观察和思考，微观视角的人类学和社会学案例研究因而成为必须。[12] 在一般意义上，若干抱持人类学和民族志视角的学者已经围绕中国城中村进行了一系列个案研究，早已将非虚构写作转化为有效的学术声音，这些成果构成一个初具规模的田野调查档案网络——包括王春光、王汉生、项飙、张鹂等对北京浙江村的调查，李培林、蓝宇蕴、储冬爱、班志远等对广州城中村的调查，黄韵然对深圳大芬村商品油画的艺术社会史研究、王大威对深圳城中村的专题著作等。[13]

另一方面，2005 年以来，一批关注城市更新议题的建筑师、规划师和城市研究学者围绕深圳城中村展开设计研究，试图提供推倒重来之外的小尺度设计干预策略，如李津逵《城中村的真问题》（2005）等辨析城中村正面价值的文章、都市实践编著的《村·城 城·村》（2006），城村架构（和林君翰）的《中国乡村改造》（2013），史蒂芬·艾尔编著的《华南城中村指南》（2014），张宇星的《城中村作为一种城市公共资本与共享资本》（2016）等。从人类学家们的在地观察，到设计师们的渐进式改造提案，再到杜鹃的深圳起源神话去魅，二十年来分散在不同领域的各类努力渐渐显露出某种共性：主张不再将城中村视为需要“解决”的“问题”，而是面向未来的一种可能的答案，自下而上，以微视角的关切和小尺度的干预，扭转城市规划热衷于大拆大建的地产思维惯性。对此，一个较早提出的重要看法来自深圳规划师黄伟文，在一篇题为《城市规划与城中村：谁来改造谁？》（2011）的文章中，他指出城中村是深圳事实上的保障房，不仅帮助深圳弥补了城市规划与管理的不足，更可谓不可或缺的城市基础设施。[14] 黄伟文认为：

> *出问题的，不是城中村，而是当下的城市规划；需要改造的，不是城中村，而是当下城市规划。而可以启发和改造城市规划的，也恰恰是被城市规划威胁的城中村。……城中村不仅仅为规划建设公众参与的、宜居宜业的、和谐低碳城市的样板，还是城市不可缺少的基础框架。如果任由当下的城市规划以改造之名行消灭城中村之实，则伤害的不仅仅是点状散落的样板区域，而是在损害这个城市的支持系统。*[9]

这样的看法此后逐渐成为知识界的共识，关于城中村的官方表述也缓缓松动，开始承认城中村对深圳的“贡献”。2017 年，由深圳市政府作为主办方，建筑师刘晓都、孟岩和艺术评论家侯瀚如策展的第七届深双以“城市共生”为主题，史无前例地把主展场设置在城中村南头古城。在这一背景下，深圳市规土委下辖的深圳市城市设计促进中心展开了《城中村：消失中的城市》一书的编辑工作。这本书旨在“记录快速发展的深圳背后多元而富有层次的社会含义，也在更长的时间与更丰富的空间层面细腻展现城市的历史切片”，已于 2020 年面世。同在 2020 年，《向深圳学习》中文版出版，但深圳还有一部分在暗处。

来福认为：“如果用自上而下的规划视角去看这些村子，则必然只能看出其毫无秩序。但用自下而上的视角去观察，会发现，在这种无序和混乱之中所产生的人口、资本和观念的流动，才是深圳经济特区得以‘成功’的原因。”[10]20 年前，这样的认识是不可思议的; 20 年后，在一大批学者、建筑师、艺术家、媒体人士和独立写作者的共同努力下，类似的看法已被广泛接受; 2019 年，“丰富城中村综合整治内涵，实现多元更新目标”被写进深圳城中村综合整治总体规划，明确划入分区的城中村居住用地在规划期内不得纳入拆除重建类城市更新单元计划、土地整备计划及棚户区改造计划，为部分城中村留下了喘息的时间。[11] 这些变化当然无法扭转“深圳的城中村时代”走向终结的历史进程，但即便如此，逆转视角方向、缩小观察尺度、祛魅神话叙事的工作不是太多，而是太少。城中村是深圳城市化的真正起点，是移民者在这座巨型移民城市的最早落脚点，研究深圳城市化也就不可能无涉城中村的历史。而城中村的历史也远远没有结束，只要迫于无奈突破规范的即兴生活还存在，城中村就必然以另外的方式延续。[15]

4 中观：“南方”作为方法，或“城中村”作为概念

“深圳的建设是‘北方性’对‘南方性’的一种征服; 不过，南北宗的持续博弈既改造了‘北方性’，又表现出一种新的差异。”在发表于 2009 年的文章《差异的南方性：广东的空间史与珠三角建筑生产》中，[12] 建筑评论家冯原指出，近代以来的珠三角演变史表明，广东不间断地身处外部经济与内部权力、地理边缘与国家样板、来自北方的政治想象与来自香港的资本想象之间，被多重两面性持续塑造，并常常提供生动的独创性，冯原将这种混合状态称为“差异的南方性”。[16] 这一看法本来并未触及城中村问题，因为城中村显然既不是深圳或珠三角的发明，也不是地理意义上南方的地域特色。“南方”和城中村之间的关联在 2017 深双得以凸显，这次聚焦城中村的双年展特别将“南方”策划为一个板块主题，

12 关于该书题目的翻译问题及中英文标题的微妙差异，可参看马立安接受“中间地带”播客的访谈：https://www.xiaoyuzhoufm.com/episode/5f8d4e2483c34e85dd66ea42.

13 这些著作包括：Da Wei David Wang, Urban Villages in the New China: Case of Shenzhen, 2016 (Palgrave Macmillan); David Bandurski, Dragons in Diamond Village: And Other Tales from the Back Alleys of Urbanising China，2016 (Penguin Books China); Winnie Wong, Van Gogh on Demand: China and the Readymade, 2014 (University of Chicago Press); 储冬爱，“城中村”的民俗记忆：广州珠村调查，2012（广东人民出版社）; 邵媛媛，转型中的实践：对一个“城中村”社区的人类学研究，2012（中国社会科学出版社）; 蓝宇蕴，都市里的村庄：一个“新村社共同体”的实地研究，2005（生活·读书·新知三联书店）; 李培林，村落的终结——羊城村的故事，2004（商务印书馆）; Li Zhang. Strangers in the City: Reconfigurations of Space, Power, and Social Networks Within China's Floating Population, 2002 (Stanford University Press); 项飚，跨越边界的社区：北京“浙江村”的生活史，2000（生活·读书·新知三联书店），等等。

14 来自黄伟文的类似论断还见于黄伟文 . 深圳：如何快速塑造一座城市？ [J]. Domus 中文版 . 2011；黄伟文 . 城市的下半身 . 黄伟文的新浪博客 , 2006-4-13。

15 如 Fulong Wu, Fangzhu Zhang 和 Chris Webster 于 2013 年指出，通过城中村再开发，政府试图消除城中村的非正规性，并通过正规的土地开发创造更多的治理空间；但由于这种做法未能从根本上解决对非正规生活和工作空间的需求，村庄的重建只会导致非正规性在更偏远的农村、其他城市街区以及在一定程度上在重建后的街区的复制。见 WU F, ZHANG F, WEBSTER C. Informality and the Development and Demolition of Urban Villages in the Chinese Peri-urban Area [J]. Urban Studies, 2013, 50(10): 1919-1934.

16 冯原对于岭南的文化建构和空间政治的研究还包括：冯原 . 南下：珠三角的空间政治 [J]. 城市中国 , 2006(13); 冯原 . 机器、制度与自梳女——从容桂镇的“标本”看珠三角工业化的性别史 [J]. 美术学报 . 2009(1); FENG Y. Transcending Geo-culture [EB/OL]. e-flux Architecture, 2018-4-24 等。

作出一次反思观察视角和扩大观察尺度的尝试。该届深双的总策展人之一刘晓都写道：

> *南方也不仅仅是一个地理概念。在中国，南方也是一个文化概念。……改革开放以来，以深圳为代表的"南方"被推到了对外开放的前沿地带，整个珠三角作为"开放的实验室"将地域化与全球化、传统与未来、迷信与幻象、落后与超前共冶一炉，建构起了南方所特有的想象力与复杂的现实。现实的"南方"不仅仅指代着五岭以南的广袤地域，更表征着开放性与可能性、创造力与效率、现实欲望和未来梦想……由此，中国的"南方"不仅表述为地理概念，还历史地与"北方"和"中心"相关联，并进而成为亚文化与另类思想的喻指。……深圳的城中村则几乎是唯一遗存的南方原生的城市形态，也成为抵抗机制仅存的滋养之地。*[13]

在这一表述中，"南方"与地理和气候的具体限定脱钩，而被视为一种关于南北差异的观念和方法，用来处理中心与边缘、影响与抵抗、地区与全球等一系列复杂问题，反抗北方中心的文化霸权，深圳城中村则是适用这一方法的典型研究案例。把深圳城中村称为"抵抗机制仅存的滋养之地"算得上是惊人之语，尤其在这一"仅存的滋养之地"正加速消失的今天。换言之，如果说 2017 年策展人试图提出一种概念化的"南方"方法论来考察城中村的具体问题，那么今天我们面临的则是一个更加概念化、更缺乏具体参照的"城中村"。把南方作为观念或方法并不是新鲜看法——如与 2017 深双同年举办的第十四届卡塞尔文献展就以"南方作为一种思想状态"作为场刊标题——但城中村成为脱离具体空间的抽象概念却是史无前例的新状况。[17] 城中村如果（在可以预见的将来）不再以具体的形式存在，那么现在就必须借助文献思考，作为概念的城中村为何不应该，也不会被"非正规性"等已被广泛讨论的概念稀释，以及该如何继续讨论城中村才是有意义的、作为方法的南方又会对此有哪些助益。[18]

若从刘晓都的概要表述出发去找具体的例证，那么城中村的的确确（曾经）为某种"开放性与可能性、创造力与效率、现实欲望和未来梦想、亚文化与另类思想"提供政治层面、经济层面和社会层面的空间庇护，而且城中村的"开放性""创造力"和"梦想"与改革开放以来国家叙事中的"开放性""创造力"和"梦想"相比，确实是"另类"的。在发表于 2020 年的论文《中国的非正规生产空间》中，乔纳森·巴赫和斯特凡·艾尔探讨了"开放"的、充满"创造力"的中国城中村非正规制造业。[14] 关于城中村非正规经济的文献已有很多，这些文献表明，城中村的集体土地上，不仅仅有外来人口密集居住的非正规住房，还有各种小规模的非正规工厂、非正规劳动和非正规劳务关系，以小尺度（官方常使用"三小场所"的说法——小档口、小作坊、小娱乐场所）存在于法规的灰色地带，居住、生产、仓储、营销高度混合。[19] 巴赫和艾尔试图进一步指出，是局部地区非正规制造业与全球正规制造业的共生关系——城中村的非正规经济早已高效地嵌入城中村之外的正规制造业生态之内，极小尺度与极大尺度以极富想象力的方式平行互联（而非竖直的等级关系）。深圳一座握手楼的一个小房间里，工人或许正在用来自中国台湾的"公版"自制无牌手机并即刻发售印度；[20] 又如吴玲在关于城中村非正规经济的博士论文（2013）中，[15] 详细调查了一座广州城中村如何在晚上变成一座非正规加班工厂：工人白天在城中村附近的正规工厂里缝衣服，晚上在城中村给衣服上纽扣，这些服装产品第二天就会销往世界各地。[21] 在类似的议题上，有两个已经稍有时代感的术语常被提及，巴赫和艾尔却并未使用："山寨"和"低端全球化"。

在一篇题为《山寨：一种南方理论？》（2020）的文章中，去殖民框架下的比较文学学者向在荣将山寨和低端全球化扩展为一种关于南方的理论思辨。值得我们注意的是，向在荣顺带提到了城中村。他写道：

> *山寨出现在"反帝反资本主义阶级斗争"这样的话语疲弱无能的时代，且与全球资本主义保持着十分暧昧的关系。近来对山寨的理论热情，最具代表的就是韩裔德国哲学家韩炳哲的《山寨：中国的解构》。这本极具洞见的哲学著述却难掩其东方主义，其中对"山寨"的讨论过度强调了所谓的东西之别，又屈服于"原创性"的魔咒，例证大多以此为山寨辩护和正名，甚至从山寨的哲学背景逐渐延伸到一种"民主希望"的大调上。这样的做法恰恰消解了山寨最本质的特点：暧昧，不具合法性，又对任何宏大叙事漠不关心。……各地的调研让我发现山寨并没有那么标新立异，"原创性"在大生产中体现得十分牵强，并没有韩氏书中描述得那般精彩，也*

17　关于第 14 届卡塞尔文献展和 "South as a State of Mind"，可参看 https://www.documenta14.de/en/south/。关于南方作为方法，学术界已有大量讨论，参看 GARDNER A, GREEN C. South as Method? Biennials Past and Present [C]// Making Biennials in Contemporary Times: Essays from the World Biennial Forum no. 2. São Paulo, 2014: 28-36; PAPASTERGIADIS N. What is the South?[M]// GARDNER A, ed. Mapping South: Journeys in South-South Cultural Relations, Melbourne: The South Project, 2013. Anthony Gardner 和 Charles Green 在他们关于双年展的南方方法的文章中写道："We see 'South' as a loose working concept for that which tries to resist easy assimilation within hegemonic global currents generated outward from the North Atlantic. Instead, 'South' emphasises two things at once: it asserts a rich history generated from long-standing unease with North Atlantic hegemony, whether through the lingering legacies of colonial and neo-colonial violence, or the struggles for decolonisation and deimperialization figured through the nonaligned politics of the Third World during the Cold War; and, secondly, an awareness that, as Nikos Papastergiadis argues, 'survival requires a coordinated transnational response' through which that hegemony might be displaced." 刘晓都的表述与这些既有看法一脉相承，同时强调了"南方"在中国的特异性。

18　关于"非正规性"概念及其历史，文献不胜枚举，此处只举几例：王晖，龙元．第三世界城市非正规性研究与住房实践综述 [J]. 国际城市规划，2008, 23(6): 65-69; HERNÁNDEZ F, KELLETT P, ALLEN L K. Rethinking the Informal City: Critical Perspectives from Latin America [M]. New York：Berghahn Books, 2010; WU F, ZHANG F, WEBSTER C. Informality and the Development and Demolition of Urban Villages in the Chinese Peri-urban Area[J]. *Urban Studies*, 2013, 50(10): 1919-1934. 学界已逐渐形成共识，非正规城市不再被认为是发展中国家才有的现象。有关发达国家城市的非正规性，见 MUKHIJA V, LOUKAITOU-SIDERIS A, eds. The Informal American City: Beyond Taco Trucks and Day Labor [M]. Cambridge, Mass.: The MIT Press, 2014.

19　关于城中村的非正规经济，已有来自相当多学者的研究成果，如薛德升、闫小培、蓝宇蕴、吴玲、周大鸣等。

20　巴赫和艾尔援引了 Silvia Lindtner, Anna Greenspan 和 David Li 关于深圳山寨电子产品的研究：LINDTNER S, GREENSPAN A, LI D, Shanzhai: China's Collaborative Electronics-Design Ecosystem[EB/OL]. The Atlantic, 2014-5-18. https://www.theatlantic.com/technology/archive/2014/05/chinas-mass-production-system/370898.

21　关于中国非正规劳动的整体状况，另可参看 SWIDER S, Building China: Informal Work and the New Precariat [M]. Ithaca, NY: Cornell University Press, 2016; AL S. ed. Factory Towns of South China: an Illustrated Guidebook[M]. Hong Kong: Hong Kong University Press, 2012.

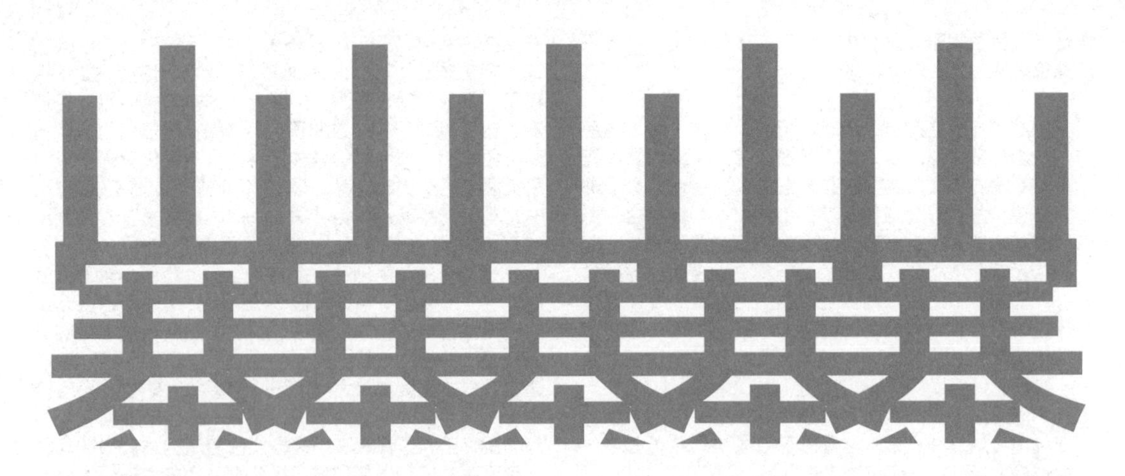

没有什么特别的反抗精神。……当“原创性”在世界各地与城市市绅化沆瀣一气的同时，深圳一个个被拆除的城中村，作为这个未来大都会的城中山寨，迅速成为历史烟云。而庶民的历史从来都是另一个节奏。[16]

全球资本主义需要山寨，正如城市需要城中村，尽管它们纷纷以“低端”之名对山寨或城中村嗤之以鼻。不妨在向在荣顺带一提的“城中山寨”城中村这里稍作补充：如果我们把文中关于“山寨的南方理论”提及的全部论据——假钞、假货、盗版、违章、“低端人口”、低端全球化、山寨手机、山寨小商品——都加起来，然后想象一个空间状态，不难得出“城中村”这个答案(当然，也可能是转型前的华强北)。这么说不是为了在“山寨”和“城中村”之间强行建立某种对应关系，而是希望借助已有的对于山寨的理论思辨来启发对于城中村的思考。“山寨”这个词已经透出某种年代感，正如“城中村”一词也注定将在不太久远的将来失去所指。倘若读者接受我的武断想象，那么一个个具体的城中村“迅速成为历史烟云”就有可能造成一个关键的历史状况。具体的城中村以多么迅速的方式消失，抽象的“城中村”就以多么迅速的方式走向一种概念，一种关于“庶民的历史”的概念。城中村的物理实存或许难逃被轰出官修历史的命运，“城中村”这朵“烟云”却仍会在庶民的历史中逡巡。关于这朵“烟云”的可能形貌，向在荣继续提醒我们：

山寨的南方理论（或者：南方理论作为山寨理论）不可能服从理论和实践的边界——包括其他人为的二元边界（南北、东西、男女、神俗、高低、真伪）。这种不服从不是抵抗和打破，而是源于它平庸的繁盛；不是英雄式的反抗，而是一种翻转，一种内在涌动：万物相通的理论 - 实践反过来生发万物皆可的伦理。[16]

此时如果再回看2017深双策展人关于“影响与抵抗”的慷慨陈词会发现，不论是握手302、白石洲小组，还是2017深双，都希望赋予城中村和城中村的居民们某种抵抗的主体性，然而事实上城中村从来没有主动承担过“抵抗机制的滋养之地”的角色，也就无所谓仅存不仅存。城中村的真正姿态从来都不是主动抵抗，更从未主动滋养过任何抵抗机制，而是不合作、不合规、不认同、不关心任何宏大叙事、游离、野生、即兴、临时、暧昧、撤退、弃权、身不由己、权宜之计、得过且过，是赫尔曼·梅尔维尔笔下抄写员巴特比的“我倾向于不”。在谈及巴特比的消极政治学时，斯洛文尼亚哲学家齐泽克认为，“弃权行为比政治内否定和投不信任票走得更远：它所否定的是整个决策框架。”[22] 与之类似，此处认清“城中村”概念的非抵抗性绝不意味着一种对现实的逃避，恰恰相反，它蕴含一种远远更为有效的路径，哪怕仅仅作为一种想象中的力量。

5 宏观：一个语言的难题

1996年春天，荷兰建筑师雷姆·库哈斯带着哈佛大学设计研究生院的学生来到珠三角，以深圳特区为城市调研的第一站。他在教学笔记中写道：[23]

在一方面，可信且普世的学说乏善可陈；而另一方面是前所未见的生产模式和发展力度，这就导致了一个非常拧巴的状况：这种城市状态正臻于顶峰，但此时也恰恰是最难以理解它的时候。……事实上，整个业界都不具备足够的语汇来探讨这个领域中最贴切、最急迫的现象，更别说用任何概念性思维来确切地描述、诠释和理解那些能够重新定义和复苏城市的力量。[17]

库哈斯指出了一个关于语言的难题，对他来说，当时的珠三角城市化现象——他称之为一个“现代化的大旋涡”“一边是混乱，一边是歌颂，中间一片空无”——正处于一个概念、术语和理论的真空之中，超出了既有

22 齐泽克写道："Why is the government thrown into such a panic by the voters' abstention? It is compelled to confront the fact that it exists, that it exerts power, only insofar as it is accepted as such by its subjects-accepted even in the mode of rejection. The voters' abstention goes further than the intra-political negation, the vote of no confidence: it rejects the very frame of decision." 见 ŽIŽEK S. Violence: Six Sideways Reflections [M]. New York: Picador, 2008: 216.

23 中文翻译见雷姆·库哈斯. 加剧差异化的城市 ©. 刘宇扬，译，附录于王辉. 误读“大跃进”[J]. 时代建筑, 2006(5): 35-39. 对刘宇扬的译稿有改动。

给《城市中国》：

当时我预料到的现象：

——香港的相对不景；
——深圳的力量；
——澳门的愈发纸醉金迷；
——广州的持续活力。

我没有预料的：

——深圳走出粗放开发的状态，而变得更为成熟；它如今是中国经营得最为谨慎的区域之一；
——深圳也是中国和"外国"的建筑师将明显变得平起平坐的地方。
——深圳也是第一个拥有（相对）老区的新城；在它建成之前，这个城市都仍要进行自我再造。

惟一仍可能对这个城市（以及中国）造成破坏的东西就是金钱。中国想用有限的预算去实现复杂的野心。如今既然有了超常的财富，就应当投资于真实的质量。

深圳可能是中国的第一个……

瑞姆·库哈斯
2006年10月23日

图 1 雷姆·库哈斯，《珠三角 / 深圳：10 年之后》，《城市中国》，2006(13): 13.

城市和建筑研究的思辨范围。他当时的策略是发明术语，为新术语标记版权并编纂词典：

这些研究叙述了珠三角的城市状态，我们称之为 CITY OF EXACERBATED DIFFERENCE©（"加剧差异化的城市 ©"），或 COED© 的一种城市共存新模式。……除了发掘到各状态的独特性外，《大跃进》同时也介绍了多个"版权所有"的专有名词，进而呈现和开拓一个新的思维来叙述并解释当代城市的状态。珠三角现象有如彗星一般从天而降，一朵"不明就里的烟云"在珠三角的实存和功能的周遭，鬼鬼祟祟地缭绕，这一切都证明了平行世界的存在，彻底否定了那种认为全球化等同于全面的知识的看法。[17]

这段写于 1996 年的锐利文字，触发来自历史的厚重回响（图 1）。偷偷自赋"版权所有"的编词造句已经开上了某种山寨式的玩笑，尽管此时"山寨"这个词尚不存在，"城中村"这个说法也才刚刚进入学术界的视野（据说国内第一篇以城中村为主题的论文是杨安的《"城中村"的防治》）。[24] 到了 20 年后的 2017 深双，OMA（Office for Metropolitan Architecture，字面意思即"都市建筑办公室"）换成了都市实践；"CITY OF EXACERBATED DIFFERENCE©"换成了展览城中村的大字标题"CITIES, GROW IN DIFFERENCE"（2017 深双的主题"城市共生"的官方英文翻译）。虽仅两字之差，但从来自外部的、对爆燃的差异的惊异，变成来自本土的、在官方的督导之下对差异和例外的小心维护，意义已完全调转（在回答建筑历史学者朱涛关于双年展主题英文译法的疑惑时，孟岩说："咱们整一个特别'Chinglish'那种词，就是中式英语，让外国人听着别扭、觉得这个词不是一个很顺的意思"）。[25]"南大门"门口，内外之间，一系列镜像、颠倒、误译和本土化过程将一个语言的难题排演成一出洋泾浜戏剧。

在 2006 年的文章《南下：珠三角的空间政治》中，冯原论述了五岭之南的珠三角历史悠久的地缘矛盾性：

岭南，或者说珠江三角洲的使命注定是如此，它在内陆的政治征服和海洋经济贸易的反复博弈中度过了两千多年，现在又迎来了新的千年。……站在珠江口放开思绪还是有价值的，那仍然是一个能够唤起历史想象的节点，无论是放眼向南还是转过头来向着北方，这里一直是南方的尽头；又是向着更远的南方的开头。[18]

十年后的深双确实看向了比南方更南的远方，与城中村并置在南头古城的展厅里的，有内罗毕的贫民区、圣保罗的法维拉、哈瓦那的城市更新案例，另有印度和东南亚地区的非正规建筑、亚非拉贫民窟共陈一室，与 6 公里外的"世界之窗"主题公园用上百座微缩古迹打造的后现代山寨奇观形成强烈的对照。在英文版介绍词中，该板块被称为"Global South"，但中文介绍词却

24 这个观察来自周毅刚．两种"城市病"比较：城中村与百年前的西方贫民窟 [J]. 新建筑，2007 (http://wen.org.cn/modules/article/view.article.php/2208). 杨安 1996 年的文章见杨安．"城中村"的防治．城乡建设，1996(8): 30-31.

25 关于 2017 深双英文主题的确定过程，可参看载于都市实践网站的记录：http://www.urbanus.com.cn/events/english-theme/.

没有“全球南方”这个对应的术语翻译，而是在“世界”和“南方”之间用一条竖线分割。[19] 中英文之间的“创造性误读”又一次显露了对内对外的两面性，按冯原的说法，这种珠三角特有的两面性已经在这片土地的地缘政治史上延宕了两千多年。

从法国社会学家涂尔干的19世纪说法“原始”与“传统”，到1950年代由法国人口学家阿尔弗雷德·索维提出并在万隆会议后被广泛采纳的“第三世界（主义）”，再到时兴的“全球南方”“后殖民主义”和“热带”，已有一系列用来标记贫困地带、殖民主义历史和地区之间不平衡发展的词汇，指示着思想界的动态前沿。术语向着远离欧洲中心主义视角的方向进化，在逐渐强调跨地域、跨洲际的团结和联系的同时，也难免稀释不同地区各有千秋的政治、经济和社会特异性。就“全球南方”而言，该词汇诞生于冷战结束后全球化背景下学界对“南方”的研究兴趣。它不是一个地理方位的描述，也不仅仅是针对贫困和不平等问题的归纳性提法，而是用来指示“第三世界主义”和去殖民化运动在全球化时代的延续，意在强化各个曾被殖民群体的凝聚力，并进一步挑战和颠覆西方中心的偏狭视角。[26] 换言之，“全球南方”不仅仅是对贫困地带和地区之间发展不平衡现象的概括；从根本上说，它是一种以推动去殖民化进程为目标的政治动员。[27] 当下的学界对这一术语的态度大致如加拿大学者马莱亚·克拉克在其综述文章中所说的：“对于有志于理解和揭示南北不平等的人们、试图参与或支持庶民群体、就一系列激进主张和实践进行组织动员的人们、想要向支配群体和不均衡发展发起挑战的人们来说，这个术语有力且有用。一言以蔽之，‘全球南方’的说法可能多少有些含糊不清，但我仍欣然采纳它。”[28]

显然，“世界之窗”的“世界”和“第三世界”的“世界”不是一个“世界”，“全球资本主义”的“全球”和“全球南方”的“全球”也不是一个“全球”，舶来的术语与中文的积习常常在岭南激荡碰撞，制造有意无意的误差，发明从未有过的语言。不论具体出于何种考虑，2017深双对“全球南方”表述的回避在事实上达成一个更多描述性、更少政治性的呈现（城中村议题也确实从未真正融入“全球南方”的政治议程），呼应了被库哈斯早早点破的语言难题。在25年大量的研究和讨论之后，我们还是观察到，“城中村”依然处于一个概念和理论的空乏状态之中，难以被“全球南方”研究或“后殖民主义”或“热带”研究等任何学术思潮真正捕获。在物理实存被消灭之后，它至多成为一种概念、一种理论、一朵“烟云”，去偷偷推倒挡在“世界”和“南方”之间的那条竖线。25年前，“加剧差异化的城市©”被汪洋恣肆的荷兰建筑师暗搓搓地自制盗版并写进了美国的大学教案；25年后，已经到了不得不为作为概念的“城中村©”抢赋版权的历史时刻。之所以需要抢，是缘于这一历史状态的紧迫性：残砖断瓦不会为城中村的复杂历史画上一个轻易的句号，自有秃鹫一般的资本和权力环伺在侧，争相把这朵残存的“烟云”定义成他们想要的样子。一种关于“城中村©”的理论思辨因而迫在眉睫，甚至比其他任何时候都更加迫在眉睫。

6 结语

尽管“后城中村时代”还未真正到来，为“城中村©”抢赋版权的争夺战已经开始。2017深双之后，双年展主展场南头在房地产开发商万科的主导下经历改造，做实了创意古城的时尚标签，升级换代的城中村招徕入时的游客赏游打卡的同时，旧有居民和商户仿佛在无声无息之间已被大面积置换。2021年4月底，由时尚传媒公司栩栩华生旗下的卷宗书店和万科联合主办的“友谊书展”在南头古城落地，该书展以“我们的城市，我们的村”为主题，使我们无法无视南头在短短四年之间从下里巴人到阳春白雪的剧烈士绅化过程（尽管“古城”的身份使得这种剧烈的士绅化没有在南头的空间形态上显现出来），并反问“我们”究竟是谁们、“我们”之外的“他们”又安家何方。[29]

2021年5月9日，在南头古城的大家乐舞台——这里正是2017深双闭幕式举办地——书展发起“南城论坛”，政府官员、地产商、建筑师和学者的话语在4年后又一次罕有地共熔一炉。紧接着南山区副区长发言的万科城市研究院院长钱源毫不讳言“南头古城”的虚构性：“所有人都说它是一个古城，哪里古了呢？所以我们要创造一个‘古’的意象……‘无中生有’的创作。这件事不光要改房子还要做内容，不仅要招商还要创造内容。”[20] 地产商绕过南头自身数百年的具体生命历史，凭空制造了一座时尚古城。“风貌”被“安排”，“场景”被“设计”，十几年来在村城之间举棋不定的南头改头换面，被虚构的“古意”装点为一座主题公园，“村”则被彻底架空。而这，已经是十几年来大量建筑师和知识界人士共同努力反抗城中村改造中推倒重来的开发商逻辑的最“成功”案例。如孟岩在随后的发言中所指出的，士绅化远远不是一个新问题：“所有参加双年展的以及参与活动的所有人都意识到士绅化是势不可挡的，我们希望能够做到减缓这个过程，当时的想法是螳臂当车（固然）做不到，但是希望把这个过程延长。”[20] 而今天，这个曾努力“延长”的过程终于还是接近了尾声。在“深圳的城中村时代”行将结束（甚至已经结束）的当下，必须抛出新的理论问题、给出新的历史判断、刺激新的历史想象；毕竟在阵地的另一边，线性发展主义叙事早已备好了它的答案。

2011年，南方报业传媒集团将一系列旗下报刊对

26 如 Nour Dados 和 Raewyn Connell 所说，"The term Global South functions as more than a metaphor for underdevelopment. It references an entire history of colonialism, neo-imperialism, and differential economic and social change through which large inequalities in living standard, life expectancy, and access to resources are maintained." 见 DADOS N, CONNELL R. The Global South [J]. Context, 2012, 11(1): 12-13.

27 关于作为一个政治计划的“全球南方”概念，见 BALLESTRIN L. The Global South as a Political Project [EB/OL]. https://www.e-ir.info/2020/07/03/the-global-south-as-a-political-project/.

28 "For those interested in understanding and unmasking this inequality, and participating or supporting subaltern groups that are organising around sets of radical claims and practices to challenge dominant groups and uneven development, the term is powerful and useful. In short, the term global South might be ambiguous and vague, but I embrace it." 见 CLARKE M. Global South: what does it mean and why use the term?[EB/OL]. https://onlineacademiccommunity.uvic.ca/globalsouthpolitics/2018/08/08/global-south-what-does-it-mean-and-why-use-the-term/.

29 关于2021友谊书展，可参看微信公众号“卷宗 Wallpaper”发布的介绍：“欢迎来到‘我们的城市，我们的村’！”https://mp.weixin.qq.com/s/nC5SOFEKgALcHS9K6pfMrA.

城中村改造问题的报道文章合集出版，以《未来没有城中村：一座先锋城市的拆迁造富神话》为题（图2）。十年后，书名已经成真，但该书结尾处留下的问题——“没有城中村，他们怎么办”依然有待回答，而且有待回答的问题没有减少而是增多了。

图2《未来没有城中村：一座先锋城市的拆迁造富神话》（南方都市报编著，中国民主法制出版社，2011）封面。
Figure 2 Front cover of *There Will Be No Villages in the City*, edited by Southern Metropolis Daily, 2011.

作为一次文献调查，本文的片面性不仅仅在于其有限的视角自带的必然的偏狭，更在于城中村自身历史上尚存的空洞。仍然要不停地问，有哪些声音、哪些人群、哪些视角、哪些历史至今仍然被忽视、被排除、被压抑、被躲闪，又是哪些人还没有准备好接受这些声音、这些人群、这些视角、这些历史。它们像朵朵烟云，深潜在庶民历史的阴影处，并一定会在未来的某个时候漂游而出。

A Literature Review on the Urban Village

An Tairan
An Tairan is a Ph.D. candidate in History and Theory of Architecture at Princeton University and he was a literature researcher of the 2017 Bi-City Biennale of Urbanism\Architecture (Shenzhen).
The article was translated from Chinese to English by Chen Yiwei.

1 Prologue: A Writing Experiment on the Question of Scale

More than three years ago, in the South Building of the main exhibition site—Nantou (literally "South End"), Nanshan (literally "South Mountain") District, Shenzhen—the "Global South" section of the 2017 UABB, which laid out the horizon and viewpoint for the entire Biennale, was on show. The overlapping "Souths" in place names, directions, and research subjects may not have been merely coincidental; rather, implicit at the point of intersection where the "Souths" meet is a question of scale: Lingnan (literally "South of the Southern Mountains") is in the southern part of China while Shenzhen is located on the southern edge of Lingnan; Nantou, as a part of Shenzhen, is narrowly classified into the tropics in geographic terms with a latitude at 22°N. To be sure, under the rubric of "Global South" which is also a popular academic field, the tiny village of Nantou situated between Yangtai Mountain and Lingdingyang Estuary provides a valuable "case." However, regarding the "urban village" as a subject of urban studies, the specific relationships between the "case" and the region, the part and the whole, and the local and the global have yet to be clarified. These relationships were only briefly touched upon in the 2017 UABB; in this sense, the exhibition is unfinished (and it does not have to be finished—as one of its chief curators, Meng Yan, said—"'Cities, Grow in Difference' is an 'ongoing' exhibition").[1] Through the writing of a literature review on the urban village as typology, the present essay endeavors to make a preliminary yet belated clarification of these relationships. By stretching the scale of observation from micro to macro, this review brings China's urban village phenomenon—with Shenzhen as a case—into the horizon of the "Global" and the "South" while also using the urban village to question the nature of these terms and why the urban village has become and still remains a matter of serious concern. One by one, urban villages are disappearing. Before they vanish into history and memory once and for all, I ask: what new questions have this particular urban form and way of life raised for urban studies? In the academic history of urban village studies, specifically, what reflections and concerns has the urban village stimulated and what investigation and research methods has it induced and raised awareness of? At a time when the axis of urbanization is fundamentally shifting globally from the North to the South and from the West to the East, a research framework spanning across both geographical and temporal scales may have to take the first step in the form of a "literature review." In this sense, this essay, like the "Global South" section of 2017 UABB, is an experiment of cross-scale observation.

As a result of the large span in the scale of observation—from the tangible minutiae of day-to-day life to the academic frontier of Global

South studies, three hypotheses may first be proposed here and examined in subsequent writings. First, literature materials that have documentation value for cross-scale observations should also be "cross-genre;" in other words, the concept of the "literature" needs redefinition. Besides the usual genres of formal literature such as academic writing, journal, book and exhibition review, some non-academic writings and other types of practice should also be included in the category of the "literature"—including but not limited to media writings such as news reports, surveys, and interviews, as well as independent research, independent writing, oral history writing, and art making by scholars, independent writers, and artists. Second, the diversity of perspectives and scale of observation inevitably imply differences in investigation methods, disciplinary traditions, and research strategies; that is, a cross-scale literature investigation will unavoidably lead to cross-disciplinary outcomes while how these different disciplines intersect with or inspire one another is a subject for further examination. Third, based on the assumption that a literature review should not and cannot set off to be all-encompassing and perfectly neutral (just like any other form of documentation), I may as well honestly confront the limitations of the starting point, the biases of the current moment, and the idiosyncrasies of local conditions. As a belated—or, as far as the 2017 UABB is concerned, a retroactive—literature review, this essay covers the period beginning with the Biennale exhibition (December 15th, 2017-March 15th, 2018) to the narrative present in 2021. During these years, many urban renewal policies directly related to the urban village have been implemented in Shenzhen and their corresponding initiatives have already started taking place and making a substantial impact on the city. These drastic short-term changes mean that the retroactive literature investigation is inevitably at the same time a rewriting of its bibliography. I narrate the massive, yet slow transformation of the informal city in reverse chronological order to allow readers to find clues for evaluating related literature. I begin my review in the time period between 2017 to 2021 and work my way across a specific geographical segment of Shenzhen. In other words, the present literature review preexists the bibliography and not vice versa; a brand new yet incomplete bibliography is gradually developed through the process of reviewing the literature.

2 Introduction: The Forgotten Baishizhou as an Allegory of an Era

On March 23, 2021, Mary Ann O'Donnell, an American anthropologist who has been a long-term resident in Shenzhen, updated her personal blog, "Shenzhen Noted," with an article titled "Is the era of Shenzhen urban villages over?" In the article, she made a bold speculation: "I think the era of urban villages in Shenzhen has ended. I'm dating the beginning of the urban village era to 1992, when rural urbanization began in the inner districts. [And] I'm dating the end of the urban village to now-ish, with the demolition of Baishizhou."[2] In O'Donnell's perspective, as Shenzhen's largest and last major urban village which stands along Shennan Boulevard, Baishizhou's demolition and gentrification since 2019 have marked the end of an era.[1] The replacement of the forgotten Baishizhou with high-rise buildings and towers is merely a matter of time. As early as 2005, Baishizhou was included in the "Old Village Renovation Plan" by the Nanshan District Government. In July 2014, the Baishizhou project was renamed "Shahe Five Villages Urban Renewal Unit" and included in the Shenzhen Urban Renewal Unit Plan. By the end of 2018, the Baishizhou revitalization project was sanctioned by Shenzhen's Municipal Urban Planning, Land & Resources Commission, and hence the official launch of its demolition and village eviction process.[2] In the following years, the urban village landlords successively signed contracts with real estate developer, Lvgem China, while leaseholders gradually moved out and all the handshake buildings were being pulled down if not already demolished. The redevelopment project is scheduled to be completed in 2029 and the site eventually will see the rise of 31 residential towers ranging from 49 to 65 stories, 21 condominiums, and 4 office towers. On this site that is roughly 0.6 square kilometers in area with more than 2,500 rental buildings, 50,000 units (suites), around 2,310 stores, and a total population of more than 140,000 (127,000 of which are migrants), the full-scale demolition of Baishizhou, which has long been on the city's planning agenda, finally began.[3] Yet it should be mentioned that Baishizhou will be reduced not only to a

1 In this short blog, O'Donnell did not provide detailed evidence for her assertion - as she stated at the beginning of her writing, this is a "speculative post."

2 The timeline of Baishizhou's revitalization agenda can be seen in, project developer, Lvgem China's report: http://www.lvgem-china.com/report/2020/05/27/6110.html.

physical ruin but also to the debris of discourse. If we put aside O'Donnell's specific periodization of "the era of urban villages in Shenzhen" for the moment and believe that this speculation about "the end" is insightful rather than nonsense, then we may ask a few questions about the connotation of her assertion: looking retrospectively from "the end," what topics and concerns that should not be forgotten and what can we pick up from the vast research, practice, and public discussion around Baishizhou during the past 20 years? The reason for using Baishizhou as an introduction to this essay is because of its importance as a case and O'Donnell's radical conjecture which is in line with this essay's concern for scale: Baishizhou is admittedly only one example among hundreds of urban villages in Shenzhen, but a focused literature investigation on this subject matter may enlighten a holistic diagnosis of the present and the past.[4] By starting from Baishizhou, I retrace a series of academic studies, public discussions, and independent observations on urban villages since 2005 (also since the first UABB); my purpose is not only to reveal a silhouette of an era, but also to call attention to some irreconcilable contradictions and to dig out some questions that have never been answered from the debris of language and words. Born out of Shahe Farm and five surrounding old villages, the old Baishizhou is about to become history as it accelerates forward into a post-urban village era, but by then, all conflicts and problems will be buried, eliminated, and forgotten.[5]

Since 2019, the mass demolition of Baishizhou has triggered a vast amount of documentation, investigation, and critical writing from both mainstream media and individuals.[6] Among them, an article by independent writer Yuan Aijia, titled "The 'valuable' Baishizhou: The Developmentalism of the Urban Village" and published on the official WeChat account of "Shenzhenyeren," provides a way of thinking about Baishizhou. According to Yuan, the history of Baishizhou has shown that this massive urban village is no less than a battleground for two distinctive ideas about the city: On the one hand, there is the linear, forward-looking developmentalism that is subservient to the logic of real estate development, while on the other hand, there is the intellectual resistance to linear developmentalism; unfortunately, the latter was never given a fair chance to confront the former. "The staying or going of communities like Baishizhou could have sparked a social debate on 'what cities exist for'." Yuan provided two specific examples of the intellectual endeavors to help Baishizhou speak for itself: The first being the aforementioned Mary Ann O'Donnell and her public art and community practice project named "Handshake 302" (established in 2013) which was stationed in Bashizhou for many years. She made unremitting efforts to cultivate a kind of self-consciousness of the urban village and to convey a message: "We did

3 According to Shenzhen Nanshan District Goverment's census report, there are a total of 110,000 tenants residing in Baishizou; however, other sources reveal that the total number may actually be close to 150,000. Similar to the count of total population in other urban villages, the actual number of population in Baishizhou will never be exact as the sources vary. The number listed here came from a research conducted by the Baishizhou Team in 2016, which has since been widely accepted as the closest count.

4 The methodology used in this review was inspired by Italian historian, Carlo Ginzburg's series of important theories on the relationship between the specificity of (the case) and the generalization of (the theory). Ginzburg points out that the micro-analysis into a specific case can inspire an understanding of regularity, and not vice versa. See GINZBURG C. Microhistory: Two or Three things That I know about it [J]. TEDESCHI J, TEDESCHI A C, trans. *Critical Inquiry*, 1993, 20(1): 10-35.

5 There have been many literature works on and documentation of the transformation history of Baishizhou. For instance, O'DONNELL M A. Laying Siege to the Villages: The Vernacular Geography of Shenzhen [M]// O'DONNELL M A, WONG W, BACH J, eds. *Learning from Shenzhen: China's Post-Mao Experiment from Experiment to Model City*. Chicago: University of Chicago Press, 2017: 120; O'DONNELL M A. Of Shahe and OCT [EB/OL]. Blog "Shenzhen Noted," 2010-1-31. http://maryannodonnell.wordpress.com/2010/01/31/tangtou-baishizhou. And the historical review on UABB can be accessed in Shenzhen Biennale of Urbanism\Architecture Organizing Committee Office. *(IN)VISIBLE CITIES: 10 Years of UABB Research*, Vol.1 [M]. Beijing: China Architecture & Building Press, 2017; HUANG W. Invisible Cities: 9 Views of UABB in 10 Years [J]. *Time+Architecture*, 2014: 48-56.

6 In July, 2019, the tenants of Baishizhou consecutively received the notice to evict from their homes by the end of August or mid-September. The abruptness of this eviction notice caused thousands of migrant children to face the problem of being out of school. This had attracted the attention of many domestic newspapers including *Nanfang Metropolis Daily*, *JIEMIAN*, *The Paper*, *Caixin Weekly*, and international newspaper like *South China Morning Post*, as well as other self-media platforms. For example, see Liang Zhou, "Shenzhen's largest urban village 'Baishizhou's' tenant clearance dilemma: the degree that is discontinued," *JIEMIAN*, https://m.jiemian.com/article/3282575.html?from=timeline&isappinstalled=0; Cai Yuqing, "The first stop for 'Shenzhen drifters' faces redevelopment, landlords are evicting tenants, who also encounter their children's schooling and other problems," *Nanfang Metropolis Daily*, https://m.mp.oeeee.com/a/BAAFRD000020190703178296.html?&layer=6&share=chat&isndappinstalled=0; Phoebe Zhang, "Migrant workers forced out as one of Shenzhen's last 'urban villages' faces wrecking ball," *South China Morning Post*, July 22, 2019. Curator Zheng Hongbin, architect Zhang Xing, and artist NUTBROTHER, who have long been concerned with Baishizhou and urban renewal issues, launched the "Shenzhen Dolls" project, collecting 400 dolls from the children of Baishizhou and using a giant 29-ton claw machine to pass them one by one from the inside of Shenzhen to outside the border to Huizhou on August 4. This artistic initiative, for the first time, brought attention to these children, who were facing education difficulties due to the city's policy changes, and brought Baishizhou into the spotlight for a short time in July and August, 2019. For a detailed description of the "Shenzhen Dolls" project, see the article "Shenzhen Dolls Project is collecting" by *ARThing*, https://www.arthing.org/archives/2019/07/post2410857.html. Other responses were not found in mainstream media but scattered on various self-media platforms. These fragments of voices were widely disseminated and stimulated a great deal of online discussion, and their documentary value should not be overlooked. For example, *NGOCN*, a self-media focused on civil society issues, published an article titled "The evicted 'Shenzhen Dolls'" in August 2019, and the left-wing platform Margaux published an article titled "Non-violent demolition and eviction as a management and governance method: a belated note on Baishizhou," and the serial "The invisible Shenzheners" by the "Baishizhou Group" on the official WeChat account. These articles provide an overview of the difficult negotiations taken by parents of Baishizhou children regarding the education of their children from their own perspectives. Other related independent writings worth mentioning include the non-fiction serial on Baishizhou published by "Wenshanhu" on the creator platform of *The Paper*, including "Leaving from Baishizhou under the eviction order, a collective farewell of Shenzhen drifters," "The long-staying Shenzhen drifting tenants, building their own Jianghu in the largest urban village" and "The fading Baisizhou, the refuge of many Shenzhen drifters."

not help Baishizhou, but Baishizhou helped us."[3] She rejected viewing Baishizhou as a slum which awaited "saving" by urban renewal plans and instead believed in Baishizhou's potential to provide an alternative "Shenzhen living experience." The second endeavor—"Baishizhou Team"—was established in June, 2015 for the same purpose of cultivating the conscious expression of Baishizhou's own experience.[7] Yuan argues that through developmentalism which pursues the logic of real estate development, the community-level, inward-oriented residential and commercial ecosystem of the urban village will inevitably be replaced by the city-level, outward-looking economic demands in the most relentless way.[8]

The pessimistic tone of this article is necessarily based on the fact that since 2005, even the combined "intellectual" efforts of a host of architects, urban planners, artists, writers, scholars, social workers and other independent observers in and around the urban village have not prevented the logic of developmentalism from bulldozing Baishizhou. As successful precedents, neither the minor achievements of the "Hubei 120 City Public Plan" (where the final Hubei renewal plan was partially approved and some parts of the old town were preserved) in 2016, nor the "Cities: Grow in Difference" concept of the 2017 UABB at Nantou Old Town was applied to or replicated in Baishizhou. In the cases of Nantou and Hubei, it was precisely the contradiction between the subject of history as historic old town and the subject of history as urban village which helped save them from being uprooted. (However, they were only partially preserved, at the expense of Nantou. As URBANUS researcher Luo Yiqian has said, "[bearing the] requirements of 'creative culture' [...] while the place is being quickly filled up by outside cases [and] the local culture of Nantou [...] ignored. "[4]) Baishizhou, which does not have the historic old town card to play, is doomed to be replaced. "Why was it possible for 'Hubei 120' to be somewhat successful but not Baishizhou?" In response to this question, Architect Duan Peng states, "Perhaps, fundamentally, 'Hubei 120' is a plan to defend a heritage site, while Baishizhou was one to defend its residents."[9] Indeed, the partial transformation of Hubei into a heritage site and the reshaping of Nantou old town have not prevented the villages' original tenants from leaving. The subtext of "upgrading" the "demographic structure" and "industrial structure" is actually to replace one residential and commercial ecosystem with another, and to replace one demographic with another. In terms of defending the residents of Baishizhou, it is not only the radical activists in the traditional sense who have lost, but also the centrist players who have been trying to "reconcile the conflicts between the reconstruction and preservation of the urban villages."[10] Among which, the URBANUS team which curated the 2017 UABB in Nantou has long been concerned with the issues of Baishizhou (at one point, URBANUS even considered placing the main exhibition venue of the Biennale in Baishizhou despite the resistance from various sides that prevented it from happening) as they have always been working toward "[balancing] the real-world economic parameters with the idealistic goals of urban design and preservation" in order to turn the "intractable sociological [problem]" into a "[possible] design problem."[5-6] However, even these pragmatic attempts may seem idealistic in face of the myth of urban modernization and the ruthlessness of reality.

Now—if we look back on O'Donnell's assertion that "the era of urban villages in Shenzhen has ended"—perhaps the end is near not only for the informal urban form of the urban village and the informal urban lifestyle it shelters, but also the important ground that Shenzhen intellectuals

7 The "Baishizhou Team" was established in June, 2015 by architect Duan Peng and others. They specifically created an event, "The Invisible Shenzheners," to invite Baishizhou residents to hold up a white board with each of their name, sex, profession, age, and hometown to stand on the street of the urban village for a picture.

8 According to Lai Fu, *Shenzhenyeren*'s editor, the article was written in 2018 for a publication "to document Shenzhen's 'multifaceted' urban space and historical context." The description of "intellectual" efforts in the review echoes O'Donnell's perspective which categorizes a group of architects, artists, scholars, and representatives from different fields into "the second generation of Shenzhen's intellectuals." Contrary to the first generation of Shenzhen intellectuals who built the city yet still identify emotionally with their own hometown, the second generation of Shenzhen intellectuals identify as "Shenzheners" who "endeavor to expand the public discourse on Shenzhen's urban form in an attempt to surpass the voices of the government and developers." See O'DONNELL M A. Heart of Shenzhen: The Movement to Preserve "Ancient" Hubei Village [M]// BANERJEE T. LOUKAITOU-SIDERIS A. eds. *The New Companion to Urban Design*. London and New York: Routledge, 2019: 480-493.

9 This quote is a recitation by Duan Peng from Shen Yuyuan's "A Sample of Life in Shenzhen 01 | Architects created their city, but also became its tools," last modified January 5, 2021, https://matters.news/@zscliterary/ 深圳人生样本 01-建筑师建造了他们的城市 - 也被异化为工具 -bafyreifahmdynathzum5mxgfbhqokjvxyraxwhxirea2ut7q7vbfzupmkq; similar to what Liu He, a member of Handshake 302, said: "We are not closely 'connected' to the residents, because I am not different from any of the residents. Just like every resident in Baishizhou, we live or work here, and we all have close neighbors and see strangers on the street." Handshake 302 is a public art project established in 2013 by members of the "Urban Village Special Forces," including Mary Ann O'Donnell, Zhang Kaiqin, Lei Sheng, Wu Dan, Liu He and others in Unit 302, Building 49, Baishizhou. Based in the urban village, they conducted various spatial experiments and performance art, while encouraging the residents of handshake buildings to express their opinions. More details about the "Handshake 302" project, see the interview with O'Donnell: https://soundcloud.com/uncuttalks/mary-ann302; O'DONNELL M A. The Handshake 302 Village Hack Residency: Chicago, Shenzhen, and the Experience of Assimilation [M]// FORREST R, REN J, eds. *The City in China: New Perspectives on Contemporary Urbanism*. Bristol: Bristol University Press, 2020: 125-140.

10 See URBANUS' project description on "Baishizhou 5 Villages Urban Regeneration Research, Shenzhen 2013," http://www.urbanus.com.cn/projects/baishizhou/?lang=en.

have been using to resist the monotonous urban development model governed by real estate speculation. This contest around the urban village has become a wreck, and we must now look at what the urban village has left us other than a ruin of language, images, and memories.

3 The Microscopic: Field Perspective, Design Research, and the Case of Shenzhen

In 2005, Juan Du, an architect working at the time for Atelier FCJZ in Beijing, went south to Shenzhen and came cross a busy night market filled with smoke from local food stands. This experience marked her first visit to an urban village. She subsequently learned that this urban village was Baishizhou.[7]4-5,[8] The first Shenzhen Biennale in 2005 conducted a specified observation on the urban village. As the assistant curator of the 2005 exhibition, Juan Du has since visited Baishizhou and other Shenzhen urban villages numerous times. In 2020, she published a monograph on the urban village phenomenon titled, *The Shenzhen Experiment: The Story of China's Instant City.* This book, which disenchants the myths around Shenzhen's urban history, begins with her first encounter with Baishizhou fifteen years ago and concludes with its eviction and demolition. At the end of the main body of the text, she points out that:

> *Rather than a demonstration of problems, the urban villages are sources of past knowledge and future solutions to improve the city's economic vitality and social resilience. Shenzhen's success cannot be simply attributed to policies or central planning. Some of the city's most remarkable innovations are the result of local responses to top-down planning, responses that sometimes directly contradicted those plans.*[7]304

One could argue that "directly [contradicting] those plans [of the city]" is not only reflective of the creative survival experience in Shenzhen's urban villages, but also the primary purpose of her monograph. In her writing, Juan Du attempts to reveal that through constructing a series of myths including those of development, instantaneousness, virgin land, and a fishing village, the narrative of Shenzhen's urban history serves the political agenda of reform and opening up. In other words, officials told an origin story for the new city. The story is an urban planning fantasy in which a leader drew a circle in the South China Sea, transforming a tiny fishing village into a "Southern Gateway" for China's modernization and internationalization. Thus, the disenchantment of this story would in fact be the dispelling of the top-down and north-to-south dream-making machine—urban planning. What this story has less focused is not only the complex overlaying of the city and the village over the centuries, but also the adaptation, endeavor, and marvelous ability to create, and the great contribution seen in and from the millions of migrant workers who have come to this city over the decades. Mining in the historical view of urban planning that is also a dream-and-history-making machine, observing from a microscopic scale starting in the South, it is to provide another narrative through anthropological watching and non-fictional writing. Juan Du shows that Shenzhen has its own long and tumultuous life history which stands far beyond the rigid limitations of any planning blueprint. As an informal urban space that is mired in the city's complex history, the Shenzhen urban village is not an other to Shenzhen's history. Rather, the urban village's spacial, political, and cultural histories constitute the city's past. To study the urban village is to study the city, and the investigations of urban villages are important supplements to urban studies.

For more than twenty years, a host of anthropologists, sociologists, and urban researchers have endeavored to de-other Shenzhen's urban villages in the hope of bringing them into the scope of the Shenzhen experience. Edited by Mary Ann O'Donnell, Johnathan Bach, and Winnie Wong (2017; Chinese edition in 2020), *Learning from Shenzhen: China's Post-Mao Experiment from Special Zone to Model City* is a major collection of essays which introduces the urban research conducted on Shenzhen to the English-speaking world. Within the book, authors such as O'Donnell, Bach, and Wong have all, in their respective essays, provided nuanced critiques on the urban village and the migrant workers whom it shelters.[11] The title of the book not only alludes to the slogan of "Learn from Daqing in industry, learn from Dazhai in agriculture" followed by Chinese peasants and workers before the Cultural Revolution, but also refers to

11 Aside from studeis in Chinese, an earlier academic contribution to the research on Shenzhen includes Linda Vlassenrood, *Shenzhen: From Factory of the World to World City* (Rotterdam: nai010 publishers, 2016). This book was written and edited based on a seminar of the same name held by The International New Town Institute (INTI) on December 12, 2014.

the famous book, *Learning from Las Vegas*, by Robert Venturi, Denise Scott Brown, and Steven Izenour (1972). In the English context, the resemblance is self-explanatory. The Chinese translation of the title, however, needs to be clarified. Similar to how Las Vegas' consumerist aesthetics has disturbed the purism narrative of modernist architecture, Shenzhen's informal urbanization also triggered an inversion of the high and the low as local and vernacular experiences are given the same level of attention as formal sites. Anthropological and sociological case studies from a microscopic perspective have thus become necessary correctives.[12] In a more general sense, scholars with anthropological and ethnographic perspectives have already conducted a series of case studies on Chinese urban villages, turning non-fictional writings into effective academic voices. With these efforts, a considerable archival corpus of field research—including the research on Zhejiang Village in Beijing by Wang Chunguang, Wang Hansheng, Xiang Biao, Zhang Li and others, the research on urban villages in Guangzhou by Li Peilin, Lan Yuyun, Chu Dongai, David Bandurski and others, the research on the social history of oil paintings which are for sale in Shenzhen's Dafen Village by Winnie Wong, and the monograph on Shenzhen's urban village by Da Wei David Wang.[13]

In addition, since 2005, a group of architects, planners, and urban researchers concerned with the issues of urban renewal have conducted design research around Shenzhen's urban villages in their attempts to intervene in the demolition and total reconstruction plan by providing alternate small-scale strategies. These endeavors are exemplified by Li Jinkui's (2005) "Realist Issues in UEVs" and other articles which provide positive evaluations of urban villages, *Village/City City/Village* compiled and written by URBANUS (2006), *Transforming the Chinese Countryside* by Rural Urban Framework (Joshua Bolchover and John Lin, 2013), Stepan Al's (2014) *Villages in the City: A Guide to South China's Informal Settlements*, as well as Zhang Yuxing's (2016) article "Urban Village As a Form of Urban Public Capital and Shared Capital." From the anthropologists' on-site observations to designers' incrementalist renovation plans to the disenchantment of Shenzhen's origin myth in Juan Du's writing, a commonality has gradually surfaced among these endeavors which have been scattered across different fields over the past twenty years. They advocate no longer perceiving the urban village as a "problem" to be "solved," but rather as a possible answer for the future. By following bottom-up strategies, observing from a microscopic perspective, and intervening only at a small scale. The goal of these efforts would hopefully be to reverse the inertia of the avid real estate mindset of urban planning—to demolish and to build. An early example of one such intellectual endeavor is an important opinion from urban planner Huang Weiwen, who, in an article titled "Urban Planning Vs. Urban Villages, Who Should Transform Whom?" pointed out that urban villages are Shenzhen's *de facto* affordable housing which not only make up for Shenzhen's urban planning and management deficiencies, but also serve as an indispensable form of urban infrastructure.[14] To illustrate, Huang Weiwen argues that,

> *The issue at hand is not the urban village, but rather the current urban planning agenda. What needs to be changed is again not the urban village, but the city's current urban planning agenda. And it is precisely the urban villages which are threatened by the urban planning agenda that can inspire and transform urban planning. [...] Urban villages are not only a model for planning a livable, harmonious, and low-carbon city that encourages public participation and promotes healthy work environments, but also an indispensable infrastructural framework for the city. The current urban planning agenda has been razing urban villages to the ground in the name of city redevelopment, if allowed, it will not only cripple the scattered sample areas, but also the support system of the city.*[9]

Such a view has become the consensus of the intellectual community. At the same time,

12 The translation of the title of the book and the difference between its English and Chinese sub-titles are further explained in O'Donnell' s interview with "Middle ground" FM, https://www.xiaoyuzhoufm.com/episode/5f8d4e2483c34e85dd66ea42.

13 These works include Dawei David Wang. *Urban Villages in the New China: Case of Shenzhen* (London: Palgrave Macmillan, 2016); David Bandurski. *Dragons in Diamond Village And Other Tales from the Back Alleys of Urbanising China* (Penguin Group Australia, 2015); Winnie Wong. *Van Gogh on Demand: China and the Readymade* (University of Chicago Press, 2014); Chu Dongai. *Urban Village's Cultural Memory: A Research on the Villages in Guangzhou and the PRD* (2012); Shao Yuanyuan. *Practice in Transition: An Anthropological Study of a "Village in a City" Community* (2012); *Lan Yuyun. Village in the City: A Field Study of a "New Village Community"* (2005); Li Peilin. *The End of the Village: The Story of Yangcheng Village* (2004); Zhang Li. *Strangers in the City: Reconfigurations of Space, Power, and Social Networks within China's Floating Population* (2002); Xiang Biao. *Communities across Borders: The Life History of the "Zhejiang Village" in Beijing* (2000).

14 Similar deductions from Huang Weiwen can also be found in Huang W. Shenzhen: The Rapid City-making [J]. *Domus*, 2011; Huang W. The Bottom Half of the City [EB/OL]. Huang Weiwen's sina blog, 2006-4-13.

the official attitude of the government has slowly unwound as it began to acknowledge the contributions urban villages have made toward the prosperity of Shenzhen. In 2017, the 7th UABB, organized by the Shenzhen Municipal Government and curated by architects Liu Xiaodu, Meng Yan, and art critic Hou Hanru, was held under the theme of *Cities: Grow in Difference*. Unprecedentedly, the main exhibition venue was placed in the urban village of Nantou. In this context, the Shenzhen Center of Design, superintended by the Municipal Urban Planning, Land & Resources Commission, started editing a new book: *Urban Villages: The Disappearing City*. It aims to "record the multifaceted and layered social meanings behind the fast-developing Shenzhen and present a delicate slice of the city's history over a longer period in a richer spatial dimension," released in 2020. In the same year, necessary technical treatments were applied in the Chinese version of *Learning from Shenzhen*. Occurrences like these remind us that some parts of Shenzhen remain unseen.

Lai Fu concludes: "If these urban villages are perceived through the lens of top-down planning, one is bound to see nothing but disorder. However, if these villages were inspected from a bottom-up perspective, one will find it is exactly the flow of people, capital, and ideas emerged from this kind of disorder and chaos that has contributed to the 'success' of the Shenzhen Special Economic Zone."[10] Such an understanding would have been considered impossible twenty years ago. But twenty years later, with the joint efforts of a large group of scholars, architects, artists, media professionals, and independent writers, similar concepts have become widely accepted. In 2019, the slogan of "Enriching and improving the content of urban villages comprehensively and realizing diversified renewal goals" was written into the *Shenzhen Urban Village (Old Village) Comprehensive Remediation Plan (2019-2025)* which specifies that urban village land used for residential purposes shall not be included in any urban renewal unit plans, land preparation plans, and shelter zone transformation plans purposed for demolition and reconstruction.[11] These changes certainly cannot reverse the city's historical course toward the end of "the urban village era in Shenzhen." Nevertheless, the work that has been done to reverse perspectives, to reduce the scale of observation, and to dispel mythical stories is far from enough. Urban villages are the real origin of Shenzhen's urbanization and the earliest landing place for migrants in this megacity of migrants. Thus, the investigation into Shenzhen's urbanization progress is inevitably intertwined with the history of the urban village. Moreover, the history of the urban village is far from coming to an end. As long as an informal life improvised under pressure still exists, the urban village phenomenon is bound to continue in alternative ways.[15]

4 The Mesoscopic: "South" as Method, or the "Urban Village" as a Concept

"[...] Through the construction of Shenzhen, 'Northernism' subjugated 'Southernism.' However the constant competition between the two schools innovated 'Northernism' while catalyzing new differences." In his 2009 article "Southernism of Differences: Spatial History in Guangdong and Architectural Construction in Pearl River Delta,"[12] architectural critic Feng Yuan points out that the recent history of the evolution of the Pearl River Delta shows that Guangdong has been incessantly situated between the external economy and internal power, geographical periphery and national model, the Northern political imagination and the Hong Kong (Southern) financial imagination, and continuously (re)-shaped by multiple dualities while often providing vivid originality. Feng Yuan calls this hybrid state "Southernism of Differences."[16] This observation did not touch on the issue of urban villages in particular, as the urban village is clearly neither an invention of Shenzhen nor of the Pearl River Delta, nor even a feature of the Global South, specifically. That said, the connection between the South and urban villages was highlighted at the 2017 UABB. To illustrate the connection, the Biennale showcased a section designated for the South, attempting to reflect on the perspective and expand the scale of observation. Liu Xiaodu, one of the chief curators of the Biennale, wrote:

15 As Wu Fulong, Zhang Fangzhu, and Chris Webster pointed out in 2013, the redevelopment of urban villages is "an attempt to eliminate their informality and to create more governable spaces through formal land development; but since it fails to tackle the root demand for unregulated living and working space, village redevelopment only leads to the replication of informality in more remote rural villages, in other urban neighborhood and, to some extent, in the redeveloped neighborhoods." See WU F, ZHANG F, WEBSTER C. Informality and the Development and Demolition of Urban Villages in the Chinese Peri-urban Area [J]. *Urban Studies*, 2013, 50(10): 1919-1934.

16 Feng Yuan's research on the cultural construct and spatial politics of Lingnan also includes FENG Y. Down to the South: Geopolitics of PRD [J]. *Urban China*, 2006(13); FENG Y. Machines, Institutions and Self-Comb Women: Looking at the Gender History of the Industrialization of the PRD a from the 'Specimen' of Ronggui Town [J]. *Art Journal*, 2009(1); FENG Y. Transcending Geo-culture [EB/OL]. *e-flux Architecture*, 2018-4-24, etc.

The South is not just a geographical concept. In China, the South is also a cultural concept. [...] Since China's Reform and Opening Up Policy began in the 1980s, Shenzhen, as the representative of the South, has become the shining example of this policy. The Pearl River Delta has become an "open lab" ever since. It has established the unique southern imagination and complex reality by melding together localization and globalization, tradition and the future, and superstition and illusion. The "South in reality does not only refer to the south of the Five Ridges expanse, but more a characterization of the openness and possibility, creativity and efficiency, practical desire and dreams for the future of the region. [...] Thus, the South of China is not only expressed as a geographical concept, but also associated with the North and the Center in history, and is, therefore, a metaphor for subculture and alternative thought. [...] The urban village is almost the only remaining original urban form of the South in Shenzhen, and it is the only place where the resistance mechanism exists. This is of great significance to the future of Shenzhen's urban development, so it deserves our careful study.[13]

In this formulation, the South is decoupled from the specific qualifications of geography and climate, and is instead identified as a concept and method to distinguish differences between the North and the South. It is used to approach a series of complex issues—such as the relationship between the center and the periphery, influence and resistance, as well as regionalism and globalism—in an attempt to resist the cultural hegemony of the northern center. The Shenzhen urban village is a typical case study for this approach. Calling Shenzhen's urban villages "the only place where the resistance mechanism exists" is a bold and startling statement, especially today when "the only place" is disappearing at an accelerated rate. In other words, if the curators had tried to come up with a conceptualized Southern methodology to examine the specific issues of the urban village three years ago, then the urban village we are faced with today would be even more abstract and less concrete. Treating the South as a concept or a method is not a new idea—for instance, the documenta 14 in Kassel, held in the same year as the 2017 UABB, had *South as a State of Mind* as the title of its exhibition journal—but it was unprecedented for the urban village to be extracted from concrete space and viewed as an abstract concept.[17] If urban villages were to no longer exist in concrete form (in a foreseeable future), then it is important to consider, with the help of existing literature, questions such as: why the urban village, as a concept, should not and will not be diluted by the widely discussed concepts like informality? How it is meaningful to continue discussing urban villages? And how can the South as a method contribute to these conversations?[18]

If we were to depart from Liu Xiaodu's claim and search for concrete examples, then it is true that the urban village (once) did provide a spatial refuge for "a characterization of the openness and possibility, creativity and efficiency, practical desire and dreams for the future of the region [and] a metaphor for subculture and alternative thought" on political, economic, and social levels, and the "openness," "creativity" and "dreams" encompassed by the urban village are indeed "alternative" compared to the "openness," "creativity" and "dreams" in the national narrative since reform and opening up. In their co-authored "Spaces of Informal Production in China" published in 2020, Jonathan Bach and Stefan Al explore the "open" and "creative" informal manufacturing industry in China's urban villages.[14] There has already been an extensive amount of literature on the informal economy model extant in urban villages. These studies

17 The details of documenta 14 in Kassel and *South as a State of Mind* can be found on https://www.documenta14.de/en/South/. There have been many academic discussions on viewing South as a method, see GARDNER A, GREEN C. South as Method? Biennials Past and Present [C]// *Making Biennials in Contemporary Times: Essays from the World Biennial Forum no. 2.* São Paulo, 2014: 28-36; PAPASTERGIADIS N. What is the South? [M]// GARDNER A, ed. *Mapping South: Journeys in South-South Cultural Relations*, Melbourne: The South Project, 2013. In Anthony Gardner and Charles Green's writing on the South method of the biennale, they "see 'South' as a loose working concept for that which tries to resist easy assimilation within hegemonic global currents generated outward from the North Atlantic. Instead, 'South' emphasizes two things at once: it asserts a rich history generated from long-standing unease with North Atlantic hegemony, whether through the lingering legacies of colonial and neo-colonial violence, or the struggles for decolonization and decriminalization figured through the nonaligned politics of the Third World during the Cold War; and, secondly, an awareness that, as Nikos Papastergiadis argues, "'survival requires a coordinated transnational response' through which that hegemony might be displaced." While Liu Xiaodu's expression coincides the above statements, it also emphasizes the uniqueness of the concept of "South" in the context of China.

reveal that the collective land in urban villages is not only densely populated with informal housing, but also with a variety of small-scale informal factories, informal labor, and informal service relations which exist on a micro-level (the officials often use the term "'three kinds of small places' *(san xiao changsuo)*"—small shops, small workshops, and small recreation sites—to describe these places) in the gray areas of law. In the urban village, housing, production, storage, and marketing are fused.[19] What Bach and Al attempt to further point out is the symbiotic relationship between local informal manufacturing and global formal manufacturing. The informal economy within the urban village has long been efficiently embedded within the formal manufacturing ecology outside the urban village in highly imaginative ways situated on the two ends of the spectrum of scale. At small or at large, they are interconnected in a parallel manner (rather than in a vertical hierarchical relationship). Within a small room in a handshake building in Shenzhen, workers may be making unbranded cell phones using reference cards from Taiwan that would be immediately sent to India for sale.[20] In her doctoral dissertation on the informal economy of urban villages (University of Hong Kong 2013),[15] Wu Ling investigates in detail how a Guangzhou urban village is transformed into an informal overtime factory at night. Workers sew clothes in a formal factory near the village during the day and attach buttons to them in the village at night, and these garments are sold around the world the next day.[21] On similar topics, two terms which are often mentioned and somewhat outmoded, however, are not used by Bach and Al, shanzhai and low-end globalization.

In an article titled, "Shanzhai: A Theory of the South?" Xiang Zairong (2020), a comparative literature scholar working in the decolonial framework, extends shanzhai and low-end globalization to a theoretical discourse on the South. It is also worth noting that Xiang mentions urban villages in passing. He writes:

> *In truth, the rise of shanzhai occurred in an age in which the language of "anti-imperialism," "anti-capitalism" and "class struggle" had lost almost all of its power. Shanzhai's relationship with global capitalism is far more ambiguous. Perhaps the most famous work to contribute to the recently burgeoning research around shanzhai theory is Byung-Chul Han's Shanzhai: Deconstruction in Chinese. An insightful philosophical work, it nonetheless at times struggles to conceal its Orientalist tendencies. There is an over-emphasis on the East-West divide in its considerations of shanzhai, and Han succumbs to certain myths about "originality" and "innovativeness" to mount a defense of shanzhai, to the extent that his discussion departs from its philosophical basis to end on the saccharine note of "democratic hope." Such an approach robs shanzhai of its most essential features: ambiguity, illegitimacy, and indifference to any metanarrative. [...] As I began looking into more cases I discovered there was little that could be described as groundbreaking going on in shanzhai. Innovation is hard to find in mass production and there is little in the way of the flair that Byung-Chul Han describes. Nor is there much in the way of a true spirit of resistance. [...] As "innovation" and gentrification takes root in cities across the world, Shenzhen's urban villages, the urban shanzhais of this metropolis of the future, will become nothing more than footnotes [(literally "smoke clouds")] in its history.*[16]

Global capitalism needs shanzhai, just as cities need urban villages, even if they both turn their noses up at shanzhai or urban villages as being "low-end." We can add a side note to the urban shanzhais which Xiang mentions only in passing. If we add up all the arguments mentioned in Xiang's article about the southern theory of shanzhai—counterfeit money, fake products, piracy, violations, low-end population, low-end globalization, shanzhai phones, shanzhai goods—and then imagine a spatialize form, it is not difficult to imagine the urban village. (Of course, it could also be Huaqiangbei before its

18 There is a vast extant literature on the concept of "informality," for instance, WANG H, LONG Y. A Brief Review on the Urban Informality Studies and Housing Practices in the Third World [J]. *Urban Planning International*, 2008, 23(6): 65-69; HERNÁNDEZ F, KELLETT P, ALLEN L K. eds. *Rethinking the Informal City: Critical Perspectives from Latin America* [M]. New York: Berghahn Books, 2010; WU F, ZHANG F, WEBSTER C. Informality and the Development and Demolition of Urban Villages in the Chinese Peri-urban Area[J]. *Urban Studies*, 2013, 50(10): 1919-1934. It has also become widely accepted in academia that informal cities are no longer a phenomenon only extant in developing countries. On the informal cities in developed countries, see MUKHIJA V, LOUKAITOU-SIDERIS A, eds. *The Informal American City: Beyond Taco Trucks and Day Labor* [M]. Cambridge, Mass.: The MIT Press, 2014.

19 There have been an extensive amount of research and fruitful results on the informal economy of the Urban Village from academics such as Xue Desheng, Yan Xiaopei, Lan Yuyun, Wu Ling, and Zhou Daming.

20 Bach and Al cited the research by Silvia Lindtner, Anna Greenspan, and David Li on Shenzhen's shanzhai electronics, see LINDTNER S, GREENSPAN A, LI D, Shanzhai: China's Collaborative Electronics-Design Ecosystem[EB/OL]. *The Atlantic*, 2014-5-18. https://www.theatlantic.com/technology/archive/2014/05/chinas-mass-production-system/370898/.

21 On the overall informal labor market state of China, see SWIDER S, *Building China: Informal Work and the New Precariat* [M]. Ithaca, NY: Cornell University Press, 2016; AL S. ed. *Factory Towns of South China: An Illustrated Guidebook*[M]. Hong Kong: Hong Kong University Press, 2012.

transformation). The aim here is not to establish a correspondence between shanzhai and the urban village, but to use the existing theoretical discussion of shanzhai to inspire thinking on the urban village. The word shanzhai is somewhat dated, just as the term urban village is destined to lose its signified in the not-too-distant future. If the readers accept my arbitrary association, then the fact that specific urban villages are, one by one, rapidly becoming smoke clouds of history points to a critical historical situation. That is, concrete urban villages are disappearing as quickly as the abstract urban village is becoming a concept—a concept of the history of common people. While the physical existence of the urban village may not escape the fate of being blown out of official history, the smoke cloud of the urban village will still hover in the history of ordinary people. Regarding the possible shape and appearance of this smoke cloud, Xiang Zairong reminds us:

> *In this sense, the shanzhai theory of the South (or Southern theory as shanzhai theory) cannot be subject to the distinction between practice and theory—or other binary distinctions such as North-South, East-West, man-woman, secular-religious, high-low and real-fake. This "refusing to be subject to" does not equate to resistance or destruction but is rooted in an ordinary abundance. It is not a heroic act of rebellion but a kind of inversion, an undercurrent: the ethics internal to the theory-practice of all-things-in-connection is that of the all permissible.*[16]

At this point, if we look back at the curators' impassioned statements on influence and resistance three years ago, we will find that whether it is Handshake 302, Baishizhou Team or the 2017 UABB, they have all hoped to give urban villages and their residents some kind of subjectivity of resistance. However, in fact, urban villages have never actively assumed the role of the nurturing place of resistance mechanism. Thus, there is no such question about only surviving or not, let alone whether it is the only place or not. The real gesture of the urban village is never active resistance, and it has never actively nourished any resistance mechanism. Rather, it is marked by the qualities of non-cooperation, non-compliance, nonconformity, indifference to any grand narrative. It strays, is wild, and improvises. It is temporary, ambiguous, a retreat, an abstention, a dispossession, an expedient measure, and a way of getting by. It is scrivener Bartleby's "I would prefer not to" by Herman Melville. Speaking of Bartleby's pessimistic political outlook, the Slovenian philosopher Slavoj Žižek believes, "The voter's abstention goes further than the intra-political negation, the vote of no confidence: it rejects the very frame of decision."[22] Similarly, recognizing the non-resistance of the "urban village" concept here in no way implies an apolitical escapism. Rather, it contains a far more radical political force, even if only as an imagined one.

5 The Macroscopic: A Linguistic Predicament

In the spring of 1996, Dutch architect Rem Koolhaas brought his students from Harvard University's Graduate School of Design to the Pearl River Delta, making the Shenzhen Special Economic Zone the first stop of his urban research trip. He wrote in his teaching notes:

> *The absence, on the one hand, of plausible, universal doctrines, and the presence on the other, of an unprecedented intensity of new production, create a unique wrenching condition: the urban condition seems to be least understood at the moment of its very apotheosis. [...] In fact, an entire discipline does not possess an adequate terminology to discuss the most pertinent, most crucial phenomena that occur within this domain, no conceptual framework to describe, interpret and understand exactly those forces that could help to redefine and revitalize it.*[17]

Koolhaas pointed out a predicament about language. For him, the Pearl River Delta's urbanization scene at the time—a "maelstrom of modernization" and "there [was] nothing left between Chaos and Celebration"—was a vacuum of concepts, terminologies, and theories that existed beyond the speculations of both urban and architectural research. His strategy, at the time, was to invent terminologies, copyright them, and compile a new dictionary:

> *Together, these studies describe a new urban condition, a new form of urban coexistence that we have called CITY OF EXACERBATED*

22 Žižek wrote, "Why is the government thrown into such a panic by the voters' abstention? It is compelled to confront the fact that it exists, that it exerts power, only insofar as it is accepted as such by its subjects-accepted even in the mode of rejection. The voters' abstention goes further than the intra-political negation, the vote of no confidence: it rejects the very frame of decision." see ŽIŽEK S. *Violence: Six Sideways Reflections*. New York: Picador, 2008: 216.

DIFFERENCE©, or COED©. Beyond the particularities of each condition that we found, the COED© project introduces a number of new, copyrighted concepts, that, we claim, represents the beginning of a new vocabulary and a new conceptual framework to describe and interpret the contemporary urban condition. The emergence of the PRD with the suddenness of a comet, and the present 'cloud of unknowing' that creates a kind of stealth envelope around its existence and performance, are in themselves proof of the existence of parallel universes that utterly contradict the assumption that globalization equals global knowledge.[17]

This poignant passage written in 1996 triggers a heavy echo from history. (Figure 3) Surreptitiously assigning "all rights reserved" to made-up words and sentences seems to be a shanzhai style joke, albeit the word shanzhai did not yet exist at the time and the term urban village had only just entered the academic world (it is said that the first paper on the subject of urban villages in China was Yang An's "The Prevention and Control of Urban Villages,"1996)[23] Twenty years later at the 2017 UABB, OMA (Office of Metropolitan Architecture, literally the Office of Urban Architecture) is replaced by URBANUS (literally "Metropolitan Practices"), and "CITY OF EXACERBATED DIFFERENCE©" was replaced with the theme of the exhibition, "CITIES, GROW IN DIFFERENCE" in the urban village. The difference between the two slogans is much greater than it seems. The external astonishment at the exacerbated difference of 1996 has become the careful maintenance of differences and exceptions under internal and official supervisions.The meaning has been turned upside down (in response to architectural historian Zhu Tao's doubts on the English translation of the Biennale's theme, Meng Yan answered: "Let's make a special 'chinglish' term that will seem awkward to foreigners.")[24] In front of the South Gate and between the inside and the outside, a series of mirrorings, inversions, mistranslations, and localizations have turned a linguistic predicament into a pidgin drama.

23 This observation came from the article, ZHOU Y. The Comparison between Two "Urban Diseases": Urban Villages in China and the Western Slum Problem around 19th Century [J]. *New Architecture*, 2007(2): 27-31. http://wen.org.cn/modules/article/view.article.php/2208.Yang An's 1996 article, see Yang A. The Prevention and Control of "Urban Villages"[J]. *Urban and Rural Development*, 1996(8): 30-31.

24 The confirmation process of the English translation of the theme of the 2017 UABB is recorded on URBANUS' website: http://www.urbanus.com.cn/events/english-theme/.

In his 2006 article "Down to the South: Geopolitics of PRD," Feng Yuan discusses the centuries-old geopolitical ambivalence of and toward the Pearl River Delta south of the Five Ridges (*Lingnan*):

Lingnan, or rather the Pearl River Delta, was destined to be this way. Having spent more than two thousand years in the repeated games of political conquest within the mainland and economic trade by sea, it is now stepping into a new millennium. [...] It is still valuable to stand at the mouth of the Pearl River and let go of one's thoughts to evoke historical imagination, whether looking south or turning one's head toward the north. The Delta region is always the end of the south, and the beginning toward a further south.[18]

Ten years later, the UABB indeed looked past the South and toward a further south. Showcased in Nantou Old Town's exhibition venue were the slums of Nairobi, the favelas of São Paulo, the urban renewal cases of Habana, and the informal architecture of India and Southeast Asia. Coexisting in one room, the slums from Asia, Africa, and Latin America staged a sharp contrast with the adjacent Window of the World, a post-modern shanzhai style theme park consisting of hundreds of miniature historical landmarks and monuments. In the English version of the introduction, the section was called "Global South," but the Chinese introduction did not use the standard translation of the term. Instead, a vertical line was placed between "World" and "South," separating the two concepts.[19] This creative misinterpretation between Chinese and English once again reveals the two-faced nature of the internal and external worlds, which, according to Feng Yuan, has lingered in the geopolitical history of the land for more than 2,000 years.

From 19th-century French sociologist Émile Durkheim's terms, primitive and *traditionnelle*, to French demographer Alfred Sauvy's proposal of Third-World[ism] (*tiers monde*), which became widely accepted after the Bandung Conference of 1955, to popular recent concepts such as Global South, post-colonialism and tropics, there has been a series of terms to indicate the poverty areas, colonialist histories, and the imbalanced development between regions, marking the dynamic frontiers of the realm of thought. The evolution of terminology is moving away from a Euro-centric perspective. While gradually emphasizing solidarity and connections across geographic and continental boundaries, the manifold political, economic, and social specificities of different regions are inevitably diluted during the process. The term, global South, for example, emerged from scholarly interest in the South in the context of globalization after the end of the Cold War. It is not considered a geographical description, nor is it merely a generalized reference to poverty and inequality. Rather, it is used to indicate the continuation of Third-Worldism and the decolonization movements in the era of globalization, to strengthen the cohesion of formerly colonized groups, and to further challenge and subvert the parochial perspective of the Western center.[25] In other words, the global South is not only a metaphor for property areas and imbalanced development between regions. Fundamentally, it is a political project aimed to move forward the course of decolonization.[26] The current academic attitude toward this terminology is best summarized by Canadian scholar Marlea Clarke, "For those interested in understanding and unmasking this inequality, and participating or supporting subaltern groups that are organizing around sets of radical claims and practices to challenge dominant groups and uneven development, the term is powerful and useful. In short, the term global South might be ambiguous and vague, but I embrace it."[27]

Clearly, the "world" in the Window of the World theme park and the "world" in Third World are not the same world, just as the "global" in global capitalism and the "global" of the global South are not the same global. These imported terms and Chinese habits often collide in Lingnan, creating both intentional and unintentional errors and inventing languages that have never been spoken before. Whatever the specific considerations, the avoidance of the expression global South in the 2017 UABB, in fact, achieves a more descriptive and less political presentation (it is also true that the issue of urban villages has never really been integrated into the political agenda of the global South), echoing the linguistic conundrum which Koolhaas had already identified. After twenty-five years of extensive research and discussions, we still observe that the urban village remains in a conceptual and theoretical void, difficult to be truly captured by any scholarly trend such as global South studies, postcolonialism, or tropical studies. After the eradication of its physical existences, the urban village may at most turn itself into a concept, a theory, a smoke cloud that could covertly trim away the vertical line between the World and the South. Twenty-five years ago, "CITY OF EXACERBATED DIFFERENCE©" was pirated by a Dutch architect and written into American university textbooks. Twenty-five years later, the historical moment to fight for the copyright of the "Urban Village©" as a concept has arrived. The need to fight is due to the urgency of this historical state. The remnants of bricks and tiles will not bring the complex history of the urban village to an easy end. There vulture-like capital and power abound, vying to define this remaining smoke cloud. A theoretical reflection on the urban Village© is therefore urgent, more so than at any other time.

6 Epilogue

Although the post urban village era is yet to come, the battle for the copyright of the urban village© has already begun. After the 2017 UABB, the main exhibition venue of the Biennale, Nantou, underwent a comprehensive transformation, which was led by real estate developer Vanke and solidified a fashionable label for the creative old town. While the upgraded urban village beckons tourists to visit and enjoy the tour, a large number of the old residents and merchants seem to have been replaced without any sound. At the end of April 2021, the "Eureka! Touring Library" co-organized by JuanZong Books, a bookstore owned by fashion media company CMC Inc., and Vanke landed in Nantou Old Town. The book fair used "Our City, My Community" as its theme, making it impossible for us to ignore the drastic gentrification process of Nantou—from the rustic and poor to the elegant and aesthetically appealing—in just four years (albeit the identity of the old town prevented the drastic gentrification process of Nantou from being revealed spatially)

25 As stated by Nour Dados and Raewyn Connell, "The term global South functions as more than a metaphor for underdevelopment. It references an entire history of colonialism, neo-imperialism, and differential economic and social change through which large inequalities in living standard, life expectancy, and access to resources are maintained." See DADOS N, CONNELL R. The Global South [J]. *Context*, 2012, 11(1): 12-13.

26 On the concept of "Global South" as a political project, see BALLESTRIN L. The Global South as a Political Project [EB/OL]. https://www.e-ir.info/2020/07/03/the-global-south-as-a-political-project/.

27 "For those interested in understanding and unmasking this inequality, and participating or supporting subaltern groups that are organising around sets of radical claims and practices to challenge dominant groups and uneven development, the term is powerful and useful. In short, the term global South might be ambiguous and vague, but I embrace it." see CLARKE M. Global South: what does it mean and why use the term?[EB/OL]. https://onlineacademiccommunity.uvic.ca/globalsouthpolitics/2018/08/08/global-south-what-does-it-mean-and-why-use-the-term/.

and to question who "we" are and where "they" live, apart from "us."[28]

On May 9th, at Nantou Old Towns's Dajiale public stage—where the UABB's closing ceremony was held three years ago—the book fair launched the "NAN City Forum." It was a rare occasion because the words of government officials, real estate developers, architects, and scholars once again conflated. Qian Yuan, President of Urban Research Institute of China Vanke spoke immediately after the Deputy District Mayor of Nanshan, frankly admitting the fictionality of Nantou Old Town, "Everyone says it is an ancient city, but how is it ancient? Therefore, we need to create an image of the 'ancient' [...] and 'to create something out of nothing.' This is not only about renovating the houses but also about the content; not only about attracting investments, but also about creating the content."[20] Real estate developers bypassed the hundreds of years of specific life history of Nantou and created a fashionable old town out of thin air. The style is arranged, and the scene is designed. For more than a decade, Nantou had vacillated between the village and the city. Now, it has been transformed into a theme park decorated with fictional antiquity, even as the village has been hollowed out. Nevertheless, Nantou is already the most successful case in which numerous architects and intellectuals have worked together, for more than a decade, to resist the developer's logic in the reconstruction of urban villages. As Meng Yan pointed out later in his speech, gentrification is far from a new issue, as "everyone who has participated in the Biennale and its activities have realized that gentrification is unstoppable and what we hope to do is to slow down the process. At that time, the idea (of course) lacked the strength to put a stop to it, but we hope to prolong the process."[20] Today, the process of prolonging is finally coming to an end. At a time when the era of urban villages in Shenzhen is coming to an end (or even has already ended), new theoretical questions must be thrown out to form new historical judgments and stimulate new historical imaginations. After all, on the other side of the battlefield, the linear developmentalist narrative has already prepared its answer.

Pearl River Delta / Shenzhen: 10 Years Later

Oct-23. 2006
To Urban China

Phenomena that I expected:
- The relative decline of Hong Kong.
- The strength of Shenzhen.
- The further corruption of Macao.
- The continuing vigor of Guangzhou.

What I did not expect:
- the maturity of Shenzhen from an urban wild west; it is now one of the most carefully managed sections of China.
- Shenzhen is also the territory where the equivalence of Chinese and 'foreign' architects will become apparent and play itself out.
- Shenzhen is also the first new city that has a (relatively) old quarter; before its completion, the city has to reinvent itself.

The only thing that could still sabotage the city - (and China) - is money. China wants to realize sophisticated ambitions with limited budgets. With its exceptional wealth, it should now invest in real quality.

Shenzhen could be the first . . .

Rem Koolhaas

Figure 3 Rem Koolhaas, "Pearl River Delta / Shenzhen: 10 Years Later," in *Urban China*, 2006(13): 13.

There Will Be No Villages in the City (Figure 2). In 2011, Nanfang Daily Newspaper Group Co., Ltd published a series of articles on the transformation of urban villages as an edited volume. Ten years later, the title has come true, but the question—"Without urban villages, where can they go?"—at the end of the book remains unanswered, and there are not fewer, but more other questions to be answered.

As a literature review, the one-sidedness of this essay lies not only in the inevitable parochialism due to its inherently limited perspective, but also in the remaining gaps and omissions in the history of the urban villages themselves. We still have to keep asking which voices, which groups of people, which perspectives, and which histories are still ignored, excluded, suppressed, and evaded, and who are not ready to accept these voices, these groups of people, these perspectives, and these histories. They are like clouds of smoke, deeply hidden under the shadow of the history of common people, and will surely surface sometime in the future.

28 On the 2021 "Eureka! Touring Library" book fair, see the introduction, "Welcome to 'Our City, My Community'!" published on the official WeChat account [卷 宗 Wallpaper] for more details: https://mp.weixin.qq.com/s/nC5SOFEKgALcHS9K6pfMrA.

[1] 孟岩 . 城市即展场，展览即实践：第七届“深双”从“城市共生”的宣言到“城市策展”的现场 [J]. 时代建筑 , 2018(4): 179.
[2] O'DONNELL M A. Is the era of Shenzhen urban villages over? [EB/OL]. Blog "Shenzhen Noted," 2021-3-23. https://shenzhennoted.com/ 2021/03/23/is-the-era-of-shenzhen-urban-villages-over/?blogsub=confirming.
[3] 沈於淵 . 深圳人生样本 01 ｜建筑师建造了他们的城市，也被异化为工具 [EB/OL]. https://matters.news/@zscliterary/ 深圳人生样本 01- 建筑师建造了他们的城市 - 也被异化为工具 -bafyreifahmdynathzum5mxgfbhqokjvxyraxwhxirea2ut7q7vbfzupmkq.
[4] 罗祎倩 . 迈向“城市共生”[J]. 建筑实践 , 2020(11): 157.
[5] BUNT T. 白石洲五村城市更新研究 [J]. 城市环境设计 , 2018(6): 295.
[6] 孟岩 , 林怡琳 , 饶恩辰 . 村 / 城重生：城市共生下的深圳南头实践 [J]. 时代建筑 , 2018(3).
[7] DU J. The Shenzhen Experiment: The Story of China's Instant City [M]. Cambridge, Mass.: Harvard University Press, 2020: 4-5.
[8] MCHUGH F. In Shenzhen, "urban villages" like Baishizhou have been lost to the megacity myth [J]. South China Morning Post Magazine, 2020-2-16.
[9] 黄伟文 . 城市规划与城中村：谁来改造谁 [J]. 住区 , 2011(5): 102-105.
[10] 来福 . 口岸、城中村与深港关系 [EB/OL]. “野人”微信公众号 . 2019-1-19.
[11] 深圳市城中村（旧村）综合整治总体规划（2019—2025）. 深规划资源〔2019〕104 号 .
[12] 冯原 . 差异的南方性：广东的空间史与珠三角建筑生产 [J]. 建筑与都市 . 2009. 27(6): 127-131.
[13] 刘晓都 . 世界 | 南方：影响与抵抗 [EB/OL]. 都市实践 . http://www.urbanus.com.cn/writings/global-south.
[14] BACH J, AL S. Spaces of Informal Production in China [M]// RAPPAPORT N, LANE R N, eds. The Design of Urban Manufacturing. New York: Routledge, 2020: 151-160.
[15] WU L. Migrant Workers and Informal Economy in Urban China: An Ethnographic Study of a Migrant Enclave in Guangzhou [D]. Hong Kong SAR: University of Hong Kong, 2013.
[16] 向在荣 . 山寨：一种南方理论？ [J]. 时代美术馆期刊 , 2020(1).
[17] KOOLHAAS R. Pearl River Delta [M]// Studio Works 5: Harvard University Graduate School of Design. Princeton Architectural Press, 1999: 134.
[18] 冯原 . 南下：珠三角的空间政治 [J]. 城市中国，2006(13): 17-20.
[19] 刘晓都 . 世界 | 南方：影响与抵抗 [EB/OL]. 都市实践 . http://www.urbanus.com.cn/writings/global-south.
[20] “南城论坛”现场速记稿 [R]. 2021-5-9.

摄影 PHOTO

城·村——一个变迁时代的症候
Urban · Village: The Symptom in the Age of Change

陈东　策展
Curated by Chen Dong

湖贝村掠影——民俗乐事，湖贝古村，曾向强，2016
Hubei Glimpse, The Joy of Folklore, Old Hubei Village, Zeng Xiangqiang, 2016

中医师和他的朋友，坳下村，彭涛，2016
A Traditional Chinese Medicine Doctor and His Friend, Aoxia Village, Peng Tao, 2016

坐玩具车的兄弟俩，大芬村，许宇涵，2016
Brothers on a Toy Car, Dafen Village, Xu Yuhan, 2016

擦肩而过的两个人："狭路相逢，勇者为王"，湖贝村，廖俊鸿，2013
Two Men Brushing Past Each Other, "Brushing past the alleyway, the king grows stronger," Hubei Village, Liao Junhong, 2013

下沙村客家盆菜宴，下沙村，刘伯良，2012
Hakka Poon Choi Feast in Xiasha Village, Xiasha Village, Liu Boliang, 2012

水贝村最后的盆菜宴，水贝村，方展帆，2015
The Last Poon Choi Feast in Shuibei Village, Shuibei Village, Fang Zhanfan, 2015

最后的家园，岗厦村，叶伟明，2008
The Last Home, Gangxia Village, Ye Weiming, 2008

建设中的城中村，黄贝岭村，郑丽萍，2012
Urban Village under Construction, Huangbeiling Village, Zheng Liping, 2012

岗厦村，岗厦村，廖鹏，2010
Gangxia Village, Gangxia Village, Liao Peng, 2010

来了，就是深圳人。到了深圳，第一个问题，就是落脚何处？

在深圳，城中村以相对低廉的房租，遍布巷道的各式餐饮店和修雨伞、裤子锁边、配钥匙、修理家电等小摊贩，为初到城市者提供了必需的庇护和生活的便利——这，是一个梦想开始的地方。

同时，深圳的城中村多脱胎于历史悠久的原生村落，承载了城市的原生文化。当城市的迅猛增长和旺盛的租房需求使村变为城中村，村民逐渐离开，城中村的宗祠和传统仍然在精神上维系着村落的人际体系——这，也是一个寄寓记忆的地方。

本次展览中的黑白影像，是近年来深圳多个城中村的真实记录，并不系统，更无关权威，甚至还带着一点荒诞：这是深圳吗？

是的，这些生活的悲喜、人情的冷暖都是深圳，它是一个关于寻梦者与城中村的来与去的故事——城中村因为城市的发展而出现，也因为城市的发展面临旧城改造。

今天，这些影像中的一些场景已经在城市更新中一去不复返，其中是否也有你熟悉的场景？有你曾经的回忆？

▷展览：你住过 / 住在城中村吗？　▷参展人：陈东

Once you're here, you're a Shenzhener. The first question when you arrive is: where to stay?

In Shenzhen, people are attracted by the relatively low rent, different stores and restaurants all over the streets as well as small vendors like seamstresses, locksmiths, umbrella tailors and household appliance repairmen, which provides necessities and convenience for the newcomers—this is where dreams get started.

Meanwhile, urban villages in Shenzhen are mostly born from the historical native villages that carry the native culture of the city. When the city besieges the village with exponential growth, the skyrocketing rental demand propelled the village to become an urban village with villagers leaving the village gradually. Nevertheless, the ancestral halls and traditions of the urban village still maintain the spirit of the village's network—this is also where memories are kept.

The black and white images in this exhibition, are the authentic records of Shenzhen urban villages in recent years. Not at all systematic, not even authoritative, it is even a little absurd: is this Shenzhen?

Yes, these human relations and emotions are Shenzhen, it is a story about the comings and goings of dreamers and urban villages—urban villages took shape because of the city's development, as well as reforming through the city's development.

Today, many scenarios among these images are gone with the wind due to the reformations. Are there any familiar scenarios that remind you of something? Any old memories that belong to you?

▷Exhibition: Have You Lived / Are You Living in an Urban Village?　▷Participant: Chen Dong

来源：深圳城市\建筑双年展组织委员会 . 城市共生：2017 深港城市\建筑双城双年展 [M]. 广州：岭南美术出版社，2021: 274-275.

福田中心区，2018 © 都市实践
CBD in Futian District, 2018 ©URBANUS

政策
POLICY

⊙ 政策 | 深发〔1992〕12号

关于深圳经济特区农村城市化的暂行规定

（1992年6月18日发布，自发布之日起施行，现行有效）

第一章 总则

第一条 深圳经济特区（以下简称特区）建立以来，特区农村的经济发展、村镇建设、生活环境都发生了重大变化，为了适应这种变化和进一步发展经济，提高人民群众生活水平，加速特区社会主义现代化建设，实现把深圳建设成为外向型、多功能的国际性城市的战略目标，特制定本规定。

第二条 本规定所指特区农村和农民系深圳特区范围内 68 个村委会、沙河华侨农场和所属持特区内常住农业户口的农民、渔鱼和蚝民。

第三条 实施农村城市化（即实现农村转化为城市、村民转化为城市居民）的基本原则是：尊重历史，面对现实，着眼未来，平衡过渡；兼顾国家、集体与个人的利益，兼顾近期利益与长远利益，有利于促进集体经济的进一步发展和人民群众生活水平的提高，有利于加强基层社会主义精神文明建设，有利于城市的总体规划和我市未来发展目标的实现：在实施过程中，不损害群众的正当利益，不降低集体分配水平，不平调集体财产。

第二章 管理体制

第四条 对特区农村现行的管理体制按照职能分解、归类理顺、分步实施、不断完善的原则进行转变。即将原村民委员会的发展集体经济和组织村民自治的两大职能分开，分别由新的集体经济组织、居民委员会和街道办事处承担。

第三章 集体经济组织

第六条 在农村城市化过程中，应建立和健全集体经济组织，使之成为自主经营、自负盈亏的城市集体所有制企业（以下简称集体企业）。实施城市化政策后，原集体财产仍属于原全体村民集体所有，由集体企业代表原村民行使集体财产的所有权。集体企业接受市、区政府有关部门的计划指导和宏观管理。市、区有关部门根据集体企业的规模、产业结构和管理水平、按照对城市集体企业的统一政策进行管理，并在人才招聘、技术改造、设备引进和经营范围等方面给予必要的支持和优惠。

第四章 土地、房屋与公用设施

第九条 根据《中华人民共和国宪法》有关规定，对现有特区内农村的土地采取如下办法实现国有化：

（一）特区集体所有尚未被征用的土地实行一次性征收，土地费的补偿办法，按照“深府〔1989〕7 号”文（《关于深圳经济特区征地工作的若干规定》）执行。

（二）按“深府〔1982〕185 号”文《深圳经济特区农村社员建房用地暂行规定》和“深府〔1986〕411 号”文《关于进一步加强深圳经济特区农村规划工作的通知》已划给原农村的集体工业企业用地和私人宅基地，使用权仍属原使用者。集体企业与个人应按市有关规定分别与市、区国土管理部门签订土地使用合同，办理有关房地产手续。

第十条 特区农村城市化应与全市旧村改造工作紧密结合，以利特区总体规划的实现和城市建设管理水平的提高。在原各村红线范围内的集体企业用地和个人宅基地，在旧城改造中空出的土地，经区政府有关部门批准，集体企业可以开发建设。原村民在政府划定的宅基地上合法建筑的房产，

为了加快我市城市建设步伐，根据《中华人民共和国宪法》关于城市的土地属于国家所有的规定，经研究，决定对特区内可供开发的属于集体所有的土地，由市政府依照法律的规定统一征用。

转为居民后，其产权不变。今后私人需在市政府原划定的宅基地上建房，应按市有关规定执行。转为居民后，集体企业职工和在市内其他单位就业人员的住房，按照市住房改革政策规定解决。

第十一条 原各村在红线范围内投资建设的公用设施，在符合城市规划的情况下，仍归原投资者使用，街道办事处和居委会应配合集体企业进行管理。今后城市各小区的市政公用设施，均应按城市统一规划的要求由市、区组织投资和建设。

第五章 人口管理

第十二条 将特区内原农民全部一次性转为城市居民。对于原特区农民配偶为特区外农业户口者，在本规定公布之前结婚的，可不受指标限制迁入特区并转为居民；在本规定公布之后结婚的，执行全市统一的入户政策。原农民一次性转为城市居民的具体方案由市公安局提出，市政府批准后实施。

（节选）

Note: The above are excerpted from *Interim Provisions of Urbanization of Villages of Shenzhen Special Economic Zone* (1992). The goal of the provision was to construct Shenzhen as an open and multifunctional international city. It impacts 68 village committees, Shahe Overseas Chinese Farm, and residents with rural hukou living in SEZ. Eight clauses in the provision stated the urbanization approaches including the transformation of the governance system, collective economic organization, land use and facilities investment, hukou registration, employment, social security and benefits.

来源：http://www.sz.gov.cn/cn/xxgk/zfxxgj/zcfg/content/post_9048582.html.

⊙ 政策 | 深府〔2004〕102 号

深圳市宝安龙岗两区城市化土地管理办法

（2004 年 6 月 26 日发布，自发布之日起施行，现行有效）

第一条 为依法推进我市宝安、龙岗两区（以下简称两区）城市化进程，根据《中华人民共和国土地管理法》、《中华人民共和国土地管理法实施条例》、《中共深圳市委深圳市人民政府关于加快宝安龙岗两区城市化进程的意见》（深发〔2003〕15号），制定本办法。

第二条 根据《中华人民共和国土地管理法实施条例》的相关规定，两区农村集体经济组织全部成员转为城镇居民后，原属于其成员集体所有的土地属于国家所有。

第十四条 继受单位和个人非农建设用地按下列标准予以确定：

（一）工商用地，按100平方米/人计算；

（二）居住用地，按100平方米/户计算，建筑面积不超过480平方米；

（三）道路、市政、绿地、文化、卫生、体育活动场所等公共设施用地，按200平方米/户计算。

非农建设用地的位置、面积及坐标，由国土管理部门确定。

本条涉及的户籍户数、人口数以批准两区镇村建制转变之日的户籍数和人口数为准。

对已按1993年《深圳市宝安、龙岗区规划、国土管理暂行办法》（深府〔1993〕283号）第四十八条标准或按本条规定划定非农建设用地的继受单位和个人不再另行划

定非农建设用地。

第十五条 依照本办法第十四条划定的用地，按照"工业进园，农业进基地，住宅进区"的原则，统一规划和建设。

（节选）

为加快我市建设国际化城市的步伐，实现深圳经济特区内外经济社会全面协调快速发展，市委、市政府决定加快宝安、龙岗两区城市化进程

Note: The above are excerpted from *Management Measures on Urbanized Land in Bao'an and Longgang Districts of Shenzhen* (2004). It declared the urbanization of Shenzhen went beyond the SEZ.

经过特区内外的两次城市化行动
深圳成为全国第一个没有农村的城市
THE TWO URBANIZATION PROVISIONS MADE SHENZHEN THE FIRST CITY IN CHINA WITHOUT VILLAGES.

来源：http://www.sz.gov.cn/zwgk/zfxxgk/zfwj/szfwj/content/post_6578010.html.

⊙ 规划 | 深城改办（2005）

深圳市城中村（旧村）改造总体规划纲要（2005—2010）

为全面完成城中村（旧村）（以下统称“城中村”）改造，加快城市化步伐，推动特区内外一体化建设，实现城中村地区与城市其它地区的统筹规划和协调发展，促进国际化城市、现代化中心城市建设，构建和谐深圳、效益深圳，特制定本纲要。

二、指导思想和基本原则

坚持积极稳妥、有序推进的原则。坚持政府主导、市场化运作的原则。坚持统筹兼顾、综合改造的原则。坚持区别对待、因地制宜的原则。

三、城中村改造目标与策略

我市城中村改造在未来五年的工作目标是：以空间形态改造为重心，以综合整治为突破口，全面推进，突出重点，逐步实现城中村生活环境的普遍改善，促进城市产业结构提升和空间布局优化，推动特区内外一体化建设，使特区外城市化水平和城市面貌与特区内接近。

具体应当实现以下目标：

全面完成特区内各级城市中心区、重点产业片区、重点景观地区、重点水源保护区、重大基础设施建设范围等对城市整体利益具有重要影响的区域内的城中村，以及存在严重安全隐患的城中村的全面改造，争取在2010年前特区内城中村的拆除重建规模达到总量的20%。

全面完成特区内其它城中村的综合整治，较大程度地改善城中村的居住环境和建设面貌。

重点推进特区外城镇中心区、重点产业片区、重点景观地区、重大基础设施建设区内城中村以及存在严重安全隐患、居住环境极差的城中村的全面改造，争取在2010年前特区外城中村的拆除重建规模达到总量的5%。

重点开展特区外城镇中心地段、各类规划重要功能区及城市干道两侧城中村的综合整治，争取在2010年前特区外城中村综合整治的规模达到总量的20%。

五年后，一方面要根据建设国际化城市的要求，继续强化对城中村空间形态改造的力度，同时要在总结近期工作经验的基础上，将工作重点转入社会形态改造方面，使城中村逐步在产业发展、社会管理、文化心理等方面完全融入城市，为实现规划的长远目标打下基础。

实现我市城中村改造目标的总体策略是：

规划先行。突出重点。整体开发。分期推进。分类指导。分区平衡。政策配套。

四、总体部署

（一）着重抓好位于重点地区的城中村的全面改造，完善城市空间整体结构，提升城市功能，加快国际化城市建设。

（二）以城镇中心地段城中村改造为重点，全面推进特区外旧围、旧屋村改造，促进产业结构升级，加快城市化进程。

（三）完善城中村的公共服务体系，全面消除严重安全隐患，保障基本安全，改善人民群众的生活环境。

（四）保护城中村的珍贵文化遗存，继承优秀地方文化传统，促进文化融合，构筑和谐社会。

（五）引导城中村经济的多元化转型，加大对原村民的生活保障力度，促进城乡经济统筹协调发展。

（六）完善法制秩序，加强行政管理，促进社区组织体系创新，把城中村改造成为现代化的城市社区

五、改造规划引导

（二）城中村改造的模式引导

我市城中村改造包括全面改造和综合整治两大类型，应加强对城中村改造的分类指导。

1、城中村的全面改造

全面改造是指通过对现状建筑物的拆除重建，较为彻底地改变城中村的空间形态，调整用地

功能，消除城中村与城市用地布局和景观面貌的严重冲突，全面提升空间环境质量。应当优先对城市近期建设重点地区内与城市规划冲突严重的城中村、特区内安全隐患等问题较为突出的城中村、特区外居住环境质量极差的旧围和旧屋村开展全面改造。

根据城市规划要求及城中村的具体情况，全面改造可以采取异地重建、整体拆建、局部拆建等模式。

城中村的全面改造可以根据具体情况，分别采取村内股份合作企业独立建设、股份合作企业联合其它开发机构合作建设、开发机构通过竞标实施建设等不同的组织形式。

城中村的全面改造一般应根据有关规定进行市场化运作，改造项目完成后，符合有关规定的建筑物和附着物可由建设单位取得完全产权，并可自由转让；特区外较为偏僻、缺乏市场动力的部分城中村改造项目，在统一规划、统建上楼的前提下，可以按照有关规定，采取鼓励措施，由村内股份合作企业组织实施村民自改，所建建筑物为自住自用，拥有不可自由转让的合法产权。

2、城中村的综合整治

综合整治是指在现状建筑空间形态不发生根本变化的基础上，采取各种手段对城中村居住环境的净化、美化、优化改造。综合整治项目中，除了极少量的市政、公用设施及环境小品外，基本不涉及房屋拆建，主要是要实现居住环境品质的提升。在近五年内，各区应积极采取有效的扶持措施，大力鼓励和重点推进综合整治的改造模式。

综合整治的对象主要为建筑布局较整齐、居住环境较好、基本能够适合中低层次居住需要的城中村。特区内所有近期不开展全面改造的城中村都应该积极开展综合整治工作。特区外应优先对位于各街道中心地段、区和街道近期重点发展的产业园区以及城镇主要道路沿线的城中村开展综合整治工作。

（五）城中村改造的功能引导

1、全市城中村改造的总体功能引导

我市城中村改造应当在合理维持和有序过渡现有低收入者住宅及低端都市加工业、服务业功能的基础上，结合全市产业结构调整和总体规划布局的要求，积极为城市支柱产业和优势传统产业发展提供配套服务，并根据项目区位不同，因地制宜确定合适的规划功能。在优势区位条件下，应积极发展商务办公、配套公寓、旅游服务等职能，在丰富城中村产业构成和经济来源的同时，促进城中村与城市功能布局的融合。

（节选）

Note: The above are excerpted from the *Master Planning Outline for the Redevelopment of Urban Villages (Old Villages) in Shenzhen (2005-2010)*. It encouraged the redevelopment of urban villages through a government-led and market-oriented (implemented by developers) approach. It was the first time "comprehensive remediation" was proposed as an alternative, nevertheless, the document focused on demolition and rebuilding.

深圳市城中村（旧村）改造暂行规定

深府〔2004〕177号

深圳市人民政府关于深圳市城中村（旧村）改造暂行规定的实施意见

深府〔2005〕56号

来源：http://law168.com.cn/doc/view?id=178953.

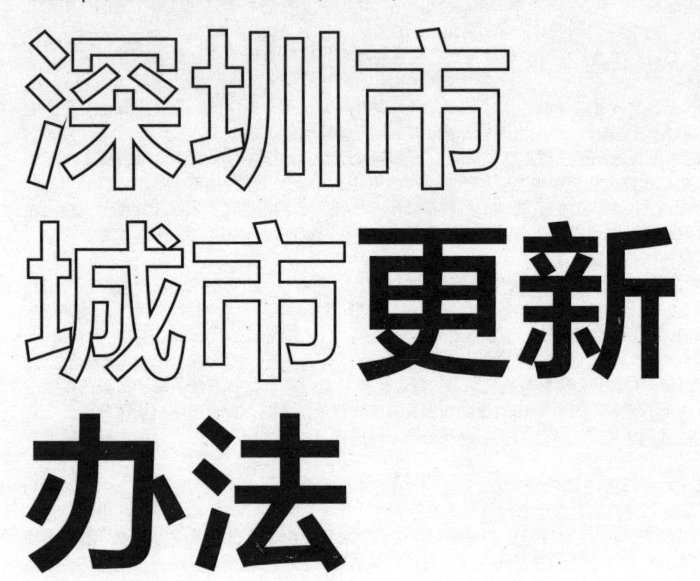

（2009 年 10 月 22 日发布，自 2009 年 12 月 1 日起施行，已修订）

深圳
在全国率先提出城市更新概念

Shenzhen
first proposed the concept of urban renewal
in China.

第一章 总则

第一条 为规范本市城市更新活动，进一步完善城市功能，优化产业结构，改善人居环境，推进土地、能源、资源的节约集约利用，促进经济和社会可持续发展，根据有关法律、法规的规定，结合本市实际，制定本办法。

第二条 本办法适用于本市行政区域范围内的城市更新活动。

本办法所称城市更新，是指由符合本办法规定的主体对特定城市建成区（包括旧工业区、旧商业区、旧住宅区、城中村及旧屋村等）内具有以下情形之一的区域，根据城市规划和

本办法规定程序进行综合整治、功能改变或者拆除重建的活动：

（一）城市的基础设施、公共服务设施亟需完善；

（二）环境恶劣或者存在重大安全隐患；

（三）现有土地用途、建筑物使用功能或者资源、能源利用明显不符合社会经济发展要求，影响城市规划实施；

（四）依法或者经市政府批准应当进行城市更新的其他情形。

第三条 城市更新应当遵循政府引导、市场运作、规划统筹、节约集约、保障权益、公众参与的原则，保障和促进科学发展。

第三章 综合整治类城市更新

第十九条 综合整治类更新项目主要包括改善消防设施、改善基础设施和公共服务设施、改善沿街立面、环境整治和既有建筑节能改造等内容，但不改变建筑主体结构和使用功能。综合整治类更新项目一般不加建附属设施，因消除安全隐患、改善基础设施和公共服务设施需要加建附属设施的，应当满足城市规划、环境保护、建筑设计、建筑节能及消防安全等规范的要求。

第四章 功能改变类城市更新

第二十三条 功能改变类更新项目改变部分或者全部建筑物使用功能，但不改变土地使用权的权利主体和使用期限，保留建筑物的原主体结构。功能改变类更新项目可以根据消除安全隐患、改善基础设施和公共服务设施的需要加建附属设施，并应当满足城市规划、环境保护、建筑设计、建筑节能及消防安全等规范的要求。

第五章 拆除重建类城市更新

第二十八条 拆除重建类更新项目应当严格按照城市更新单元规划、城市更新年度计划的规定实施。

第二十九条 根据城市更新单元规划的规定，城市更新单元内土地使用权期限届满之前，因单独建设基础设施、公共服务设施等公共利益需要或者为实施城市规划进行旧城区改建需要调整使用土地或者具备其他法定收回条件的，由市规划国土主管部门依法收回土地使用权并予以补偿。

第三十条 除依法应当收回的外，市政府可以根据城市更新的需要组织进行土地使用权收购，城市更新单元内的土地使用权人也可以向市规划国土主管部门申请土地使用权收购。土地使用权收购的程序、条件、价格按照土地储备和土地收购的有关规定执行。

第三十一条 除鼓励权利人自行改造外，对由政府统一组织实施城市更新的，可以在拆迁阶段通过招标的方式引入企业单位承担拆迁工作，拆迁费用和合理利润可以作为收（征）地（拆迁）补偿成本从土地出让收入中支付；也可以在确定开发建设条件且已制定城市更新单元规划的前提下，由政府在土地使用权招标、拍卖、挂牌出让中确定由中标人或者竞得人一并实施城市更新，建筑物、

构筑物及其他附着物的拆除清理由中标人或者竞得人负责。

第三十二条 拆除重建类城市更新项目范围内的土地使用权人与地上建筑物、构筑物或者附着物所有权人相同且为单一权利主体的，可以由权利人依据本办法实施拆除重建。

第三十三条 拆除重建类城市更新项目范围内的土地使用权人与地上建筑物、构筑物或者附着物所有权人不同或者存在多个权利主体的，可以在多个权利主体通过协议方式明确权利义务后由单一主体实施城市更新，也可以由多个权利主体签订协议并依照《中华人民共和国公司法》的规定以权利人拥有的房地产作价入股成立公司实施更新，并办理相关规划、用地、建设等手续。

第三十四条 同一宗地内建筑物由业主区分所有，经专有部分占建筑物总面积三分之二以上的业主且占总人数三分之二以上的业主同意拆除重建的，全体业主是一个权利主体。城中村、旧屋村拆除重建的，应当经原农村集体经济组织继受单位股东大会按照有关规定表决同意。本办法所称城中村是指我市城市化过程中依照有关规定由原农村集体经济组织的村民及继受单位保留使用的非农建设用地的地域范围内的建成区域。

第三十六条 拆除重建类城市更新项目中城中村部分，建筑容积率在 2.5 及以下部分，不再补缴地价；建筑容积率在 2.5 至 4.5 之间的部分，按照公告基准地价标准的 20% 补缴地价；

建筑容积率超过 4.5 的部分，按照公告基准地价标准补缴地价。

城中村依本办法补缴地价进行拆除重建后，符合有关规定的建筑物和附着物均可以由建设单位取得完全产权，并可以自由转让。

城中村拆除重建项目补缴地价可以按照市政府有关规定返拨所在区政府，作为城中村基础设施和公共服务设施建设费用。

（节选）

Note: The above are excerpted from *Shenzhen Municipal Urban Renewal Measures* (2009). Shenzhen government further encouraged redevelopment plans from developers. Meanwhile, it provides general guidelines for "comprehensive remediation" to improve the infrastructure, facilities and environment of urban villages.

深圳市人民政府关于修改《深圳市城市更新办法》的决定

市政府令第 290 号（2016）

一、删除第二十七条、第三十六条至第三十九条。

二、删除第三十五条中“本办法”字样。

三、在第四十八条后增加一条：“城市更新项目地价计收的具体规定，由市政府另行制定。”

四、在第五十条后增加一条：“为了推进强区放权，加快城市更新实施，市政府可以根据工作实际，调整职责分工，创新工作机制，并向社会公布。”

（2016年11月12日发布，自发布之日起施行，现行有效）

来源：http://www.sz.gov.cn/szsrmzfxxgk/zc/gz/content/post_9453090.html,
http://www.sz.gov.cn/zfgb/2016/gb982/content/post_4945589.html.

深圳市城中村（旧村）综合整治总体规划
（2019—2025）

第一章 总则

第六条 规划目标

全面推进城中村有机更新，逐步消除城中村安全隐患、改善居住环境和配套服务、优化城市空间布局与结构、提升治理保障体系，促进城中村全面转型发展，努力将城中村建设成安全、有序、和谐的特色城市空间。

第二章 综合整治分区划定

第八条 分区划定对象及其规模

综合整治分区划定的对象是全市城中村居住用地，扣除已批更新单元计划范围内用地、土地整备计划范围内用地、棚户区改造计划范围内用地、建设用地清退计划范围内用地以及违法建筑空间管控专项行动范围内用地。综合整治分区划定的对象总用地规模约99平方公里。

第九条 分区划定原则

综合整治分区应当相对成片，按照相关规划与技术规范，综合考虑道路、河流等自然要素及产权边界等因素予以划定。综合整治分区内单个地块面积原则上不小3000平方米。位于基本生态控制线、紫线、历史风貌区、橙线等城市控制性区域范围内的城中村用地，市、区城市更新“十三五”规划明确的不宜拆除重建城中村用地，不符合拆除重建标图建库政策的城中村用地，以及现状建筑质量较好、现状开发强度较高的城中村用地，原则上应当划入综合整治分区。

第三章 分区管理及调整

第十二条 综合整治分区管理要求

综合整治分区空间范围执行刚性管控要求，除法定规划确定的城市基础设施、公共服务设施或其他城市公共利益项目的用地、清退用地及法律法规要求予以拆除的用地外，综合整治分区范围内的用地不得纳入拆除重建类城市更新单元计划、土地整备计划及棚户区改造计划。

第四章 实施保障机制

第十七条 健全城中村更新管理机制

强化城中村更新统筹。以有机更新为理念，以综合整治分区为抓手，统筹安排城中村拆除重建和综合整治，科学、规范、有序地指导全市城中村更新工作的开展。积极推进城中村综合整治类更新，重点关注城中村安全隐患消除、居住环境和配套服务改善以及历史文化特色保留。

统筹多种存量开发实施手段。梳理城市更新、土地整备、棚户区改造之间的衔接关系，建立协调互促的城中村存量用地开发管理机制。以综合整治分区为抓手，通过协调多种存量用地开发手段的实施时序，控制城中村改造节奏，促进城中村可持续发展。

第十八条 优化城中村综合整治实施

强化政府主导和统筹作用。一是由市城管部门牵头，按照《综合治理行动计划》及其实施方案对全市城中村开展综合治理工作。二是积极鼓励各区在综合整治分区内开展城中村综合整治类更新工作，涉及城中村综合治理的，区政府应加大统筹力度有效衔接城中村综合整治类更新。

完善市场主体参与机制。鼓励市场主体参与城中村综合整治类更新，由政府制定相关规则规范市场行为，实施过程中加强政府监督和规划统筹。

强化城中村综合整治质量把关。加强城中村城市设计和建筑风貌控制，强化城中村现有建筑、新建建筑与周边景观环境、风貌、尺度、文化之间的协调。完善公共空间组织，注重开敞空间的互联互通、开放共享。

加强经费支持。按照公平、公开、公正、效益的原则，由各区政府负责牵头制定相关规定，出台资金扶持政策，对综合整治分区内由政府主导实施的综合整治类更新工作予以资金支持。扶持资金应做到计划管理，专款专用，切实实现资金保障目标。

第十九条 强化城中村租赁市场管理

加强城中村租赁市场监管。政府相关部门应加强城中村租赁管理，要求企业控制改造成本，并参照租赁指导价格合理定价。改造后出租的，应优先满足原租户的租赁需求，有效保障城中村低成本居住空间的供应。加强市场秩序整治，严厉打击城中村租赁市场违法违规行为，将违法违规信息纳入信用信息共享平台。

引导城中村存量房屋开展规模化租赁业务。政府相关部门应明确城中村规模化改造的要求和流程，通过计划引导、规划统筹、价格指导等手段，引导各区在综合整治分区内有序推进城中村规模化租赁改造，满足条件的可纳入政策性住房保障体系。经政府统租后实施综合整治类更新的城中村居住用房纳入政策性住房保障体系，进行统筹管理。

第五章 配套政策建议

第二十一条 构建政府主导管理机制

各区政府根据辖区城市更新五年规划和本规划的要求，制定城中村综合整治类更新单元年度计划，按照“一村一规划”的原则，组织编制更新单元规划，科学制定更新单元实施方案，消除城中村消防隐患，打通交通微循环，完善公共配套设施，增加公共开放空间，提升环境品质。

第二十二条 鼓励多方参与综合整治

综合考虑城中村现状建设情况及经济利益因素，通过允许局部拆建、降低更新单元计划合法用地比例的准入门槛、建立政府专项扶持资金等方式，要求各区政府联合原农村集体经济组织继受单位或鼓励有社会责任感、有实力的大型企业，不以追求利润为目的，积极参与城中村综合整治，以及参与综合整治后物业的统一经营管理，确保城中村的可持续发展。

（节选）

Note: The above are excerpted from *Shenzhen Urban Village (Old Village) Comprehensive Remediation Plan (2019-2025)*. It prioritized "organic renewal" which put emphasis on the "comprehensive remediation" of residential land use in urban villages. It also encouraged large-scale leasing of urban village buildings through comprehensive remediation and included them in the housing security system of Shenzhen City.

来源：http://www.sz.gov.cn/cn/xxgk/zfxxgj/ghjh/csgh/zxgh/content/post_1316996.html.

⊙ 政策 | 市人大常委会（2020）

深圳市第六届人民代表大会常务委员会
第四十六次会议通过

（2020 年 12 月 30 日发布，自 2021 年 3 月 1 日起施行，现行有效）

第一章 总则

第一条 为了规范城市更新活动，完善城市功能，提升城市品质，改善人居环境，根据有关法律、行政法规的基本原则，结合深圳经济特区实际，制定本条例。

第三条 城市更新应当遵循政府统筹、规划引领、公益优先、节约集约、市场运作、公众参与的原则。

第四条 城市更新应当增进社会公共利益，实现下列目标：

（一）加强公共设施建设，提升城市功能品质；

（二）拓展市民活动空间，改善城市人居环境；

（三）推进环保节能改造，实现城市绿色发展；

（四）注重历史文化保护，保持城市特色风貌；

（五）优化城市总体布局，增强城市发展动能。

第七条 城市更新项目由物业权利人、具有房地产开发资质的企业（以下简称市场主体）或者市、区人民政府组织实施。符合规定的，也可以合作实施。

第十一条 城市更新应当加强对历史风貌区和历史建筑的保护与活化利用，继承和弘扬优秀历史文化遗产，促进城市建设与社会、文化协调发展。

第十二条 市城市更新部门、各区人民政府应当建立健全城市更新公众参与机制，畅通利益相关人及公众的意见表达渠道，保障其在城市更新政策制定、计划规划编制、实施主体确认等环节以及对搬迁补偿方案等事项的知情权、参与权和监督权。

第三章 拆除重建类城市更新

第二十五条 申报拆除重建类城市更新单元计划时，拆除范围内物业权利人更新意愿应当符合下列要求：

（一）用地为单一地块的，应当经全体共同共有人或者产权份额四分之三以上的按份共有人同意，建筑物区分所有权的，应当经专有部分面积占比四分之三以上的物业权利人且占总人数四分之三以上的物业权利人同意，其中旧住宅区所在地块应当经专有部分面积占比百分之九十五以上且占总人数百分之九十五以上的物业权利人同意；

（二）用地包含两个或者两个以上地块的，每一地块物业权利人更新意愿应当符合前项规定，且物业权利人同意更新的面积不少于总用地面积的百分之八十；

（三）用地属城中村、旧屋村或者原农村集体经济组织和原村民在城中村、旧屋村范围以外形成的建成区域的，应当经原农村集体经济组织继受单位的股东大会表决同意进行城市更新，或者符合前两项规定，经原农村集体经济组织继受单位同意。

申请将旧住宅区纳入拆除重建类城市更新单元计划，自发布征集意愿公告之日起十二个月内未达到前款物业权利人更新意愿要求的，三年内不得纳入城市更新单元计划。

第二十八条 拆除重建类城市更新单元规划经批准后，物业权利人可以通过签订搬迁补偿协议、房地产作价入股或者房地产收购等方式将房地产相关权益转移到同一主体，形成单一权利主体。

属于城中村拆除重建类城市更新项目的，除按照前款规定形成单一权利主体外，原农村集体经济组织继受单位可以与公开选择的单一市场主体合作实施城市更新，也可以自行实施。

第三十二条 属于旧住宅区城市更新项目的，区人民政府应当在城市更新单元规划批准后，组织制定搬迁补偿指导方案和公开选择市场主体方案，经专有部分面积占比百分之九十五以上且占总人数百分之九十五以上的物业权利人同意后，采用公开、公平、公正的方式选定市场主体，由选定的市场主体与全体物业权利人签订搬迁补偿协议。

第三十六条 旧住宅区已签订搬迁补偿协议的专有部分面积和物业权利人人数占比均不低于百分之九十五，且经区人民政府调解未能达成一致的，为了维护和增进社会公共利益，推进城市规划的实施，区人民政府可以依照法律、行政法规及本条例相关规定对未签约部分房屋实施征收。

城中村合法住宅、住宅类历史违建部分，已签订搬迁补偿协议的物业权利人人数占比不低于百分之九十五的，可以参照前款规定由区人民政府对未签约住宅类房屋依法实施征收。

第三十七条 依照本条例规定对未签约房屋实施征收的，可以不纳入全市年度房屋征收计划，由区人民政府参照国家和本市国有土地上房屋征收与补偿的有关规定，依法分别作出征收决定。

被征收人对征收决定或补偿决定不服的，可以依法申请行政复议或者提起行政诉讼。

被征收人在法定期限内不申请行政复议或者提起行政诉讼，又不履行征收决定确定的义务的，区人民政府应当自期限届满之日起三个月内向人民法院申请强制执行。

第四章 综合整治类城市更新

第四十八条 实施综合整治类城市更新不得影响原有建筑物主体结构安全和消防安全，原则上不得改变土地规划用途。涉及加建、改建、扩建、局部拆建的，所在区域应当未被列入土地整备计划、拆除重建类城市更新单元计划。

第五十三条 市城市更新部门负责组织划定城中村综合整治分区范围。划入综合整治分区范围的城中村，不得开展以拆除重建为主的城市更新活动。

城中村内的现状居住片区和商业区，可以由区人民政府组织开展综合整治类城市更新活动。

第五章 保障和监督

第五十七条 实施主体在城市更新中承担文物、历史风貌区、历史建筑保护、修缮和活化利用，或者按规划配建城市基础设施和公共服务设施、创新型产业用房、公共住房以及增加城市公共空间等情形的，可以按规定给予容积率转移或者奖励。

（节选）

Note: The above are excerpted from the *Urban Renewal Regulations of Shenzhen Special Economic Zone* (2021). The regulation stated the urban renewal planning and further articulated the regulation in two categories, demolishing and rebuilding as well as comprehensive remediation. English version of the regulation can be found at http://sf.sz.gov.cn/fggzywyb/content/post_9907361.html.

来源：http://www.sz.gov.cn/szsrmzfxxgk/zc/gz/content/post_9453864.html.

卓越
大中華

非常经验
TOOLKITS

触发城市的另一种可能
Catalyzing Alternative Urbanism

数十年间，笼罩在单一的现代城市图景下的城市建设，既带来城市面貌的巨变和经济的腾飞，也触发城市同质化和社会分化的危机。而城中村所孕育的城市性——多元、差异、杂糅、抵抗——已自发地矫正和补充这幅单一图景的盲区。时至今日，摒除单一的历史观与发展观，根植于非正式空间的研究与实践已呈现出多种可能的城市发展方向。其工作包括对城中村的认识、认同、利用、优化、辐射，从而建立规划与自发、介入与内生的沟通与联系，最终生成新的城市机制，是为非常经验。积累此类经验的过程挑战了现有城市观念，也为其更新迭代提供了借鉴。

《城市即展场，展览即实践》从总策展人的视角阐明“城市共生”（在南头古城举办的2017深双的主题）作为城市观念和行动宣言，提出以城市策展推动城村重生计划。《城中村自建房调查》从观察者的视角记录广州赤沙村和小洲村自建房的建造逻辑和使用状态，从空间博弈中发现城中村建筑的生成机制和居民的生存方式，探讨个体建造的权利。《家园营造：香港都市中的社区参与和设计》展开建筑师的长期社区参与，在租赁关系、资金和面积的限定下，通过设计改善香港劏房租户的生活环境。《城中村家具交换计划》关注个体差异，将城中村居民制作的家具视作生活的文献，又作为将展览植入居民日常生活的媒介。

For decades, the single-minded vision of the modern city has shrouded urban construction, bringing revolutionary changes to the urban landscape and economy while provoking crises of urban homogenization and social differentiation. The diverse, different, hybrid and unyielding urbanity that emerged in urban villages has overcome the blind spots of that single-minded vision. Today, interconnected studies and practices of informal urbanism allow us to revise the single-minded view of history and development, and suggest many possible futures for urban development. These works understand, recognize, utilize, optimize, and influence urban villages which have established communication and interconnectivity between planning and informality, where intervention and endogenous growth have led to a new urban mechanism. The accumulation of such experiences challenges the existing concept of what a city is and provides references for its regeneration.

"City as Biennale, Exhibition Is Action" elaborated the theme of the 2017 UABB held in Nantou Old Town—*Cities, Grow in Difference*—from the perspective of the chief curator. As a concept of cities and a manifesto for action, the article locates urban curation as a site for revitalizing cities and villages. "An Investigation into Self-built Buildings in Urban Villages" records the observation of the logic of construction and space utilization of self-built buildings in Chisha and Xiaozhou Villages in Guangzhou. Through the lens of the game of space, the team uncovers the formation of buildings and inhabitants' ways of living in urban villages, exploring individuals' right-to-build. "Home Improvement: Community Engagement and Design in Hong Kong" recounts the architects' unfolding, long-term community engagement, which has used design to improve the living environment of tenants with limited lease, budget, and area in subdivided units. "Urban Village Furniture Exchange Program" takes the homemade stools of urban village inhabitants both as evidence of everyday life and as a vehicle for bringing the exhibition into inhabitant's everyday life.

城市即展场，展览即实践

City as Biennale, Exhibition Is Action

孟岩
URBANUS 都市实践建筑设计事务所创建合伙人、主持建筑师，2017 年深港城市 \ 建筑双城双年展（深圳）总策展人。
本文首次发表于《时代建筑》，2018(4): 174-179；由陈一苇中译英；照片 © 都市实践 ©UABB。

Meng Yan
Meng Yan is the Principal Architect and Co-founder of URBANUS, and he was the chief curator of the 2017 Bi-City Biennale of Urbanism\Architecture (Shenzhen).
The article was first published in *Time+Architecture*, 2018(4): 174-179; translated from Chinese to English by Chen Yiwei; image ©URBANUS ©UABB.

1 序言

在一个无时无刻不被技术与媒体充斥的当下生活现实中，双年展为什么还值得我们亲临现场？它怎样才能从一场短暂的学术与媒体盛宴中自我突围？双年展能够在日常的喧嚣琐碎之中创建一个理想的“乌托邦”吗？或许能像它所承诺的，成为一片观念上的“飞地”，或是在多重现实中实现一种新的“公共领域”？双年展所开启的一场持久的城市介入行动能否如期而至？在策划第七届深港城市 \ 建筑双城双年展（深圳）的两年间，笔者的这些疑问和自省一直盘绕着。2017 年第七届深双无疑是一次雄心勃勃的展览，是一次集大成的实验，尤其突出的是借助城市策展手段介入城中村的复杂现实。然而在深双的历史语境下，它并不是一次突变，而是承袭了历届深双传统并将其推向极致，在保持一个年轻双年展生猛无畏的探索精神的同时，更试图建立一种展览现场与介入实践合体共生的另类展览模式。如今深双已落下帷幕，喧嚣归于平静，而这些追问仍萦绕不去。当一个展览直面当下城市问题并触及当代文化的症结，它可以成为一道劝谕、一声呐喊、一份宣言和一次行动，能够对当今城市的空间、社会和文化进行反思、批判和重建。

本文回溯 2017 深双主题——“城市共生”提出的背景和意义，策展人一反国际双年展日趋宽泛、中性和政治正确的主题范式，针对当今世界的文化境况和中国城市化的严峻现实，鲜明地推出既是城市观念更是一项行动宣言的“城市共生”主题。文章首先分析了这一主题如何通过盘点和反思中国近四十年快速城市化

图 1 2017 深双主展场南头古城
Figure 1 Nantou Old Town, main venue of the 2017 UABB

1 Preface

In the current reality of life that is perpetually inundated with technology and media, why is the Biennale still worthy of our presence? How can it extricate itself from this short-lived academic and media feast? Could the Biennale still create an ideal "utopia" amidst the mundane hustle-and-bustle of modern life? Perhaps it could become a conceptual "enclave" as promised or —within multiple realities—progress into a new type of "public domain?" Could a long-lasting urban-intervention movement initiated by the Biennale kick off as expected? These questions and introspections of the author have circled around the two-year planning regime of the 7th

的历史遗产，适时回应了当下迫切的城市问题；随后阐述了在这一主题下的城市策展理念与介入手段，解读了如何在极端复杂和不确定性的城中村铺设多重展览叙事结构，展览设计如何凸显城/村共生并应对多元混合的观展人群，最终在3个月展期内编织了展览与日常生活同构的混杂现实。文章最后以“后策展时间”为题，用未完待续式的插叙展示了深双鲜为人知的另一侧面：自发的双年展、隐形的双年展和未完成的双年展。在此之上，2017深双最重要的成功是助推了南头城/村重生计划的启动，用自身的学术平台和展览实践试图寻找中国城市化的另类样本和未来模式（图1-图2）。

2 建筑展览——从抽离基地到回归现场

也许从来不存在理想的展览模式，但对于策展人来说，建筑展无疑是最难做的展览。展览并不是建筑学与生俱来的工具和语汇，具有自觉意义的独立建筑展的历史也并不遥远。由于在西方传统中建筑与艺术之间的历史勾连，在画廊和美术馆中展览建筑变得顺理成章。专业的建筑展览兴起于启蒙运动时期，然而从一开始它就面临着一个悖论，因为建筑本身无法与建成环境分离，如何解决展示主体不在场的困境？倘若如此，我们究竟展的又是什么？建筑展呈现的大多是建筑物建成以后或未建成之前的拟像、替代物、实物片段和模拟再现等多种表达媒介。前纽约现代艺术馆的建筑策展人巴瑞·博格多就曾撰文关注“在多大程度上展示建筑的现象改变了建筑的特性和可能性”。他指出自启蒙时期以来，展览作为一种有效手段，对于发展现代建筑那些最为鲜明的特征起到关键的作用。同时他也认为，抽离现场的建筑展览促生了对现代建筑的公共属性和责任的批判性思辨，也开启了一段寻求建筑含义以及塑造国家、区域和地方特征的建筑历史。[1]

埃娃·丽萨·派柯南于《在悖论中开掘》一文中铺陈了建筑展览的众多面相，展览作为建筑思想的实验室，在历史上甚至催生了“建筑电讯派”和“超级工作室”这类激进的建筑思潮和建筑团体；同时，展览让建筑师暂时逃离日常事务的琐碎，可以采用极端思辨的立场开辟建筑学新的可能性。除去纯粹建筑学领域的探索之外，城市、社会与公众参与也是建筑展览历史上的长久话题。特别是20世纪60年代晚期动荡的社会现实、激进变革的影响也曾深深植入同一时代的展览现场，政治宣言和社会理念回荡在展览与周遭现实的激烈互动之间。总之，建筑展览始终是建筑学的历史与未来的重要组成部分，“建筑展览不只是宣传建筑师和建筑物，并且时刻提醒我们建筑学能够和应该成为什么的全部可能性。”展览作为媒介不仅可以再造一个沉浸式的模拟性体验，更能够成为鼓励公众参与的界面，甚或是一场社会动员和行动，“展览也迫使建筑师和策展人一次次

Bi-City Biennale of Urbanism\Architecture (Shenzhen) (UABB) in 2017. The 7th UABB was undoubtedly an ambitious exhibition and an epitomizing experiment especially as it intervened the complex reality of the urban village through the means of urban curation. However—in its historical context—the 7th UABB was not a break with the past but rather it inherited the traditions of all previous Biennales and pushed them to the extreme. While maintaining the fearless exploration spirit of a youthful exhibition, the 7th UABB furthers its evolution from the past by attempting to establish an alternative exhibition model where the site and the intervention practice are symbiotic. Now that the curtain has come down on the exhibition, the public uproar over the Biennale has quieted down. Nevertheless, the author's inquiries still linger. As an exhibition confronts the urban problems of the city and touches the crux of contemporary culture, it also becomes an exhortation, a cry, a manifesto and a movement which reflects on, criticizes and reconstructs the city's present-day spaces, society, and culture.

Retrospectively, the following essay discusses the theme of the 2017 UABB—*Cities, Grow in Difference*—as well as its background and significance. In contrast to the increasingly vague, neutral and politically correct thematic paradigm of international biennials, the curators proposed the theme, "Cities, Grow in Difference" not only as an urban concept, but also as a manifesto in direct response to the current global cultural climate and the reality of rapid urbanization in China. In the first section of the essay, the author provides an in-depth analysis on how the theme countered the pressing issues of the city, taking stock of and reflecting on the legacy of the past forty years of rapid urbanization in China. Following this historical analysis, he expounds on the urban curatorial concepts and interventions that were implemented under the theme and explains how the multi-function exhibition narrative was laid out in the extremely complex and volatile environment of the urban village. At the same time, the author elaborates on how exhibition design was able to highlight the coexistence of the city and urban village while addressing a diverse audience. He also looks at how the exhibition was woven into the daily life of the urban village as an intricate reality over the three-month span of the exhibition. The essay concludes with a section titled, "Post-Curatorial Time"—a cliffhanger to showcase other aspects of the Biennale that are less known. It is spontaneous, invisible, and unfinished. Indeed, the most prominent achievement of the 2017 UABB is that it contributed to the revitalization project of Nantou City/Village, while using its own academic platform and exhibition practice in an attempt to find an alternative prototype and a future development model for China's urbanization (Figure 1-2).

图 2 深双开幕式，南头古城报德广场，2017
Figure 2 UABB opening ceremony, Baode Square, Nantou Old Town, 2017

地重新定义建筑学的范畴以及边界。” [2]

建筑展衍生于艺术展，而远为庞大、复杂的建筑和城市存在于我们周边且不断演化，与人们直面相遇，本无需依托展览媒介而存在。最理想的建筑展也许只有真实的建造展，就像 1927 年德国斯图加特“魏森霍夫”展，以及 20 世纪 80 年代的柏林住宅展实现了足尺的城市建筑实验。除却这种最极端的建筑实践展以外，建筑展更广泛地囊括了研究与知识生产、观念陈述与社会宣言的内容。自 1980 年第一届威尼斯建筑双年展起，国际大展围绕一个统一的主题展开成为新的传统。为了迎合一个日益媒体化的社会，一方面展览主题必须足以引人关注且又能够包罗万象，越来越多明星建筑师代替职业策展人策划威尼斯和其他国际双年展；另一方面近年来除了库哈斯在第 14 届威尼斯建筑双年展采用了强势策展方式高度整合展览之外，更多策展人采用的是在一个并不针对特定问题的宽泛主题下呈现汇编和集锦式的“来自前线的报告”。正是在此大趋势下，15 年前才加入双年展大家庭的深双开始突显其另类本色：与其做一场包罗万象的展览，不如径直反向地邀请双年展的观众参加一趟“去往‘前线’的探险”。因为深双作为聚焦城市化主题的年轻双年展不仅是一个展览，同时也是 20 世纪和 21 世纪最剧烈的城市化现场。观众的观展体验是从亲历或见证这一急剧城市化的真实进程开始的，事实上从深圳到珠三角，城市本身才是深双最大的展场。历届深双都试图紧扣城市化进程中最为迫切的城市问题，因而展览主题往往也更具问题导向且立场鲜明，可以说七届深双主题的演变可以清晰地勾勒出深圳城市和文化转型的历史。

3 “城市共生”作为普世价值、城市观念和宣言

紧锣密鼓筹备第七届双年展的 2016 年所面对的恰是一幅危机四伏、充满不确定性的世界图景。它如同大自然美丽表象之下无时无刻的暗流涌

2 The Architectural Exhibition: From Extracting the Base to a Return to the Site

Even if there does not exist an ideal exhibition model, nevertheless, architectural exhibitions are undoubtedly the most difficult for curators. Exhibition is not a tool or a vocabulary term that exists within the innate capacity of the architecture discipline. Historically, the conscious awakening of the independent architectural exhibition occurred in the not too distant past. In the history of traditional western culture, architecture and art have long been closely intertwined. Exhibiting architecture in galleries and museums thus became sensible. The first initiative toward an independent architectural exhibition was observed during the Enlightenment. However, since architecture itself cannot be separated from its built environment, from the beginning, there has existed a paradoxical difficulty: How can we fix the dilemma of displaying architecture in the absence of its context? Generally, architecture exhibitions today present the buildings either before or after they are built with expressive mediums such as simulacrum, substitution, physical fragmentation, and simulation. Barry Bergdoll, former Chief Curator for Architecture and Design at the Museum of Modern Art (MoMA), once undertook a study on "to what extent has exhibiting architecture changed the characteristics and possibilities of architecture?"[1] He points out that exhibitions, as an effective measure, have been the driving vehicles for the innovation of the most distinctive characteristics of modernist architecture since the Enlightenment. Simultaneously, Bergdoll believes architectural exhibitions—which have extracted architecture from its site and moved it into the space of the gallery—have been instrumental in the debate on modern architecture's responsibility to engage with a larger public and to set off another path, in the history of architecture, which seeks the meaning, the cultural context, and the vernacular characteristic of architecture.

In "Mining the Paradox," Eeva-Liisa Pelkonen lays out the many facets of architectural exhibition. As the experimental lab of architectural ideals, exhibition as a medium has facilitated, even expedited the emergence of radical architectural movements and groups exemplified by Archigram (*The Living City*, 1963) and Superstudio (*The Continuous Monument*, Trigon '69) in the 1960s. At the same time, exhibitions have not only provided architects with a temporary escape from the minutiae of everyday affairs, but also a platform for new speculations and progressive inventions within the field of architecture. In addition to the investigation on the subject of architecture, urban, social and public partici-

动，矛盾丛生：一方面，世界经济发展不平衡，政治单边主义抬头，国际恐怖主义升级，宗教冲突和难民问题，民粹主义、保守主义和狭隘的民族主义兴起；另一方面，全球主义、消费主义、信息与媒体技术又对既有世界秩序、经济格局以及人的生活方式实施颠覆，不断将人们的思维和行为方式重新格式化。旧有平衡已经打破，新的游戏规则正待形成，未来变得更加难以预测。

正是这一充满矛盾与冲突的背景使得带有鲜明后现代文化色彩的“共生”理念呼之欲出。迈克·费瑟斯通在《消费文化与后现代主义》的最后一章中谈到“多样性的全球化进程”时认为，后现代主义可以“是一种导向型工具，因为它在形成和解构文化时、在提高有鲜明个性的艺术与知识生活之秩序时，具有重要地位”，进而指出，建立“新框架需要更富弹性的原生性结构，需要认可、容忍广泛的差异”[3]。“共生”理念的重新发现表明一种态度，即对个体、社会和文化多元性的认同，对“另类”“他者”与“不同”的包容。而实现它的途径即反对单一价值取向、反乌托邦的发展方式，主张杂糅、差异与抵抗。“共生”是从根源上对文化层面、社会层面、空间层面上不同起源、不同状态、不同价值观的认同与包容，是对文化中心主义的反叛，也是对正统现代主义美学和单一“进步”大历史观的修正。

反观中国近四十年来的造城运动，在权力和资本裹挟下经历了高速发展，城市渐趋同质和单一。近年来如火如荼的城市更新进一步把层积丰富的历史街区和多样杂糅的城市生活彻底清除，代之以全球化、商业化的标准配置。城市人赖以生存发展的自由开放空间正在被规范化、机构化和标准化的空间实践挤压和清除，按既有模式建造的城市已经难以容纳更深远的创新活力了。“共生”可以是一种发展模式——“城市共生”强调城市生态系统之平衡，如同自然界物种之共生与动态平衡，城市中的矛盾混杂不应被肆意打破而失衡。“共生”更是一种城市观念——我们今天的城市应自觉地反抗单一和理想化的未来图景，因为城市本质上是一个高度复杂的生态系统。“仰观宇宙之大，俯察品类之盛”，今日之城市应是多元价值体系平衡的结果，是人们身处同一世界，心存不同梦想的高度异质性和差异化的文明共同体。城市本应无所不包、和而不同，它的生存与繁荣恰恰在于能最大限度地对“差异性”的文化认同。[4]

4　城市策展：一次异质与共生的策展实践

4.1　城中村中的双年展

深圳是自上而下理性的规划与自下而上自发生长的共生与合体，是乌托邦与异托邦、秩序与

pation has also been a long-standing topic in the history of architectural exhibitions. In particular, the exhibition site of the late 1960s was indeed reflective of the turbulent social reality and radical changes of its time as it circulated contemporaneous political statements and social ideologies. It suffices to say that architectural exhibitions will always be an important part of the history and future of architecture as "Architectural exhibitions not only promote architects and buildings, but also serve as a constant reminder of all the potential possibilities of what architecture can and should be." The exhibition as a medium can not only recreate an immersive and simulated experience for the audience, but can also become an interface that encourages public participation, or even a social action; "the exhibition [will thus] force architects and curators into actively redefining the scope of architecture and its boundaries."[2]

The architectural exhibition derives from the long-established art exhibition. However, as colossal and complex beings that continuously evolve in our presence, the existence of architecture and the city does not need to be solidified by the medium of exhibition. Perhaps the most ideal architectural exhibition would be one which physically exhibits the building of architecture. Exemplified by the Weissenhof Estates which was built for the Deutscher Werkbund Exhibition in Stuttgart in 1927 and the International Building Exhibition Berlin in the 1980s, these two German projects illustrated an ideal image of architectural experiment in urban context. In addition to showcasing architecture in the most literal and practical sense, architectural exhibitions further encompass the content of research, knowledge production, conceptual statement as well as social manifesto. In 1980, the 1st Venice Biennale of Architecture set a new tradition for large-scale international exhibitions: to have a unifying theme. On the one hand, in adaptation to a society immersed in mass media, the exhibition theme has to be capable of attracting attention while also being inclusive. Moreover, in the context of biennales, more iconic architects are now stepping in to replace professional curators in the planning of the exhibition. On the other hand, with the exception of Rem Koolhaas' strong curatorial approach in organizing the 14th Venice Biennale of Architecture, more curators have taken the method of presenting a compiled collection of "front-line reports" under a wide-ranging theme rather than directing the exhibition toward one specific problem. The UABB which joined the biennale family merely fifteen years ago revealed its alternative nature to the influence of this trend, as it directly invites the audience onto "a front-line adventure" rather than establishing itself as an all-inclusive exhibition. Therefore, as a youthful exhibition which focuses on the theme of urbanization, UABB does not only serve as an exhibition, but also presents an image of the most drastic urbanization progress of the 20th

混乱的合体与共生。城中村是城市与村庄的合体与共生，它仍处于城市中的“未完成”状态，持续演进、自我繁殖、自我更新、投机钻营与民间智慧无处不在。城中村的发展是借力与自助结合，既灵活又善变，它的空间生产策略在其他城市地区无法应用和复制。正是在清晰与混沌之间、合法与非法之间，在一个非黑即白的价值判断体系里，城中村这一灰色地带被保育成当下最具潜力的城市实验场和另类新生活的培育基地。城中村一栋栋密密匝匝的“握手楼”排布出“迷你街区”，滋养了灵活多样的自发经济模式，使无数年轻的外来者在拥挤狭窄的出租屋和内巷中暂栖身心，也容纳了梦想，成为新来者真正的落脚城市。城中村永不落幕的“未完待续”状态使持续的创造成为可能，人们不断把自身的生活理想和未来想象投射到这里并持续实验，以一种相对低廉的成本、在相对自由的机制中不断生发出种种新的可能性。

深双是游牧式和实质性城市介入的展览，于 2005 年为应对最为迫切的城市问题而生。它始终坚持在城市中四方游走，直面当下，切入现实。2005 年恰逢深圳城中村改造办法公布，在城中村首次面临大规模拆除的关头，首届深圳双年展第一次展出城中村改造研究的板块，在当时引发了广泛争议和共鸣。其后经历 2010 上海世博会深圳馆大胆地将城中村中普通人的生活境遇与梦想作为核心叙事，2017 年深圳双年展首次走进城中村，这十余年间，政府、公众、学界和媒体对城中村问题的包容和态度发生了巨大的转变。然而在深圳面临土地资源枯竭转向对建成区全面改造的当下，城中村再次成为深圳城市化向内深耕的最新前沿，城中村的未来命运也必将触及城市平衡发展的最后底线。2017 深双再次直面困局，直接介入城中村改造并以展览作为探索新的共生发展模式的契机。

4.2 另一种双年展的可能性

最重要的事件正是那些发生在街道上的微不足道的小事情。

——乔治·布雷希特

2017 年第七届深双是寻找新的理论、新的实验和新的实践模式的平台，南头古城主展场和位于四个城区、五个城中村中的分展场均紧扣“城市共生”主题，同时又是一系列各不相同、各自独立的展览，在城市之中处处开花、全城互动。南头主展场也呈现为多个展览的共生同体，它们似乎是环环相扣、一个套一个相互叠加的双年展。展览模拟城市体验、融入城市本身，既有严密的整体构想又可以随机应变，充分应对展场位于城中村的种种不确定性。南头主展区包含三个大的展览：第一部分是城 / 村介入实践展，展示了两年以来南头古城改造实践的阶

and 21st centuries. In fact, the audience's involvement with the exhibition begins from witnessing and personally experiencing the drastic urban changes that are concurrent with the exhibition, making the city itself the grand venue of UABB. UABB themes have always been issue-oriented and distinctive as they endeavor to tackle the city's most acute problems. It suffices to say that the conceptual evolution of the 7th UABB has clearly outlined the history of Shenzhen's urban and cultural transformations.

3 "Cities, Grow in Difference" as a Universal Value, Urban Concept and a Manifesto

In 2016, preparations for the 7th UABB took place in a world-scape of crisis and uncertainty. This image resembled nature, where under its beautiful facade there exists turbulence and conflicts. On the one hand, global economic development was unbalanced, political unilateralism was surfacing, international terrorism was escalating, while religious conflicts and refugee problems, populism, conservatism and narrow nationalism were all on the rise. On the other hand, globalism, consumerism, and media hegemony were upending the existing world order, economic system, and lifestyles as they constantly reformatted people's ways of thinking and acting. The previous balance had been broken. New rules of the game were yet to be formed, and the future had become more difficult to predict.

It is precisely this ambivalent background which has urged the distinctive post-modern concept of "symbiosis" to emerge. In the last chapter of *Consumer Culture and Postmodernism*, Mike Featherstone argues that postmodernism "as a means of orientation is to emphasize its place within the processes which form and deform the cultural sphere and give rise to distinctive artistic and intellectual life orders" in the context of "the globalization of diversity." He then points out the need to establish "a new frame which entails a more flexible generative structure within which a wider range of differences can be recognized and tolerated."[3] The rediscovery of the concept of symbiosis shows an attitude, that is, the recognition of individual, social and cultural diversity and the tolerance of the alternativeness, otherness and difference. To realize this new regime is to oppose the dystopian mode of development following a single value orientation. Moreover, it is to advocate hybridity, difference and resistance. Symbiosis is based on a fundamental recognition and tolerance of different origins, states and values on the cultural, social and spatial levels. It is not only a rebellion against cultural centralism, but also a revision of orthodox modernist aesthetics and the tradition of focusing on one monolithic form of progress at a time.

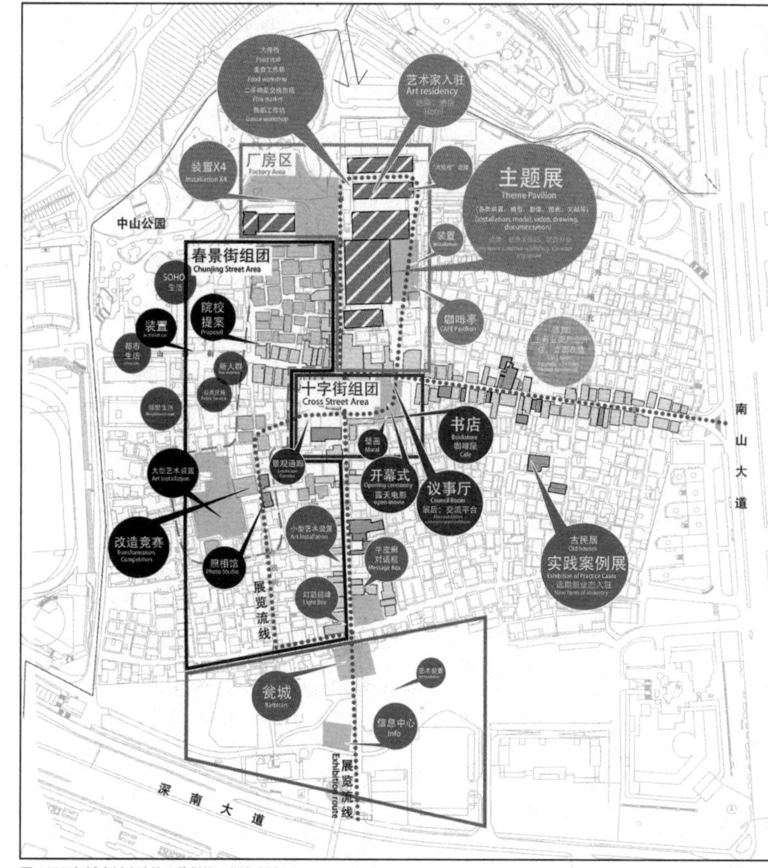

图 3 深双与城中村改造的无缝衔接，都市实践
Figure 3 Cohesion of UABB and urban village regeneration, URBANUS

段性成果，它不仅仅是建造，更是一项以公共空间为线索的本土文化复兴和社区重建计划；[5] 第二部分是位于古城北部厂房区内的主展场，囊括了“世界 | 南方”“都市 | 村庄”和“艺术造城”三大板块的核心部分；第三部分是从古城南门起到古城中心广场沿线及散布于古城各处的小型展览和街道、社区的微介入计划(图3)。这三大展览在结构上是真正意义上的“展中展”，在“展览”之外的南头古城 / 城中村才是第七届深双最大的展场。让人们进入城中村亲身感受和参与才是最重要的观展体验，城中村里众声喧哗、活色生香的生活现场才是真正的主展。在城中村和展览之间的穿越模糊了展场与现场，构成一次次叙事的重叠与交织，它们相互合成并互为印证。展厅内部浓缩了城中村的历史，映射了生活现实，也涵盖了对未来的想象，而包裹展区的城中村本身成了展览的主体，同时也是展览注视的对象、佐证，有时甚至构成对展览本身的反证和批判。此次双年展吸引了政府决策者、专业人士和媒体之外的很多城中村居民和城市年轻人群的广泛参与。它给大量从未走进过城中村的人们提供了一次近距离关注同一座城市中不同生活现实的契机，而进入城

In contrast, for forty years, the city-making movement in China has experienced rapid development under the coercion of power and capitalism, and cities are becoming homogeneous and unitary. In recent years, urban renewal has further eliminated rich historical blocks and diverse urban life, replacing them with a new configuration that is standardized, globalized, and commercialized. The free, open space available for the survival and development of the urban population is being squeezed and eliminated by standardized and institutional spatial practice. Cities built according to this model have become incapable of accommodating further innovation vitality. Symbiosis can be a development model. *Cities: Grow in Difference*, for example, emphasizes a balance of the urban ecosystem that is similar to the symbiosis and dynamic balance between species in nature. The mixed contradictions in the city should not be recklessly broken to cause more disparity. More importantly, symbiosis is an urban concept. Our cities today should consciously resist the single-ended and idealized image of the future because each city is essentially a highly complex ecosystem itself. "Facing upwards to the blue sky, we behold the vast immensity of the universe." Contemporary cities should be seen as a product of multiple balanced value systems in a highly heterogeneous, harmonized, and tolerating society. The city should be all-embracing as its survival and prosperity lies in, to the greatest extent, the cultural recognition and tolerance of difference.[4]

4 Urban Curation: A Heterogenous and Symbiotic Curatorial Practice

4.1 An Urban Village Biennale

Shenzhen's urbanization progress is marked by the coexistence and symbiosis of top-down rational planning and bottom-up autonomous growth. This city encompasses not only utopian ideals and order, but also dystopian scenes and chaos. The urban village is a

图 4 南头古城工厂区展场
Figure 4 Venue in the Factory Zone, Nantou Old Town

中村的双年展提供了环视和反思当下城市问题的另一种批判性视角。

4.3 展览的多重叙事结构

位于北部工厂区主展场（图4）的南楼为“世界|南方”板块，首层和二层与中部的“都市|村庄”板块在空间上衔接，随后连接到北侧的“艺术造城”板块。“世界|南方”板块“以‘影响与抵抗’作为副题，着眼于探讨地方性与全球化的权力、资本与文化动力之间的广泛博弈与丰富表现，以及它们如何通过自身条件的适应与改造，导致‘差异化的现代性’后果的实现”。[6]

“世界|南方”垂直叠加了最大差异化的展览呈现。一层是“南方南方”，介绍拉美地区特定政治、经济背景下非正式建筑的历史、现状、建造模式和新的城市策略（图5）。二层以“漂变的珠三角”为题聚焦岭南、客家与深圳，采用几组当代艺术装置寓言式地呈现珠三角移民的历史、客家迁移和社会文化变迁，试图找到差异的现代性的历史基因。三楼“他者南方”对两个南方——江南和岭南做建筑类型学上的对比研究，以及地缘文化和空间的差异分析。四层以何志森指导的“图绘南头”工作坊成果回到南头古城。一到四层的展览从南美大城市中民间的愉悦和创造氛围转到珠三角遭遇现代性的从容与荒诞，展览体验也经历了从江南的轻盈细腻到岭南当下的纷繁杂陈。王昀的聚落研究项目从历史的维度展示世界聚落从自然到人工的聚居生长机制图谱，意在表明城中村乃是世界聚落在当代的一种呈现形式，具有其必然性。这一作品在空间上衔接了南方板块与二层核心的“城中村档案馆”，加强了两大板块间的叙事联系。

“都市|村庄”板块包含中部主厂房首层的主题展、二至三层的研究展、展中展和特别展，以及散布城中各处的“城中展”，总体呈现为多个展览的重叠与共生。“都市|村庄”板块以“杂糅与共生”为副题，是第七届双年展的主题板

symbiosis of the village and the city. While it continues developing, self-multiplying, and self-renewing, speculations are commonplace within this unfinished social environment full of folk wisdom. The development of the urban village is a combination of both the processes of appropriation and self-sustainment. It is flexible and changeable. These unique characteristics of the city's space-making strategy cannot be applied and duplicated elsewhere. Within the black and white dichotomy of values, the urban village is, however, a gray area, comprising both clarity and chaos as it travels between the two ends of the spectrum of legality. Thus, the urban village is conserved as the most embryonic present-day experimental field for city development as well as a base for cultivating alternative lifestyles. The densely laid out "handshake buildings" within the urban village formed "mini blocks" that would nurture a spontaneous and versatile free market economy, providing a place for countless migrants to inhabit in. Between the narrow alleys and within the cramped rental rooms, it accommodates the dreams and a sense of belonging for these newcomers. The never fixed, unfinished status of the urban village urges the emergence of more possibilities. The low-cost spontaneous economic order within these encompassing dwellings has provided grounds for people to not only project but also achieve their dreams and imagination.

Established in 2005 to tackle the most critical problems of the city, UABB is both a nomadic exhibition and a substantial procedure that attempts to reveal and improve the reality of the city. The same year that UABB was established, the transformation measures for Shenzhen urban villages was issued; urban villages faced demolition. It was at this crucial moment that the first UABB shed light on this critical matter, sparking controversies and evoking resonation. It wasn't until the EXPO 2010 Shanghai where the Shenzhen Pavilion fearlessly showcased the life and stories of ordinary urban villagers in pursuit of their dreams that people came to embrace the coexistence of the urban village and a rapidly developing city. In 2017, UABB finally stepped foot into the urban village. The consistent endeavor to present the lively image of this urban informality over the past decade has changed the controversial impression previously held by the government, public, academia, and media. As Shenzhen faces the shortage of land and turns toward the inner transformation of its existing fabric, the urban village has, once again, become the frontier of Shenzhen's further cultivation of its core. Inevitably, the fate of the urban village will run parallel to the future of Shenzhen as they coexist to achieve balance in this city's develop-

块。展览叙事主线以“城中村”这一深圳和珠三角地区特殊的城市现实为样本，以历史研究、现实观察、介入实践与未来想象等多元视角回应与呈现“城市共生”主题。

位于首层的“都市 | 村庄”主题馆是一系列自由散落的半封闭展区，类比于故事屋、游戏室、照相馆、图画室、录像厅、音乐室、实验室，以及未来馆等都市民间的日常空间，内容取材于丰富的城中村研究与观察，展示方式力求新颖引人。两条十字交叉的内街模拟了城市街道的体验，嵌入墙面的大型摄影作品和隐藏在楼梯间的霓虹灯装置（图 6）犹如巨型广告牌侵入城中村的日常；迷你街区之间是纵横编织的街巷，充满了不同色彩和质感，游荡其间便会撞见随处嵌入的微型影像展，就像与街道涂鸦的不期而遇。策展人为主题展定制了一系列特殊的空间地形，为参展项目制造限制条件的同时也鼓励展览即兴发挥和不断生长。于是展览成了一场不断添加、无休无止的实验，观众的观展体验也不再是一览无余，而是不断发现、遭遇惊喜和主动参与。

位于二层和三层的展览叙事呈现密集、重叠与多重讲述，首尾相接又变幻莫测。主题板块下设置有“城中村档案馆”“城中村演武场”和“城中村实验场”三个子板块，从城中村的知识库、改造工具手段和空间实践三个方面聚焦城中村。二层正中杜鹃的档案墙《世纪变革：深圳转型过程》全景式地呈现深圳城中村的正式和非正式的历史演变，由这里串联出的“城中村文献库”和“城中村观察站”，囊括了众

图 5 圣保罗贫民窟艺术实践，富兰克林・李、吉尔森・罗得里格斯
Figure 5 Paraisópolis Art Practice, Franklin Lee and Gilson Rodrigues

图 6 变迁——城中村，刘晓亮
Figure 6 Transformation - Urban Village, Liu Xiaoliang

ment agenda. The 2017 UABB confronted this challenge by directly participating in and exhibiting the reconstruction of the urban village in search for new opportunities of its harmonious integration within the city.

4.2 An Alternative Biennale

"The most important things are the insignificant ones happening in the street."

- George Brecht

The 7th UABB offered a platform which sought new theories, experiments and new ways of practice. With Nantou Old Town as the main site and other subsections of the exhibition scattered in five urban villages across four districts, *Cities, Grow in Difference* comprised a series of distinct and independent exhibitions that all reflected the same theme yet interacted with the city differently. Nantou Old Town, as the main venue, was a heterogeneous symbiosis presented as a juxtaposition of numerous overlapping biennales. With an overall guiding concept and the ability to adapt, the exhibition was capable of facing all the uncertainties that exist in the complex environment of the urban village. The main exhibition site was divided into three sections. The first, City/Village Intervention Practice Exhibition showcased the milestones which the Nantou Old Town transformation project achieved from 2016 through 2018. It was not only a renovation project, but also a cultural and communal revitalization plan that followed the lead of public space.[5] The second section of the exhibition included three subsections, "Global South," "Urban Village," and "Art Making City" while the third is a collection of all the smaller exhibitions scattered around the alleys and communities within the Nantou Old Town (Figure 3). Structurally, these three subsections were embedded into the main exhibition as the largest site of the 7th UABB, the Nantou Old Town/Urban Village which exists outside of the exhibition. To have people come into the urban village to experience and participate in the bustling and lively local environment was, in fact, the main purpose of the exhibition. The interaction between the urban village and the exhibition blurred the boundary between the exhibition site and the local scene, constituting numerous overlapping and interweaving narratives that synthesized and corroborated each other. Inside the exhibition venue existed the condensed history, the reflection of reality, and the imagination for the future of the urban village. The urban village which wrapped around the exhibition site became the main body of the exhibition while also being the main subject for discourse, the evidence, and even criticism on the concept of the exhibition itself. The 7th UABB attracted the attention and involvement of political decision-makers, professionals, media personalities and many other local residents and young people.

图 7 深圳案例，一个城中村的再生故事，史建、黑一烊
Figure 7 Shenzhen Case: the Regeneration of an Urban Village, Shi Jian, Hei Yiyang

图 8 城市印迹之城中村系列，谢卉、杨秋华
Figure 8 Urban Villages, a Marking from Shenzhen's Urbanization Effort, Xie Hui, Yang Qiuhua

图 9 一楼宇 · 动物园，尹烨敏、郑楚玲
Figure 9 One House Zoo, Yin Yemin, Zheng Chuling

多参展人用极其多样的媒介手段全方位呈现的对城中村演进的追踪、生存机制的记录，以及真实与虚构交织的未来想象（图 7- 图 9）。三层的展览将国内外学术机构、建筑师、艺术家、地产商、独立研究者和民间组织针对城中村未来的提案共冶一炉，成为一场众声喧哗的集体发声（图 10）。

本届深双在城市和建筑板块的基础上，第一次引入当代艺术板块，国际策展人和评论家侯瀚如策划的“艺术造城”板块为双年展增添了关注城市现实与未来更加多元的视角。“艺术造城”的“终极目的是在经典的建筑和城市之上，加上一个更有活力、更有创意、更开放的系统。在高速的经济和基础设施的发展之上，引进一个可以不断延续和发展的软架构。”[7] 事实上展场内艺术、建筑和城市之间剪不断理

It gave people who had never been inside the urban village a chance to closely examine this alternative urban lifestyle, even as the Biennale provided critical reflections on the city's urban problems.

4.3 The Multi-narrative Structure of the Exhibition

Located in the Factory Zone to the north of the site (Figure 4), the main exhibition venue was divided into three different sections, displayed south to north: "Global South," "Urban Village," and "Art Making City." Through the first and second levels of the main exhibition space, all three sections were spatially connected. With "Influence and Resistance" as a subtitle, the "Global South" section focused on discussing the extensive interplay between regionalism and globalized power, capital, and cultural impetus. It also examines how they can reconcile the consequences of "differentiated modernity" through self adaptation and self-retrofitting.[6]

The "Global South" section was vertically overlaid to present the most juxtaposed exhibitions. On the first level, the "'South' South" introduced the informal architectural history, present state, building typologies, and new urban strategies under the specific political and economic environment of Latin America (Figure 5). On the second level, an exhibition titled "A Drifting PRD " revealed the connection—PRD immigration history, Hakka and social culture migration—between Lingnan, Hakka, and Shenzhen by integrating art and history through an installation. On the third level, "The Other South" presented a comparative analysis of architectural typology as well as the geographical, cultural and spatial differences between China's "two Souths" Jiangnan and Lingnan. On the fourth level, the visitors returned to Nantou Old Town with a mapping workshop project, *Mapping Nantou Old Town* which was directed by Jason Ho. From the venue's first to fourth level, this vertical exhibition experience began with the joyful and creative atmosphere of large, Latin American cities moving toward the ease and absurdity of differentiated modernity in the PRD. The experience continued through the lightness and delicacy of Jiangnan to the complexity of contemporary Lingnan. Wang Yun's study of *World Settlements* aimed to show the inevitability of the urban village as it illustrates, from a historical perspective, an atlas of the formation mechanism of human community adaptation. This work in "Global South" spatially echoed the "Urban Village Archive" on the second level, enhancing the connection between the narratives of these two sections.

The "Urban Village" section contained a series of themed exhibitions, research presentations, on-site exhibitions which were scattered throughout the old town in a coherent overlapping manner to reveal the symbiotic feature of the exhibition. Subtitled "Hybridity and Coexistence," "Urban Village" was the main themed section

还乱的“挑逗性关系”[8]，甚至是试图互相消解和批判的对垒，使展览更为立体丰满。艺术家何岸的装置作品（图 11）试图解构建筑的日常理性——在两座厂房之间的空隙里建造的一片倾泻而下的墙体，原本功能性的管线、墙上的小便斗和广场边的路灯同时失去本来意义，几近捉弄和反抗建筑和城市通常所呈现的过度理性和僵化。林一林在拥挤的街道上的行为同样消解和打断了日常生活，正如策展人所言，艺术是非功利的，转瞬即逝但是足以供人们日后怀念。迥异于艺术家们处置现实的方法，建筑师们则是通过精心设计的空间装置介入城中村，如都市实践的瓮城（图 12）、张永和的信息亭（图 13）、NADAAA 的膜结构凉亭（图 14）以及犬吠工作室的超级烧烤巨构等都表达了建构本身的理性、节制和思辨，它们或承载记忆、挖掘历史，或关注空间修辞与城市戏剧场景的重塑。

图 10 湖贝生死进行时——城市更新迷宫的第三方公共参与策略，湖贝古村 120 城市公共计划策展小组
Figure 10 Life and Death of Hubei - the labyrinth of urban renewal, Hubei 120 City Public Plan Team

图 11 是永远不是，何岸
Figure 11 It is Never, He An

of the Biennale. The exhibition narrative used the urban village, a unique urban phenomenon in Shenzhen and throughout the PRD region, as a case to conduct historical analysis, on-site observations, practical interventions and future imaginations in relation to the theme of the Biennale, "Cities, Grow in Difference." The theme venue of "Urban Village" offered a series of semi-closed exhibition spaces scattered on the first level of the building, including a story den, a game room, a photo lab, a drawing studio, a videography studio, a music studio, a laboratory, a future gallery, and other everyday spaces occupied by ordinary people. The content of these rooms derived from observations and studies conducted in the urban village, while the exhibition was purposefully arranged to attract attention. The two criss-crossed alleyways simulated the experience of walking down an urban street, and the large-scale photography work embedded on the walls and the neon light installations hidden within the stairwell (Figure 6) functioned as commercial billboards that had seemingly invaded the daily affairs of the urban village. The miniature streets and blocks were full of colors and textures and the space was designed so that people would encounter the randomly placed micro photography exhibitions as they strolled around the village just as they would see graffiti in the city. The curator tailored a unique spatial terrain for the themed exhibition, setting restrictions while encouraging improvisation and growth over time. Thus, the exhibition became an endless experiment that would keep multiplying, while and the audience's experience was no longer merely that of being a spectator, but also that of being an active participant who discovered the city and encountered surprises.

The exhibitions located on the second and third levels were narrated in a dense and overlapping manner, echoing each other while showcasing unpredictable changes in their content. There were three subsections under the umbrella of the main theme,"Urban Village Archive," "Urban Village Armory" and the "Urban Village Laboratory." At the center of the second level, Juan Du's *Massive Change: Centuries of Shenzhen's Transformation* unfolded panoramically, displaying the informal and formal historical transition of the Shenzhen urban village. "Urban Village Document Room" and "Urban Village Observatory" radiated from there, including the tracking of the urban village's evolution, documentation of its survival mechanism, an imagination of its future interlaced with fiction and reality as they were illustrated in various mediums and methods by the exhibition participants (Figure 7-9). On the third level, the exhibition brought together different proposals for the future of the urban village, these proposals were offered by domestic and international academic institutions, architects, artists, real-estate developers, independent researchers and civil society organizations, into one collective voice (Figure 10).

图 12 瓮城，都市实践
Figure 12 Barbican, URBANUS

5 后策展时间：一项长期计划的开始

5.1 隐形的双年展

“城市共生”是一个“进行时”的展览。展前策展团队发起了绘本城中村、城中村影像馆、城中村材料库、城中村图书馆、城中村信息中心等一系列公开征集与合作计划；展览期间由策展人委托平面设计师参与的“城中村布告栏”、“牛皮癣对话框”、小海报、招贴画不断涌现。随着时间的推移，展览渐渐演化成一座人们游走其间的多重迷宫，它模糊了现实与乌托邦，它是工厂、实验室、工作坊、图书室、画廊、游乐园，也是探向未来的窗口。更重要的是展前六个月策展人发起了一系列“共生实验室”(图15)，以期开放策展过程、持续城村共生的学术探讨。“城中村故事会”增强了当地居民的参与分享；在古城中心广场设置的“南头议事厅”试图进一步打通展览与生活现实的界面，提供一处城市决策者、专家与当地居民共同参与古城保护与更新的议事空间与公共平台。

“城中村档案馆”是在展览之前已先行启动的一项长期研究计划。策展人通过委托专项研究、合作计划及公开征集，对国内外近十几年来与城中村相关的研究进行海量收集与整理，形成城中村历史与现实生态的庞大数据库。它同时也是不同参展人以多元视角和丰富的视觉语言呈现的展览现场。这里汇集了个体经验、观察模型到大数据收集，理性逻辑的分析研究与众生嘈杂的个性解读喧嚣并置。它还是一处活跃的知识生产空间，展期内不断与观众和当地居民互动，展后也希望为未来城中村的相关研究留下一份珍贵的档案。[9]

5.2 自发的双年展

第七届深双是注重“发现”而非预先设定的交流平台。在清晰的主题策展结构框架下，在深双历史上首次公开征集部分参展人，极大调动了年轻建筑师、研究者和艺术家的参与热情，由于大量应征者缺乏大型国际展览的参展经验，

The 7th UABB introduced a contemporary art section for the first time. The "Art Making City" section by international curator and critic, Hou Hanru added a more diverse perspective on urban reality and the city's future to the Biennale. The ultimate goal of "Art Making City" was to "append on top of classic architecture and the city a system of greater vitality, creativity, and openness. Above and beyond breakneck developments in the economy and infrastructure, a continually extensible and developing 'soft structure' (*ruanjiagou*) can be ushered in."[7] Within the exhibition site, a complex, entangled "teasing relationship" among art, architecture and the city existed.[8] Their attempt to reconcile with while criticizing each other made the exhibition more three-dimensional and all-embracing. The installation artwork by He An was an attempt to deconstruct the mundane rationality of architecture (Figure 11). The pouring wall built between the gap of two factory buildings displaced the original functional pipelines, urinals and the street lights on site to tease and rebel against the overly rational and rigid presence of extant architecture in the city. Lin Yilin's performance on the crowded streets similarly cleared up and interrupted people's daily life. As stated by the curator, art is non-utilitarian, it is transient yet suffices to create a sense of nostalgia for its audience. Unlike the ways artists have dealt with reality, architects implanted meticulously designed spatial installations into the urban village. URBANUS' Barbican (Figure 12), Yung Ho Chang's Information Pavilion (Figure 13), NADAAA's membrane structure Pavilion (Figure 14), and Atelier's Bow-Wow's Fire Foodies Club expressed the rationality, control and speculations of architecture, performing diverse functions, including the transmission of memory, the excavation history, the display of the fluidity of space, or the reconstruction the dramatic urban scene.

5 Post-curatorial Time: The Beginning of a Long-term Plan

5.1 An "Invisible" Biennale

Cities, Grow in Difference has been an ongoing exhibition. The curatorial team initiated a series of open calls for exhibits and collaboration prior to the opening, including the "Urban Village Picture Book," "Urban Village Photo and Video Room," "Urban Village Databank," "Urban Village Library," and "Urban Village Info Center." During the exhibition, the curators involved graphic designers in the project to design an urban village bulletin board, a Post-it Whole notice board, small posters, and advertisement posters. Gradually, the exhibition evolved into a multi-layered maze filled with visitors wandering

也极大增加了策展团队的工作量和展览呈现的不确定性。难得的是经过三次公开征集和与策展团队多次交流，城市和建筑板块有大约 1/4 参展作品出自这些年轻的团队，他们作品的最终呈现也格外出色。更令人欣喜的是由于本届深双的主题和选址的特殊性激发了社会各界的参与热情，公开征集、自发组织的微型设计介入和大量活动也吸引了本地居民的广泛参与，双年展真正实践了一次自发的自下而上的“自我组织”和“自上而下”的统一策划的共生，以此形成对其自身机制的批判和变革。

5.3 未完成的双年展

这届深双是一个不断生长的双年展也是一个没有完成的双年展。原本设想“为城中村的未来献上一条计策”是线上展览和展览现场的互动计划，而未实现的《城中村之声》小报则可以称为对非正式的媒体小报的回归。“城中村群艺馆”原本是深双的特别活动策划，强调居民参与，将一系列拍摄、放映计划和戏剧活动嵌入城中村的日常生活之中，通过一系列的公共活动回馈南头居民。这些舞蹈、戏剧展演、电影拍摄与露天电影放映计划原本是本届深双极其重要的展映单元，最终因故取消无疑是巨大的遗憾。

“城市共生”成就了一场遍布全城的展览。相对于集锦式的主题展，古城中散落的历史建筑、民居、街道、空地，甚至烂尾楼中间也嵌入了与当地居民互动的作品和活动现场。以展

图 13 信息亭，张永和
Figure 13 Information Pavilion (Brick Pabilion), Yung Ho Chang

图 14 柱廊华盖：南头城中村意象，纳迪尔 · 特拉尼（NADAAA）
Figure 14 Zhulang Huagai (Colonnade with Fantastic Canopy): A Figure for the Nantou Urban Village, Nader Tehrani (NADAAA)

around. It blurred the boundary between reality and utopia as it is a factory, a lab, a workshop, a library, a gallery, a playground, a theme park, and a window to the future. More importantly, the curators launched a series of CGD Lab (Cities, Grow in Difference Lab) (Figure 15) six months prior to the exhibition, opening up the curatorial process to the public and continuing the academic debate on the urban village's coexistence with the city. The "Urban Village Stories" enhanced the participation and sharing amongst local villagers, and the "Nantou Forum" set up in the center of the Old Town furthered the curators' attempt to unify the exhibition and reality, providing a public platform to allow decision-makers, experts, and local residents to work together for the preservation and revitalization of the Old Town.

On the one hand, the "Urban Village Archive" is a long-term research project which was launched before the exhibition. Through commissioning research programs, cooperation plans, and accepting open calls, the curators have compiled a comprehensive collection of studies related to both domestic and foreign urban villages, forming a massive database on the formation and current condition of the urban village. Simultaneously, the "Urban Village Archive" is also an exhibition that has brought out, using various visual languages, the different perspectives of its participants. It has put together individual experiences, observation models and a massive database that has presented a juxtaposition of logical analysis and the clamoring subjective interpretations of all participants. The archive also became an active production space of knowledge as visitors who came to the exhibition interacted with local residents. These valuable interactions were then recorded and archived for future studies on the urban village.[9]

5.2 A Spontaneous Biennale

The 7th UABB drew attention to discovery rather a preconceived communication platform. Under the framework of a clear curatorial theme, UABB, for the first time, adopted an open-call strategy for some of the exhibition participators. This strategy stimulated the enthusiasm of young architects, academic researchers and artists to participate. As many participants lacked the experience of partaking in large-scale international exhibitions, this drastically increased both the workload of the curatorial team and uncertainty about the quality of the exhibition. Fortunately, through three open-calls and countless back and forth communications provided by the curatorial team, about a quarter of the great works shown in the Urbanism and Architecture section were from these young teams. As a result of the theme and the particularity of the chosen site, significant public responses from all sectors of the community emerged in interaction with the Biennale. These self-organized micro design interventions and activities also attracted the participation of

图 15 共生实验室，2017
Figure 15 Cities, Grow in Difference Lab, 2017

览为契机实际介入城中村改造、提升公共空间品质，以此回应“双年展可以为城中村做些什么而不是城中村为双年展做些什么”的问题。最后，一届在城中村举办的双年展注定是与众不同的，“众生共荣”的城中村是年轻移民的落脚社区在当下城市中的最后堡垒。保留城中村不仅在于保留城市档案和城市记忆，也为筹划一个更加健康的城市未来。这个未完成的双年展“隐含了一种劝谕，保留城中村，并由这里出发审视、反思和重建一种杂糅与共生的城市多元主义体系”[10]。

6 结语

2017 深双无疑是一场多面向、另类和不拘一格的展览，但更是一种文化与社会的观念陈述、阐释和宣言。展览是一个社会平台来构建或重建作品同社会语境之间的关联，提供多重可能性，让奇观主导下的现实世界得以发声，展览成为一

落脚城市

local residents. This exhibition truly experimented with the symbiosis between top-down rational planning and bottom-up autonomous growth in order to criticize and reform the systematic approach of the Biennale.

5.3 An "Unfinished" Biennale

The 7th UABB was an unfinished and growing exhibition. "An idea for the future of urban village" could have provided interaction between the online and the on-site exhibitions, while the unrealized *Voice of Urban Villages* was intended to revive the unofficial tabloid newspaper. Similarly, the "Urban Village Club" was specially planned for the UABB which would have called for the participation of all local residents. In turn, the documentation of the activities such as filming, screening, and drama performances held by the club were to have been reinserted into the daily life of the urban village as a measure to give back to the local residents. These dance performances, drama festivals, filming events and drive-in theatre setups were meant to be a very important part of the Biennale. Their cancellation due to management issues was undoubtedly a shame.

Cities: Grow in Difference achieved popularity all over the

个替代的、抵抗型的空间。在这里，合法的与被边缘化的、无法获得表达的空间与社会群体得以展示自身、共存共生。[11] 正如策展人小汉斯·奥布里斯特所声称的，“展览不是解决方案，不是一个有着大团圆结局的制作，而是一个出发点。它是起点、转折和探险。”[12] 随着双年展的结束，对城中村未来的讨论与空间博弈正进入一个全新的阶段，一项以南头改造与重生计划起始的城村共生实验也刚刚开始。

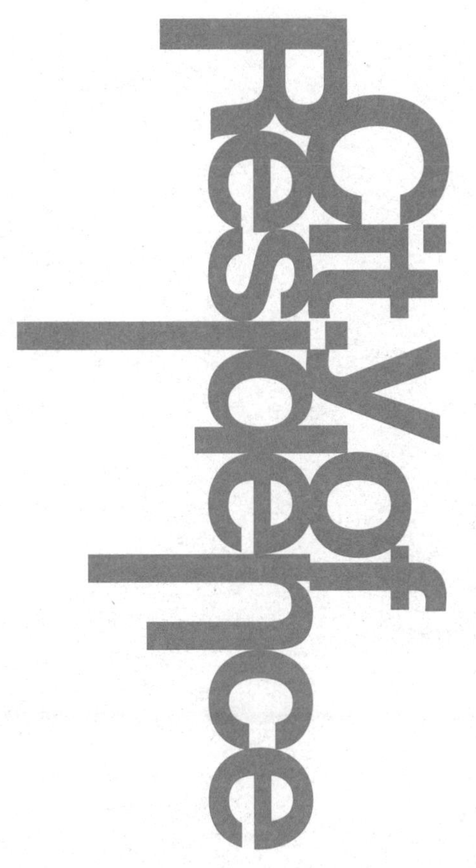

city. Different from a synopsized themed exhibition, historical buildings, residential areas, streets, open spaces, and even unfinished buildings scattered throughout the Old Town were turned into venues to house interactive works and events. Utilizing the exhibition as an opportunity, the curators intervened in the renovation of the urban village to help improve the quality of its public space. This gesture responded to the question "what should the Biennale do for the urban village?" rather than answering the question, "what can the urban village do for the Biennale?" Last but not least, a biennale hosted in an urban village was destined to be different from other exhibitions. The dynamic and vibrant environment of the urban village is the last haven for young migrants who have just arrived in the city. Retaining the urban village form is necessary not only to preserve the history and memories of the city, but also to plan for a better and healthier urban future. This "unfinished" exhibition also "implies an advocation to preserve the Urban Village and to start reviewing, reflecting on and rebuilding a hybridized and symbiotic urban system."[10]

6 Epilogue

The 7th UABB was undoubtedly a diverse, distinctive, and innovative exhibition. It manifested cultural and social concepts, interpretations and declarations. An exhibition is a social platform that constructs or rebuilds the connection between displayed works and their social context, providing infinite possibilities while revealing the reality led by wonders and spectacles. The exhibition can be an alternative spatial resistance which allows both prominent and marginalized spaces and social groups to emerge and coexist.[11] As stated by Hans Ulrich Obrist, "An exhibition [is] not a solution, a production with a happy ending, but a point of departure. It's a beginning, a shift, an adventure."[12] As the curtain fell on the 2017 UABB, the discourse about the urban village's future and spatial rearrangement was only marching into a new phase. A coexistence experiment, which originated from the Nantou Old Town Renovation and Regeneration project, was just beginning.

[1] BERGDOLL B. Out of Site/In Plain View: On the Origins and Actuality of the Architecture Exhibition [M]// PELKONEN E-L, CHAN C, TASMAN D A, eds. Exhibiting Architecture: A Paradox?. New Haven: Yale School of Architecture, 2015: 14.
[2] PELKONEN E-L. Mining the Paradox [M]// PELKONEN E-L, CHAN C, TASMAN D A, eds. Exhibiting Architecture: A Paradox?. New Haven: Yale School of Architecture, 2015: 9.
[3] 费瑟斯通 . 消费文化与后现代主义 [M]. 刘精明 , 译 . 南京 : 译林出版社 , 2000: 209.
[4] 侯瀚如 , 刘晓都 , 孟岩 . 城市共生：从城中村开始⋯⋯[R]// 城市共生 Cities, Grow in Difference. 城市 + 建筑分册 . 深圳城市 \ 建筑双年展组织委员会 , 2017.
[5] 孟岩 , 林怡琳 , 饶恩晨 . 村 / 城重生——城市共生下的南头实践 [J]. 时代建筑 , 2018(5): 58-64.
[6] 刘晓都 . 世界南方：影响与抵抗 [R]// 城市共生 Cities, Grow in Difference. 城市 + 建筑分册 . 深圳城市 \ 建筑双年展组织委员会，2017: 109.
[7] 侯瀚如 , 杨勇 . 艺术造城：现场 [R]// 城市共生 Cities, Grow in Difference. 深圳城市 \ 建筑双年展组织委员会 , 2017.
[8] PHILLIPS A. Exhibitions Matter [M]// PELKONEN E-L, CHAN C, TASMAN D A, eds. Exhibiting Architecture: A Paradox?. New Haven: Yale School of Architecture, 2015: 196.
[9] 孟岩 . 都市村庄：杂糅与共生 [R]// 城市共生 Cities, Grow in Difference. 城市 + 建筑分册 . 深圳城市 \ 建筑双年展组织委员会，2017: 178.
[10] 朱大可 . 村史、城史、国史：一部田野调查报告的诞生 [M]// 南方都市报 , 编著 . 未来没有城中村：一座先锋城市的拆迁造富神话 . 北京 : 中国民主法制出版社 , 2011: 7.
[11] 侯瀚如 . 日常奇观 [M]// 侯瀚如 , 奥布里斯特 . 策展的挑战：侯瀚如与奥布里斯特的通信 . 顾灵 , 译 . 北京 : 金城出版社 , 2013: 112.
[12] 奥布里斯特 . 平行现实 [M]// 侯瀚如 , 奥布里斯特 . 策展的挑战：侯瀚如与奥布里斯特的通信 . 顾灵 , 译 . 北京 : 金城出版社 , 2013: 92.

南头古城，张超摄于 2017 ©UABB
Nantou Old Town, photograph by Zhang Chao, 2017 ©UABB

宣言
MANIFE-
STO

城市共生：从城中村开始……
Cities, Grow in Difference: Starting from Urban Village...

侯瀚如、刘晓都、孟岩（按姓氏拼音排序）
2017 深港城市 \ 建筑双城双年展（深圳）总策展人
由孟岩起草

Hou Hanru, Liu Xiaodu, and Meng Yan (in alphabetic order)
Chief curators of 2017 Bi-City Biennale of Urbanism\Architecture (Shenzhen)
Drafted by Meng Yan

2017 年深港城市 \ 建筑双城双年展（深圳）以"城市共生"作为展览主题。"城市共生"问题的提出不仅是我们对于当今世界和中国城市化现实的批判性解读，也是提出另一种未来城市图景的尝试。

我们正生活在一个危机四伏且充满不确定性的世界之中：经济发展不平衡、文化冲突、价值相左、矛盾丛生；与此同时，全球主义、消费主义和媒体霸权又主导着既有世界秩序、经济格局以及人的生活方式，并不断将人们的思维和行为方式重新格式化。回望中国，当代的城市化进程在权力和资本裹挟下经历了三十多年的高速发展。在原有苏俄式现代主义和市场主导的彻底功利主义双重规划模式驱动下，我们生活的城市概无例外地变得趋同和单一：这不仅可见于京沪粤深等一线城市，而且在二、三线城市甚至乡镇也越演越烈。近年来旨在提高生活品质的"城市更新"往往更进一步地清除层积丰富的历史街区和多样杂糅的城市生活，代之以全球化、商业化的标准配置。城市的士绅化正在光鲜的表面下制造着社会分化、生活乏味的城市病症。面对这样的现实，我们呼唤一种多元"共生"的城市模式。

我们认为应该自觉反抗单一和理想化的未来图景，因为城市本身是一个高度复杂的生态系统。今日之城市应该是多元价值体系平衡的结果，是人们身处同一世界、心存不同梦想的高度异质性和差异化的文明共同体。城市本应无所不包、和而不同，它的生存与繁荣在于能最大限度地对"差异性""另类""他者"的包容和文化认同。我们以"城市，因异而生"的直接表述作为"城市共生"的英译，旨在强调城市这一复杂系统之中文化、社会、空间和日常生活的多层面共生：它们是多重身份和多重视角的叠加，包含对社会多元价值和生活方式的认同，也包含对空间和时间的多样性与不确定性的认同，并且以“混杂与共生”强调我们的城市观念应走向多元、差异、杂糅和抵抗。

“城市共生”即是从根源上对文化层面、社会层面、空间层面不同起源、不同状态、不同价值观的认同与包容，是对主流文化中心主义的反叛。我们必须尊重城市生态系统的平衡，如同自然界物种之共生与动态平衡，城市中的矛盾混杂不应被肆意打破而失衡，而对于它者的尊重则是对城市包容性的考验：它是反纯粹的，杂糅的，也是对正统现代主义政治－空间图景及其美学的抵抗，对单一“进步”大历史观的修正；它是对城市人想象力、自由的心灵和创造激情的承载。

创意和想象力需要在城市中不断寻找新的栖居地，这一次它游走到本届展览的主展场：南头古城。南头古城的辖区自晋代以来就包括今天的香港、澳门、东莞、珠海等广大地区。1840 年鸦片战争之后，香港被从新安县割去。近百年间古城不断消退而村庄不断膨胀，随着深圳城市化的加剧，最终形成城市包围村庄，而村庄又包括古城的城 / 村环环相扣，古城时隐时现的复杂格局——“城中村，村中城”。南头古城是当代城中村与历史古城的高度融合，它既拥有超过 1700 年的古城遗址，也有在城市化进程中不断出现的自发性和另类性的当代空间。作为深双主展场，它全光谱式地展示着从近代到当下城村演变丰富的空间样本：中西共生，古今共荣。作为全球唯一以“城市 \ 建筑”为固定主题，以关注城市和城市化为使命的双年展，深双关注中国当下与每一个人休戚相关的城市议题。与世界上其他双年展不同的是：深双不仅是展览现场，也同时身处 20 世纪和 21 世纪最剧烈城市化的现场。从深圳到珠三角，城市本身既是最大的展场，同时又是事件的发生地。而深双恰恰是不断应对当下最紧迫城市问题的交流平台，同时也是城市建筑和日常生活实质性改善的试验场。

城中村作为当代城市另类模式，以特殊的方式体现着城市长期演变的未完成状态。它自身在被外力逼迫下自发形成并持续演进。自我繁殖和自我更新是它的立命之本。它既是深圳城市化向内深耕的最新前沿，同时也是触及城市平衡发展的最后底线。作为经济特区城市，深圳正处于“后城中村”的时代，也在经历二次城市化的浪潮，不断加大的空间密度亦在拷问城中村的生存与未来，深双的介入恰逢其时。在加速发展的城市化进程中，城市既有自上而下的城市规划，也有自下而上的自发推动力，“城中村”恰恰位于两种力量之间。“城中村”是中国市场经济时期的快速城市化与计划经济时期遗留的城乡二元结构相碰撞的矛盾产物，现今自发地成为爆炸式增长的城市新移民的“落脚城市”。城中村面积约占深圳总面积的六分之一，在深圳两千多万人口里有约 900 万人住在城中村，即城中村以 16.7% 的空间容纳了深圳 45% 的城市人口。作为一种自发形成并相对独立自治的“飞地”，城中村的空间与社会价值在于其包容性和多元性。相比于今日大量在高速都市化和全球化条件下被设计出来的布景式城市，城中村中密度极高的物理空间形态出奇生动，显示着人类营造自己家园中的人性魅力。它处于传统和现代之间、清晰与混沌之间、合法与非法之间，在非黑即白的价值判断体系之外，城中村的意义恰恰在于因其所处的灰色地带而被保育和发展出蓬勃的、自下而上的自发潜力。由此，城中村可以成为另类城市新生活的培育基地，可以成为新来者的落脚城市，可以以其永不落幕的未完待续状态使得持续创造成为可能。

在"城市共生"的主题之下，本届深双是寻找新理论、实验和行动方式的平台，是拒绝预设、强调自发的交流空间。这是一个不断生长的双年展。本届深双作为城市介入的手段与古城更新计划合体并进，展览将遍布南头古城的大街小巷、绿地广场、住宅和厂房之中。它有时仅仅是轻轻的城市介入，让展览植入日常生活，鼓励漫游式的发现和惊喜，路边不经意走过时的一瞥却别有深意。展览在此呼应乔治·布雷希特所声称："最重要的事件正是那些发生在街道上的微不足道的小事情。""城市共生"也是一系列各不相同、各自独立的展览，它们在城市之中处处开花、全城互动。展览就像城市本身一样结构复杂，它既有整体构想又可以随机应变，充分应对展场位于城中村的种种不确定性。它还是一次自发的自下而上，自我组织的草根想象和自上而下的官方规划的共生，借以形成对双年展自身机制的批判，体现对不同的认同。

在以往"城市\建筑"领域的基础之上，本届深双第一次明确把艺术作为一个板块凝聚进来，这是更加全面理解城市发展的一个很重要的变量。它构筑出一个艺术家与规划师、建筑师、设计师跨界合作的图景，进而在城中村的场地空间之中，以非常规的手段和对城市文化的想象为展览注入新的活力，促进新的从内到外的社会关系形成。艺术计划以种种令人意想不到的方式出现在历史建筑、工业厂房、街市摊档，甚至普通的出租屋中。这种结合城中村特殊空间、以日常生活主题为切入点的介入实验能够唤起人们对城中村、对城市与公共空间的重新思考，在留下视觉记录的同时，为社区营造、邻里互动、专业人士与市民的对话和交流产生长远的影响。最终，艺术的介入在城中村高密度的城市形态中会创造出新的公共领域，或者说，艺术也是一种不可或缺的造城力量。

在"城市共生"的总主题之下，展览从内容结构上分作三个有机联系的板块："世界|南方""都市|村庄"和"艺术造城"。"世界|南方"板块为展览主题"城市共生"提供了进行讨论的背景、视野和立场，即本届深双立足于以深圳为代表的珠三角地区，从全球经验的视野来探讨超越东西对立的过时模式；"南方"为我们观察和探讨城市发展提供了更符合当代世界现实和未来走向的新范畴。"都市|村庄"板块作为城中村研究及与此问题相关展览内容的主板块，从城中村的知识库、改造工具手段和介入实践的实验场三个方面向城中村学习，在打破固有的城乡区别的偏见同时，激发居民参与，共同创造城中村未来的可能图景。展览由此融入城中村的日常生活之中，建构展览与城中村生活的嵌入式互动关系，激发城中村的潜在活力。"艺术造城"板块在展示当代世界各地艺术家对街道、家居、公共空间以及社会生产问题的种种批判性探索同时，引入一系列"艺术介入城市"的直接行动，利用当代艺术的活力和创造性的表达介入城市生活，激发多元的试验，尤其强调与当地社区居民的互动合作，由此衍生更有活力和创意、更开放、民主的城市生活系统。

我们希望在展后为深圳城中村的发展留下一份丰厚的城市档案和范例，同时也把城市问题的讨论引向更广的范围。本届深双既是首次进入城中村这一特殊的城市地区，也是以城中村作为起点，反观今日的城市，更是对未来城市策略的讨论。我们并不认为本届双年展将主展场选定为城中村，它就只属于城中村的双年展。就展览的叙事策略而言，城中村只是我们探讨"城市共生"的起点。它以一个社会平台构建或重建作品与社会语境之间的关联，提供多重可能性。同时，展览本身形成一个针对任何规划强权的抵抗和替代空间：合法的与被边缘化的、无法获得表达的空间与社会群体得以展示自身，通过展览及其引发的城市改造尝试共存共生。由此，2017 年第七届深港城市 \ 建筑双城双年展（深圳）不仅是一个展览，同时也是一场介入计划，一次直接的造城行动。

As the theme of the 2017 Bi-City Biennale of Urbanism\Architecture (Shenzhen), "Cities, Grow in Difference" represents not only a critical interpretation of the current urbanization in today's China and the world but also an attempt to envision alternative models of future cities.

Today, we are living in a world of turbulence and uncertainties, such as unbalanced economic developments, cultural conflicts, divided views, and illogicalities. Meanwhile, globalism, consumerism, and media hegemony are dominating the existing world order, economic system, and lifestyles, constantly reformatting people's ways of thinking and acting. Prompted by power and capital, urbanization in China has gone through rapid booms for over 30 years. Driven by the combination of two main planning models, namely the former Soviet-style modernism and the market-oriented total utilitarianism, the cities we live in tend to be homogeneous and generic. This is not only the case of first-tier cities like Beijing, Shanghai, Guangzhou, and Shenzhen but also of second and third-tier cities, even towns and villages. In recent years, urban renewal, pretending to improve quality of life, has further swept away the time-honored historical areas and a rich urban life based on diversity and hybridity. They are now replaced by a globalized and commercialized standard configuration. Underneath the flashy gentrification, social polarization and lusterless life rear their ugly heads. Faced with this reality, we call for a new concept of urban diversity to promote "coexistence" of differences, and growth through difference.

We are convinced that we should consciously resist the promise of a mono-culture and idealistic vision of the future city. The city is a highly complex ecosystem. Today's city should be a manifestation of a balanced coexistence of different value systems. It should be a civilized community with maximum heterogeneity and diversity in which people coexist in one domain with different dreams. A city should be all-embracing, and seek harmony through negotiating with differences. Its survival and prosperity lie exactly in being more inclusive and tolerant toward "difference," "alternativeness" and "otherness." This makes up a real urban cultural identity. *Cities, Grow in Difference* emphasizes coexistence at various levels of culture, society, space and daily life in the complex system of a city: it embraces multiple identities and perspectives, and acknowledges the diverse values and lifestyles in society as well as the variety and uncertainty of space and time. Like a "jungle grown out of difference," it stresses the importance of diversity, differences, hybridity, and resistance in our conception of a city.

Cities, Grow in Difference fundamentally signifies a recognition and inclusion of things of different origins, status and values at social, cultural and spatial levels. It is a revolt against the mainstream culture ruled by "centralism." We must respect the balance of the urban ecosystem like we should do with the delicate balance of nature, in which the ecological system exists in a dynamic equilibrium. On one hand, the balance of contradiction and hybridity in the city shouldn't be broken arbitrarily. On the other hand, respecting otherness is a test of the degree of tolerance of a city. Opposing purity and advocating hybridity are resistance to the orthodox modern political-spatial vision and its aesthetics, it is a revision of the single-minded "progressive" view of history. The hybrid urban ecosystem is where the imagination, free minds, and passion for the creation of the city dwell and reside.

Creativity and imagination always need to be relocated and renewed across the city. This time they find their new habitat in the main exhibition venue, Nantou Old Town. Since the Jin Dynasty, a vast area consisting of today's Hong Kong, Macao, Dongguan, and Zhuhai had been under the administration of Nantou Old Town, until Hong Kong was separated from Xin'an County after the end of the Opium War of 1840. In the past century, the ancient town has gradually vanished while the village was constantly expanding. The exacerbation of urbanization in Shenzhen has resulted in an intertwined layering of the historical town embedded in the urbanized village, which is again encircled by the city. Nantou Old Town is a combination of a modern urbanized village and an ancient town. It has both the ruins of a 1,700-year-old ancient city and the constantly growing, spontaneous and alternative space during the course of urbanization in modern times. As the main exhibition venue, it displays in a full spectrum a complete set of spatial evidence of the urban village transformation in recent history. It fully embraces its Chinese origin and Western influence, as well as the hybrid coexistence of the past and the present. As the one and only biennale of urbanism\architecture in the world, and with the mission to focus on cities and urbanization, UABB devotes attention to the discussion on urban issues which concern every individual now in China. Different from other biennales, UABB is not only an exhibition but also the site witnessing the most drastic urbanization during the twentieth and twenty-first centuries. From Shenzhen to the Pearl River Delta, cities in themselves are the biggest exhibition venues that accommodate the transformations. UABB is a platform for continuous discussion on the most pressing urban issues, and a laboratory for actual improvement of urban architecture and daily life.

Urban village, an alternative model of the contemporary city, manifests the ongoing status of the evolution of a city in a unique way. Under the pressure of external forces, it is forming spontaneously and evolving continuously, steered by its own self-reproduction and self-regeneration. Urban village is the last frontier of Shenzhen's urban renewal campaign, and also the bottom line of a balanced urban development. As an economic powerhouse, Shenzhen is currently undergoing the "Post-Urban-Village" era and going through a second urbanization. The ever-increasing spatial density of the city raises concerns about the survival and future of the urban villages, and UABB is born to enter the discourse. In the course of urbanization, the development of a city is propelled by both top-down urban planning and bottom-up growth. Under the influence of the two forces, the urban villages emerged as a result of the collision between the rapid urbanization in the current dynamism blending market economy and the urban-rural dual structure inherited from the planned economy period in China. They have now become the "Arrival City" where the exponentially growing number of new immigrants to a city reside. The urban villages covering about one-sixth of the total land area of Shenzhen house approximately nine million of the over twenty million people in Shenzhen. In other words, they accommodate 45% of the population with 16.7% of the space. As an extra-territory spontaneously formed and self-governed in a sense, the urban villages are inclusive and diverse in terms of space and community. Compared with a large number of cities designed like "stage-sets" in the context of fast urbanization and globalization, the urban villages are surprisingly dynamic due to the high-density physical space, manifesting the charm and creativity of human beings building their own homes. Residing in between past and present, order and chaos, legal and illegal statuses, and outside of the all-or-none system of value judgment, the urban villages are valuable for the bottom-up spontaneous potential preserved and developed from the gray zones where they are located. That's why the urban villages become the incubator of alternative new life in a city, and a place for newcomers to stay, nurturing indigenous creativities with their ever-lasting "incompleteness."

Themed on *Cities, Grow in Difference*, 2017 UABB presents a platform to seek new theories, experiments and ways of action and offers a space for spontaneous exchanges without presumption. This is a constantly growing exhibition. 2017 UABB will be implemented concurrently with the old town regeneration plan. The exhibitions will be spread all over the lanes and alleys, parks and squares, residences and plants in Nantou Old Town. Sometimes, they are just subtle urban interventions, which infiltrate into our daily life and encourage unexpected encounters and discoveries. A glimpse during a walk, a moment, or an encounter may bring about profound understanding. It is in this way the exhibition responds to what George Brecht once said, "The most important things are the insignificant ones happening in the street." *Cities, Grow in Difference* consists of a series of distinct and independent exhibitions scattering over the city and echoing with each other. The exhibition structure will be as complex as the city itself. With an overall guiding concept and the ability to adapt, it should be able to fully cope with all the uncertainties in the urban villages where it is held. Moreover, it is also a coexistence of a spontaneous bottom-up "self-organization" and top-down official planning, which serves as a critique of the biennale's own mechanism and shows recognition of the juxtaposing differences.

Based on the previous fields of urbanism\architecture, 2017 UABB for the first time integrates an entire art section, making art as a key variable for a full picture of urban development. It will set up a platform where artists work alongside with planners, architects, and designers. With unconventional means and imagination reflecting the urban culture, it will inject new vitality into the exhibitions in the urban villages to facilitate the formation of new social relationships among the inhabitants as well as with the outside world. Art projects will be introduced to historical buildings, industrial plants, market stalls, and even ordinary rental apartments in unexpected ways. Such interventions and experiments that integrate the specific spaces in urban villages and are inspired by the daily life of ordinary people will evoke our reflections on the significance of urban villages, the city, and public spaces. While leaving visual testimonies for UABB, they will also exert far-reaching influences on community building, neighborhood interaction, as well as dialogues and communications between professionals and the public. Eventually, the art interventions will create new public realms in the high-density urban villages. In other words, art is also an indispensable force in city-making.

Under the theme of "Cities, Grow in Difference," three interconnected sections in exhibition narrative are presented, namely "Global South," "Urban Village" and "Art: Making City." The first section "Global South" sets out the background, horizon, and stance for the discourse. Specifically, based on PRD region represented by Shenzhen, 2017 UABB will debate the obsolete east-west model through the perspective of global experiences. The "South" defines a new position for us to observe and explore urban development, which better interprets the world's current reality and its future. As the themed exhibition, "Urban Village" section focuses on the research on urban villages and related issues it draws on experiences of

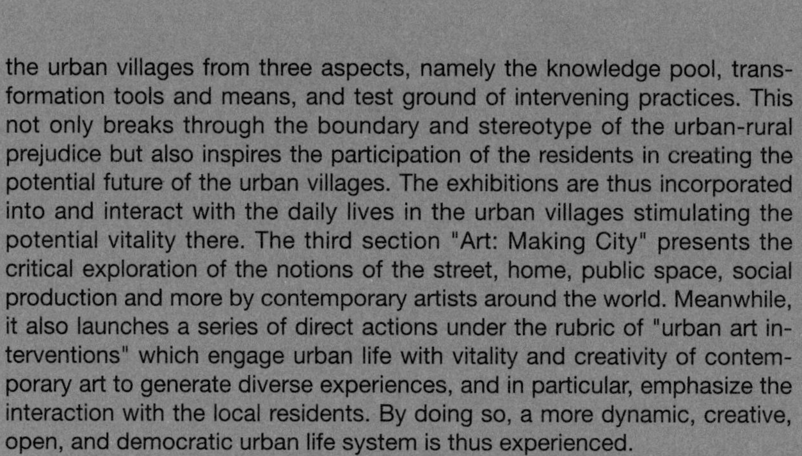

the urban villages from three aspects, namely the knowledge pool, transformation tools and means, and test ground of intervening practices. This not only breaks through the boundary and stereotype of the urban-rural prejudice but also inspires the participation of the residents in creating the potential future of the urban villages. The exhibitions are thus incorporated into and interact with the daily lives in the urban villages stimulating the potential vitality there. The third section "Art: Making City" presents the critical exploration of the notions of the street, home, public space, social production and more by contemporary artists around the world. Meanwhile, it also launches a series of direct actions under the rubric of "urban art interventions" which engage urban life with vitality and creativity of contemporary art to generate diverse experiences, and in particular, emphasize the interaction with the local residents. By doing so, a more dynamic, creative, open, and democratic urban life system is thus experienced.

After 2017 UABB, we hope to leave behind a comprehensive urban archive and cases for the future development of urban villages in Shenzhen, meanwhile carrying the discussion further on broader urban issues. UABB for the first time takes place in an urban village, a special type of urban area, from where we can reflect on today's city, and more importantly, discuss future urban strategies. Yet, a UABB with its main venue in an urban village doesn't mean the exhibition is only about the urban villages. In terms of the narrative strategy of the exhibition, the urban village is just where we start to explore *Cities Grow in Difference*. 2017 UABB is a social platform to build and/or rebuild the connection between the exhibition and its social context, stimulating diverse possibilities. Meanwhile, the exhibition itself forms an alternative space of resistance against authoritarian planning, allowing the marginalized spaces and neglected voices of communities to emerge. 2017 UABB triggers an experiment of urban coexistence promoting the future of urban regeneration. Therefore, the 7th Bi-City Biennale of Urbanism\Architecture (Shenzhen) in 2017 is not just an exhibition, but also an urban intervention and a city-making action.

来源：深圳城市\建筑双年展组织委员会 . 城市共生：2017 深港城市\建筑双城双年展 [M]. 广州：岭南美术出版社，2021: 44-50.

城中村自建房调查

An Investigation into Self-built Buildings in Urban Villages

许志强

广州大学建筑与城市规划学院教师。他在广州美术学院取得学士和硕士学位。城中村自建房调查小组成立于 2016 年，小组成员不固定，根据项目更替人员，主要以城市、建筑空间为研究对象，以呈现当下的生存空间，发现未被定义的空间类型。

本文于 2017 年 3 月 30 日首次发表于“澎湃新闻”网站；由莫思飞编辑和中译英；图片 © 城中村自建房调查小组。

Xu Zhiqiang

Xu Zhiqiang is a lecturer at the College of Architecture and Urban Planning, Guangzhou University. He received his bachelor's and master's degree from Guangzhou Academy of Fine Arts. In 2016, the investigation team on self-built buildings in urban villages was established. The team members are not fixed, constantly changing according to different projects. They take urban and architectural space as the research object to present our current living space and discover undefined space types.

The article was first published on website *The Paper*, March 30, 2017; edited and translated from Chinese to English by Mo Sifei; image ©The Urban Village Self-built House Study Group.

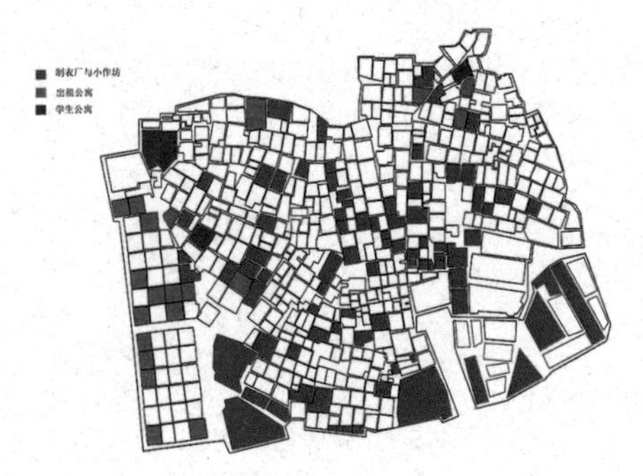

图 1 赤沙村制衣厂、小作坊、公寓分布图
Figure 1 Mapping of garment factories, small workshops, and apartments in Chisha Village

1 住宅工厂里的生活

1.1 城中村里的制衣厂

广州海珠区的服装生产业非常有名。以中山大学对面的“中大布匹市场”为中心，周边大片城中村成为服装加工厂、小作坊以及连带相关服务配套的聚集地（图 1），形成一个庞大的服装产业链。“去中大”成了服装批发和加工从业者的口头禅。赤沙村（图 2）处于该生产链的边缘，以制衣与服装加工为主，形成工厂与城中村自建房结合的两种生产模式：①城中村外围以大型制衣厂为主，面积较大，可满足大订单的需求；管理、消防、通风系统完善。②城中村内部则以小作坊为主，租金低廉但消防和管理系统复杂多样，一般处理约几十件的小订单，雇佣的十几个工人大多是老板的老乡，也不乏家庭小作坊。

1 Life at the Residential Factory

1.1 Garment Factory in the Urban Village

Haizhu District, Guangzhou is famous for its clothing industry. The "Zhongda cloth market," which is situated opposite the main entrance of Sun Yat-sen University is its center, while surrounding urban villages have become magnets for clothing processing factories, small workshops and related services, forming a huge clothing industry chain (Figure 1). "Going to Zhongda" has become the catchphrase of clothing wholesalers and processing workers. Located at the edge of the production chain, Chisha Village (Figure 2) mainly makes and processes garments, informing two construction strategies for integrating factory and self-built building: (1) large garment factories are primarily located at the urban village periphery. These buildings have enough space to meet large orders, while management, fire protection and ventilation system are well in place, and; (2) cheap and small rental workshops are primarily located inside the urban village. These buildings handle small orders of several dozen pieces, while fire protection and management systems are unstandardized and vary widely. Most workers are fellow townspeople of the owner, and there are also many small family workshops.

1.2 Large Garment Factory: Mixed Services

Most of the large-scale, self-built factories are mixed-use. In one three-story self-built factory, for example, a garment factory, supermarket, health care center and

图 2 赤沙村 2017 年鸟瞰图
Figure 2 Bird's eye photo of Chisha Village, 2017

1.2 大型制衣厂：多种服务叠加

大型自建厂房大部分混合了多种使用功能。其中一栋三层高的自建房工厂，一楼是制衣厂、超市、保健馆和诊所；二楼是三家制衣厂；三楼是两家制衣厂，其中一家已搬离。工厂的三层是在原有的建筑基础上，用钢构桁架与铁皮外墙搭建起来的，钢桁架减少了结构承重柱的数量，让室内更开阔。三层的屋顶则安装上吊顶，兼具美观和隔热功能（图 3）。

根据制衣厂工作流程顺序形成裁剪一缝制一熨烫一包装的空间序列：以主干道连接两头的活动空间和生产空间。入口处是开阔的活动空间和人流交汇处。中间交通主流线两边的结构柱列自然分隔形成左右两个堆积着产品的生产空间。最后方是熨烫和包装空间，一圈靠墙的工作桌围合着包装用的大桌子，形成一个相对松散的空间，这里是制衣厂次级的活动空间（图 4）。

该厂房属于相对正规的厂房，约 500 平方米的面积，月租金是 20 元 / 平方米，比工业区正式厂房约 28 元 / 平方米的价格低，且允许员工居住。从业近二十年的工厂老板透露，由于制衣工作辛苦，待遇低，做这一行业的人逐年减少，使得招聘困难，再加上订单也减少了，整体利润降低，只能勉强维持生活。在赤沙村的调研中，我们发现很多制衣厂都已关张，并贴出转租广告。

1.3 住宅工厂：工厂与家庭的叠加

城中村内部有大量自建住宅楼中的制衣作坊，可分为两类：一是楼下生产，楼上住人；二是

clinic are located on the first floor, three garment factories are located on the second floor and two more are located on the third floor, one of which has recently moved out. The third floor of the factory is an add-on based on the original building with steel trusses and metal weatherboard exterior walls. The steel trusses reduce the structural load-bearing and open up the interior. The roof of the tin house is installed with a suspended ceiling, which has the double function of beauty and heat insulation (Figure 3).

The factory's spatial order derives from its workflow, moving from cutting, sewing, and ironing to packaging. The open space by the entrance serves as the main space for activities and circulation. At the center, the main circulation artery connects the activity space and the production space at the two ends. The colonnade of the central circulation artery divides the space into two production spaces with piles of products. A large table for packaging at the end of the space is encircled by tables against the wall, forming a relatively unstructured secondary activity space within the garment factory (Figure 4).

The factory belongs to the category of "relatively formal," with an area of about 500 square meters. The monthly rent is 20 yuan per square meter, which is lower than the roughly 28 yuan per square meter rent in a formal factory in the industrial zone. It also serves as a dormitory for the workers. The owner of the factory, who has been in the business for nearly 20 years, revealed that the number of workers in this industry has decreased

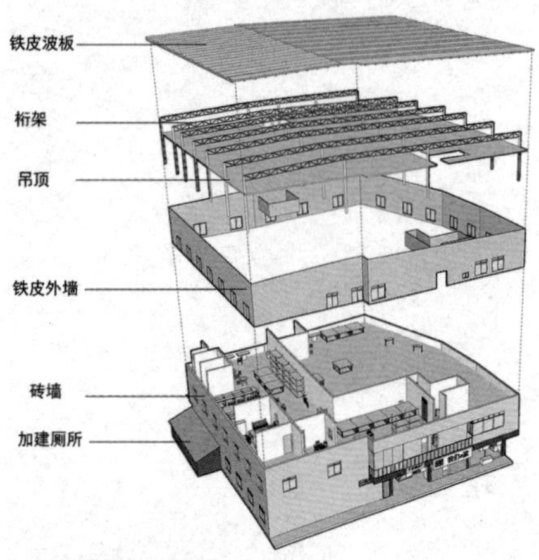

图 3 制衣厂建筑结构分解图
Figure 3 Exploded diagram of the architectural structure of a garment factory

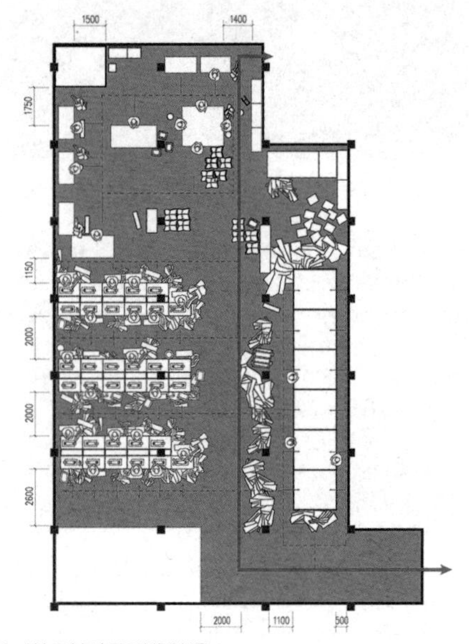

图 4 制衣厂空间序列及流线分析图
Figure 4 Layout and circulation analysis of the garment factory

生产与居住混合。后者是所谓的“三合一”住宅（图 5- 图 9），即居住、生产、存储设置在同一空间内；每一层的混合方式都不尽相同，生活与生产模式呈现出复杂的多样性，成为展现城中村现实的缩影。

虽然“三合一”住宅存在消防隐患，但对于房东而言，出租给家庭作坊做工厂，租金来源稳定，且一家作坊的面积就相当于好几户人，管理方便。房东会要求小工厂购买消防器材，而楼梯也可直达天台作为逃生通道。由于认定违规的标准模糊，执法人员检查时一般关门即可，只有出现事故才会加强管理。

夹在城中村建筑群中，只有内阳台且每个窗都装有防盗网的自建房，室内光线非常昏暗。从日照分析图（图 10）可以看出，每层楼获得

over the years due to its labor-intensive nature while low-wage offers making recruitment increasingly difficult. In addition, a decrease in orders has reduced overall profits, leading to survival challenges. During our survey of Chisha Village, we found that many garment factories were shut down and advertising for sublease.

1.3 Residential Factory: The Superimposition of Factory and Family

Many garment workshops that are located in a self-built residential building of an urban village can be divided into two categories. The first type situates the production on the lower floors and the living functions on upper floors, while the second type mixes production and living functions. The latter is the so-called "three-in-one" residence (Figure 5-9), where residential, production and storage functions are programmed in the same space. Every floor exhibits a unique mix and match of functions, presenting a complex and diverse model of production and living arrangements. In turn, each floor seems to be a miniature of the urban village.

Although the "three-in-one" residence poses fire risks, landlords prefer this arrangement because it is easier to manage, the rent from family workshops is guaranteed, and a dedicated workshop space would displace potential living space for several families. The landlord requires small workshops to have fire protection equipment and to keep the staircase to the roof clear as an escape route. Since the standards for identifying violations are vague, it is often enough to simply close doors during an inspection. It is only when accidents occur that enforcement is strengthened.

Squeezed inside the urban village, the natural lightning condition of building interiors is poor because self-built houses have inward-facing balconies with security bars installed. It can be seen from the sunlight analysis diagram (Figure 10) that even at high noon, when sunlight is strongest natural light can barely reach the interior of each floor, except for the fifth floor (the fourth and fifth floors were added on to the original three-and-a-half-story building). Consequently, artificial lights need to be turned on during the day, making electricity bills expensive.

1.4 The Superimposition of Factory and Ancestral Hall

During site research, we found a garment production factory built on top of a traditional ancestral hall. After the collapse of the latter part of the ancestral hall, villagers raised funds to restore the building while also adding a garment factory. The kitchen on the right side of the ancestral hall was removed and converted into a warehouse for the factory, while an office was set up on the upper floor. The plan of the garment factory corresponded to that of the original ancestral hall, preserving the

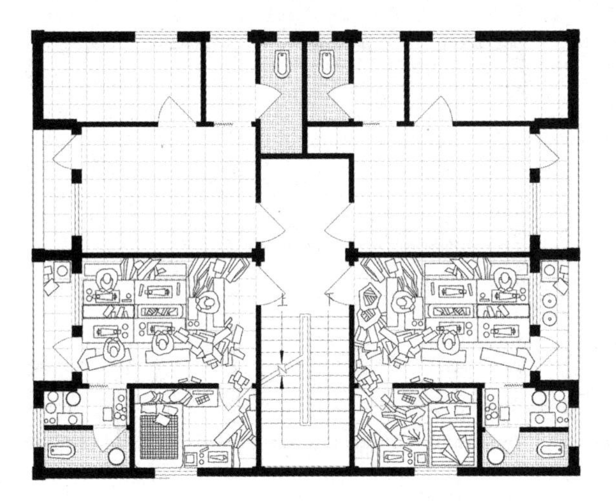

图 5 三楼
一间房子的月租金是 600 元，平常夫妻两人工作，相较工厂一天 12 小时的工作，自己单干每天需要 15~16 小时的高强度工作，才能保证收入，但相对来说时间比较自由。夫妻俩的小孩在乡下生活，忙时会有老乡来帮忙。因为房间小，平常他们都在厨房里吃饭，怕在工作台吃饭弄脏做好的衣服。整个房间充斥着布料的气味，卧室也堆满了做好的新衣服。
Figure 5 Third Floor
The couple on the third floor pays 600 yuan per month for rent and work together. Instead of a 12-hour shift in the factory, having their own business requires an intensive workload of 15-16 working hours to ensure sufficient income. Time management, however, has become more controllable. Their child lives back in their home village, while people from their home village will come to help when busy. The apartment is so small that even the bedroom is filled with finished products and the couple eats in the kitchen to keep the products clean. The odor of fabric pervades the whole apartment.

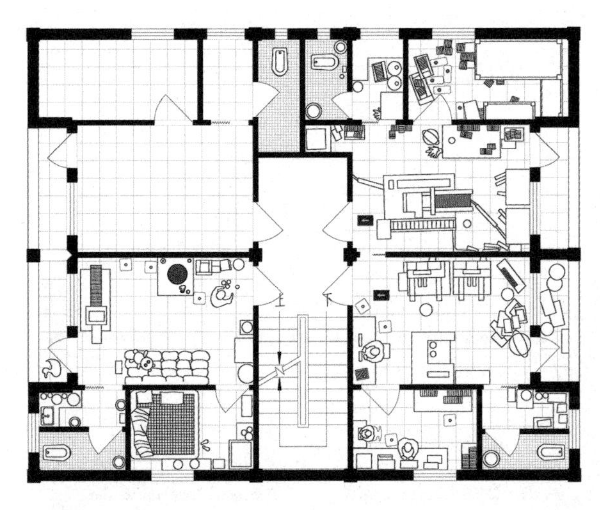

图 6 二楼
生产 LED 灯的家庭作坊，以右边两套屋子作为生产用，左边一套屋子作为生活使用。另外雇佣了一个工人，工人住右边屋子的一个房间。他们租了四年，一直就干这个。每间的月租金与居住用途的一般租客一样，是相对便宜的 500 元。他们对屋子的不满在于光线太暗，白天都要开灯，加上工作忙的时候就会连夜开着机器赶工，24 小时都要开着灯，一度电 9 毛多，同时机器和空调用电也多，一般一个月电费要 3000 元左右。电费加房租共约 5000 元 / 月。
Figure 6 Second Floor
This is a LED lighting family workshop. Production functions are located in the two rooms on the right and living sections are located in the left room. In addition, they also provide accommodations for their employee in a room on the right. They have rented the rooms for four years, doing the same productive work. The monthly rent for each room is 500 yuan, the same as for residential tenants. According to the couple, the poor natural lighting is the building's only deficiency as artificial lighting is required even during the day. During a rush, machines, air-conditioning and light need to be kept on 24 hours a day. With a unit price of 90 cents, their monthly electricity bill is about 3,000 yuan. Electricity bills and rent add up to 5,000 yuan per month.

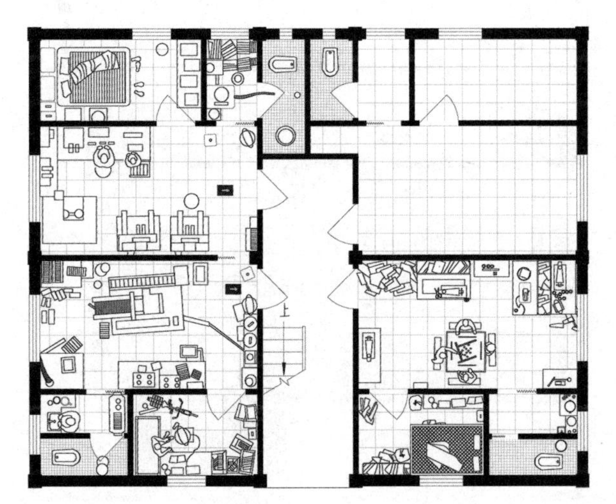

图 7 一楼
其中两套一室一厅的屋子是一户生产 LED 灯的家庭作坊，每套约 30 平方米，中间的墙打通一个小门，便于工作。两个客厅是生产车间，放置了噪声较大的自动化运作的机器和一台不停地散热的大风扇，小孩一般不会到这里玩耍。两个卧室分别为儿子的房间和夫妻二人与女儿共住的房间，其中也有工作与居住混合的区域。吃饭时，在客厅机器狭窄缝隙中摆开小桌子，一家四口围坐吃饭。傍晚小女儿放学回来，用工作台边的电脑看动画或是自己画画，母亲则会在旁边工作；隔着运转的机器，父亲在厨房中煮饭。晚饭后他们继续工作至深夜 12 点。
Figure 7 First Floor
Two of the 30-square-meter one-bedroom apartments have been combined into a family workshop that produces LED lights. A door was added to the partition wall between the original apartments for easy access. The production area is located in one of the living rooms, where a large, noisy automatic machine operates, and a large fan keeps the room cool. Children do not play in this room. The son lives in one of the two bedrooms, and the couple and their daughter live in the other. There is also a mixed-use area, where a small table is placed in between the machines in the living room during meal times. After school, the daughter watches cartoons and draws on a computer by the workstation while her mother works beside her. The running machine separates this space from the kitchen, where the father cooks. After dinner, the couple usually continues working until midnight.

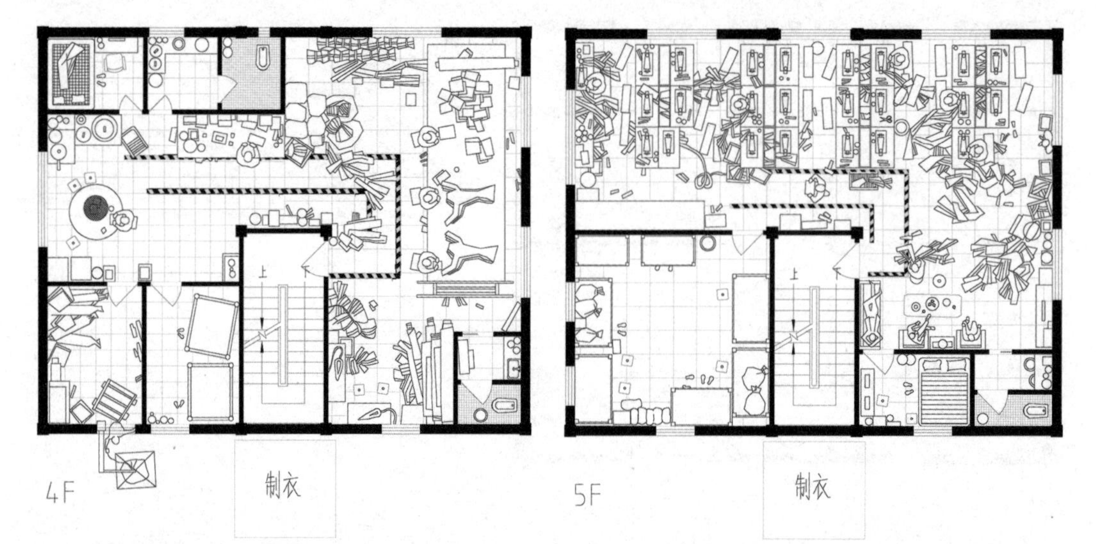

图 8 四楼、五楼是加建部分，为一家制衣厂（分两层）。这里格局不同于楼下三层，是一个连通的大空间。与城中村外围的大型制衣厂不同，这里的制衣厂分成了两层，男女工人分开工作，四楼的男工主要做裁剪、熨烫、包装，五楼的女工主要做服装缝制，完成一般制衣厂裁剪一缝制一熨烫一包装的流程。当我们看到地上贴的防火通道标识被各种布料掩盖、阻挡。提及消防安全，他们也意识到不符合规定，但这里租金便宜，只能接受现实。

Figure 8 The fourth and fifth floors comprise a two-story garment factory. These two stories have been added on to the original building, therefore the layout of these floors is open and is different from the floor plan of the lower floors. Unlike large garment factories at the urban village's periphery, this garment factory is divided into two floors where female and male workers work separately. On the fourth floor, the men cut, iron and package products, while on the fifth floor, the women sew, completing the process. The fire escape signage and route are covered and blocked by cloth. The owners acknowledge the potential fire risk but accept it due to the cheap rent.

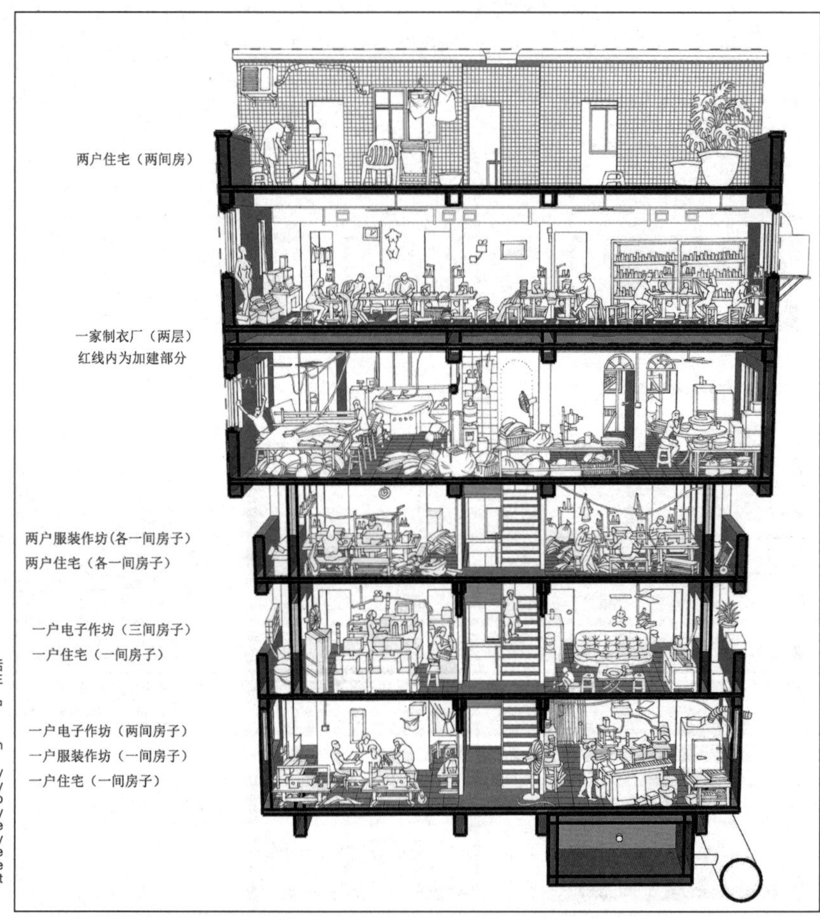

图 9 住宅与生产空间关系图
这个剖切图可以呈现五层半的建筑内人的活动，在垂直方向上可以看到每一层不同的生活情景。一楼：一户四口的电子家庭作坊、一户服装加工作坊、一户住宅。二楼：一户三口 LED 灯电子作坊，一户住宅。三楼：两户服装加工作坊，两户住宅。

Figure 9 Mapping of living and production spaces
The cross section shows the different daily activities of each floor in the five-story building. The first floor is a family workshop that produce electric components, a family garment workshop and an apartment. The second floorproduces LEDs as a family workshop, while different people use the other apartment. On the third floor are two apartments and a two family garment workshop.

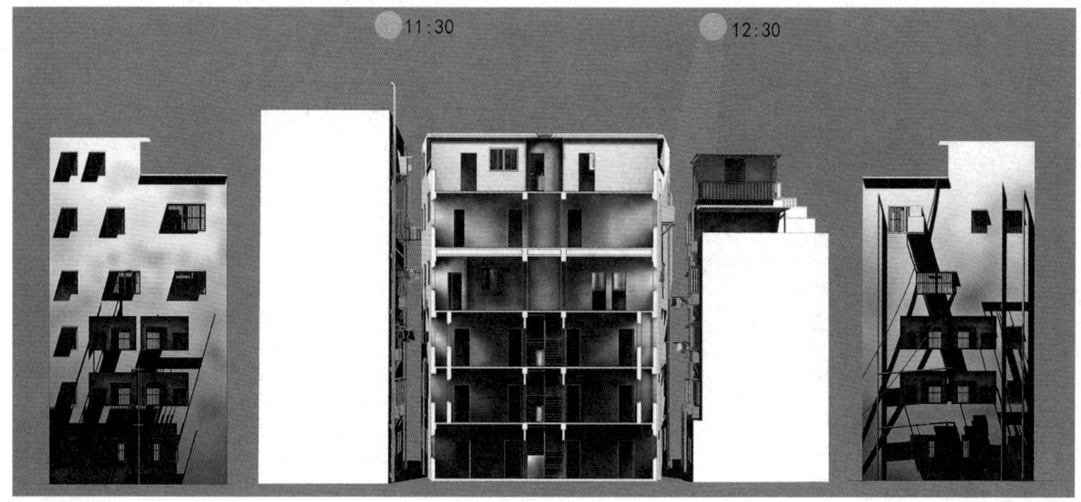

图 10 日照分析图
Figure 10 Sunlight analysis diagram

光线的差别，即使在阳光最强的中午，除了五层（原三层半的建筑加建到五层半），其他楼层几乎没有光线能进入室内，白天基本都要开灯，使得电费较高。

1.4 工厂与祠堂的叠加

在走访时无意中发现，一家制衣厂建造在一座传统的祠堂上面。由于祠堂后面部分坍塌了，村里集资修复，在修复祠堂时加建了一家制衣厂。制衣厂把原来祠堂右边的厨房部分拆除，一楼改造为制衣厂的仓库，楼上为办公室。制衣厂的平面对应祠堂原来的平面，中间的天井也留了出来，从祠堂天井往上看，是被制衣厂高高的马赛克外墙围合的天空，还有一个巨大的制衣厂的空调排气口。

制衣厂的楼梯贴着祠堂的墙壁往上，从二楼可以看到祠堂的屋顶。为了满足消防安全的需要，有另外一部楼梯从楼上插入原来的祠堂内部，再从祠堂延伸到外面。从建筑的角度来看，这样的叠加极其粗暴，作为传统象征的祠堂由于经济利益关系可能慢慢被取代，继而消失。

2 建一栋房子要花多少钱

当城中村村民转化为城市居民时，他们也就从农业耕作的生产方式转化为另一种空间生产方式：在自己的宅基地上建造房子，解决生存问题。这时他们的自建房就以一种空间生产的方式介入社会关系，改变现有的城市空间关系。因此一个城市的空间不只是由政府、地产发展商、城市居民，同时还有城中村的村民共同塑造而成。而城中村的自建房包含着复杂的现实：历史的遗留问题、国家政策问题、人与人的社会关系问题、地理位置的复杂问题等等。那么，自建房是怎么建造起来的呢?

middle courtyard. Looking up from the courtyard, the sky is encircled by the high walls of the factory with a large air vent.

The stairs of the garment factory hug the wall of the ancestral hall, connecting the first and second floors. The ancestral hall roof is visible from the second floor. In compliance with fire regulations, another staircase was inserted into the ancestral hall and extended outside. Such a superposition is rather brutal in terms of architectural design. Ancestral halls symbolizing traditional culture might gradually disappear in the urban village's drive for profit.

2 How much does it cost to build a building?

As urban villagers continuously transition into urban residents, their mode of production also undergoes a spatial transformation; instead of farming, they build houses on their housing plots to meet their needs. At this point, the self-built buildings are a form of spatial production that intervenes into social relationships, changing extant urban spatial relationships. Hence, urban space is shaped not only by the government, real estate developers and city residents, but also by the inhabitants of urban villages. The self-built buildings in urban villages actualize a complex reality, comprising historical legacies, national policies, social relations between people, and physical geographic locations, begging the question: how do self-built houses get built?

2.1 Land Acquisition

Since October 1, 2002, all local villagers in Haizhu District, Guangzhou have received urban hukou. This change in status has meant that villagers no longer receive agricultural subsidies. Nevertheless, they are

图 11 卫星图上万亩果园与小洲村旧村和新村的关系
Figure 11 Satellite image of the spatial relation between Wan Mu Orchard and Old and New Xiaozhou Villages

2.1 土地的征收

从 2002 年 10 月 1 日开始，广州海珠区所有的本地村民都变成城市居民。村民原有的农业补贴因为身份的转变而消失，却又没有城市居民的待遇。于是赤沙村和小洲村的村民都变成“还需要靠农业为生的居民”。

小洲村（图 11）位于广州市海珠区，是广州城区内发现的最具岭南水乡特色的古村寨，2000 年被列为广州市首批 16 个历史文化保护区之一，并被评为广东省生态示范村。小洲村村民世代以种果树为生，果林成片，瀛洲生态公园与附近的果林共约 2 万亩（30 公顷），素有广州“南肺”之称。由于古村落和广州“南肺”的保护，小洲村发展缓慢，工业不得进驻。古村内的建筑高度规定为两层半，外围是三层半。种植水果收入低，既没有产业，又没有房产出租的村民说：他们是海珠区最贫穷的一个村落。不少小洲村民面临生计问题，甚至有村民一年收入不到 1000 元，连小孩读书都要借钱来解决。“保得了你的肺，就保不了我的胃”，是小洲村村民经常挂在嘴边的一句话。

2004 年毗邻小洲村的大学城正式投入使用，由于小洲村的古村风貌，吸引了众多艺考培训画室的入驻。随之而来的是每年近万名艺考生短期租住古村的房子。

2008 年，中央政府提出全国各地农村土地确权、登记、颁证工作：完成集体土地所有权、集体建设用地使用权和农民宅基地使用权等三权的登记发证工作。2009 年开始土地确权试点启动，并计划于 2012 年底完成农村集体土地确权工作。土地在未确权之前，政府征地主要是跟村委商量；而确权之后，村民们拥有相应的土地证，征地时只能跟对应的村民协商。

2012 年，涉及多个村的万亩果园在土地未确权之前被征收，小洲村是被征地最多的。政府承诺每征地 10 亩，划 1 亩地给村民作商业用地，用于生活和发展，3 年内给村民的商业用地红线划定不落地，给予生活补贴。附近的大学城进驻以及征地获得补偿后，小洲村进入抢

not entitled to the same benefits as other city residents. Consequently, Chisha and Xiaozhou villagers have become "urban residents who still need to live on agriculture." (In Chinese, this statement refers to the fact that historically rural and urban residence determined livelihood. Specifically, rural residents were farmers and urban residents worked in non-agrarian jobs.)

Located in Haizhu District, Guangzhou, Xiaozhou Village (Figure 11) is not only an ancient village, but also the city's most typical example of a Lingnan water village, which is a translation of its name "Xiaozhou." In 2000, it was listed as one of the first 16 historical and cultural reserve districts in Guangzhou and rated as a Guangdong Province ecological demonstration village. Xiaozhou villagers have been planting fruit trees for generations. Yingzhou Ecological Park and nearby orchards occupy about 20,000 mu (13.3333 square kilometers) and are known as Guangzhou's "Southern Lung." Due to the protection of both the ancient village and its "Southern Lung" function, the development of Xiaozhou Village has been slow with no manufacturing industries allowed within its borders. The building height is restricted to two and a half stories in the old village and three and a half stories at the periphery. Income from fruit growing is limited, and without the option of establishing an industry or rental properties, the inhabitants describe Xiaozhou Village as the poorest village in Haizhu District. In fact, many Xiaozhou villagers face livelihood challenges. Some earn less than 1,000 yuan a year and even need to take out loans for their children's education. Xiaozhou villagers frequently say, "For your lung to be saved, my stomach is sacrificed."

In 2004, the university town near Xiaozhou Village was put into use. Xiaozhou Village attracted training studios for art examinations due to its ancient village landscape, followed by nearly 10,000 art candidates' short-term rentals every year.

In 2008, the central government proposed the confirmation of property rights, land registration and the issuance of certificates in rural areas nationwide. The goal was to complete the registration and certification of ownership of collective land, to determine collective access rights for the construction of buildings, as well as to confirm the use rights of individual homestead land. The land rights confirmation pilot project was launched in 2009 to be completed by the end of 2012. Before the confirmation of land rights, the government and village committee entered negotiations. Upon confirmation of land rights, any use of a piece of land had to be negotiated with the villager holding the relevant land certificate.

In 2012, before land rights had been confirmed,

建的黄金时期，村民开始改建房子，出租给学生和画室使用（图 12）。但等房子从两三层的瓦房建成五六层的公寓时，画家开始因为古村落变成钢筋水泥森林，没有了以往的情怀而选择离开，租房子的人减少了，使得画家和村民开始互相埋怨。

2.2 改建与抢建

2013 年底到 2015 年，海珠区对城中村自建房的管理放开，一方面默许建到五层半，同时默许在原来三层半的基础上加建。那一两年里，自建房开始疯狂生长，连同那些本来没有明确土地所属（没有宅基地证）的地方也开始建设。甚至还有超过五层半的建筑陆续出现，只要有一家这样做且没有被制止，其他人就会跟上脚步。

一般城中村改建房子，要做危房鉴定后才可申报，继而报有关部门备案，排期改建。改建规定，建筑往外出挑不能超过街道的三分之一，建筑高度不得超过 15 米（或三层半）。但那个时期是各个城中村抢建的黄金时期，有太多的房子同时想要改建，而且抢建的速度快，约半年时间就能建好，根本无法按程序与规定实施。2015 年后，由于政府对自建房违建现象的监管力度加大，失去了抢建的机会。

抢建时期，各村都处于全面施工状态，场面混乱，城管巡逻频繁。城管人手不足，就让警察盯管每条巷子，留意出现的沙堆。如果警察发现违建，先报街道办，街道办再上报城管，由后者处理。若城管到来前未完工，则拆除违建房子；但若在城管到来前完成抢建工作，并拆除排栅清理现场施工痕迹，则被视为既定事实可不予以拆除。如果严格按照规定，这些踩线的违规建筑应被拆除，但由于抢建现象太普遍，执行工作量太大，城管只能频繁巡逻以提前制止抢建和拆除明显的违建。

Wan Mu (Ten Thousand Mu) Orchard was expropriated from several villages, including Xiaozhou Village which had the most land being expropriated. The government promised to allocate 1 mu (666.667 square meters) of commercial land rights to the villagers for living and development in exchange for every 10 mu of rural land requisitioned. In addition, if the property lines of the promised commercial land had not been marked within 3 years, living subsidies were to be given to the villagers. Upon the opening of the nearby University Town and having received compensation for their land, Xiaozhou villagers rushed to take advantage of this golden opportunity, building houses to rent to students and artists (Figure 12). However, as soon as the two-to-three-story houses became five-to- six-story apartment tenements, the painters relocated because the village no longer had the same ambiance and was instead a forest of reinforced concrete. The rental property market shrank, and the painters and villagers blamed each other for the situation.

2.2 Reconstructions and a Rush to Build

From the end of 2013 to 2015, Haizhu District loosened its management of self-built housing in urban villages. Constructions of five and a half stories in height were permitted, while add-ons to old three-and-a-half story buildings were tacitly approved. During this period, buildings rose rapidly, even on land without confirmed ownership (no homestead certificate). Moreover, all it took was one successfully completed building taller than five and a half stories and others began appearing.

The normal application process to rebuild an urban village home can only begin after a building has been evaluated and condemned. Then the application is filed with the relevant department and the reconstruction is

图 12 2008 年（左）和 2017 年（右）的小洲村　Figure 12 Xiaozhou Village in 2008 (left) and 2017 (right)

2.3 建造与成本

那么，城中村村民自建房的资金来自哪里？建造成本如何？自建房又是如何建造起来的？在赤沙村，我们认识了一位特别细心的房东。2007 年，他把一栋两层带院子的旧房子改建为六层的新房子。建完后他整理并保留了当时所有的文件与费用单据。既有城管部门发出的两份《责令限期改正通知书》，也有当时购买材料的名片和清单费用，甚至细到配了多少把钥匙，花了多少钱都有单据保留。我们整理了这份清单，对每一笔支出的材料和人工费用进行了确认。还有一位在广州城中村建房十几年的建筑包工头，跟我们讲述了建房的过程与技术。

整个建造过程的第一笔支出是花费 100 元购买的报建调查审核表，隔了一个多月审批下来后，房东开始拆掉旧房子，并与建筑包工头签订建房合同，之后房子一步步建造起来，花费也越来越多。

在 2007 年，如果包工头包工包料的成本是每平方米 700 元（现在包工包料是每平方 1200 元左右）。这个 660 平方米的房子的建造成本就为 46 万元，而房东为了节省成本只让包工头负责建造，自己购买所有材料，整个建造费用约 36 万元。

除了自己采购以及选择购买便宜的建材节省开销外，节省成本主要是根据泥水师傅开出的购买清单再重新计算一遍，算一个更接近实际用料的数额。一般来说，泥水师傅怕用完了要耗时间追料，开的数目比实际用料要多一些。如果是包工包料，泥水师傅会帮忙节约材料，因为节省下来的钱可以归自己所有。

八个月的建造过程中，前期完成整个砖混结构的建造，支出的费用不到总成本的一半，但到了装修时用钱很快，几万块一两天就没有了。实际上，很多自建房的资金都是借来的，存款有十几万就敢建房子，没钱了则借钱周转，一般按进度先支付建筑工人的费用，材料可以先拿货记账，只支付部分款项，到最后再结清材料费。房子建好后一般几年就可以回本，以一栋六层的自建房为例，一层五个房间，每间月租金 400 元，（若全部出租）总租金收入为 12000 元每月，一年总收入 144000 元，按 36 万元的建造成本，3 年即可回本。

我们根据清单对其中两种材料的使用进行了详细的描述：①给排水与供电系统（图 13），②砖的成本与数量计算方法（图 14）。然后，根据清单中涉及的所有建造材料，画出一个材料汇总图（图 15）。把一栋六层的建筑

scheduled. The reconstruction provision stipulates that the overhang of new buildings shall not exceed one-third of the street and that building heights shall not exceed 15 meters (or three and a half stories). However, it was difficult to comply with these measures because too many buildings were rushed into reconstruction during the period. Moreover, construction was completed in only half a year. Since 2015, the government has tightened the supervision of illegal construction.

During the period of rushed construction, all villages were under construction, with chaotic scenes and frequent patrols by the city management bureau. As the city management bureau was understaffed, the police were requested to keep an eye on each alley and pay attention to the appearance of sand piles. If an illegal construction was found, the police reported it to the sub-district office, which reported the case to the city management bureau, which would then enforce construction regulations. Illegal constructions were to be demolished if the building was unfinished when the city management bureau arrived. However, if the building was completed, the scaffolding was removed, and there were no traces of construction onsite, then the city management bureau viewed the building as an established fact and did not demolish it. If housing regulations were strictly enforced, these buildings should have also been demolished. However, the scale of construction made it difficult to enforce regulations. As a result, the enforcement patrols only stopped the construction of buildings that were in flagrant violation of the regulations.

2.3 Construction and Its Costs

How do urban villagers fund the construction of buildings? How much does the construction cost? How did these houses get built? In Chisha Village, we met a meticulous landlord who had replaced his old two-story house and yard with a new six-story new building in 2007. He organized and kept all the documents and receipts from the construction process. The documents included two "notices to rectify and reform within a prescribed time limit" issued by the city management bureau, business cards from material suppliers, and a full list of fees of purchased materials. He even recorded the cost and number of keys for the building. We confirmed the material and labor costs on his list. In addition, a construction contractor who had worked for over ten years in urban villages in Guangzhou went through the construction process with us, sharing insights into his building techniques.

The first expenditure in the construction process

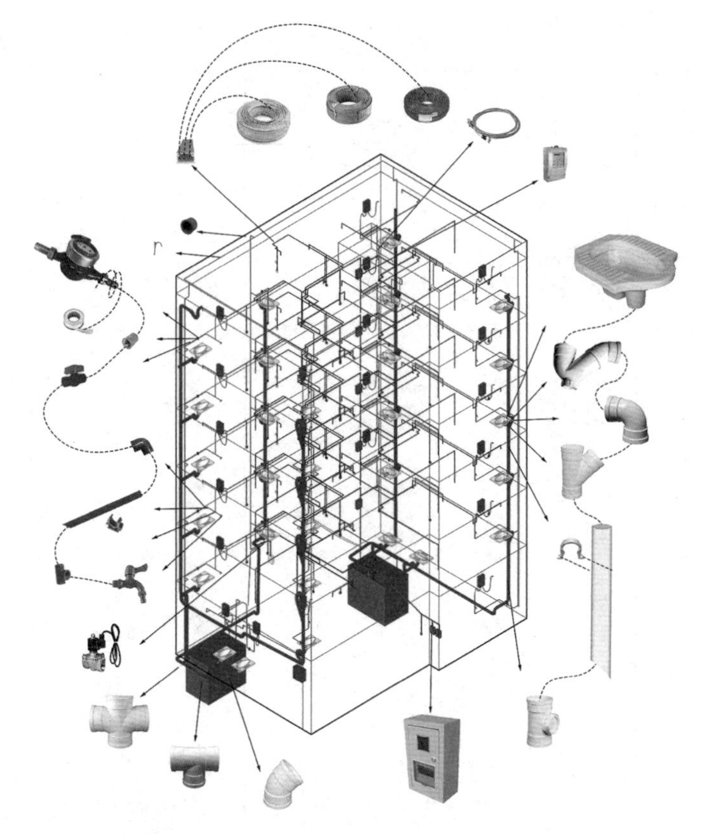

图 13 给排水及供电系统材料图
Figure 13 Mapping of water supply and drainage as well as power supply system

包工头计算红砖使用数量：92526 块

按墙面每平方 70 块砖计算
一层：211 平方米 × 70 块 / 平方米 = 14770 块
二层—六层：1032 平方米 × 70 块 / 平方米 = 72240 块
楼梯扶手：32.1 平方米 × 70 块 / 平方米 = 2247 块
天台、女儿墙：46.7 平方米 × 70 块 / 平方米 = 3269 块

清单实际使用红砖数量：82500 块

旧砖 35000 块
0.15 元 / 块 × 35000 块 = 5250 元

新砖 47500 块
0.3 元 / 块 × 47500 块 = 14250 元

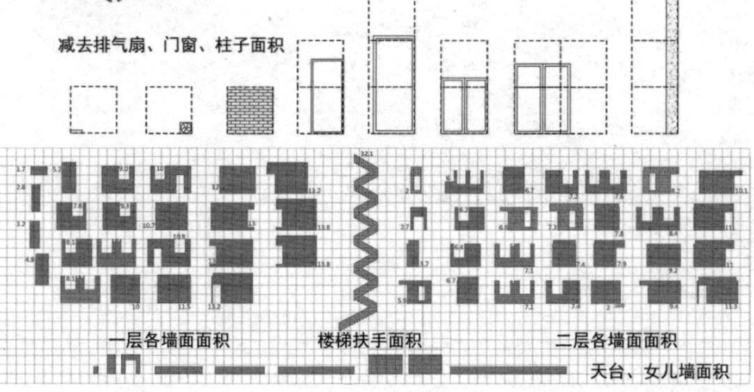

图 14 在建材汇总清单中，红砖使用数量为 82500 块，相较包工头提供的每平方米用砖 70 块的计算方法得出的 92526 块红砖，节省了大约 10000 块红砖。
Figure 14 According to the list, 82,500 bricks were used, which was 10,000 bricks less than the contractor's estimate. The contractor calculated the number of bricks based on a formula of 70 bricks per square meter, requiring 92,526 bricks to finish the job.

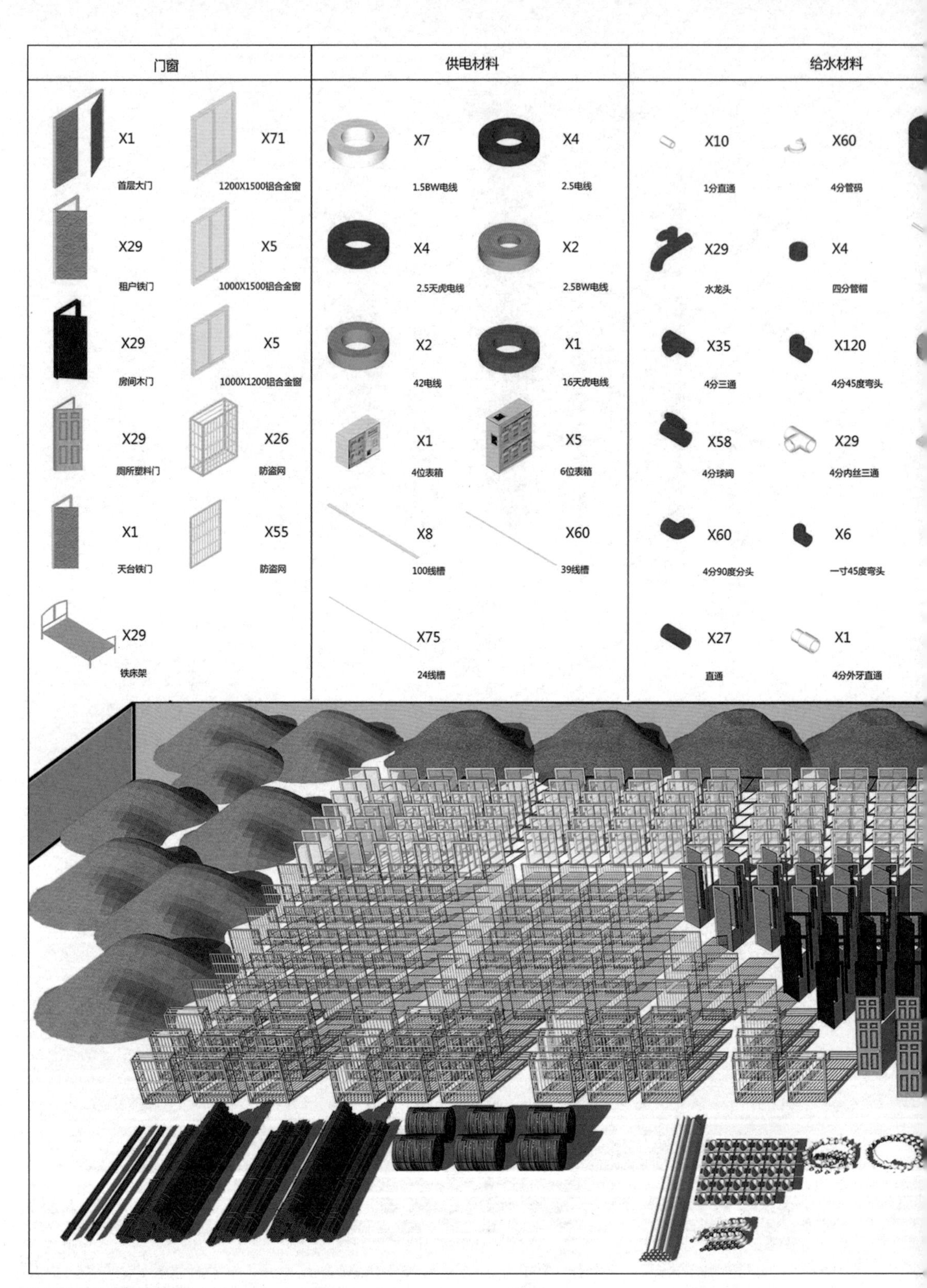

图 15 城中村自建房建筑材料汇总图
Figure 15 Illustration showing materials used in the construction of self-built buildings in urban villages

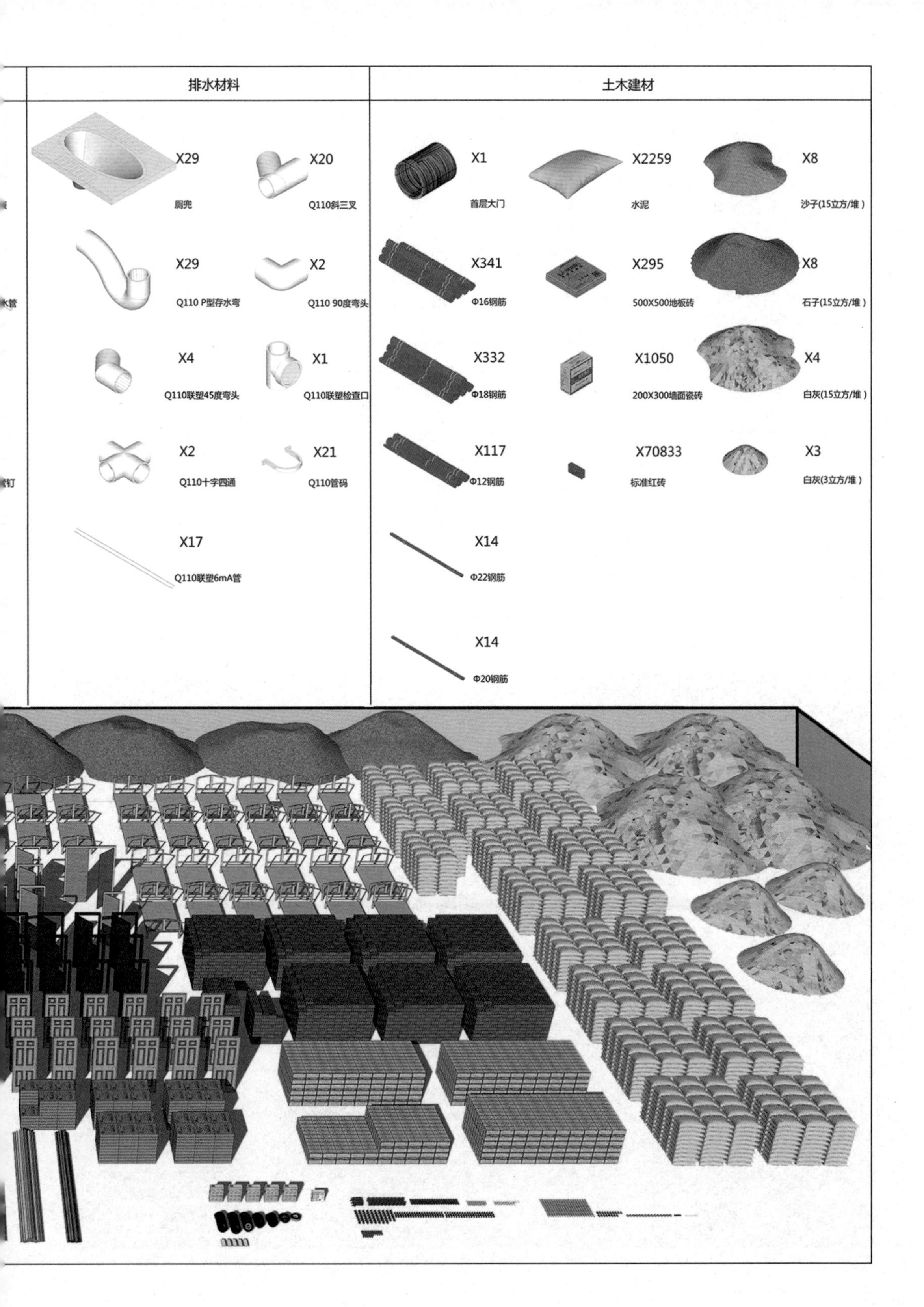

排水材料
土木建材
X29
厕兜
X20
Q110斜三叉
X29
Q110 P型存水弯
X2
Q110 90度弯头
X4
Q110联塑45度弯头
X1
Q110联塑检查口
X2
Q110十字四通
X21
Q110管码
X17
Q110联塑6mA管
X1
首层大门
X2259
水泥
X8
沙子(15立方/堆)
X341
Φ16钢筋
X295
500X500地板砖
X8
石子(15立方/堆)
X332
Φ18钢筋
X1050
200X300墙面瓷砖
X4
白灰(15立方/堆)
X117
Φ12钢筋
X70833
标准红砖
X3
白灰(3立方/堆)
X14
Φ22钢筋
X14
Φ20钢筋

还原为一个个原材料时，似乎能在材料和现实之间建立起一种联系，也会对建造有不同的认知。

2.4 空间使用与成本的关系

我们发现除了建造外，在实际的空间使用中，有一些很具体的成本使用情况与节省方法（图 16）。

比如，房子室内部分的墙壁没有铺设墙面砖，虽然省了钱但是脱落的墙皮影响房客，只好贴胶纸防止白灰的脱落。而在走廊、楼梯等公共区域铺设墙面砖则是为了便于出租人管理、清洁。另一个细节是厨房与厕所一体，没有分隔，使用厨房时厕所则无法使用。为了节省成本厨房不安装水槽，共用厕所的水龙头。

包工头给我们看了他过往参与过的自建房工程的平面图，我们发现有两个与这间房子的户型极为相似，可以看出包工头对户型布置的思考历程。城中村自建房的包工头设计户型时都非常实际，通过控制房间数量和质量，结合不同房东的需求进行户型布置，在有限的地块里，用最省钱的方式为房东获得最多的租金。包工头经常说：“这里摆个床 2 米，加一个开门的衣柜总共 2 米 5；厕所最多 1.3 米，厨房放 1 米到 1 米 3 的灶

握手楼

was 100 yuan for the review form of the building permit application. A month later, after receiving the building permit the landlord began to dismantle his old house and signed a building contract with the construction contractor. As the building went up, so did the expenses.

In 2007, contractors billed 700 yuan per square meter, inclusive of labor and materials. The budget for this 660-square-meter building would have been 460,000 yuan. (Current labor and material costs are about 1,200 yuan per square meter). To save money, the landlord only paid the contractor for labor costs, purchasing building materials himself. In this way, his total construction costs were only 360,000 yuan.

In addition to saving money by purchasing cheap building materials by himself, it was also possible to save money by re-calculating the required material lists from artisan builders, who overstated the amount of materials needed to avoid shortages and replenishment costs. In contrast, if the job included both labor and material, then the contractor and artisan builder cooperated to save on building materials, increasing their profit margin.

Construction took eight months to complete. At the early stage of construction, the concrete and brick frame structure and walls required less than half of the budget. In contrast, interior renovation costs tens of thousands

图 16 这个建筑的外墙瓷砖总共是 5000 块左右，比较便宜。而现在建房子，外墙瓷砖就要花 2 万左右。以前没什么钱，而且当时也流行这种风格的瓷砖。防盗网有两种形式，一种是价格贵一些的外凸防盗网，可以作为阳台使用（晾晒衣服），现在为了室内更大的使用面积已不再建造阳台。另一种只是纯粹的防盗网作用，价格较便宜。

Figure 16 When built, the cost of exterior wall tiles was approximately 5,000 yuan, which was relatively cheap. In contrast, the current cost of these tiles is about 20,000 yuan. In the past, villagers were not so well-off, and these tiles were popular. There were two types of security bars. One was an expensive, convex mesh that was primarily used on balconies. However, balconies are no longer constructed to fully utilize rental space. These spaces use cheap security bars to prevent burglaries.

台就可以。”通过家具的摆放和人的活动定下最小化的面积，这就是包工头的空间设计方法。

3 争夺空间的博弈

小洲村在2013年到2015年的改建、抢建时期，由于村内巷道狭小，大车无法进入，村里也没有足够的开敞空间堆放建筑材料，村口就成了被选中的空间，总是被水泥搅拌车和一堆堆的沙石占据着。村民会互相计划安排各家建筑材料的堆放空间位置与使用的时间段，以避免矛盾。但街道空间毕竟有限，对往来交通造成不便，不时会出现堵塞甚至是碰撞事故。这样的场景仿佛是城中村自建房与城市建设的规章制度之间博弈的缩影。城中村村民在自己的宅基地上进行建造，争取在城市空间中的权利，而各种出现在违章建造边缘的建造方法就是这种博弈的体现。

城乡规划规范中对建筑的间距有很详细的规定，但城中村“不适用前款兼具规定”，指出改建需要适应旧村原有的历史遗留复杂性，同时也带来模糊的博弈地带，让村民有机会出于自身利益最大化的驱使，进行空间占有。我们把城中村自建房博弈的空间占有方式分为三个部分：①建筑高度；②建筑出挑；③基地延伸。

3.1 建筑高度

关于建筑的高度，曾经规定：“建筑层数应当不超过三层，根据功能需要可以增设楼梯间和功能用房，三层部分建筑高度应当不大于11米，楼梯间和功能用房的建筑高度应当不大于14米。鼓励采用坡屋顶”。

以小洲村为例，经过抢建黄金期，早期符合规定的两层半和三层半高度的建筑几乎绝迹。基本都是三层半以上的加建，甚至是超过五层半的加建（图17-图18）。而在相邻建筑间距不及臂展的、高密度的城中村，村民在加建时利用灵活多变的排栅术（图19）应对建筑相邻过近的现实，避免影响公共巷道的日常和安全使用。得益于村民自发建造，这些不同的加建方式、立面材料的选择、加建体量的形态以及临时与永久构造的差异，使城中村立面变得丰富多元。对加建类型的选择反映出村民对原建筑结构的适用性及自身需求的思考，对加建资金与利益最大化之间的权衡，甚至是对规章制度的试探和博弈。

城管对违规加建的处理方法也是很有讲究的，一般不会去砸有证的房子的梁柱，避免建筑结构不稳或倒塌，而是把楼板砸出一个大窟窿，使违建部分无法使用。而村民的对策就是迅速补楼板。调研中有一个案例，有两栋相邻的建筑，首先改建五层的邻居因后建五层建筑完全遮挡了其采光，且还高出半层，进行举报——按规定五层半的半层只能为一个楼梯间的梯室，但该屋主多建了两间小房子。经过反

of yuan each day. Most of the building was financed through loans. Someone who had saved a little over one hundred thousand yuan would be confident enough to start building. When the money ran out, they would borrow and continue building. Construction workers were paid as they built. Materials were delivered on account with a small downpayment, and the account was settled at the end of construction. It took several years for landlords to recover their investment. For example, if a room rented for 400 yuan per month, a six-story building with five rooms per floor (if fully occupied) would provide a monthly income of 12,000 yuan, for an annual income of 144,000 yuan. Thus, it would take at least three years to recover the construction cost of 360,000 yuan.

We illustrated the following two materials: the water supply and drainage as well as the power supply system (Figure 13), and the cost and quantity of bricks (Figure 14). We also drew an illustration that summarized all construction materials on the list (Figure 15). By returning the six-story building to its raw material state in illustrations, we were able to establish a connection between the building materials and actual conditions, giving us new insights into this construction process.

2.4 Program Layouts and Costs

We discovered that in addition to savings during the construction process, costs could be cut through specific practices (Figure 16).

For example, the walls of rental rooms were not tiled to save money. The wall finish peeled off, and tenants pasted gummed paper on walls to prevent lime from peeling off. However, wall tiles were used in the corridor and staircases, making it easy for the landlord to maintain and clean common areas. Another detail was the integration of the kitchen and toilet into one space. As a result, the kitchen and toilet could not be used at the same time.

When the contractor showed us plans for the buildings he had worked on, we discovered that two of the layouts were similar to the one we had visited. The contractor was thus able to give insights into the layout. Specifically, the contractor designed the layout of self-built buildings to maximize rental income within the limitations of the site and to satisfy the specific needs of different landlords by manipulating the quantity and quality of rooms. According to the contractor, placing a bed required 2 square meters, while a bed and sliding-door wardrobe used 2.5 square meters. A toilet space takes up to 1.3 square meters, while a kitchen with a stove takes up 1 to 1.3 meters. The contractor's design approach was based on furniture placement and daily activities, with an eye to determining the smallest possible typology.

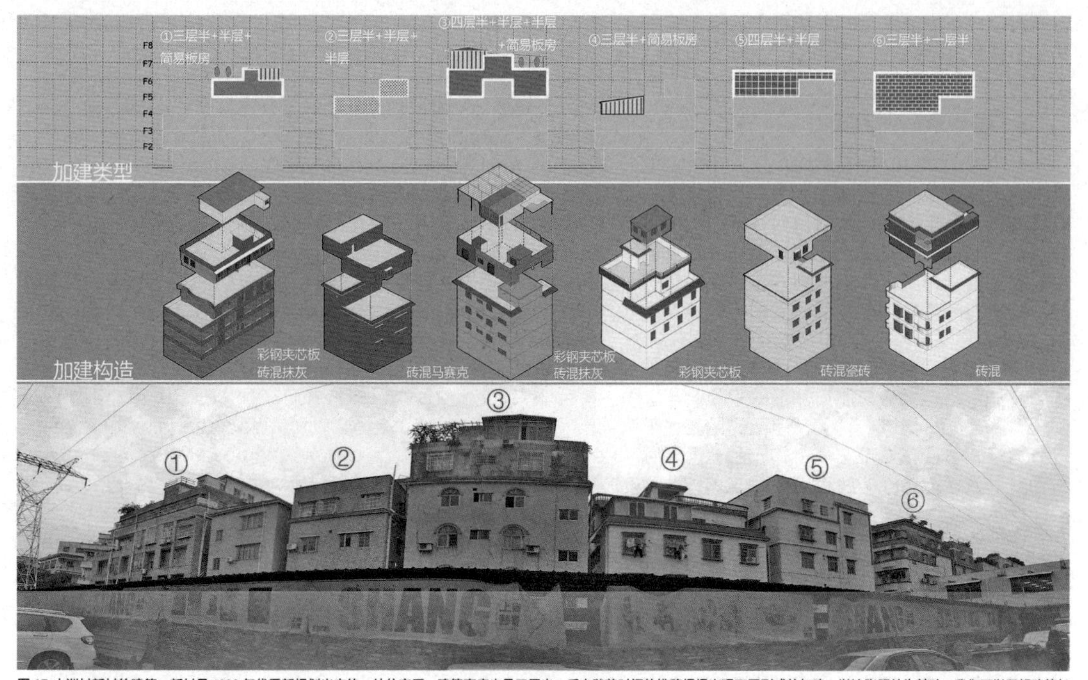

图 17 小洲村新村的建筑，新村是 1990 年代重新规划出来的一片住宅区，建筑高度也是三层半，后来随着时间的推移慢慢出现不同形式的加建。以这张照片为基础，我们可以了解建筑加建的方式。其中有六种不同的加建类型。

Figure 17 Building in Xiaozhou New Village. Planned and built in the 1990s, the new village was as a residential area of three-story buildings. Over time, different additions appeared. This photograph provides a basic understanding of the diverse approaches to building additions. We have identified six types.

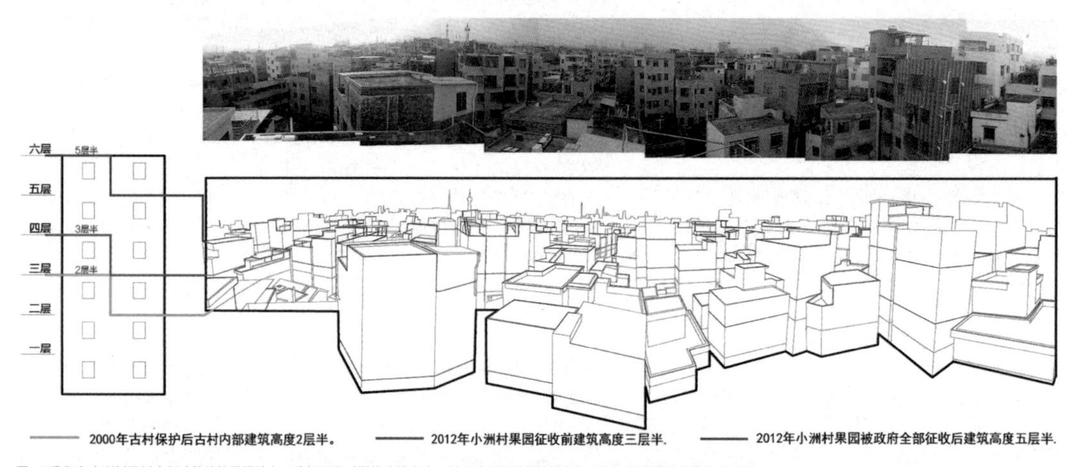

图 18 我们在小洲村旧村内部建筑的俯瞰照片上，分析不同时期的建筑高度，以三种不同颜色的线条标示出三个时期的建筑高度对比。

Figure 18 On this bird's eye photo of Xiaozhou Old Village, we have analyzed building heights over time. Three colored lines indicate the heights of buildings in the three periods.

复举报和多次砸楼板、补楼板，屋主等房子全部建好、拆掉排栅，最后才把楼板补起来。这样建成后，以既定事实为由，城管就无法再进行监管。

3.2 建筑出挑

城中村自建房常利用建筑出挑，即以悬臂梁的方式使连续梁、板，伸出基地红线范围，在建筑间距规定[1]的基础上实现空间占有最大化。一般建筑出挑部分用作阳台，而城中村自建房则把允许出挑部分直接作为建筑内部使用，而

3 The Game of Fighting for Space

Xiaozhou Village's reconstruction and rush construction period lasted from 2013 to 2015. During that time, large vehicles were unable to enter the narrow alleyways of the village and within the village itself there was no open space for the temporary storage of building materials. In the absence of centralized planning for storage, cement mixers and piles of gravel were stored at the entrance to the old village. Villagers planned and made a schedule for the use of the space to avoid conflicts. However,

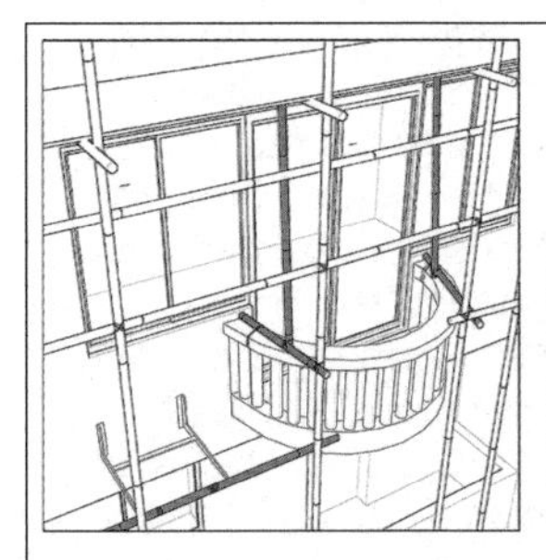

加建排栅术节点一
利用阳台护栏沿作为支撑。上部排栅重力一边由从地而起的竖直竹竿传递到地面上，一边由支撑在阳台护栏上的2根竖直竹竿传递到阳台上。通常都有一根横向竹竿绑定排山与建筑构件，诸如防盗网、窗框、阳台、空调机架等，以加强排栅水平向的稳定性。

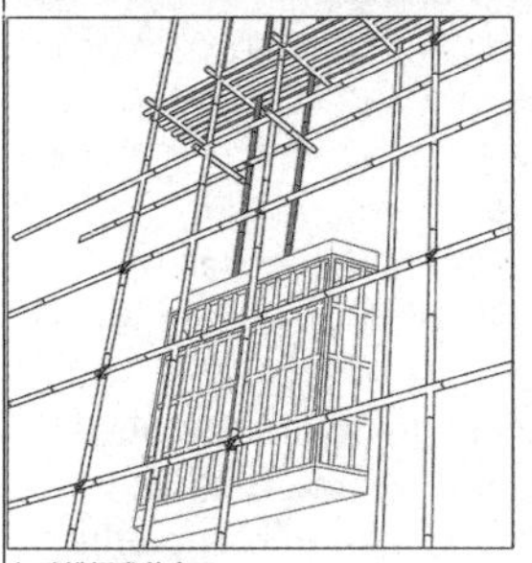

加建排栅术节点二
利用出挑窗檐作为支撑。上部排栅重力一边由从地而起的竖直竹竿传递到地面上，一边由支撑在窗檐上的2根竖直竹竿传递到窗户上。

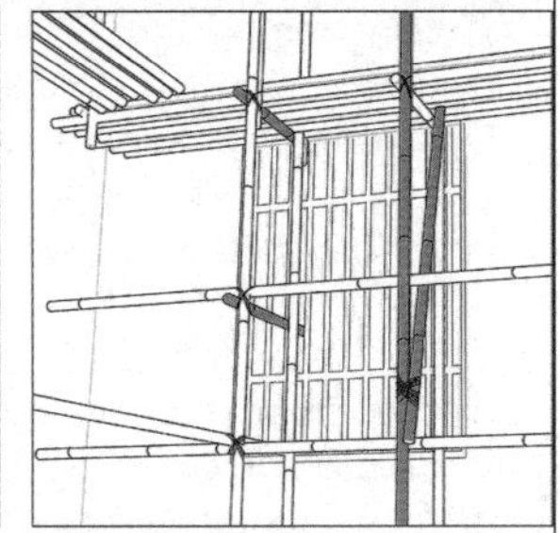

加建排栅术节点三
当没有建筑突出物作为支撑点时，利用一根竹竿作为斜撑，呈Y型把重力由从地而起的竖直竹竿传递到地面上。

图 19 不同的城中村加建方式
Figure 19 Different add-on methods in urban villages

阳台的晾晒功能则由窗户上外凸的防盗网替代（图 20）。也不乏在某一巷道集中出现因同一产权的宅基地被巷道一分为二、出挑部分用于连接两侧房子的现象。

street space was limited and vehicles passed by frequently, which made traffic jams and collisions unavoidable. This scene was a miniature of the game played between the self-built buildings and the urban regulatory framework. Villagers built on their homestead plots, securing rights within urban space. However, this process resulted in methods that bordered on the edge of illegality.

Urban and rural planning specifications include detailed provisions regarding the distances between buildings, but "the provisions do not apply to" urban villages because reconstruction must adapt to the complex legacy of the old village. In turn, this created a grey area within the game of space, giving villagers an opportunity to appropriate space for their interests. We categorized three methods of space appropriation of self-built buildings in urban villages: building heights, building overhangs, and base extensions. The former two will be elaborated.

3.1 Building Heights

The regulation regarding building heights states that "buildings should not exceed three stories, staircases and utility rooms can be added according to needs, the height of the three-story building should not be greater than 11 meters, and with the staircase and utility rooms, the total height should not exceed 14 meters. Pitch roofs are encouraged."

In Xiaozhou, the early two-and-a-half-story and three-and-a-half-story buildings that were in compliance with regulations almost completely disappeared during the rush constructions. The most common buildings exceed three and a half stories in height, with some exceeding five and a half stories (Figure 17-18). The proximity of buildings in high-density urban villages is less than an arm span, making it necessary for villagers to use flexible scaffolding techniques to ensure the safety and function of public alleyways when constructing add-ons to extant buildings (Figure 19). Their methods facilitated spontaneous construction, with each villager adopting different construction methods, facade materials, add-on forms, and temporary or permanent structures. The different types of additions reflected the villagers' ability to adapt based on their needs, to consider trade-offs between costs and benefit maximization, and to navigate the regulatory edge.

The city management bureau was very particular in its handling of illegal add-ons. In general, they avoided damaging the main beams of the legal buildings and

1 对于建筑的间距，有这样的规定：“村民住宅，主要朝向的建筑间距应不小于 6 米，次要朝向的建筑间距应不小于 2 米（建设联排式住宅的除外）。现状村民住宅改建、扩建，不得影响邻屋安全，临路方向退缩建筑间距不少于 2 米（以道路中心线为基准），其他方向退缩建筑间距不少于 0.6 米(以用地边界为基准，建设联排式住宅的除外)。最小间距范围内，不得设置阳台、飘窗、雨篷、花槽、台阶等。住宅次要朝向的外墙应当为防火墙。”

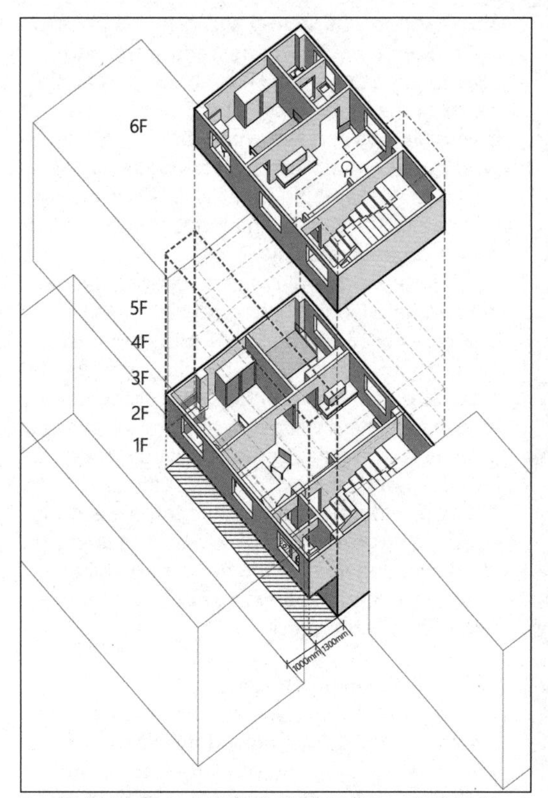

图 20 右边的白色房子是 1990 年代初建造的（从外墙瓷砖判断），早期阳台的出挑还不超出建筑基地外部，一方面离街道远一点噪声没那么大，另一方面可以有个自己的小庭院，而且当时对空间的需求没那么大。而后面的黄色房子是最近改建的，可以看出，建筑出挑已超出基地，阳台也没有了。
Figure 20 The white building on the right was built in the early 1990s (judging by the exterior tiles). The early overhang of the balcony did not reach beyond the site which allowed distance from the noisy street while creating a small front yard. In addition, the demand for space was modest at the time. In contrast, the yellow building in back was recently reconstructed. The overhang visibly exceeds the red line and there is no balcony.

图 21 这栋房子刚开始建造，遭到邻居的投诉，认为他们出挑的 1.3 米会影响自己未来加建房子的光线，于是城管禁止出挑，导致建筑每层面积减少了 10 个平方米，影响了房子的户型，房间由两房转为一房，客厅的面积也缩小了三分之一，而原本在出挑部分的厨房厕所也移到另一侧。现在从外立面上看，仅余下了一层的梁板突兀地裸露在外。
Figure 21 The landlord had just begun construction, when his neighbor filed a complaint, claiming that the 1.3 meter overhang would affect the natural light of their buildings. The city management bureau banned the construction of the overhang. As a result, the size of each floor was reduced by 10 square meters from the reconstruction plan. The two-bedroom layout became a one-bedroom layout and the living room was 30% smaller than planned. Kitchens and bathrooms that would have been placed on the overhang were relocated to other sides of the building. The remaining beams and slabs on the first floor were still visible on the facade.

preserved the structural integrity of the houses. Instead, they smashed a hole in the floor of the add-on, making it impossible to use the additional space. Villagers responded by quickly fixing the floor slab. During our research, we learned about a case involving two neighbors. The first reconstructed his house and built a five-story building. His neighbor subsequently reconstructed his house, bringing its height to five and a half stories. According to regulations, the half-floor could only be a staircase, but the second neighbor had added two small rooms. The first neighbor filed a complaint, claiming that the additional half story in height blocked his natural sunlight. Repeated complaints led to several cycles of floor smashing and fixing the floor slab. However, once the second neighbor finished his construction and removed the scaffolding, the building became an established fact, making it impossible for the city management bureau to intervene.

3.2 Architectural Overhangs

Self-built buildings in urban villages often use overhangs to maximize space. Continuous beams and plates are extended beyond the red line of the base by means of cantilever beams, and the occupation of space is maximized with respect to building spacing regulations.[1] The most common type of overhang is a balcony, but overhangs in urban villages can be programed as interior space. Instead of serving as a balcony, the convex security bars provides a place for drying clothes (Figure 20).

The Xiaozhou New Village which was planned in the 1990s complied with city regulations. The distance between two buildings facing the street is 7 meters (Figure 21-22) and 2 meters in secondary directions. As urban village housing was not subject to sunlight regulations, the length of overhangs was negotiated between the owners of adjacent buildings. If a dispute could not be resolved through mediation, the city management bureau intervened by dismantling the overhangs.

1 The building regulation with respect to the space between building is as follows. "For villagers' houses, the distance between buildings oriented in the main direction shall not be less than 6 meters, and the distance between buildings oriented in the secondary direction shall not be less than 2 meters (excepting townhouses). The reconstruction and expansion of extant villager housing shall not affect the safety of adjacent houses, and the distance between buildings retreating in the direction near the road shall not be less than 2 meters (based on the road centerline), the distance between buildings retreating in other directions shall not be less than 0.6 meters (based on the land boundary, excepting townhouses). Balconies, bay windows, awnings, flower troughs, and steps shall not be set within the minimum spacing. The secondary facing exterior wall of the residence shall be a firewall."

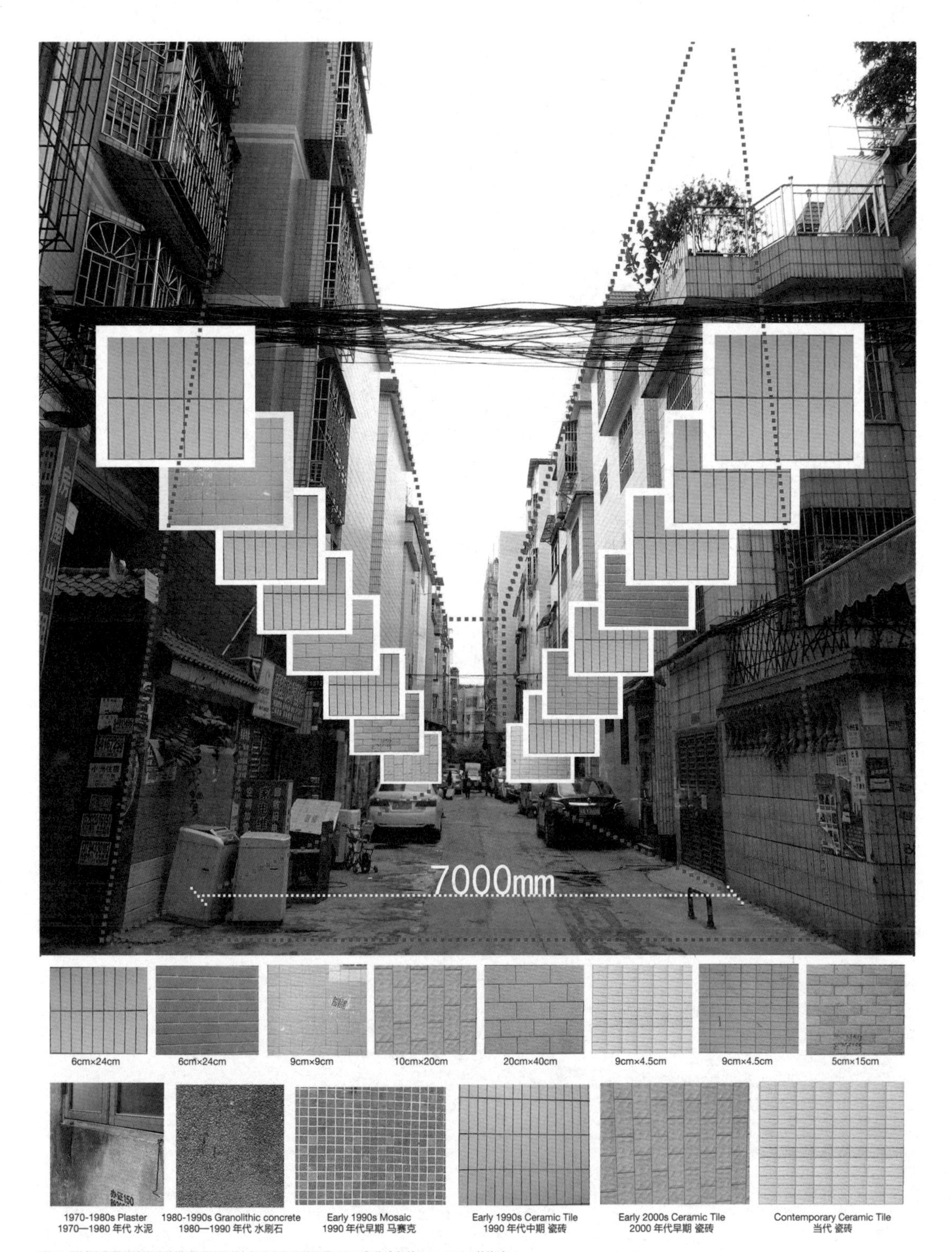

图 22 通过这张图外墙瓷砖的描述可以看到大部分房子用的都是 1990 年代流行的 6cm×24cm 的瓷砖。
Figure 22 In the photograph, the exterior wall tiles of the main buildings exemplify the 6cm×24cm tiles that were popular in the 1990s.

1990年代规划的小洲村遵循了相关的规章制度，临街朝向的建筑间距为7米（图21-图22），次要朝向的建筑间距为2米。城中村的建筑没有日照间距的规定，自建房的出挑距离大部分是由双方屋主共同协商决定，一般双方会定下相同的出挑距离，默许出挑各占街道的三分之一，留下作为采光的空间。但建筑间距的出挑与环境以及邻里关系直接相关，不同的场地问题会有不同的出挑方式。如果和邻居没有协商好出挑问题，出现矛盾时，城管会介入解决，拆除出挑。

> “‘城中村’是中国市场经济时期快速城市化与计划经济时代城乡二元体制冲突并融合的产物，城中村问题在以深圳和广州为代表的珠三角地区表现得尤为突出。”

4　正视城中村

以城中村作为展览场地的2017深双，主题为“城市共生”。建筑策展团队在“都市|村庄”板块这样描述城中村：“城中村是中国市场经济时期快速城市化与计划经济时代城乡二元体制冲突并融合的产物，城中村问题在以深圳和广州为代表的珠三角地区表现得尤为突出。”

这次城中村自建房调查持续了半年时间，城中村的建造空间让我们这些所谓的专业建筑师看到很多现实，不管从专业的建筑角度，还是从城市发展的角度，城中村都值得被给予更多的了解和正视。

注：城中村自建房调查项目人员：
许志强　广州大学建筑与城规学院教师
陈洁琦　建筑师
李文立　建造师
曾一凡　广州大学建筑与城规学院学生
李振南　广州大学建筑与城规学院学生
卢秋莹　广州大学建筑与城规学院学生
陆晓岚　广州大学建筑与城规学院学生

4　Valuing Urban Villages

Located in an urban village, the theme of 2017 UABB was "Cities, Grow in Difference." In the City/Village Intervention Practice Exhibition, the curatorial team described the urban village as follows. "The urban village is a product of the conflict between and the integration of rapid urbanization during the market economy and the residual urban-rural dual system of the planned economy. It has become a typical urban issue in Pearl River Delta cities, most notably in Guangzhou and Shenzhen."

Our investigation of self-built buildings in the urban village lasted for half a year. As so-called professional architects, we saw many actual situations in the constructed space of the urban village. From either an architectural or urban development perspective, urban villages deserve to be understood and addressed.

Note: The research was conducted by the Urban Village Self-built House Study Group. The 2017 UABB presented the results of the research.
Self-built House Study Group members:
Xu Zhiqiang, Lecturer at the College of Architecture and Urban Planning, Guangzhou University;
Chen Jieqi, Architect;
Li Wenli, Architectural Technologist;
Zeng Yifan, Li Zhennan, Lu Qiuying, Lu Xiaolan, Students at the College of Architecture and Urban Planning, Guangzhou University.

从 2016 年 9 月到 2017 年 3 月，城中村自建房小组对广州市海珠区赤沙村和小洲村自建房进行调查。城中村自建房是居民对抗城市化的自我策略，面对大资本来临时的策略。他们以最低的成本，最简单、有效的建造完成最大利益化的回报，成为城中村居民特有的生存方式，让建造作为一个人的权利得以实现。调查围绕以下三个问题展开：

（1）城中村自建房的使用需求、生活状态（制衣厂、制衣作坊与生活关系）

（2）城中村自建房的建造与成本（一栋房子的建造成本、建造过程与使用情况）

（3）城中村自建房的空间博弈（①建筑的高度；②建筑的出挑；③基地的延伸）

我们不关注建筑的自由创造，也不关注审美，只注重现实应对。这是一个来自于城中村前线的报告。

▷ 展览：城中村自建房调查 ▷ 参展人：城中村自建房调查小组

The Urban Village Self-built House Study Group has researched self-built houses in Chisha Village and Xiaozhou Village, Haizhu District, Guangzhou since September 2016. During the researching process, we found self-built house could be understood as a strategy of residents fighting against urbanism and capitalism. The residents earned their right to the city through building at minimum cost (which includes efficiency and reasonable) to pursue maximum profit.

The research unfolds with following aspects:

(1) The demands and living condition of self-built houses (clothes making company, clothes making studio and the residents inside);

(2) The tectonic and cost of self-built houses (the cost, building process, and using condition of a self-built house), and;

(3) The spatial game logic of self-built houses (① height limit; ② cantilever limit; ③ extension limit).

We only focus on reality solution rather than free creation of space and aesthetic. This would be a report from the frontline of urban village which shows how the residents build their houses under the current circumstances.

▷Exhibition: Research Toward Self-built House in Urban Village ▷Participant: The Urban Village Self-built House Study Group

来源：深圳城市\建筑双年展组织委员会. 城市共生：2017 深港城市\建筑双城双年展 [M]. 广州：岭南美术出版社，2021: 280.

摄影 PHOTO

我也不是这里的
I'm Not from Here, Either

梁永光
Mark Leong

K2

乙贰

20 世纪 90 年代，当我刚住在北京开始学普通话的时候，经常在街上迷路，所以要向路过的人问路；而他们往往是抱歉地耸耸肩，说“我也不是这里的。”

广东的城中村往往是那些“不是这里的”流动人口的一个过渡地带。它处于城乡之间、发展与建成之间。就我的纪实摄影风格而言，我长期以来都重点关注于这样的社区：它们一方面在适应新的生活方式，而同时又一小块一小块地画地为营。

近三十年间，我在中国拍摄的很多照片中都带有这种“不安定流动”的精神实质。我感觉自己也处在一个过渡地带，身处其中却并不一定有归属感：一个外乡人拍摄其他的外乡人。因此，这次展览中所展示的图片并不是想要记录时间轴线上某个单一的地理性时刻，而是希望能记录像城中村村民定义其自身过渡性空间的特定视角。

▷展览：我也不是这里的 ▷参展人：梁永光

In the 1990s, when I first started living in Beijing and learning Mandarin, I would frequently get lost in the streets and would ask passers-by for directions. As often as not, they would shrug their shoulders apologetically and tell me "Wo ye bushi zhelide"—I'm not from here, either.

Typically, the Guangdong urban village has been a transition zone for migrant people not from here, between city and countryside, between development and establishment. For my style of documentary photography, I have long gravitated toward communities like these, adapting to new ways of living while marking out small pockets of territory.

Many of the pictures I have shot over almost three decades in China share this same spirit of unsettled flux. I, too, feel like I am in a transition zone, present yet not necessarily belonging: an outsider photographing other outsiders. The idea, then, of these images shown in this exhibit is not to record a single geographical moment in time, but rather a particular point of view that, like an urban villager, may define its own transient space.

▷Exhibition: I'm Not from Here, Either ▷Participant: Mark Leong

来源：深圳城市\建筑双年展组织委员会．城市共生：2017 深港城市\建筑双城双年展 [M]. 广州：岭南美术出版社，2021: 252-253.

家园营造：香港都市中的社区参与和设计

Home Improvement: Community Engagement and Design in Hong Kong

杜鹃
多伦多大学约翰丹尼尔建筑、景观与设计学院院长、教授，香港大学建筑学院荣誉教授及城市生态设计研究室主任。在入职多伦多大学前，任教于香港大学和麻省理工学院。她是中国快速城市化研究领域的著名学者，其作品被国际期刊和媒体广泛报道。专著《深圳实验：中国速生城市的故事》由哈佛大学出版社出版并荣获 2020 年度亚利桑那州立大学人文科学研究所颁发的"跨学科研究年度最佳书籍奖"。
本文首次发表于《城市 空间 设计》，2016(5): 60-65；由王珏英译中；文字更新，2021 年。

Juan Du
Dean and Professor of the John H. Daniels Faculty of Architecture, Landscape, and Design at the University of Toronto. Du is also Honorary Professor and director of the Urban Ecologies Design Lab at the University of Hong Kong. Prior to joining the University of Toronto, she taught architecture and urban design at the University of Hong Kong and Massachusetts Institute of Technology. Du is a recognized scholar on China's rapid urbanization, and her works have been featured by international journals and media. Her book *The Shenzhen Experiment: The Story of China's Instant City* was published by Harvard University Press and received the "2020 Book of the Year Award for Interdisciplinary Research" by the ASU Institute for Humanities Research.
The article was first published in *Urban Flux*, 2016(5): 60-65; tranlasted from English to Chinese by Wang Jue; text was updated in 2021.

香港公共房屋目前需要 3~8 年的轮候时间，无法满足对可负担住房的需求。作为在香港具有代表性的非正规住房，劏（*tāng*，粤语用词，有割开、剖开的意思）房广泛地隐藏于城市中心区域，并为 21 万低收入者提供了可担负的居住空间，其在某种程度上可以而且应该为居民和城市的发展发挥作用。由于香港的公屋在短期内无法涵盖所有低收入群体，劏房将会继续存在并发挥其不可替代的住房角色。诚然，与现代公寓套间相比，劏房具有着"低标准"的生活条件和"较差"的环境质量，但其已成为低收入者的"家"，而不是一种临时"避难所"。在"宜居"和"体面"层面，我们仍然可以通过更为细致和以人为本的设计介入，对这些劏房进行改善和提升。

行走在香港街头，无论在城市中心区，还是在新城区的住宅塔楼之间，我们是很难观察或想象非正规住宅现象广泛而隐秘地存在着。尽管和亚洲大多数发达或发展中城市比较，香港有着非常完善的公共房屋政策，但在昂贵的私人住宅和公共房屋安全网之外，香港仍有很大一部分人居住于低标准的生活环境里。

由于低收入阶层的收入水平和私有市场上的高房价之间的失衡，不符合申请资格或在等候分配公共房屋的家庭就只能进入非正规住房市场，其形式如私占村地上的寮屋、顶楼加盖屋、笼屋以及劏房。此非正规住房市场带来的隐形资金交易，社会、空间网络均脱离香港法律框架的约束。执法部门对现有村地建房的严格管控，顶楼加盖屋的有限供给以及笼屋里极端糟糕的居住环境，使得分租单元——劏房成为边

With current waiting lists of 3-8 years, public housing cannot cover Hong Kong's need for affordable housing. As Hong Kong's representative informal settlements and housing 210,000 residents in urban centers, Subdivided Units (SDUs) play an important role for the people and the city. Although many units have substandard living conditions and poor environmental quality, SDUs have become homes to a significant number of residents rather than temporal and transit shelters. With understanding and adequate improvements, SDUs could continue to contribute to the Hong Kong's housing needs.

Walking on the streets of Hong Kong, whether in the central urban center or amongst the residential towers of the new towns, it is difficult to observe or imagine that there is a pervasive and hidden phenomenon of informal housing conditions. Even though compared to most developing and developed cities in Asia, Hong Kong has an excellent public housing system; there is a large population who dwell in substandard living conditions outside of Hong Kong's formal housing options of high-cost private housing as well as the safety net provided by the city's public housing system.

Due to the disproportionate high cost of housing in the private market in relation to the low-income group's earnings, many of those who do not qualify for public housing turn to the informal housing sector such as squatters on village lands, rooftop houses, cage cubicles and subdivided units. This has created invisible financial, social and spatial networks that are outside of Hong Kong legal frameworks. Stringent building control over existing village squatters, limited supply of rooftop

缘阶层最受欢迎的居住选择。

劏房指的是将正常的公寓单位分割为 2~6 个独立的居住单元，也就是说将已经非常拥挤的香港公寓套房挤入 4 个或更多的家庭，有些情况下甚至出现多代同堂的现象。这使得分租单元在几类主要的非正规住房中承载了最多的租户和边缘阶层居住人口。最新的政府报告统计显示（2021 年），香港目前有超过 9 万个家庭，或大约 21 万人居住在分租单元。但是，社区组织以及非政府机构所统计到的数字远远超过政府提供的数据——目前香港的 750 万居住人口中有约 100 万居住在分租单元。同其他城市的非正规住房不同，如中国内地的城中村和印度的贫民窟，香港公寓楼里存在的分租单元，其空间的极端密度和居住在其中的居民是不容易察觉的。

通过香港大学的社区参与项目以及与之并行开展的研究计划，我们希望能曝光广泛存在但仍普遍隐藏的香港的另一面：因贫困造成的高密度及其社会和物质条件。这个项目的长远目标是改善香港社区居民的生活条件，但这类项目一般不会出现在建筑设计师的项目列表中。

为了对当地条件和社区有更深的了解，在项目的第一阶段我们集中调研了香港仔地区。此地区有着丰富的历史文化，大量分布的老旧建筑承载了数以万计的分租单元（劏房）。此地区也是香港岛上收入中位数最低的社区之一。当前，香港仔地区老旧的无电梯公寓楼——唐楼（图 1）有着上千的居民居住其中。此种特色鲜明且颇具代表性的唐楼在香港依旧大量存

houses and the extremely poor environments of cage cubicles had made one particular type of dwelling the most popular choice for the marginal groups—SDUs.

SDU refers to rental units that take a regular apartment and subdivide them further to make 2-6 individual units, meaning the already squeezed Hong Kong-sized apartments would accommodate up to four or more families, some of multiple generations. As a result, SDUs accommodate the largest number of households and population amongst the unenviable options. The latest governmental report (2021) estimates over 90,000 families, or around 210,000 people, lives in Hong Kong's subdivided units. However, community groups and NGOs present a far greater number – one million out of the current 7.5 million residents of Hong Kong. Unlike informal housing settlements in other cities such as the urban villages in China or slums in India, the extraordinary density of space and dwellers inside builds that houses SDUs are invisible.

Through community engagement and research projects as well as parallel research seminars at The University of Hong Kong (HKU), we aim to expose the prevalent yet normally hidden aspect of Hong Kong: the social and physical conditions of poverty-based density. The project further aims to improve the quality of life for a community of residents that normally would not receive the services of architectural designers.

In order to gain in-depth knowledge of local conditions and communities, the first phase of the project concentrated on Hong Kong's Aberdeen Area, Southern

图 1 位于香港岛南区香港仔地区的一栋典型的唐楼建筑，大约建于上 20 世纪 60 年代。
Figure 1 A typical walk-up building located in Aberdeen Area, Southern District, Hong Kong Island, built in the 1960s.

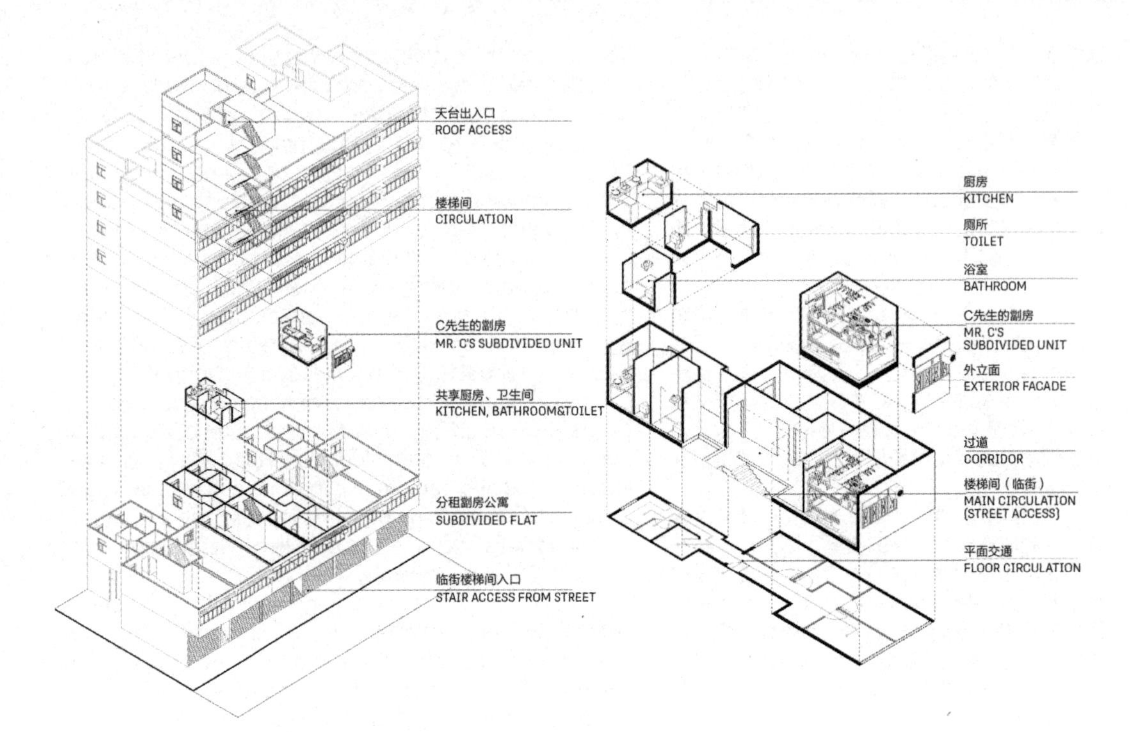

图 2 对每一个劏房进行多尺度而系统的空间现状分析，从而找出特定的空间设计问题。
Figure 2 In-depth spatial analysis of each subdivided unit (urban, building, floor, and unit scales) was conducted in order to identify the design issue.

在着，由于其租金相对低廉，居住环境较差，致使它们成为劏房的热门空间载体。

我们对唐楼以及其中的劏房以文献报告和实地考察的方式对其居住空间条件和生活状况进行调查（图 2）。通过问卷的形式，我们在征得愿意参与本项目以改善居住条件的租户同意的前提下，调研了上千个劏房家庭。通过实地探访和测绘，劏房所在建筑和室内的情况在图纸上一目了然（图 3）。此时，图纸成为特别的建筑工具——不是用来预见未来房屋的空间形式或者房地产开发建设，而是通过图纸透视城市中隐形的人居生态。

通过香港大学社区外延计划（香港大学同行计划），我们与明爱香港仔社区中心（非政府社区组织）合作，走访调查了大量的劏房，最终选定 6 个家庭的居住空间作为项目的首期研究与实践对象。社区中心的社工协助我们与这些家庭建立起联系，使我们能够和各个家庭成员进行深入沟通，了解他们亟待改善的居住区域（图 4）。我们发现，很多方面需要改善，如漏水、发霉的天花板和墙面、糟糕的通风条件、自然采光不足以及最为棘手的问题——空间的极度拥挤。当一家四口栖居于仅 13 平方米的单元之内的时候，空间开始变成最宝贵的资产（图 5）。为了能更好地理解其空间的局限性和亲密性，在一些单元的调查测绘中我们还详细地描绘出了房间内的每一件物品（图 6）。

本项目中一个非常困难的方面是我们为之工作的是分租单元里的租户而不是房东，因此

District. This area has a rich past history and a great number of older buildings currently hosting thousands of SDUs. The district also has the lowest median income of all communities on Hong Kong Island. The aged walk-up buildings (*tong lau*) (Figure 1) in the Aberdeen Area are hosts to thousands of residents living in this type of dwellings. This distinctive Hong Kong building typology exists all over Hong Kong and often hosts SDUs due to their relatively cheap rent and poor physical conditions.

We conducted literature and field research on the living situations as well as the physical conditions of the SDUs and the host buildings (Figure 2). Through questionnaire surveys, hundreds of units were investigated soliciting families who are interested in participating in this project to better their living conditions. Along with user surveys, mapping and drawings of both the building and the interior conditions were created (Figure 3). One agenda is to use the analytical drawings as an architectural tool—not to speculate for future housing and real estate development—but for making an invisible condition in the city visible.

Working with HKU's community outreach program (the WAY Project), we collaborated with Caritas Community Centre - Aberdeen, a district office of the NGO. Over the numerous home visits, we slowly narrowed down to six families for the initiation phase of the project. Social workers at the community center facilitated the home visits and allowed us to have in-depth conversations with the families on areas they would like to see

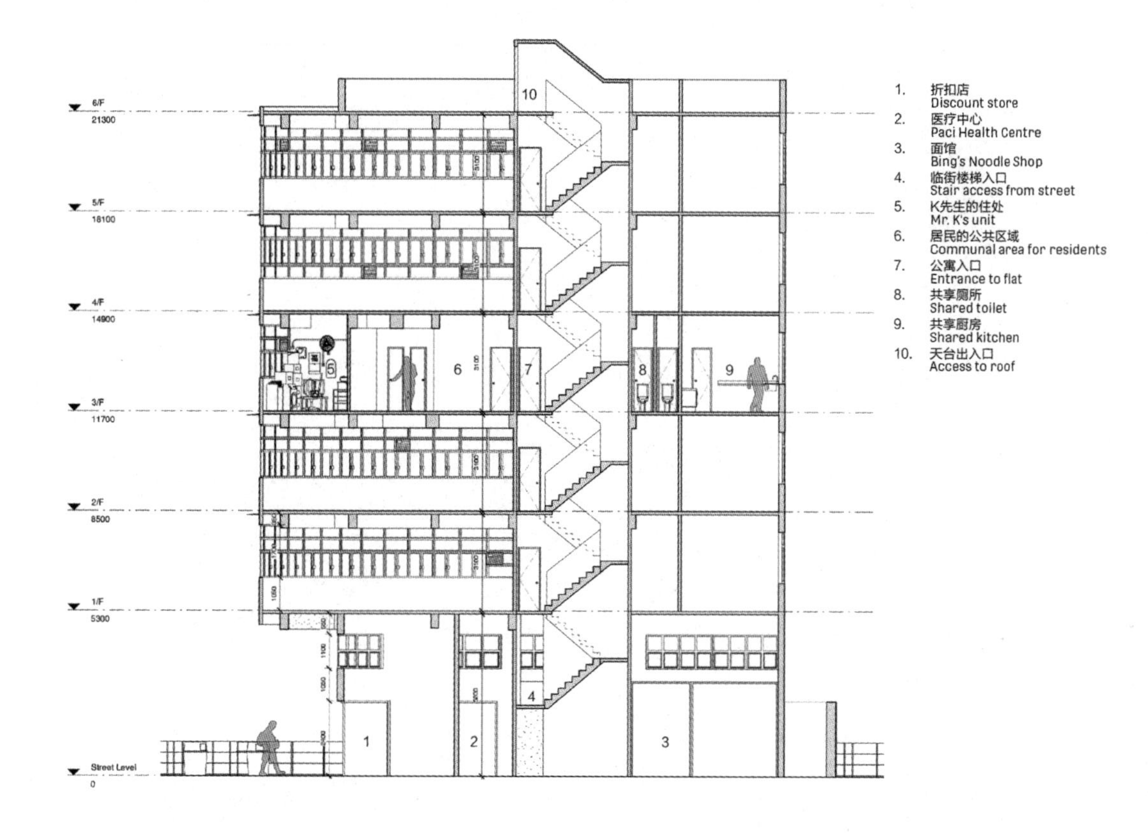

图 3 本项目中某劏房单元所处建筑的剖面图、平面图及该劏房单元的详细空间剖面图。
Figure 3 Location of a SDU in the building (building section, floor plan) and detailed section of the SDU.

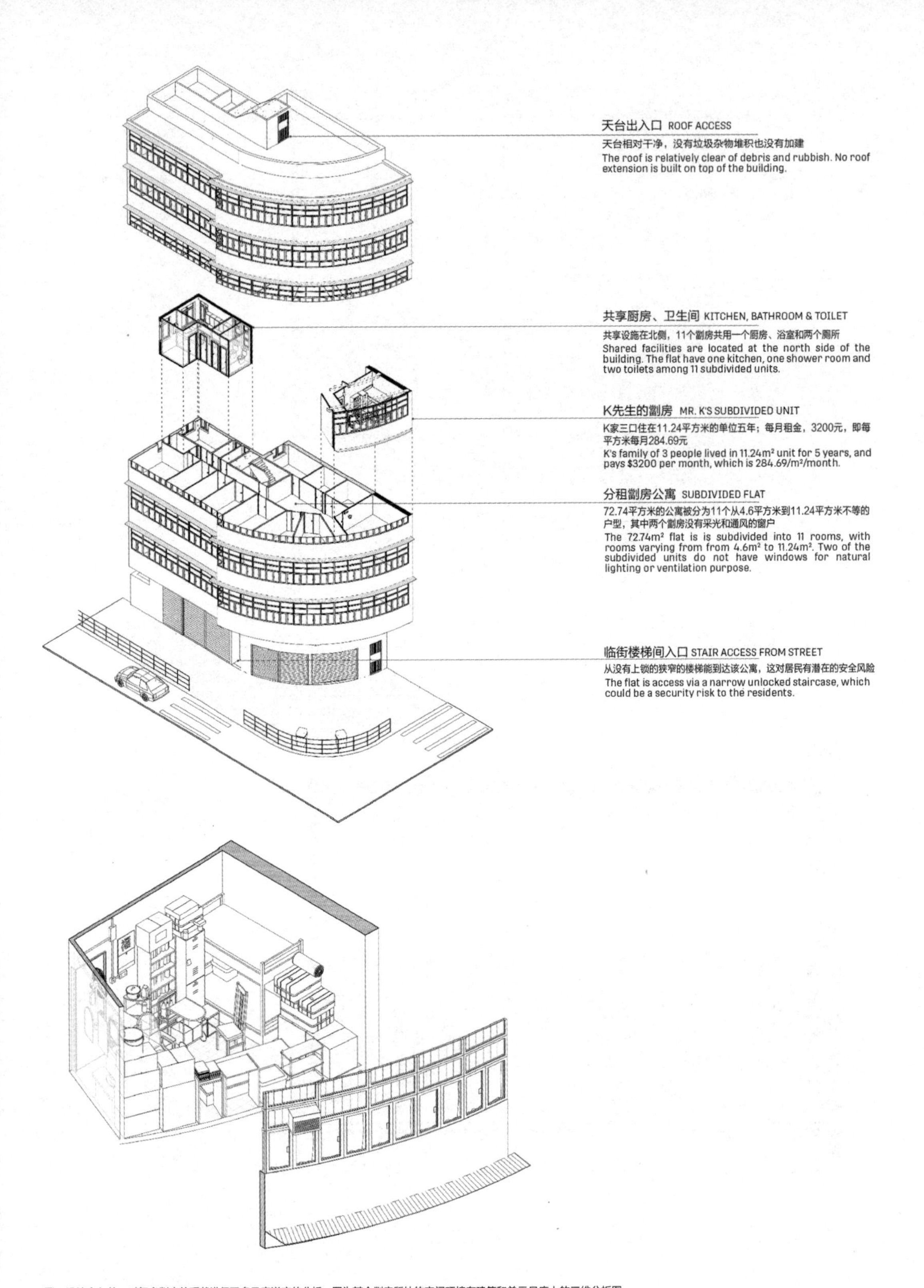

图 4 设计介入前，对每个劏房的现状进行了多尺度详实的分析。图为某个劏房所处的空间环境在建筑和单元尺度上的三维分析图。
Figure 4 Before design intervention, the urban and spatial contexts of each SDU were systematically analysed. This image shows 3D analysis and visualization of the SDU at building and unit scales.

图 5 设计介入前劏房拥挤的室内空间现状。
Figure 5 Photos of the interior "extreme-density" situation of the subdivided units.

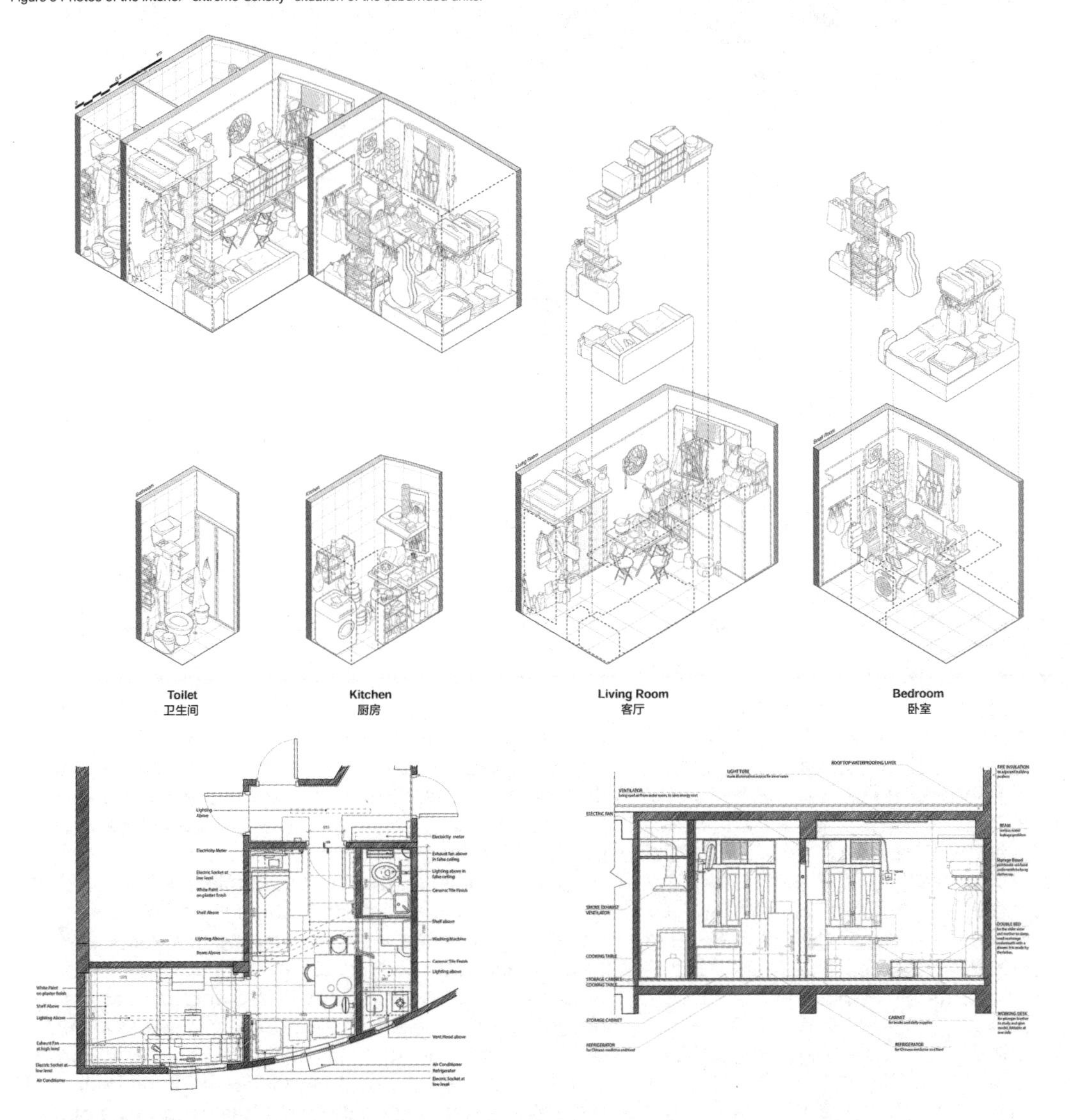

图 6 通过实地探访和测绘，劏房所在建筑和室内情况以及房间里每一件物品所处的空间状态在图纸上一目了然。
Figure 6 Along with user surveys, mapping and drawings of both the building and the interior conditions were created. In order to understand the constraints and intimacy of the space, every object in the rooms were drawn for some of the units.

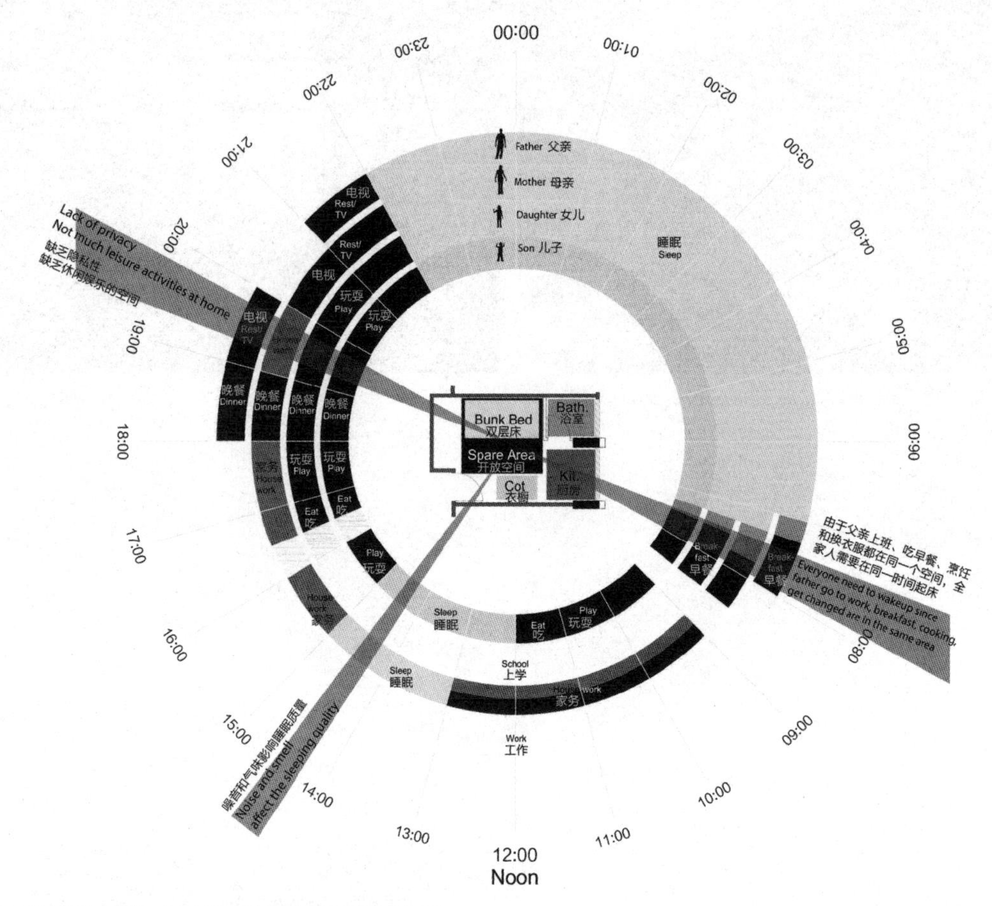

图 7 设计介入前期学生们对每一位家庭成员日常生活规律与活动轨迹的可视化分析。
Figure 7 A visual analysis of the daily routine and activities of each family member living in the subdivided unit.

在改造中不能够改动墙体结构和房屋平面。此种状况再加上社会项目非常有限的资金支持，促使我们的设计致力于以最轻型的介入获得最大的改善，尽量避免或减少对已经非常有限的空间增加负担。在与租户家庭的交流中，他们最需要得到关注并改变的是居家空间中最基本的功能区——卧室和收纳空间。在课程中学生们画出平面和分析图（图 7），制做模型，并同租户家庭进行深入交流（图 8）。从最初概念的提出到最终设计的完成，所有小组的设计方案都经历了显著的调整和改变，因为每个家庭对自己珍爱的居住空间都有着非常独特的打算。

项目的结果都是以多功能为导向解决在有限的空间及预算的情况下满足多种不同的活动需求。主要的营造尺度都是家具大小，以满足最基础的功能，如休息、吃饭、起居、学习。这些功能都在一个共享的空间内被重新构思，在一个单间内实现两代或三代同堂（图 9）。一张抬高的床在考试季兼具半私密学习空间的功能，三叠床可以转换为用于休息和看电视的沙发，双人床可以拉出变成一张书桌，一张大

improved. And there were many aspects to their living conditions that were in need of improvements (Figure 4): water damage, molded ceiling and walls, poor air ventilation and circulation, poor natural lighting condition… all these are compounded by the biggest problem of all – lack of space. When a family of four is living in 13 square meters, space becomes the most precious commodity (Figure 5). In order to understand the constraints and intimacy of the space, every object in the rooms were drawn for some of the units (Figure 6).

A very difficult aspect of the project is the fact that we were working with renters of the units so changes could only be made without changing the wall structure or general layout. This condition along with a limited funding for the social project limited and focused the design on light-weight interventions and at a scale that does not further overwhelm the limited space. In conversations with the families, the most areas to be improved were slowly focused on the most basic functions of a home—sleeping and storage. Throughout the course, the students made plans, diagrams (Figure 7), and models

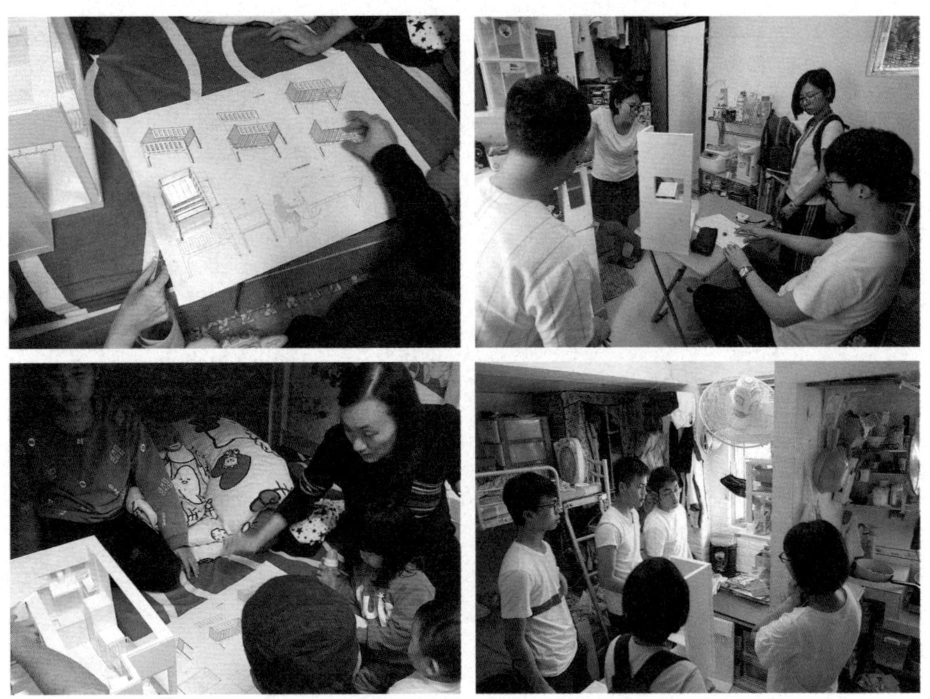

图 8 设计方案生成阶段学生们和劏房居民进行多轮深入交流和共同设计。
Figure 8 Students deeply discussed with the subdivided unit residents to develop the design in order to achieve their real needs.

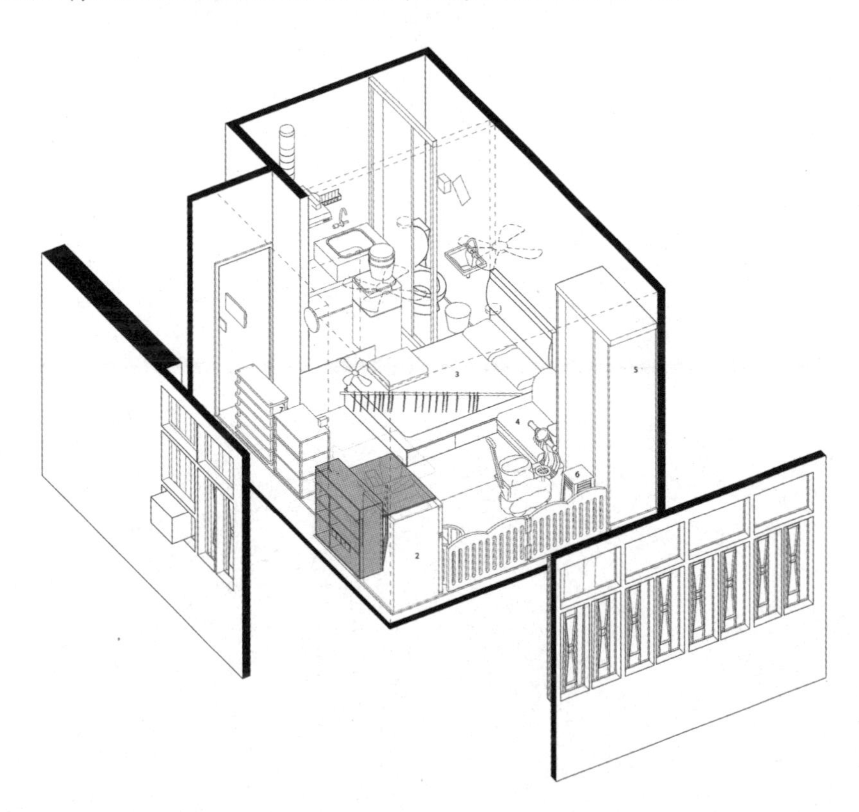

图 9 由于建筑空间极度压缩，设计需要充分考虑每个物件在空间中的角色与状态以及它们与各个空间界面和建筑要素的关系。
Figure 9 Due to the "extreme-density" of the interior space, our design intervention needed to carefully consider the role of each object and its relationship with and impact on the space and architectural elements.

床可以让父母和孩子都睡下。在周末邻居好友拜访的时间段，餐桌可以变成孩子的床；储藏空间可以作为餐厅亦可灵活地分隔为三个成年儿子的私密空间（图 10）；墙体的嵌柜可作为储藏、学习和餐饮空间，最重要的是它可以将电饭锅和电水壶等散热装置同两个年幼的孩子分隔开来。

我们的设计充分考虑了居民的日常生活习惯以及楼宇空间的限制（图 11）。此项目不仅仅是关于在极限条件下的设计介入，在和特定使用者沟通过程中，给我们提供了在极简条件下检验建筑营造最核心的内容——空间和形式。最后的建造和安装阶段也是重要的学习过程。一个特别的方面就是所有的组件都可以拆开来运输并组装，因此我们需要在安装前预先考虑所有组件的强度以及将来劏房租户对便捷搬运的需求（图 12）。通常在安装过程中，租户会有更多的参与感并表达出他们的兴奋（图 13）。虽然装置设计得较为朴实，但是它们给我们以及租户带来的却是一定程度上的新鲜感

to communicate with the families (Figure 8). From initial conception to final design, all projects went through significant revisions and changes as the families had very defined opinions on their cherished living space.

The resulting projects are all multi-functional to accommodate the various activities within the limited space and budget. Basic functions of furniture-scaled constructions—sleeping, dining, living, and studying are reconceived to adapt the unique constraints of a shared space—a single room for two or even three generations (Figure 9). A loft-bed with storage space to be used a semi-private study space during exam time; Triple deck bed that can be converted into a TV-watching sofa; Double bed that can be pulled out forming a study desk and a King-sized bed to allow parents and child to sleep together; Dining table to can be converted into a child's bed during weekend visits; Storage unit that functions as dining space as well as room divider to provide some privacy for the 3 grown sons; and wall cabinets for storage, studying, dining, and most importantly it organizes

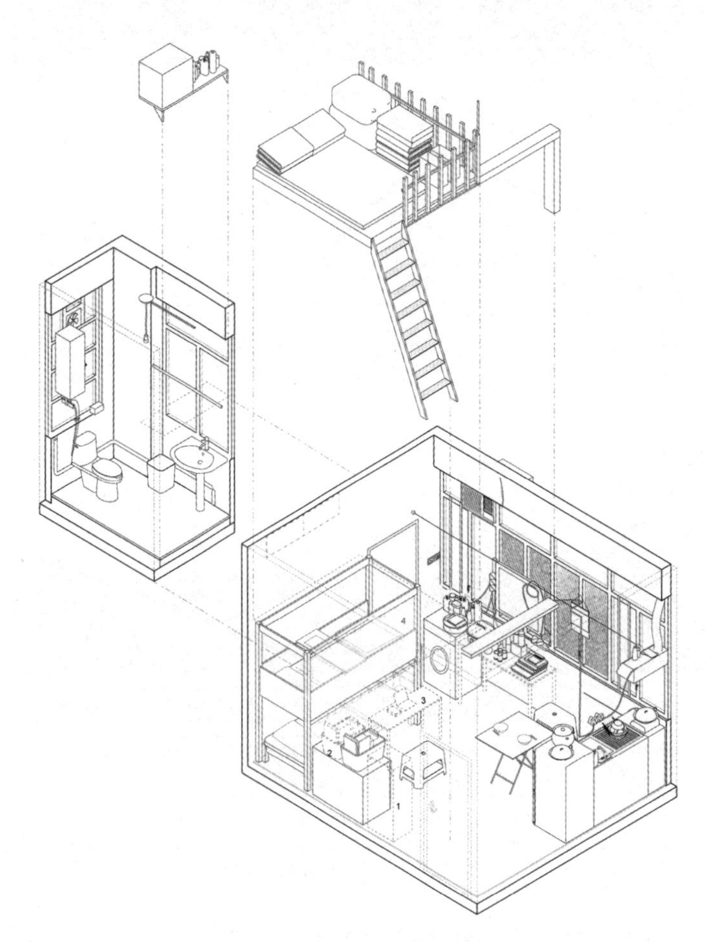

图 10 三维空间分析图显示我们的设计介入在满足居民日常生活功能需求的前提下帮助其“拓展”了空间的利用率。
Figure 10 As showed in the 3D drawing, we tried to make full use of the original space and "extend" the space for residents' daily use.

和惊喜感，以及我们和参与家庭之间的亲密感（图 14）。

在当代亚洲城市中，对建筑师和学者来说，通过居住者日常生活来细致观察多样化的城市非正规空间，逐渐成为一种有效的学习与研究过程。这些课程通常可以反映出正规规划及设计标准在满足居民日常生活所需方面的不足之处。本计划拓展了城市研究的视角和范围——尝试推出一种从开放的街道走进隐秘的室内空间的调查和研究方法。

and keeps heat-emitting appliances away from two young children (Figure 10). The designs learn from the daily living habits of the residents as well as the constraints of the building type (Figure 11). This project not only exposed the process of working with constraints, communicating with defined users, it also allowed an opportunity for us to examine the most core issue of architecture—form and space at the most pared-down and essential level. The final construction and installation phase is also a learning process. One unique aspect is for all the pieces

图 11 置入我们所设计和制造的"A 型架"装置后，劏房的室内空间变得宽敞明亮了许多，增加了室内空间的使用效率。图为该劏房居民正在使用该"A 型架"装置。
Figure 11 With the interventions of the Adobe Shelf, the interior space of the subdivided unit becomes more organised and "spacious," the interior ventilation and natural daylighting conditions were significantly improved.

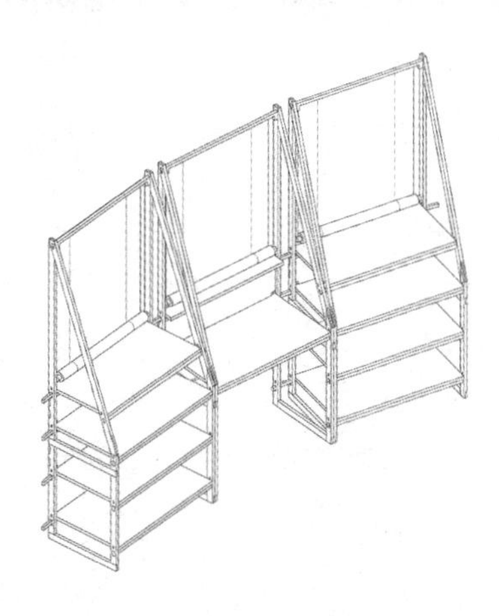

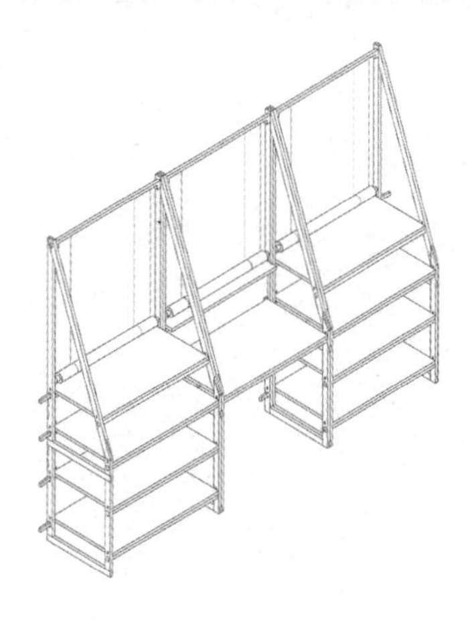

图 12 经过特别设计后可灵活组合以顺应不同墙体界面的"A 型架"装置三维模型。
Figure 12 The 3D models of the Adobe Shelf and different ways of installation and use to fit the interior walls.

图 13 香港大学建筑学系学生及劏房户居民对设计和制造完成的“A 型架”装置进行现场安装与调试。
Figure 13 HKU Architecture students worked together with the subdivided unit residents to install and test the Adobe Shelf on site.

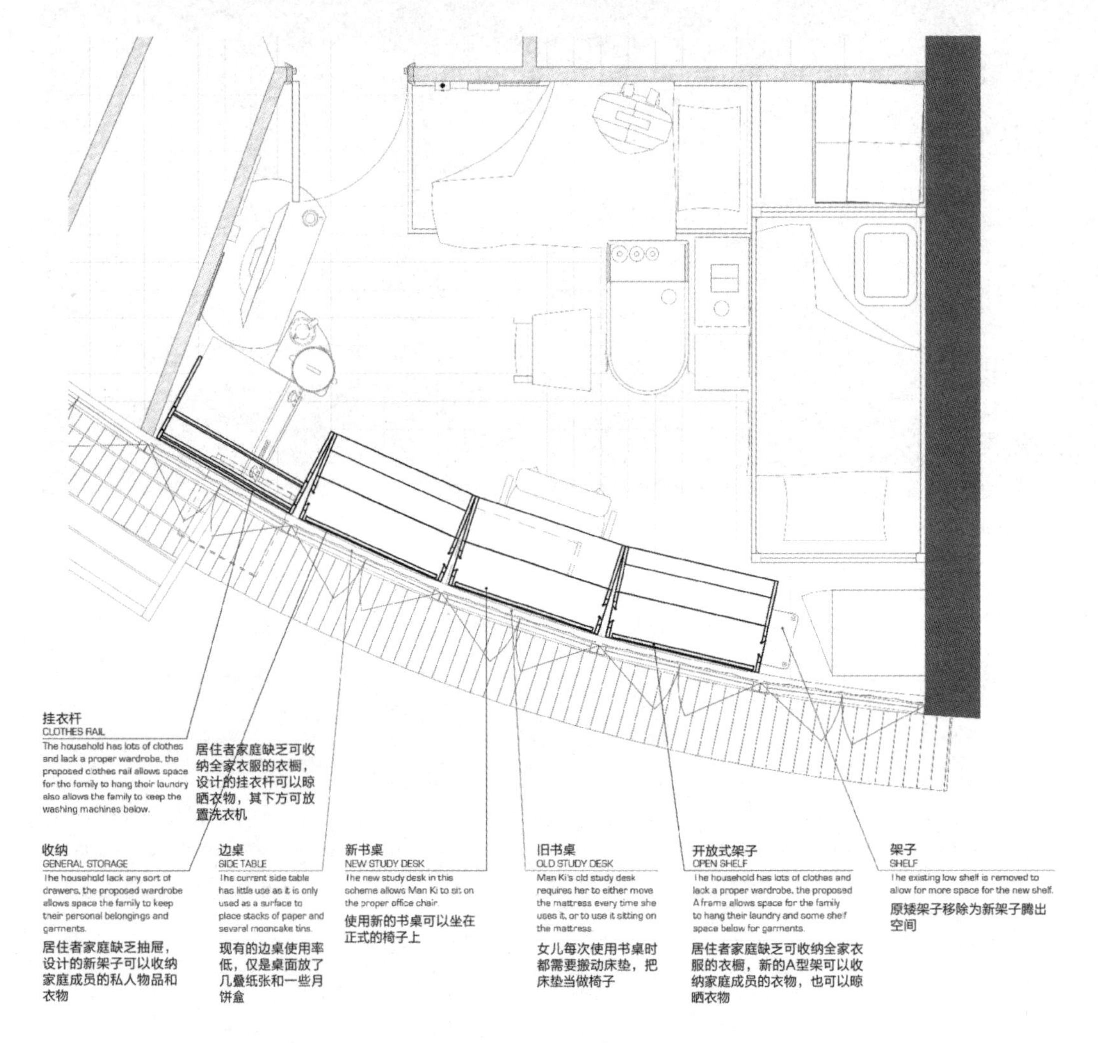

图 14 本项目某劏房空间改造后的详细平面布局（三个“A 型架”装置的介入）。
Figure 14 Detailed floor layout of the subdivided unit after our design intervention (with the installation of three Adobe Shelves).

注：本文首次发表时，作者任香港大学建筑学院副院长、副教授及城市生态设计研究室主任。继上述2015年“家园营造”的领航计划落成后，香港大学城市生态设计研究室的团队持续与明爱社区中心、香港社区服务组织以及香港仁人家园合作。在杜鹃的带领下城市生态设计研究室目前已为超过200位劏房居民改善居住环境，为有紧迫需求的低收入家庭营造过渡转型住房、为无家者营造共居单位，以及营造共享社区空间让社区工作者服务更广大的居民，如食物派发及公益咨询服务。

to be able to be taken apart and reassembled (Figure 12), this is both in anticipation that we needed to strength test every piece prior to installation, as well as the migratory tendency of residents in SDUs. Often times, it is during the construction process that the residents got even more involved and showed emotions of excitement (Figure 13). As modest as the new installations are, they bring a certain joy of newness, unexpectedness, and intimacy to the families (Figure 14).

Careful observations of the dynamic urban informality and learning from the everyday life of the occupants have emerged as a productive learning process for architects and researchers working in contemporary Asian cities. These lessons often demonstrate the deficiencies with which our formal planning and design measures have failed in meeting the everyday needs of the residents. This current research and design project on Hong Kong's Subdivided Units extends the investigation and methodology from the open streets to the hidden interiors of the city.

Note: The author served as Associate Professor and Associate Dean, and director of the Urban Ecologies Design Lab (UEDL) of the Faculty of Architecture at HKU when this article was first published. Following the completion of the pilot project "Home Improvement" in 2015, the author and UEDL have continued to work with NGO collaborators such as Caritas Community Centre, Society for Community Organization, and most recently, with Habitat for Humanity Hong Kong. These research and community design projects have improved the living conditions of hundreds of residents living in SDUs, created new transitional housing units for families in need, co-living units for the city's homeless residents, as well as provide community spaces for the delivery of social services such as food distribution and consulting services.

香港众多低收入工作家庭因为各种原因，迫不得已地居住在不符合设计规范的分租单元（又称“劏房”）里，使多代同堂居住在一个狭小房间的现象越来越普遍。通过与社会工作者、社区合作伙伴和当地的居民进行合作，香港大学城市生态设计研究室的学生及研究人员展开了一项为期两年的研究和实践项目。

▷展览：合作居住设计：小尺度的长期社区参与 ▷参展人：香港大学城市生态设计研究室（UEDL）

With multi-generations cramped into room-sized apartments, the generally unregulated Subdivided Units of Hong Kong are often the housing of last resort for many low-income working families. Working with social workers, community partners, and most importantly the residents, researchers and students of Hong Kong University's Urban Ecologies Design Lab conducted a two-year project to create small-scale upgrades and improvements to the housing units.

▷Exhibition: Co-Design: Long-Term Community Engagement through Small-Scale Home Improvements ▷Participant: HKU Urban Ecologies Design Lab (UEDL)

来源：深圳城市\建筑双年展组织委员会．城市共生：2017深港城市\建筑双城双年展[M]. 广州：岭南美术出版社，2021: 279.

摄影 PHOTO

九龙城寨
Kowloon Walled City

格雷戈·吉拉德
Greg Girard

乙贰

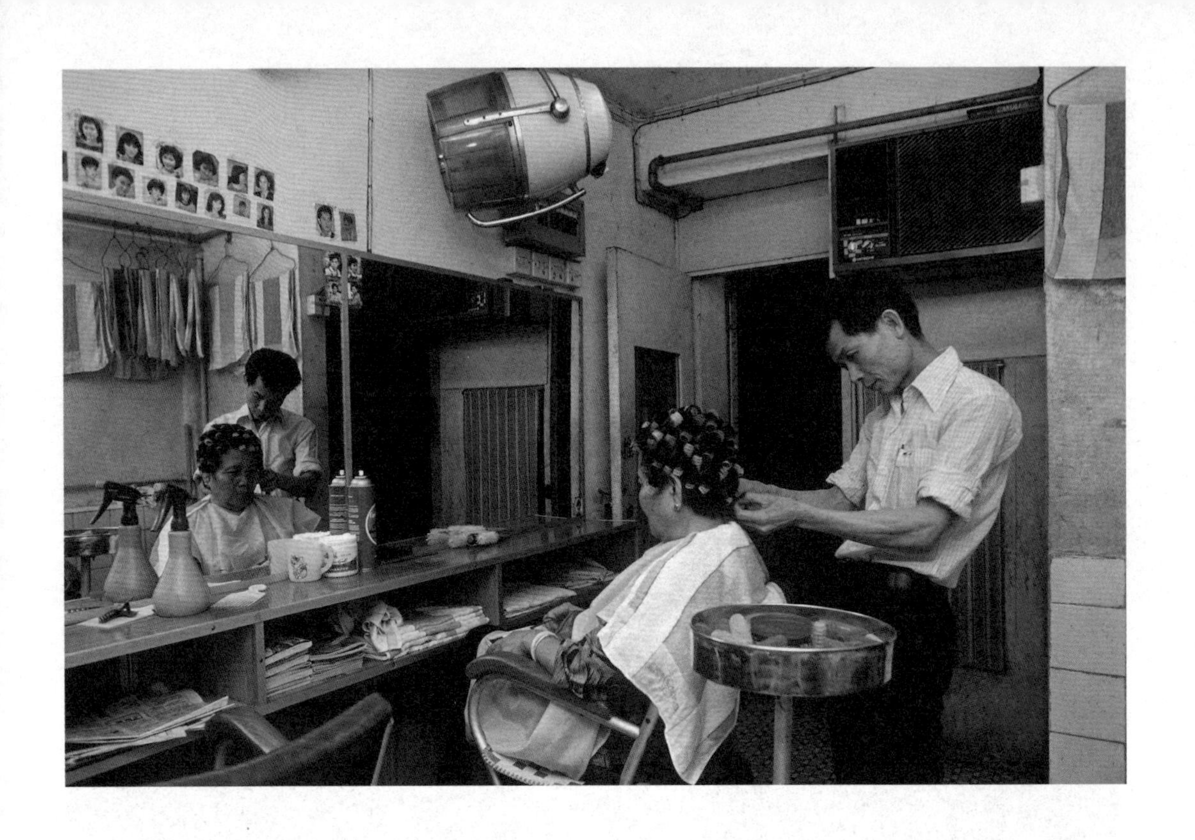

位于香港的九龙城寨，已从人们的视野中消失了近二十年。这座承载着几代人记忆的、“死去”的城中村，在由人群聚落而成的非正规建筑类型里，具有不容忽视的历史意义。

九龙寨城由于人口的高度密集，成为地球上最富“魅力”的贫民窟，也使其成为西方和日本小说、漫画、电影及游戏中出现频率最高的贫民窟。

1986 年，加拿大摄影师格雷戈 · 吉拉德被九龙城寨吸引——“这里是密集的有机建筑群，成千上百的居民会在黄昏的阳光下沐浴，活像是中世纪的场景”——吉拉德这样描述自己对城寨的第一印象。他在四年内踏进城寨不下百次，用相机为香港底层的庶民生活留下宝贵的记录。

▷展览：九龙城寨 ▷参展人：格雷戈 · 吉拉德

The Kowloon Walled City in Hong Kong has disappeared from people's sight for nearly 20 years. This "dead" urban village, which carries the memories of generations of people, has historical significance that cannot be ignored in the informal building type formed by crowds. Due to its high population density, the Kowloon Walled City became the most "charming" slum on the planet, making it the most frequently seen slum in Western and Japanese novels, comics, movies, and games.

In 1986, Canadian photographer Greg Girard was attracted by the Kowloon Walled City. "This is a densely packed organic building complex, where hundreds of residents will bathe in the evening sun, like a medieval scene," Greg Girard described his first impression of the walled city like this. He stepped into the walled city no less than a hundred times in four years, using his camera to leave a valuable record of the lives of the people at the bottom of Hong Kong.

▷Exhibition: Kowloon Walled City ▷Participant: Greg Girard

来源：深圳城市 \ 建筑双年展组织委员会 . 城市共生：2017 深港城市 \ 建筑双城双年展 [M]. 广州：岭南美术出版社，2021: 278.

城中村家具交换计划

Urban Village Furniture Exchange Program

黄河山
设计师、艺术家，毕业于清华大学美术学院，作品有《野生设计》《假宜家》《秃力富》。
本文由罗祎倩中译英。

Huang Heshan
Huang Heshan is a designer and artist, graduated from Academy of Arts & Design, Tsinghua University. His works include *Wild Design*, *Fake IKEA*, and *TooRichCity*.
The article was translated from Chinese to English by Luo Yiqian.

1　交换故事：一切都是关于"人"

在考察野生设计的时候看过很多由小板凳构成的公共空间，这种地方虽然场地条件很差，但是只要有板凳出现瞬间就会变成人群聚集气氛融洽的公共空间。像三个小板凳就是几个小朋友每天吃饭的固定地点，野生设计已经与他们的生活紧密地融合在一起，是他们不可分割的生活伴侣。

关于饼盖小板凳，小孩的父亲对我说："你是艺术家啊？我不跟艺术家说话的。"有一次交换凳子时，它的主人——2岁的小宝宝看到我要拿走他的凳子马上哭着伸手抓住不放。在项目过程中听到最多的就是不舍得，虽然这些小板凳不值什么钱，但在长期使用中人们已形成很深的感情。设计是关于人的学问，除了掌握设计技术，对人的了解和关怀同样重要。

1　Exchange Story: Everything is about People

When investigating wild design, I saw a lot of public spaces composed of small stools. Even when a site was in poor shape, as long as there were stools in place, that site immediately became a public space, where people gathered harmoniously. For example, the three small stools were located at the same place where several children ate every day. In this way, wild design has been closely integrated into their life, becoming an inseparable companion.

About the biscuit-jar stool. The child's father said to me, "Are you an artist? I don't talk to artists." When I was exchanging for the stool, its owner, a 2-year-old child, saw that I was taking away his stool. He immediately cried, and held on to it. While implementing the Exchange Program, the most common thing I found was that the owners were not willing to give their stools away. Although the small stools were not worth much, people had deep feelings for them over the course of long-term use of the stools. Design is a form of knowledge about people, thus it is important to understand and care for people along with mastering design skills.

× •••

我可能看到了一个假的宜家

记者 **黄河山** 2017-12-14

农村致富经 稿

近日记者发现，我村黄二狗和隔壁村姜铁牛借了一万块钱回乡创业，在村口小卖部旁边开了一家店。村里的妇女们议论纷纷，于是记者迅速找到他们了解相关情况。

记者：二狗，请问你开了一家什么样的店引起大家关注呢？

黄二狗：请叫我黄老板。

记者：黄老板，请问您开了一家什么样的店引起大家关注呢？

黄老板：我们是定位面向全村所有普通家庭，集生活美学和工匠精神于一体的日常家具品牌，我们把它命名为“全家家居”。不管是洗脚、喝茶、聚众打牌还是霸占车位，村民都能用到我们的家具。全家的产品比市面上出售的塑胶小板凳质量要好上几个量级，怎么坐都不会坏掉。我们立志要冲出本村，成为全镇最良心的家居品牌！

记者了解到，为了让品牌显得高端大气上档次，黄老板特意花了 200 块钱重金邀请县城最大图文打印店“先进员工奖”获得者刘哥设计了本店 LOGO。

姜铁牛：不管是在巷口写作业的狗蛋，还是在公园斗地主的欧阳，都能在全家找到最适合他们人体工学和生活习惯的小板凳，我们的口号是“为每个人而设计”。为了寻找最具匠心和智慧的设计师，我们跑遍了全城所有的城中村。这些人是生活的艺术家是技术的巨人，不但很好地控制了生产成本，还能为顾客设计出最美观的外形。有他们的加入，我们全家家居前景一片光明！

× •••

记者：请问全家家居与这些设计师将会产生什么样的化学反应呢？

黄老板：正是他们给全家家居带来了让人尖叫的设计，每一个产品都遵循低价、实用、美观相结合的原则，我将其称为“民主设计”，因为全家相信，不论贫富贵贱人人都应该拥有属于自己的好家具，全家家居就是为广大老百姓而生的！

“全家家居”与这些设计师合作推出了超级赞小板凳系列作为他们的首发王牌产品。

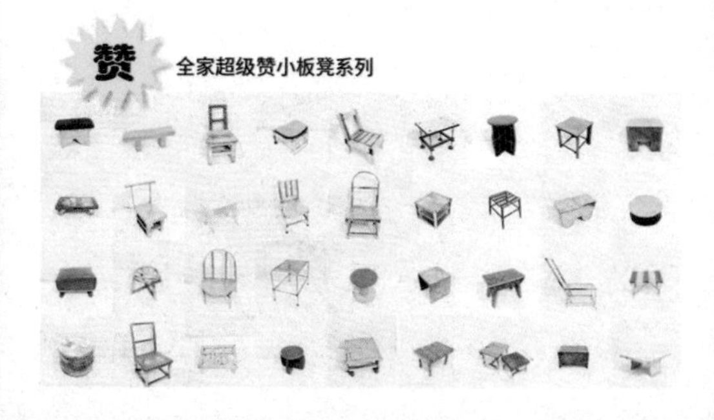

× •••

黄老板： 为了这个系列能成功上架，我们设计师倾注了大量心血设计出很多有趣又实用的产品，背后满满的都是故事。我给大家介绍一下。

IKEA
全家家居

新 STARWONG 菲利普·斯达黄
飘移凳，15x35 厘米，不锈钢
¥ 50.00

黄老板： 像这个“跑の月饼盖儿”就是出自我们湖南设计鬼才菲利普·斯达黄之手，修理自行车出身的他天生对带轮子的东西感兴趣，这个月饼盖与轮子的奇妙组合再次证明了他信手拈来的天才创意，它的可移动性也让修车变得前所未有的方便快捷。

IKEA
全家家居

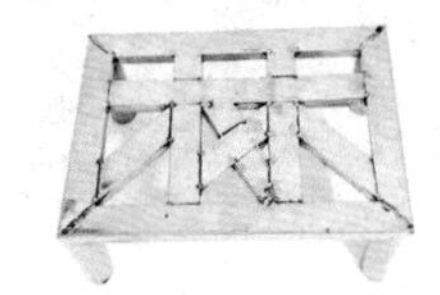

新 KOBAYASHI 小章林
定制凳，5x15 厘米，不锈钢
¥ 25.00

黄老板： 我们还邀请了著名字体设计师小章林携手全家家居推出高级姓氏定制服务，顾客提供汉字姓氏由我们来制作出精致的不锈钢姓氏小板凳。小章林这个汕头人让我们真正见识到跨界的魅力！

IKEA
全家家居

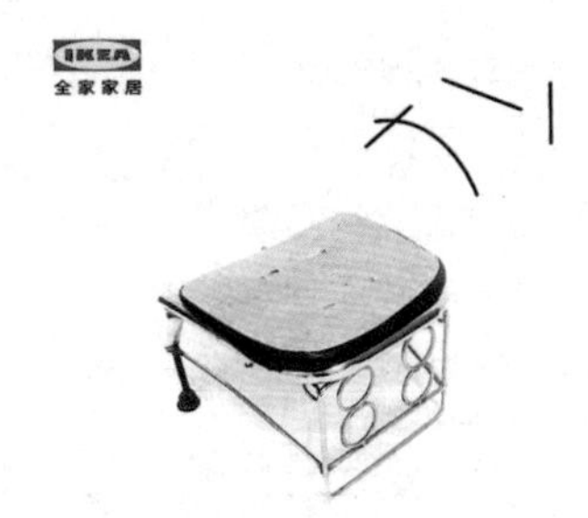

新 GEHRYZENG 弗兰克·盖曾
后座凳，15x20 厘米，不锈钢
¥ 20.00

× •••

黄老板： 另外著名的解构主义大师弗兰克·盖曾也决定与全家合作一款小板凳。此前他的设计因为成本高昂广受诟病，这次他为了困难群众一改过去奢侈作风，主动把定价从 30 元下降至 20 元。

IKEA LV
全家家居

新 LIANGCOBS 马克·梁可布
木门凳，25x40 厘米，木板
¥ 25.00

黄老板： 我们在走低价路线的同时也满足人们对高端产品的渴望，所以我们与镇上奢侈品牌路易·威丁的前总监马克·梁可布共同推出合作款小板凳，就连 LV 总店都跑出来举报：这很 LV。

IKEA
全家家居

新 MIESYU 密斯·凡·德·余
爱之凳，20x40 厘米，不锈钢
¥ 30.00

黄老板： 全家家居的产品除了经济实用以外还具有人文主义精神，比如这个由密斯·凡·德·余设计的爱之凳就带有浓烈的浪漫主义色彩，他当年就是靠给女朋友设计这个储物小板凳，在最穷的时候成功将女朋友娶回去当老婆的，十分感人泪下。

（说到这里，黄老板不禁哽咽。）

记者： 黄老板请冷静一下，还有没有更多体现全家理念的东西呢？

姜铁牛： 不好意思二狗他触景生情，诶，我们全家也非常关心儿童的健康成长，所以我们设计了一些萌萌哒小板凳，专门针对我们村里的熊孩子。

IKEA 萌萌哒系列

记者：全家似乎要采取凳海战术?

姜铁牛：对，我们采取多样化的品类策略来适应不同顾客的审美口味，我们有温暖的原木自然风，也有明快的工业金属风，数十种不同的风格总有一款适合您。不同的设计师也会有个人的风格植入到作品中，顾客也可以根据对设计师的喜好来选择产品，全家家居把最好的设计介绍给普通群众。

IKEA 款式多多

新 KOBAYASHI 小椋林
工作凳，25x35 厘米，不锈钢
¥ 30.00

新 JACQUESHERZOG 赫尔佐格
工作凳，25x45 厘米，不锈钢
¥ 50.00

新 RIETVELDGO 里特维德高
门厅凳，15x35 厘米，木头
¥ 10.00

新 LOUISL WANG 路易斯·汪
地板凳，5x15 厘米，铝合金
¥ 50.00

新 ZENGO KUMA 张研吾
修车凳，5x15 厘米，生铁
¥ 50.00

新 JEAN NOUVELW 让·努维尔黄
工作凳，35x35 厘米，不锈钢
¥ 50.00

新 SHANGHAI MAN 上海师傅
胶布凳，5x15 厘米，塑料和胶带
¥ 30.00

新 NIEMEYER RD 尼迈耶红
电工凳，5x15 厘米，不锈钢
¥ 10.00

新 NIEMEYER YE 尼迈耶黄
电工凳，5x15 厘米，不锈钢
¥ 10.00

款式多多

匠心别具，精美日常

全家与HAY携手打造全新 YPPERLIG 伊波利 系列

姜铁牛：由于现在全家的能力有限，有的优秀设计作品还没有正式上市，将来会陆续推出，敬请期待!

×

姜铁牛：全家还特别推出了“精雕细琢”系列高端产品，专门针对挑剔的高层次顾客。精细完美的做工，用过都说好，以至于其中一款凳子还没开业就被我店员工偷走了。

IKEA |精雕细琢|

IKEA |精雕细琢|

黄老板：对对对，我们强烈谴责这种行为。

记者：那么全家家居要走出本村，有没有什么宣传策略呢？

姜铁牛：我们还建立了自己的网站，全家的产品全线登陆互联网，网站现正紧张测试中，希望未来可以取代某宝某东成为最牛逼的电商网站。

×

黄老板：除了线上宣传我们还在线下搞活动，听说最近有个深圳什么双双双双展，我们会把全家的小板凳系列放到展场里给大家使用的，大家走累了可以坐在我们全家小板凳上歇会，体验一下。

姜铁牛：对，大家可以根据自己的喜好来调整小凳子的形态，可以排一排可以排一圈，不过不要偷偷拿走噢！

记者：好的，感谢二位的时间，祝你们全家生意兴隆。

黄老板：祝贵报生意兴旺，财源茂盛；祝大展宏图，财源亨通；祝事业蒸蒸日上，更上一层楼，开业大吉，恭喜发财，财源广进，福气多多，早日倒闭，吉祥如意，贵人相伴，好运常在，好事连连。

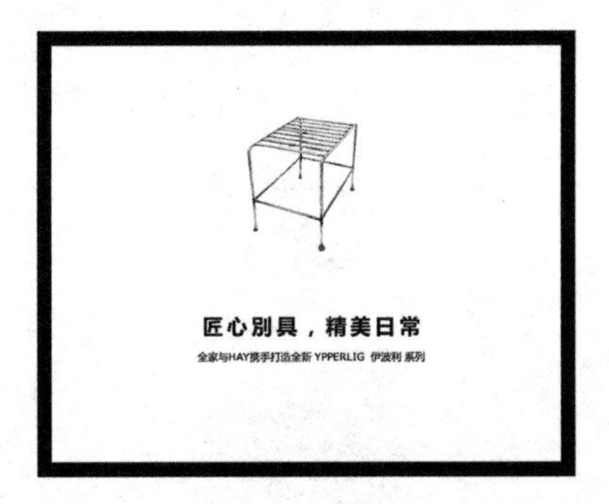

以上仅供娱乐

Note: The above is a fictional post created by the author at the beginning of the biennale. The narrative adopts the format of an interview to introduce the designers of the Shanzhai (counterfeit) IKEA. Together with the stools exhibited at the biennale, it addresses the idea that creativity is inspired by everyday life and that everyone can be a designer.

大概会有三成的人无论如何都不会交换凳子，因为对他们来说这个凳子是工作生活里很重要的一个工具。像这个修车师傅在讲起他的凳子时如数家珍，详细给我讲解怎么制作的，有什么样的功能。在我将近一个小时的软磨硬泡下，还是拒绝了我。他说这两个凳子用了十几年了，有很深的感情，不舍得送人。

About 30% of the people declined all possible proposals of exchanging their stools because the stools were very important tools in their life and work. For example, when the car repairman talked about his stool, he explained to me in detail how it was made and how it functioned. I badgered him for nearly an hour, but he still refused me. He said that he had already used the stool for more than ten years and was deeply attached to him. He couldn't just give it away.

这张板凳是我最喜欢的其中之一，皮革包着一整块大海绵，坐上去异常柔软，而且能感觉到里面空气缓慢挤出、凳面慢慢下沉的过程。让我惊讶的是，这些小板凳里，很多还是小年轻做的。他们大多是子承父业，帮家里打理店铺，通常都有自己的手艺活。跟这些年轻人打交道要明显比他们父辈容易，他们也更愿意接受新事物。

This stool is one of my favorites. It is made from a big sponge wrapped in leather. It is a soft seat, and while sitting on it, you can feel the air inside the sponge slowly escaping as the seat sinks. What surprised me is that many of these small stools are made by young people. Most of them had followed in their parents' footsteps, working in the family shop. They were usually some kind of artisan. As they were more willing to accept new things, it was noticeably easier dealing with them than with their parents.

老板自豪地告诉我，这个板凳是当年谈女朋友的时候做的，他去她家里发现没有凳子坐，回来专门给她焊了一个。他说："当年就靠这些把女朋友'骗'成老婆的。"临走时，老板还自己拿出手机拍了几张留念。我想他看着这些凳子心里应该充满了美好回忆吧。

The boss proudly told me that he had made this stool when dating his girlfriend. When he called on her, he realized that there were no stools to sit on. So he went back home and welded one for her. He said, "I used these stools to 'trick' my girlfriend into to marrying me." Before I left, he took out his mobile phone and snapped a few souvenir pictures. I think that his heart must fill with beautiful memories when he looks at these stools.

你是拿去做展览的啊？这个我自己做的，还有好几个，你要就送你啦！

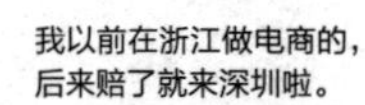

我以前在浙江做电商的，后来赔了就来深圳啦。

我这个凳子是军工材料的，坐500年都不会烂，结实的很，只有这一个，你要不要？

我做这个十几年了，这些都是我做的，你可以给我介绍一下生意啊。

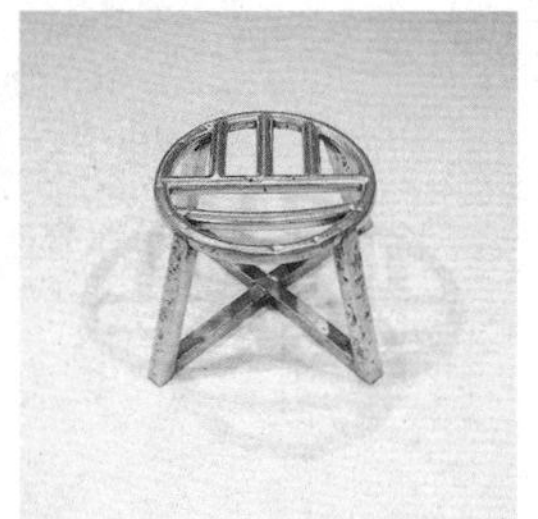

2　古城公寓 40 元 / 晚

南头古城里 40 元一晚的小旅馆，成了我临时工作室。清洁阿姨永远搞不懂为什么我每天要把被套被芯分开，还放那么多凳子。实在抱歉。

2　Old Town Hostel 40 yuan/night

A small hostel costing 40 yuan per night became my temporary studio in Nantou Old Town. The cleaning lady didn't understand why I separated the cover from the quilt every day. And then placed so many stools on it. I'm deeply sorry.

当设计开始逃离“标准、永恒、纯粹”，乌托邦圣殿向它的反面坍塌时，我们的视野会被那些混沌状态的事物所吸引。新的社会需求迫使我们不断思考如何在高度交织的社会系统中制造能量。如何在复杂的背景下做设计注定是一个庞大的问题，甚至不是一个设计问题。

在麻烦与生机共存的城中村里，居民会在生活中使用自己制作的“野生设计”。虽然工艺粗糙、材料廉价、毫无“美感”，但务实直接、简单有效。它们是见证城市变迁和底层生活有趣而真实的“文献”，但却不是生活最好的选择。所以我们实施了一个城中村家具交换计划，用双年展原本购置的新凳子交换城中村居民自己制造的“野生”凳子。通过这个交换计划我们从被换掉的“野生”凳子中选出 50 张有代表性的作品放置到展场中以供参观者使用。这些凳子实现使用功能的同时，作为文献向公众展现城中村居民的日常生活智慧，同时让更多人关注城市发展进程中底层人民的生存。这个项目通过交换凳子把双年展带入居民的日常生活中，也把双年展的理念传递给参观者。

▷ 展览：城中村家具交换计划 ▷ 参展人：黄河山、姜凡

When design deviates from "standards, eternity, pureness" and the utopia temple starts to collapse, our vision will be lured to chaotic things. Emerging social needs urge us to constantly think of ways to generate energy in a highly intricate social system. Against such a complicated backdrop, design is bound to be a great challenge, rather than a mere design challenge itself.

In urban villages where trouble and opportunity coexist, many residents use their original "grassroots' design" in daily lives. Though poorly-made from cheap materials without any sense of aesthetic beauty, these designs are economically made with simple and direct internal logics to serve very practical purposes. They are vivid and genuine "documentation" that bears witness to urban change and the life of grassroots, but they certainly are not the best choice for life. So we implemented an urban village furniture exchange program, where we exchanged new UABB stools for the stools of "grassroots design" with the residents of urban village. Through this program, 50 stools of "grassroots design" were selected as representative examples and placed in the exhibition venue for visitors to use. These stools, apart from providing practical functions, also serve as documentation demonstrating to the public the wisdom of urban village residents in their daily life, highlighting the life of grassroots in the process of urban development. The stool exchange program not only brings the UABB exhibition into residents' daily life, but also conveys the UABB concept to the visitors.

▷Exhibition: Urban Village Furniture Exchange Program ▷Participant: Huang Heshan, Jiang Fan

来源：深圳城市\建筑双年展组织委员会 . 城市共生：2017 深港城市\建筑双城双年展 [M]. 广州：岭南美术出版社，2021: 232-233.

摄影 PHOTO

第一站，白石洲
First Station: Baishizhou

张超
Zhang Chao

侨城电脑
侨城电脑

诚信 准称
峨眉山木耳素火腿

2010 年至 2013 年是我来深圳的头三年，和很多初来深圳的年轻人一样，城中村是我在这个城市落脚的第一站。三年后我搬出了城中村，本以为忙碌的生活并没有留下太多的记忆，而 2015 深双摄影工作坊的机会，让当时作为建筑摄影师的我选择重新回到这里，进行两周的拍摄记录，这让曾经在城中村生活的我有了更多的思考。希望用相机记录下我们所遇到的城中村，一个鲜活而又充满矛盾的地方。

▷ 展览：第一站，白石洲 ▷ 参展人：张超

From 2010 to 2013, I spent the first three years in Baishizhou after my graduation. It was my first station in Shenzhen. In 2015, I took part in UABB photography workshop. The living experience in Baishizhou led me to go back to this place and have a long-term shooting record in this place.

▷Exhibition: First Station: Baishizhou ▷Participant: Zhang Chao

来源：深圳城市\建筑双年展组织委员会 . 城市共生：2017 深港城市\建筑双城双年展 [M]. 广州：岭南美术出版社，2021: 272.

特别实验

TEST GROUND

探索多元改造模式
Navigating Models of Diversity

“城市共生”在宏大叙事和个人叙事、集体记忆和个人记忆中得以展演。以之为契机，发生在南头的在地实践既是个例，也是系统性思考下多个维度的在地实验。它们游击队式地侦察、行动、精准突破。在充满不确定性的城市生态中，实验行为既是主动选择，源于深圳特区词义中不可或缺的实验性；也是被动选择，以改善和解决城市问题为导向。这些积聚的微实践以多重视角、多重身份审视长期以来单一且粗犷的更新模式，在共同寻找城中村空间生产、进化与再生的新的平衡点。

《关于深圳城中村再生的多种实践模式》通过案例调查，梳理不同政策和利益相关方影响下的城中村更新模式。《超级乱糟糟 :Mapping 南头古城》从城中村微小、日常的事物切入，发掘“乱糟糟”表象下的深层逻辑及其与全球化的内在关联。《南头奶茶店：一间基于社会设计的快闪店》从社会设计的视角介入城中村更新实践，为租户赋能。《单车桌与街道生活》运用城中村的设计语言，在村里研发和制造一组供村里使用的街道家具。《房屋 17 号：全球化大都会的反叙事》由外来者借助城中村原生的关系网络，修复百年老宅，从中生成新的社会关系。《南头游乐场：城市体制与南头体质》由策展团队邀请建筑师改造一栋烂尾楼，尝试“有限度地介入”，探寻城中村空间生产背后的政策和利益博弈，虽最终未能实施，但作为深双最为激进的实验具有特殊意义。《果园城记：设计介入实验》延续至双年展后，通过设计社会交往促成社会关系的重建，激发居民自主营造公共空间。《一场展览背后：来自深双的田野笔记》记录在双年展介入项目中观念植入和使用需求的协调过程。

Cities, Grow in Difference unfolded through grand and personal narratives as well as collective and individual memories. Onsite practice in Nantou was not only a single case, but also a multi-dimensional onsite experiment with systematic thinking, a guerrilla tactic that investigated, aced, and hit the target. Living in a city of uncertainties, experimentation is an active move in terms of the SEZ's spirit, even as it is also a passive choice with an eye to improving and resolving urban issues. Looking at the long-lasting single-minded extensive urban renewal model from the perspectives of diverse individuals, the collection of small-scale practices searches for a new equilibrium in the production, evolution, and revitalization of urban village spaces.

"Practice Approaches for the Regeneration of Urban Villages in Shenzhen" uses case studies to summarize how different policies and stakeholders have shaped urban village regeneration. "Super Messy: Mapping Nantou Old Town" uses small everyday objects to explore the hidden logic of the messy phenomenon of urban villages and its imbrication with globalization. "Nantou Milk Tea Shop: A Pop-up Shop Based on Social Design" is a social design practice that intervenes in an urban village to empower tenants. "Bike-Table and Street Life" incorporates the visual language of urban villages into the design and production of urban village street furniture. A third-party restoration project of a century-old house, "House 17: A Counter-narrative of a Globalized City" utilizes the extended local social network of the urban village to generate new forms of social relationship. An unfinished building rehabilitation project by invited architects, "Nantou Playground: The Urban System and Nantou Spirit" experiments with limited interventions, speculating the policies and interests behind the spatial production of specific urban villages. Although it became an unrealized project, Nantou Playground has significance as the most radical initiative in 2017 UABB. A project that extended beyond the Biennale, "Story of Orchard City: An Intervention Experiment on Design" promotes the reconstruction of social relationships through social interactions that are designed to stimulate engagement with public space making. "Behind an Exhibition: UABB Fieldnotes" narrates the mediation process between introducing concepts and fulfilling demands in UABB interventions.

关于深圳城中村再生的多种实践模式

Practice Approaches for the Regeneration of Urban Villages in Shenzhen

莫思飞

2017 年深港城市\建筑双城双年展（深圳）策展助理，都市实践研究员。

Mo Sifei

Curatorial assistant of the 2017 Bi-City Biennale of Urbanism\Architecture (Shenzhen), URBANUS Researcher.

一刀切 One size fits all

1980 年，深圳经济特区开展土地开发及有偿转让使用权的实验和变革，为城市发展提供了坚实的物质基础。2004 年出台的《深圳市城中村（旧村）改造暂行规定》中，城中村改造延续地产思维模式，以优惠地价及提高容积率的政策鼓励地产开发商，以市场运作的方式开展旧改工作。尽管在 2005 年学界和建筑师就提出针灸式的介入模式，政策上也提出除推倒重建外的“综合整治”[1] 办法，但在深圳土地紧缩政策的大背景以及巨大资本利益的驱动下，作为城市价值洼地的城中村依旧是地产开发商博弈的前沿。“拆迁造富”的时代下催生了股份公司与地产开发主体合作的模式，共同推动城中村的“旧改”。一刀切的拆除重建模式一夜间彻底与城中村问题切割，也一夜间清除了延续的生活方式和改革开放以来的城市肌理与城市记忆。巨额成本由谁买单，“千城一面”所带来的平庸及匮乏感，高负荷的城市服务和断崖式的士绅化等城市问题随之而来。

1 2005 年 4 月 7 日，《深圳市人民政府关于深圳市城中村（旧村）改造暂行规定的实施意见》中提出两种改造模式，其一为“综合整治”。同年 10 月，《深圳城中村（旧村）改造总体规划纲要（2005—2010）》提出全面改造与综合整治的分类指导。

In 1980, the Shenzhen Special Economic Zone experimented with state-owned land development, leasing and transferring land-use rights to fund future urban development. Subsequently, in 2004, Shenzhen City promulgated the *Shenzhen Municipal Government Interim Provision of Shenzhen Urban Village (Old Village) Redevelopment* (hereafter Interim Provision). Since then, urban village renewal has followed a real estate-based development model that has encouraged developers by offering low land prices, increasing the permitted floor area ratio, and adopting market-orientated implementation for old village redevelopment. Although in 2005, academics and architects proposed small-scale intervention referred to as "urban acupuncture,"[1] nevertheless policies advocated "comprehensive remediation"[2] as an alternative to demolition and rebuilding. Mobilized by policies relating to land shortages and financial gain, undervalued urban villages have become major projects for developers. The idea of "making wealth through demolition"[3] has encouraged village stock-holding corporations and developers to collaborate on redeveloping urban villages. In turn, the complete demolition and relocation (*chaiqian*) of extant villages makes the issues of urban villages irrelevant, even as the lifestyles, urban fabric and memories of

1 Jaime Lerner coined "Urban Acupuncture" in his book in 2003. It became a popular term to describe small-scale urban interventions in China. Lerner, Jaime. *Urban Acupuncture*. Island Press. 2003.

2 On April 7th, 2005, "comprehensive remediation" was first proposed in the implementation advice of the *Interim Provision*. In October, the guideline for categorizing comprehensive remediation was outlined in the *Master Planning Outline for the Redevelopment of Urban Villages (Old Villages) in Shenzhen (2005-2010)*.

3 A common saying refered to the phonomenon of owner receiving extreamly high value compensation package for demolition and redevelopment of their properties.

2018年11月，深圳市规划和自然资源局会发布了《深圳市城中村（旧村）总体规划（2018—2025）》征求意见稿，其中划定各区综合整治对象总规模约99平方公里，规定罗湖区、福田区、南山区综合整治划定比例不低于75%。历经“中国梦想试验场”“湖贝120”和“城市共生”等公共事件的发声以及实践探索后，深圳城中村有机更新的改造观念得以政策落实，但城中村有机更新实施路径的探索才刚起步。本文将梳理具有代表性、实验性的深圳城中村再生的六种模式及其试点案例，从更新经验和观念中讨论深圳城中村再生的可能性：①以大芬村为案例的外部资源入驻引发的片区更新；②政府主导的综合整治试点（2005—2017）；③以“冈厦1980”为案例的自发改造；④以2017深双为例的第三方平台介入的有限度实验；⑤以水围柠盟人才公寓为案例的保障性租赁住房试点；⑥以万村计划为案例的政府与企业主导的长租公寓计划。

1 从资源入驻到环境和产业提升

以“深圳案例”[2]为名的大芬模式（图1）早已为城中村有机更新提供了“非常经验”。1989年以前，大芬村是一个普通的、种植水稻的自然村。1989年，专营油画外贸的香港画商黄江带着26名临摹画工，入驻邻近罗湖口岸的大芬村，以低廉的租金租用自建房，开始了名画的临摹、复制、收购及销售的产业。 随着黄江的成功，香港商人和国内一些画家开始进驻大芬村，使得大芬村逐步发展成为仿制画集散地，形成完整、标准化的生产链，从原材料到画师临摹，再到画框装裱，最后打包物流，一应俱全。到了1992年，大芬村的油画产业已经非常成熟。黄江忆述，1992年4月，400多位画工在他的召集下，用一个半月完成了36万幅油画的订单。[1]

1998年，龙岗区及布吉镇二级政府投入大量资金对大芬村进行环境整治，由镇干部做工作并带头拆除乱搭建的小院，疏通并修建大芬村内部道路，提升环境卫生。改造后，自建房的一层都改造成商铺，楼上作为出租屋、画室或是工作室（图2）。[2] 2003年，大芬油画村确定为深圳市第一届中国（深圳）国际文化产业博览交易会分会场，区、镇、村三级政府投资1000多万元，对沿街立面进行更新，并投入400多万元拆除村口自建房4栋，建设油画艺术广场。基础设施和环境的提升以及产业的引导，进一步加速了大芬村的发展。大芬村基础设施和城市管理的到位，印证了城中村的空间

reform and opening up are lost. This has been followed by the emergence of concomitant urban issues, including who pays the huge cost of redevelopment, the "all cities look the same (*qiancheng yimian*) phenomenon, over-loaded urban services and rapid gentrification.

A shift in policy occurred in November 2018, when the Planning and Natural Resources Bureau of Shenzhen Municipality issued the consultation paper, *The Shenzhen Urban Village (Old Village) Master Plan (2018–2025)*. A 99-square-kilometer area as a target area for comprehensive remediation was designated, stipulating a comprehensive remediation area of no less than 75% of extant urban villages in Luohu, Futian and Nanshan Districts. Significantly, it was only after public discussions such as "Frontier for China Dreams," "Hubei 120" and "Cities, Grow in Difference" explored development alternatives that the concept of the organic renewal of urban villages in Shenzhen manifested as policy. Yet, the search for the best organic renewal practices has just begun. This paper reviews six representative and experimental modes of urban village regeneration in Shenzhen and their pilot cases, discussing the possibility of urban village regeneration in Shenzhen from perspective of these cases: (1) Dafen Village, where external resources stimulated district improvement; (2) government-led pilot comprehensive remediation (2005–2017); (3) Gangxia 1980, a spontaneous renovation; (4) the 2017 UABB, a small-scale intervention experiment by a third party; (5) the Shuiwei LM Apartment, a pilot project for affordable rental housing, and: (6) the Wancun (Ten-thousand Villages) Plan, a government and enterprise-led rental apartment project.

1 From the Introduction of External Resources to Upgraded Industry

Known as the "Shenzhen Case,"[4] the Dafen model (Figure 1) provides alternative experiences in the organic renewal of urban villages. Before 1989, Dafen Village was a rural area where rice was grown. In 1989, Hong Kong art dealer Huang Jiang, who specialized in trading commercial oil paintings, brought 26 painters to settle in Dafen Village, which was located near Luohu Port, facilitating shipping from Shenzhen to Hong Kong. He rented low-cost self-built housing and started the industry of the copy, acquisition and sale of famous paintings. The success of Huang Jiang attracted Hong Kong dealers and Chinese painters, gradually making Dafen a distribution center for reproductions. The comprehensive and

2 2009年09月，世博会国际遴选委员会审议通过深圳提交的“深圳大芬，一个城中村的再生故事”作为2010上海世博会城市最佳实践区深圳案例馆的展览提案。由孟岩策展，深圳案例馆展示了一系列与深圳大芬村相关的故事。参见：都市实践．2010上海世博会深圳案例馆展览简介[EB/OL]. [2020-12-18]. http://www.urbanus.com.cn/writings/introduction-to-the-exhibition.

4 On September, 2009, Shenzhen's proposal, "Dafen Village-the Regeneration of an Urban Village in Shenzhen" was approved by the International Selection Committee of the World Expo and exhibited in the Urban Best Practice Area of the EXPO 2010. Shenzhen Case Pavilion was curated by Meng Yan, showcasing a series of exhibition about Dafen Village. see URBANUS. "Expo 2010 Shenzhen Case Pavilion—Introduction to the Exhibition". http://www.urbanus.com.cn/writings/introduction-to-the-exhibition/?lang=en.

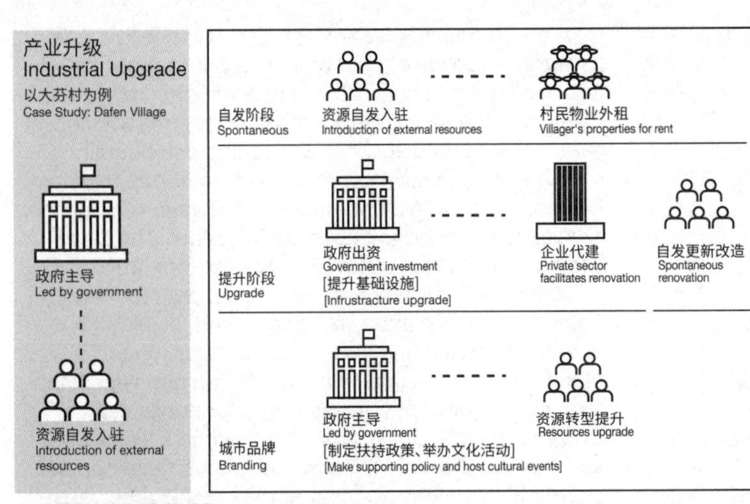

图 1 大芬村政府主导产业升级模式
Figure 1 Dafen Village government-led industrial upgrade model

standardized production chain included the production of raw materials, painting, mounting, packaging and shipping. By 1992, the oil painting industry in Dafen Village was well established. Huang Jiang recalled that in April 1992, he organized more than 400 painters to complete an order of 360,000 oil paintings in one and a half months.[1] Dafen Village was under the administration of the governments of Longgang District and Buji Town, which in 1998 invested in the environmental improvement of the village. Led by the township authorities in Dafen, illegal constructions were demolished, inner streets were connected and cleaned, and the overall environment was improved. After the renovation, the first floor of self-built buildings became shops, while the upper floors served as tenements, studios or workshops (Figure 2).[2] In 2003, Dafen village was appointed as the sub-venue for the first session of the China (Shenzhen) International Cultural Industries Fair. Three levels of government, the district, township and village invested over 10 million yuan toward the beautification of street elevations. 4 million yuan were used to demolish four self-built buildings that were situated at the entrance of the village and replace them with the Oil Painting Art Plaza. These infrastructural, environmental and industrial improvements further stimulated Dafen's growth. In 2005, worldwide 70% of commercial paintings came from China, and of those paintings, 80% were produced in Dafen Village, making it a significant global distribution center for commercial paintings. From 2005 to 2007, the gross output of Dafen Village increased from 279 million yuan to 430 million yuan.[3] In 2007, the Longgang District Government invested nearly 100 million yuan toward an internationally renowned art museum designed by URBANUS, Dafen Art Museum. The introduction of infrastructure and urban management in Dafen Village establishes an argument that the urban fabric of urban villages did not lead to urban issues, but the lack of urban infrastructure and urban management did.

The case of Dafen is one of integrated cooperation between the market, multiple levels of government and an inclusive community. The art dealer Huang Jiang settling in Dafen was a single event, but it was also a paradigm for mobilizing the establishment of a bottom-up industry and comprehensive remediation by the government corresponding to local demand. The tourist

图 2 大芬村，2021 © 都市实践
Figure 2 Dafen Village, 2021 ©URBANUS

肌理并不是引发城中村问题的原因。2005 年，全球 70% 的商品画来自中国，而其中 80% 来自大芬村，大芬村已成为全球重要“商品画”交易集散地，大芬村的油画产值从 2005 年的 2.79 亿元一路攀升至 2007 年的 4.3 亿元。[3] 2007 年，龙岗区政府委托都市实践设计，投资近 1 亿元，建设了享誉国际的大芬美术馆，作为大芬村的配套设施。

“大芬案例”是市场、政府与社区三者相互合作互动模式实践。画商黄江的入驻虽具有独特性，但其所激发和带动的自下而上的产业聚集，以及政府尊重当地需求引导的综合整治都是可借鉴的更新观念。相似的案例如大鹏新区的较场尾同样是从个体入驻，自发改造[3]，到产业入驻，进而推动政府投资综合整治[4]、引导产业提升[5]与社区自发促进社区合作。

“市场”是“大芬模式”的关键要素，2007年大芬村的油画产业为4.3亿元的产值[4]，十年间增长了将近十倍，在2017年达到41.5亿元[5]。一旦这些驱动更新改造的力量在以经济逻辑为内核的“数学”公式中缺失，“大芬模式”是否依然成立？以梧桐山为例，因与香港关口较近，服装生产行业曾经是支撑梧桐山的主要产业，后因地处水资源保护区，大部分不合标准的制衣厂被叫停。此时，梧桐山世外桃源般的环境优势和服装产业务工者的聚集吸引了服装设计师的入驻，ffiXXed Studios的菲奥娜·刘和凯恩·皮肯则是其中之一，这让梧桐山的产业转型有了新的希望。经过在梧桐山几年的努力，ffiXXed Studios成为国际知名的小众时尚品牌，出现在各种国际时尚媒体的专访中，甚至用“梧桐”作为他们一个设计系列的命名。然而梧桐山中许多优秀的时尚设计师资源却不断地在流失，ffiXXed Studios也在2017年决定搬迁至更有时尚氛围和机遇的上海。可见，在对经济指标缺乏贡献的前提下，自发力量无法形成足够的集聚效应和群体力量，变得不可持续进而不断流失。自综合整治模式的提出后，政策制定者、学者和实践者都在积极地探索可以跳脱现有的经济逻辑束缚，有效地实现有机的，可持续的微改造微更新的实施路径。

2 综合整治（2005—2017）

2005年《深圳市人民政府关于深圳市（旧村）改造暂行规定的实施意见》中，首次提出“综合整治”的改造模式。其实施路径基本由政府管理部门牵头，与股份公司及社会企业合作，采用“代建”形式进行。2005—2009年，综合整治指“基本不涉及房屋拆建的环境净化、美化项目”，[6]主要针对“面子”问题的立面美化，[7]对城中村建筑沿街外立面进行统一粉刷翻新，从外立面颜色上统一趋向城市住宅的

industry in Jiaochangwei, an urban village in Dapeng New District developed and revitalized in a similar manner. Individual settlement and spontaneous renovation[5] preceded and then stimulated government-invested comprehensive remediation,[6] leading to industrial upgrading[7] and improved self-organized community cooperation.

The market has been the key to the revitalization of Dafen, the Dafen model. Dafen's gross output of the commercial oil painting industry reached 0.43 billion yuan in 2007[4] and increased about tenfold to 4.15 billion yuan[5] in ten years. Once these revenue driving the renovation are removed from the financial equation of urban village revitalization, will the cases of regeneration still be viable? Consider the case of urban villages in Luohu District surrounded by Wutong Mountain. Wutong's proximity to the port of Hong Kong made the garment industry the area's main industry. Most substandard garment factories, however, were suspended because Wutong Mountain is a designated water conservation area. Meanwhile, the beautiful natural scenery of Wutong Mountain and the remaining garment workforce attracted fashion designers to settle there. Fiona Lau and Kain Picken from ffiXXed Studios were among those who brought hope for industry transformation to Wutong Mountain. After several years of hard work in Wutong Mountain, ffiXXed Studios attracted international attention, and they even named a collection after Wutong. However, many emerging fashion designers in Wutong Mountain are constantly relocating. ffiXXed Studios also relocated to Shanghai for more opportunities in the well-established fashion industry. These bottom-up forces failed to form collective power due to an absence of contribution to indicators of economic performance, making regeneration initiated by small groups of people unsustainable. Since the introduction of comprehensive remediation in 2005, policymakers, researchers and practitioners have been exploring a path that can break away from financial logic for a sustainable and effective small-scale organic renewal.

3 2008年，经常到较场尾附近冲浪及帆板爱好者自发改造和经营民宿，逐步吸引民宿经营者入驻。2015年，大鹏区民宿经营者成立全国第一个民宿协会。参见：訾建业，徐灵枝，刘琳琳. 2020广东乡村民宿发展白皮书 [N]. 广东建设报，2020-12-25.

4 2013年底至今，大鹏新区先后投入1.5亿元完善污水管网、电力设备、停车场等市政公共设施，使较场尾成为全市第一个接驳入户污水管的城中村，污水收集率达95%以上，海水质量明显提升。未来，较场尾还将建成滨海公园，并将与环龙岐湾片区其他景点一起打造5A级景区。参见：张妍，梁茂生，胡世民. 较场尾民宿“头号难题”破解 [N/OL]. 深圳商报，(2016-7-25) [2020-12-30]. https://finance.qq.com/a/20160725/001750.htm.

5 2015年大鹏新区颁发省内第一部民宿管理办法，由政府提供前期资金支持，鼓励民宿产品达到规范要求和标准。2015年11月，《大鹏新区民宿发展规划》发布；2016年4月6日，《深圳市大鹏新区民宿管理办法（试行）》发布；2018年10月31日，《深圳市大鹏新区民宿管理暂行办法》发布。

5 "In 2008, Surfers and windsurfers in the area spontaneously renovated self-built housing as homestay, which attracted many homestay owners. In 2015, the homestay owner established the nation's first homestay association." – YU J, XU L, LIU L. White Paper of Development of homestay in Guangdong Village[N]. *Construction Paper*, 2020-12-25.

6 "Since 2013, Dapeng New District has invested 0.15 billion to improve the sewage system, electricity system, public facilities such as public parking, making Jiaochangwei the first urban village in Shenzhen to enjoy being connected to the municipal sewage system. 95% of households have been connected to the system, which collects polluted water rather than letting it discharge directly into the ocean. In the future, Marine Park will be constructed in Jiaochangwei, making 5A tourism site with other attractions of Longqi Bay area." see, ZHANG Y, LIANG M, HU S. Solve the Number One Problem of Jiaochangwei Homestays[N/OL]. *Shenzhen Economic Daily*, (2016-7-25)[2020-12-30]. https://finance.qq.com/a/20160725/001750.htm.

7 "In 2015, Dapeng New District issued the first homestay management method in the province, and the government provided preliminary financial support to encourage homestay products to meet regulatory requirements and standards. In November 2015, *Dapeng New District Homestay Development Plan* was released; on April 6, 2016, *Shenzhen Dapeng New District Homestay Management Measures (Trial)* was released, and; October 31, 2018, *Shenzhen Dapeng New District Interim Measures for Homestay Management* was released.

色系。2009 年，深圳首次提出城市更新办法，其中关于综合整治项目的分类“主要包括改善消防设施、改善基础设施和公共服务、改善沿街立面、环境整治和既有建筑节能改造等”，[8] 而后一系列的政策和实施细则都推动城中村综合整治转向以重大安全隐患的消除、市政基础设施的完善以及治安、环境的提升等。以福田区为例，2013 年福府办 6 号文明确，区政府设立专项资金，对公共配套，如文化体育设施、道路维修改造以及水环境治理工程设补助最高可达核定工程预（概）算的 49%，按标准收费的社区服务如公共停车场和幼儿园的建设设补助最高可达核定工程预（概）算 40%，市政基础设施建设由政府 100% 出资的实施细则。6 此实施细则逐步将城中村的硬件建设纳入城市管理体系中。2014 年，克服了重重技术困难后，水围社区成为深圳首个燃气管道改造的示范项目，作为试点项目，城中村的电网改造工程也逐步得以推进。[9] 2005 年到 2015 年间，福田区政府对水围环境综合整治投入 1.3 亿元资金，包括市政道路、通信管网、供水供电、消防设施、监控系统、环境卫生等。同时水围股份公司与政府共同投资 1.5 亿元建设文化设施、修葺文物及保护百年古树。

综合整治模式在政策管理手段中也逐步地把城中村吸纳到行政管理中。2013 年《南方都市报》刊登了一篇标题为《深圳：城中村出租屋水价乱收费或将受规范》的报道，因为大部分城中村的供水、供电体系都不在城市的规范管理之内，导致城中村居民虽然租金便宜，却承担着比其他市民更高的自来水价格，城中村自来水价格管理规范受到政府的重视。2017 年《深圳市水价改革实施方案》明确，“对于暂时无法抄表到户的城中村出租屋，实行合表用水价格政策，规范收费行为”[10]。

2018 年是全面推进城中村有机更新的元年，规划要求“有序开展城中村综合整治，促进有机更新”。深圳市城市更新和土地整备局副局长王海江表示，“城中村综合整治工作进入到新的阶段，在以往工作的基础上，在更新理念、目标和机制等方面进一步拓展了综合整治的含义，其核心内容是以品质提升为目标，深入挖掘城中村的独特价值，保留并活化城中村的历史传承及现状特色空间，保持城中村活力，以微改造的方式，促进城中村转型并可持续发展。”[7] 在“微改造”“挖掘城中村独特价值”这些偏向自下而上，点状式发展的核心理念下，传统的以自上而下为主的综合整治面临着巨大

6 “深圳市福田区支持城中村市政基础设施和公共配套设施 建设专项资金管理实施细则”（福府办［2013］6 号），其中公共配套设施，是指城中村内的文化体育设施、公共绿化、美化、卫生等环境设施、治安、安防监控、交通、公共消防设施、社区公共停车场、幼儿园等公共配套设施。市政基础设施，是指符合城市规划的城中村内供水管道、排水管道、供配电管线及设备、道路、电气管线及设备、燃气管线、有线电视管线、通信管线等市政基础设施。

2 Comprehensive Remediation (2005-2017)

In 2005, the implementation advice of the *Interim Provision* proposed comprehensive remediation, a government-led project that involves working with stock-holding corporations and enterprises and is facilitated by the private sector. From 2005 to 2009, comprehensive remediation referred to sanitation improvements and environmental beautification projects that involved no demolition of buildings.[6] It mainly entailed facade beautification,[7] where the facade of urban village buildings along main streets were uniformly painted to match the color of the commercialized residential towers in the city. In 2009, Shenzhen first proposed urban renewal measures in which comprehensive remediation included the "improvement of fire protection infrastructure, urban infrastructure and public facilities, street facade, environmental remediation as well as renovation for energy efficiency."[8] Subsequent policies and detailed rules for implementation promoted the comprehensive remediation of urban villages as the elimination of safety hazards and improvement of public security as well as the environment. For example, the No. 6 document of the Futian District Government published in 2013 stated that the district government would set up special funds to subsidize up to 49% of the approved budget for the construction of public facilities, road maintenance and environmental water treatment in villages in the district. Community services, such as public parking and childcare, could receive subsidies of up to 40% of the approved budget, while the construction of municipal infrastructure would be 100% funded by the government. Such an implementation rule has integrated the construction of urban villages into the city management system.8 In 2014, Shuiwei Community overcame technical difficulties to become the first pilot renovation project involving gas pipeline construction in Shenzhen. Power grid construction in urban villages was also carried forward that year.[9] From 2005 to 2015, the Futian District Government invested 130 million yuan toward Shuiwei's comprehensive remediation, including the construction of municipal roads, pipelines, water supply systems, fire protection facilities, surveillance systems, and sanitation. Meanwhile, Shuiwei Stock-holding Corporation and the Futian District Government co-invested 150 million yuan toward cultural facilities, heritage restorations and heritage tree protection.

8 See (Fufuban [2013] No. 6), "Implementation Rules for the Management of Special Funds for Supporting the Construction of Municipal Infrastructure and Public Supporting Facilities in Urban Villages in Futian District, Shenzhen." In this document, public supporting facilities refer to cultural and sports facilities and public open space in urban villages, beautification, sanitation and other environmental facilities, public security, surveillance system, transportation, fire protection facilities, community or public parking lots, kindergartens and other public supporting facilities. Municipal infrastructure refers to municipal infrastructure that conform to urban planning such as utility system, sewage, power grid and equipment, roads, gas pipelines, television cables, and communication system.

的挑战。以综合整治规划所要求的城市设计导则为例，分片区对城中村风貌以统一的颜色、材质作为设计指导原则并不适合充满不确定性、处处是惊喜的城中村。因此，探求为城中村“量身定制”的政策、多元参与的机制和城中村介入、实践的模式，是城市的政策制定者和实践者最为迫切的使命。时任深圳市规划和自然资源局副局长王策飞就曾在 2018 年 3 月举行的“畅想南头未来——南头茶划会暨第一届南头街道民生论坛”中提出，“综合整治要把城中村内在的活力激发出来就需要长久的、可持续性的营运。”

3 内生力量驱动自发改造

城中村是借力与自助的结合，[11] 紧紧跟随城市发展和租赁市场的变化而快速生长，又时刻调整“产品类型”和策略应对外来人口居住空间需求的改变。这些居住产品类型通常都是房东和本土施工队作为设计师，通过口耳相传，经验累积的方式，结合租赁市场和租金收入为考量不断改良而成的。随着城市的转型，人口结构的变化，年轻白领、创业者逐步成为租赁市场的主力军，其中也不乏自发改造的新力军。在住客逐渐提高居住需求的情况下，城中村的居住产品也需要在保持经济性的情况下，更人性化。2015 年到 2016 年，“我们家”“蜂巢

自发更新
Spontaneous Regeneration

村民/业主主导
Led by villager/owner

中小企业实施
Implemented by enterprise
[设计、施工、运营或由业主自行外租、管理]
[Design, construct, manage/lease and manage by Owner]

租赁协议
Rental agreement

租户主导
Led by tenant
[自住或经营]
[Live/manage]

中小企业实施
Implemented by enterprise
[设计、施工]
[Design and construct]

图 3 冈厦 1980 自发更新模式
Figure 3 Gangxia Village spontaneous regeneration model

公寓”“You+ 公寓”“魔方公寓”“鲤鱼公寓”等长租公寓品牌在不同的城中村落户，2017 年万科集团宣布以“万村计划”进入多个城中村，形成以长租公寓结合城中村的改造模式。在长租公寓市场的冲击下，自我开发的产品竞争力下降成为房东们自发更新改造的驱动力。“冈厦 1980”则是一个由城中村二代业主（指改造其父辈拥有的城中村物业产权）自发改造的案例之一（图 3）。物质需求得到满足后的城中村二代大部分更向往的是精神层面的追求，寻找一种归属感，他们是社会各行各业的中坚力量，

Comprehensive remediation has also included the integration of the administrative management of urban villages and the city. In 2013, the *Southern Metropolis Daily* published the story "Shenzhen: arbitrary water rates and fees of tenements in urban villages would be regulated."[9] Due to the lack of formal utility services provided by the city, urban village residents paid affordable rent while being charged higher water bills than other city residents. The government attached great importance to regulating the water rate in urban villages. The 2017 "Shenzhen Water Price Reform Implementation Plan" asserted that "In the interim, a shared water meter policy under a regulated price plan would apply to tenements in urban villages without individual water meters."[10]

2018 was the first year to advance the organic renewal of urban villages in planning, with calls for "a comprehensive remediation of urban villages in an orderly manner to facilitate organic renewal."[9] The deputy director of the City Renewal and Land Development Bureau of Shenzhen Municipality, Wang Haihong, stated that "As the comprehensive remediation of urban villages enters the new era, based upon past work, the definition of comprehensive remediation is expanded in terms of goals and mechanisms. Today, it aims at improvement, discovering the unique value of urban villages, preserving and revitalizing the historical heritage and fabric of urban villages, maintaining the vitality of urban villages

图 4 “城中村爆改户” 冈厦 1980 改造 © 城中村爆改户
Figure 4 Glocal Gangxia 1980 renovation © Glocal

and facilitating the regeneration as well as sustainable development of urban villages through small-scale renovation."[7] The core concept of small-scale renovation and the discovery of the unique value of urban villages relying on strategies of bottom-up urban acupuncture has challenged traditional top-down comprehensive remediation. For example, urban design guidelines under

9 *Shenzhen Urban Village (Old Village) Comprehensive Remediation Plan (2019-2025)*, released in 2018.

图 5 由厘米制造设计的新洲村十点创意饮品店改造前，陈丹平摄于 2017 © 厘米制造
Figure 5 10. Creative Drink before renovation designed by CM Design, Xinzhou Village, photograph by Chen Danping, 2017 ©CM Design

图 6 由厘米制造设计的新洲村十点创意饮品店改造后，张超摄于 2017
Figure 6 10. Creative Drink after renovation designed by CM Design, Xinzhou Village, photograph by Zhang Chao, 2017

也是一股在城中村更新的新阶段中需要动员的新锐力量。

2016 年，自称“城中村爆改户”的是一个由 6 位土生土长、留洋归国的 80 后“深二代”组成的团队。团队成员中不少是在城中村生活过的二代业主，他们对城中村和深圳本土文化有着特殊的情感，希望通过改造使得城中村既能跟上城市发展的步伐，又能重现记忆中的邻里文化。[12] 冈厦 1980 是一栋藏匿于岗厦村的一栋建于 1980 年的房子，房东文先生是“城中村爆改户”成员之一文子杰的父亲。文子杰说服父亲，把这栋老房子拿出来作为改造的示范点，由学习建筑学的成员谢克非负责设计，改造过程中保留了许多承载着他们童年记忆的老物件和建筑语言（图 4）。改造后的成果得到文子杰父辈亲戚朋友们的一致认可，租金也从原来的单间 400~500 元，上升到 3000 元以上。后期的运营也是由“爆改户”团队负责，从报名入住的房客中面试，以求入住的租客认同并共同促进公寓内的邻里文化。意识到城中村的改造对周边产生影响的可能性，2017 年，冈厦 1980 结合福田区海绵城市绿色屋顶示范项目，建设了对外开放的屋顶花园，举办公共活动以反哺周边社区。

comprehensive remediation require unitary colors and materials for urban village renovations, rejecting the uncertainties and diversity of urban villages. Hence, the pressing mission for policymakers and practitioners is to explore tailored policies, community engagement tactics and the paradigm of interventions and practices for urban villages. During "The Future of Nantou-Nantou Planning Symposium with Tea Reception and the First Session of the Nantou Street Livelihood Forum" held in March 2018, the deputy director of the Planning and Natural Resources Bureau of Shenzhen Municipality, Wang Cefei indicated that long-term sustainable operation is a key for comprehensive remediation to activate the internal vitality of urban villages.

3 Spontaneous Regeneration by Internal Forces

The development of urban villages involved a combination of leveraging and self-sufficiency,[11] growing rapidly to adapt to the changes in urbanization and the demands of the rental market. Its products and strategies were adjusted to respond to the changing demands for migrant living spaces. The typologies of residential products co-designed by the landlords and contractors constantly improved through word of mouth, experience and profit maximization within their market. As the city has upgraded its economic base, concomitant demographic changes have meant that the rental market has shifted from migrant workers to junior white-collar workers and start-up entrepreneurs, creating new forces of spontaneous renovation. Demands for improved quality of life have made not only affordability, but also livability two of the main residential products in the urban villages. From 2015 to 2016, for example, rental property management companies, including Warm+, Beehive, You+, Mofang Apartment, and Liyu Apartment emerged in urban villages. In 2017, Vanke announced the "Wancun Plan" in several urban villages, establishing a model of urban village renovation of integrated long-term rental apartment management, in which a company rented most of the urban village buildings in one area, modified them, and then brought in property management and then leased the upgraded apartments (rental apartment model). This emerging rental stock outperformed many self-developed housing products, which in turn mobilized the landlords to renovate their stock. "Gangxia 1980" was a case of spontaneous renovation (Figure 3) by the sons/daughters of property owners of urban villages, or "Village 2s" as they are colloquially known. Village 2s are economically secure and they yearn for personal growth and a sense of belonging. They are the backbone of society and a new force to be mobilized in the new era.

In 2016, six second-generation Shenzheners, or "Shen 2s" returned from study abroad to establish

从改造的结果而言，冈厦 1980 与通过统租物业实现的“长租公寓模式”相同，针对年轻社群，通过提升居住空间品质同时推动租金上涨。但由于二代业主作为改造的实施主体（或是实施主体中的一员），在无需考虑支付统租租金成本的情况下，加上独栋改造投入规模小，有效地避免了“长租公寓模式”面临的效益问题。相较于万村计划片状式的改造，冈厦 1980 倾向

"Chengzhongcun Baogaihu (urban village house flippers)." Many of them grew up and have lived in urban villages, valuing the local culture of urban villages and Shenzhen. They hope their renovations will allow urban villages to keep up with the pace of urban development while continuing to nourish the neighborhood culture of their memories.[12] Gangxia 1980 is a building located in Gangxia Village, Futian District. Built in 1980, it was owned by the father of Wen Zijie, a member of the house flippers group. Wen Zijie convinced his father to have the building renovated as a pilot project. Designed by Xie Kefei, a flipper who had studied architecture, the renovated self-built house included many objects and architectural echoes from their childhood memories. (Figure 4) The parental generation praised the renovation, which increased rent by 400-500 yuan per apartment, bringing monthly rentals to over 3,000 yuan a month. Baogaihu manages the building, interviews tenants, and promotes community building among tenants. Realizing the possible impact on the neighborhood, Gangxia 1980 was included in the Futian District Sponge City Green Roof Pilot Project, in which an open rooftop garden was built to hold regular public events to benefit the community.

Targeting junior white-collar workers, Gangxia 1980's redevelopment led to an increase in rent and improved quality of life, which is similar to the "rental apartment model" but without a single corporation renting buildings. As a second-generation owner facilitated the renovation, no rent was paid to acquire the property. In addition to improving cost efficiency, a small-scale investment in a single building avoids the issues of low-profit margins faced by the "rental apartment model." Compared to the large-scale Wancun Plan, Gansha 1980 instead adopted a small-scale spontaneous renovation model. Rent has increased significantly, yet an incremental renovation on a single building would not affect the rental market of the area all at once. The small shops and accommodations of low-income groups are protected against issues of large population displacement. The resulting renovation project also provides a feasible approach for the inhabitants of the community to stimulate future spontaneous regenerations. The participation of multiple parties also allows diversity to go beyond architectural style.

Certain challenges in adaptability and

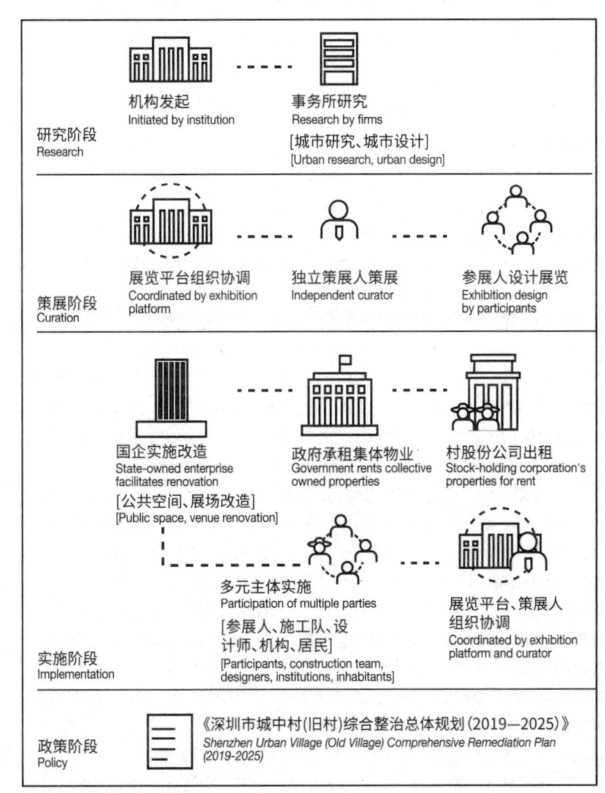

图 7 2017 深双城市策展模式
Figure 7 2017 UABB urban curation model

图 8 南头古城，张超摄于 2017 ©UABB
Figure 8 Nantou Old Town, photograph by Zhang Chao, 2017 ©UABB

点状式的自发改造模式，虽然改造后的独栋公寓租金成倍数上涨，但对整个片区租金的影响微乎其微，从而原有的小商户业态、低收入群体的居住都得到保护，避免大规模人口结构置换的问题。改造成果也为社区中的居民提供了一个可参照的成果，引导更多的自发更新改造，也只有在多主体的自发更新下才能使“多元”不仅停留在建筑风貌上，而是从内而外地实现多元包容。

多主体自发更新的普及和实现存在着一定的挑战。其一，城中村业主及本土施工队大多缺乏建筑、结构等专业知识，在消除消防隐患、结构安全加固以及居住空间品质提升等方面都有一定的局限性。近年来，通过大众媒体、社交媒体的传播，专业人士介入城中村改造的优势及必要性逐步打破专业圈层，进入民众的视野；但消防安全及生活品质的意识仍有待提高。而城中村业主与专业人士对接平台的缺乏，也导致有效的、专业的自发改造受限。事实上，许多自发改造的实施案例，如冈厦 1980、新洲村十点创意饮品店（图 5- 图 6）、官湖村艺栈和梧桐山赤水洞村 ffiXXed Studios 等，都离不开业主或是店主本身所具备的建筑设计和建造的资源。其二则是精细化的政策、机制和实施路径的缺失，无法如成片试点项目得到与政府职能部门沟通的平台，导致城中村业主虽有自发修缮、改造的意愿，但却无章可循、无法可依，进而出现禁止城中村修缮、改造一刀切的现象。

4 “非正规军”的弱介入

2016 年 4 月，孟岩、刘晓都、侯瀚如提交 2017 深双策展提案，并于 10 月正式成为 2017 深双总策展人。策展提案确认首次以城中村为主展场，以展览为契机介入城中村改造、提升公共空间品质。这是把深双展览模式推向极致，同时也是具有跨时代意义的一场城中村介入实践（图 7）。深双展览平台与南山区政府共同建

realizing multi-party spontaneous self-built development exist. First, there are limitations to the elimination of fire hazards, structural reinforcement, and spatial improvement, as property owners and contractors are not professional architects and structural engineers. Recently, the necessity of having professionals involved in urban village renovation has been promoted through public and social media, as awareness of fire protection and quality of life has been neglected under the earlier self-built model. Moreover, the lack of a platform for owners to meet professionals has also limited the effectiveness of spontaneous renovations. In fact, many spontaneous regeneration projects, such as Gangxia 1980, 10. Creative Drink in Xinzhou Village (Figure 5-6), Artinn in Guanhu Village and ffiXXed Studios in Chishuidong Village, have all relied on the property or shop owner's resources to complete architectural design and construction.Unlike large-scale projects, which have been supported by government departments, small-scale regeneration has lacked policies, mechanisms and implementation paths to support owner-led regeneration. As a result, there is a one-size-fits-all ban on independent repairs and transformation of the urban villages.

4 Informal Gentle Interventions

In April 2016, Meng Yan, Liu Xiaodu, and Hou Hanru submitted a proposal for the 2017 UABB and in October were appointed the chief curators. The 2017 UABB was the first biennale that situated its main venue in an urban village, with satellite venues in other villages throughout the city. The exhibition provided an opportunity to renovate urban villages and improve the quality of public spaces. It aimed to push for an ultimate UABB model, while representing a significant urban village intervention practice across time (Figure 7). The UABB Platform and Nanshan District Government jointly established a UABB committee to coordinate relevant departments, stimulate participation and enthusiasm for all sectors and organize community events as well as small-scale renovation projects based on research, resources and inspiration from the curators. The Nantou Sub-district Office, the curatorial team and the Shum Yip Group, which facilitated construction and coordinated communication with the residents of Nantou Old Town, invited them to participate in the activities of the biennale and the small-scale-renovation plan to try to build a sustainable platform to support future spontaneous micro renewal.

This was a renovation model that integrated multiple aspects, including top-down and bottom-up forces for

立双年展指挥部，协调相关职能部门，借助策展人的研究、资源及号召力，激发社会各界的参与热情，组织社区介入活动和微型设计改造。南头街道办、策展团队及改造代建方深业置地有限公司协同，在不同层面与南头古城居民沟通，邀请居民参与到双年展的活动和微改造计划，试图为自发的微更新微改造建立一个可持续的平台。

这是一次自上而下和自下而上结合、多方位结合的改造模式，实现有限度的介入计划：一是以公共空间为叙事线索，塑造、延续及复兴城中村公共生活，在南头古城高密度的城市肌理中创造和提升不同类型的公共场所（图8）；二是以古城北部厂房区作为主展场的轻度改造，清理厂房的立面，在完整地保留厂房原貌的基础上以局部空间重塑、建筑及艺术装置置入的方式为其带来新生；三是散布在南头古城各处的小型展览以及引导与自发相结合的社区微改造项目。点状式的更新方式使其具有试错的空间又能达到实验的目的。与介入计划交织的实施路径可概括为三类：一是由策展团队及都市实践主导的，自上而下的展览介入、展场及公共空间改造；二是以展览为平台通过邀请和征集的方式，调动青年建筑师和专业团体以及南头居民参与热情，实现社区共建（值得一提的是，在双年展与居民合作的共建项目中，并不是以金钱利益作为协商的筹码，而是从居民自身的改造需求出发，双年展以一个平台的身份，制订规则，匹配及协调资源）；三是双年展计划之外的，主动参与微介入和社区营造的社会力量。

2017 深双是一个未完成的双年展，一方面因为介入的城中村的复杂度，有一系列未能实现的项目；另一方面双年展仅仅是渐进式介入城中村改造的开端，而非大团圆结局 。

随着双年展的闭幕和《深圳市城中村（旧村）总体规划（2018—2025）》征求意见稿的发布，对城中村未来的讨论与空间博弈进入一个全新的阶段。以双年展起始的南头改造与重生计划需要一个挣脱经济资本逻辑，以文化资本和社会资本累积为目标，可持续、长期稳定的平台，建立和完善共建的游戏规则，有限度地引入外部资源，进行本地培育，不断地引导更新。所以“2017 南头模式”是结合多种模式构建的非常规介入方式，而非建立一套可复制粘贴的方法论。“城市策展”是一种改造的思维模式，量身定制的点状式、渐进式、引导式的改造是它的策略，个体价值的体现和多元生活方式的延续是它的内核。

a self-constrained intervention. First, interventions were introduced based on public space as a spatial narrative to shape, inherit and revitalize public life in urban villages. This was done with the aim of creating and improving different types of public space in the high-density urban fabric of Nantou Old Town (Figure 8). Secondly, a gentle renovation of the factory zone to the north of the old town was done, preserving the factory and cleaning up the facade while reshaping the interior space and inserting architectural and art installations to give the factory zone new life. Thirdly, small-scale exhibitions as well as guided and spontaneous community renovation projects were scattered throughout Nantou Old Town. Such an acupuncture approach to regeneration tolerates errors while enabling further experimentation. The above plan included three types of implementation paths. First was the top-down exhibition interventions as well as exhibition venues and public space renovation led by the curatorial team and URBANUS. Second was community building through open calls for the participation of young, professional teams and inhabitants. Third was collaboration with social forces that actively participate in micro intervention and community building outside the biennial plan. It is worth mentioning that instead of negotiating for monetary benefits, cooperation with inhabitants was based on their renovation needs. The biennale was a platform for making rules and coordinating resources.

The 2017 UABB was an unfinished biennale. On the one hand, there were several unrealized projects due to the complexity of urban village interventions, while on the other, the biennale was a starting point rather than a happy ending. With the closing of the biennale and the release of the draft of *The Shenzhen Urban Village (Old Village) Master Plan (2018–2025)*, the discourse and future plans for urban villages have entered a new stage. The revitalization of Nantou that started with the biennale requires a sustainable platform that is free from the economic logic of capital and directed more toward the accumulation of cultural and social capital. It is about establishing and improving rules for community building, selectively introducing external resources for local cultivation and constantly guiding regenerations. Instead of building a replicable methodology, what might be called the "2017 Nantou Model" was an unconventional intervention that adopted and integrated different models. Hence, urban curation becomes a renovation approach with an acupuncture style, a strategy of incremental and guided renovations and an ideology of individual values and inherited diverse lifestyles.

5　城中村与住房保障体系结合

2016 年 6 月，《深圳商报》发布新闻，称福田区政府拟“返租”水围村 35 栋共 600 套房源，作为城中村与保障房结合的试点进行改造，用

5　The Integration of Urban Villages and the Housing Security System

In June 2016, the *Shenzhen Economic Daily* released a press statement, indicating that the Futian District

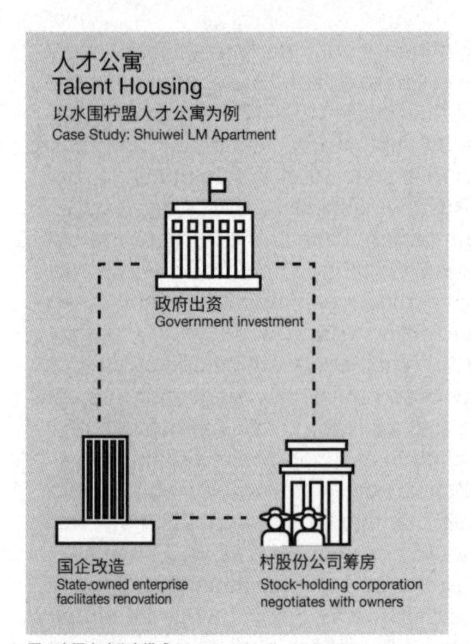

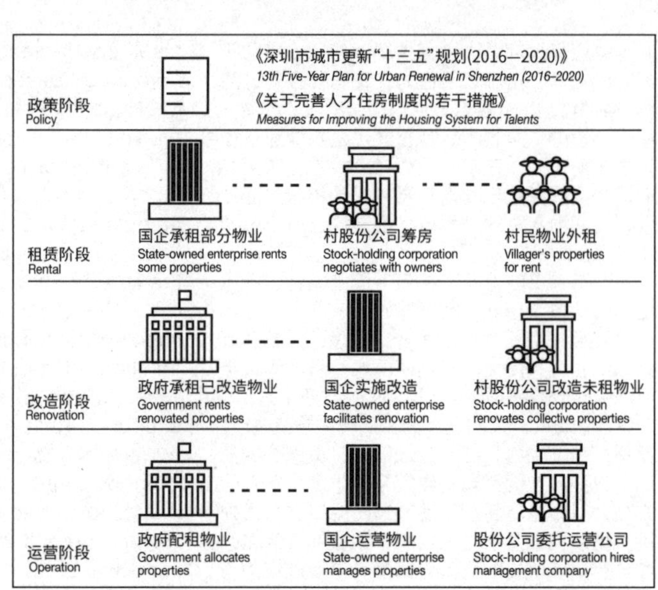

图 9 水围人才公寓模式
Figure 9 Shuiwei Village talent housing model

图 10 由创始点设计的水围柠盟人才公寓改造 ©IVY Photography
Figure 10 Shuiwei LM Apartment renovation designed by DOFFICE ©IVY Photography

作人才公寓，提供给深圳高端人才、引进人才和重点企事业单位职工。[13] 2016 年 7 月，中共深圳市委、深圳市人民政府印发《关于完善人才住房制度的若干措施》（简称《措施》），其中提到"支持原农村集体经济组织继受单位依法将已建成、审批手续不完备的住房改造为租赁型人才住房"。《措施》中还提出，所在区政府应给予适当的改造和运营资金补贴。2016 年 11 月，《深圳市城市更新"十三五"规划（2016—2020）》正式出台，提倡有机更新并明确指出："城中村以完善配套和改

Government intended to "rent back" a total of 600 units in 35 buildings in Shuiwei Village as a pilot project for the integration of urban villages and affordable housing, providing apartments for high-end professionals, and recruiting talent and staff from key public institutions and enterprises.[13] The following month, the Shenzhen Municipal Committee of the Communist Party of China and the Shenzhen Municipal People's Government issued the *Measures for Improving the Housing System for Talents* (hereafter the Measures), which mentioned "supporting the inherited parties of original rural collective

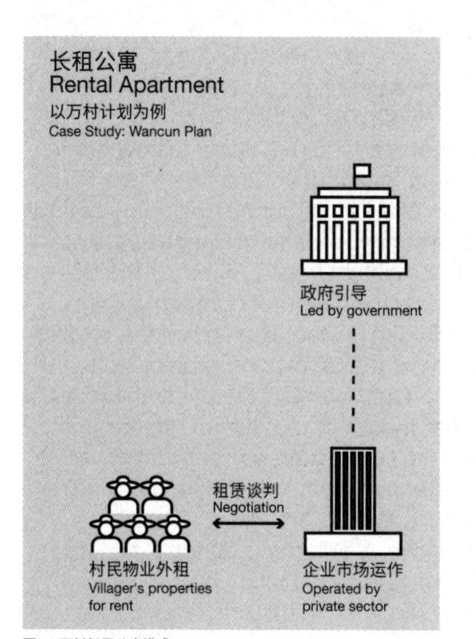

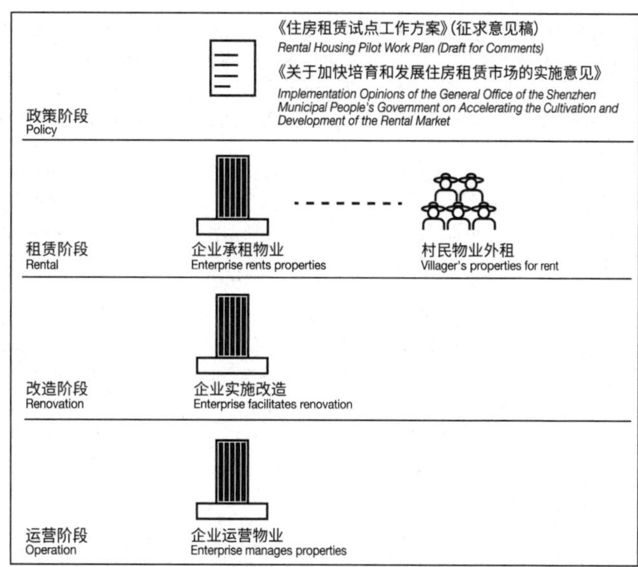

图 11 万村长租公寓模式
Figure 11 Wancun rental apartment model

善环境为目标，以综合整治为主，拆除重建为辅。”[14] 深圳在“十三五”规划期间，需提供人才住房和保障性住房共 35 万套。在建设用地紧缺的情况下，作为存量用地的城中村居住条件的优化，结合住房保障体系的模式，成为有可能解决两大城市问题的出路。

“水围模式”（图 9）的探索从 2015 年开始，2018 年落成，深业集团有限公司（简称深业集团）作为纽带，从事协商、施工筹建等事务，创始点建筑事务所作为改造设计团队介入，福田区政府及区住建局参与协调市政基础设施改造所涉及的相关各行政职能部门。水围股份公司与 35 栋约 500 多户业主进行协商，以股份公司每月向业主支付约 73 元 / 平方米的租金，每两年 6% 的增幅，与 29 栋业主签订了 10 年的租赁协议，实现了物业使用权的转移；再由深业集团以每月 70~80 元 / 平方米的租金与水围股份公司签订新村 29 栋 10 年的返租协议。[15] 一至二层底层商业空间与部分三层保留水围实业有限公司的使用权并自行改造，其余居住部分由深业集团按照人才住房标准改造（图 10），公共配套和基础设施（燃气管道、给排水管网、供电系统等）由区政府投资进行综合整治，[16] 福田区政府整体改造投资为 1 亿元。改造后由深业集团以每平方米 150 元 / 月租给区政府，区政府再以约每平方米 75 元 / 月的人才住房租金配租给辖区产业人才（相当于政府贴补每月 75 元 / 平方米，租金补贴比例 1:1）。[16]

在“水围模式”中，人才公寓与城中村改造二合一的经验及成果是城中村改造实践的关键节点，从政策、行政管理以及改造规范上都是在非正式社区改造的一次具有“实验精神”的重要突破。政府职能部门以信函代替工程审批证书，以会议纪要的方式进行决策，以领导

economic housing with incomplete approval procedures to transform it into rental housing for talents." The Measures also proposed that the district government should provide appropriate funding for renovations and operations. In November 2016, the *13th Five-Year Plan for Urban Renewal in Shenzhen (2016-2020)* was officially released, which encouraged organic renewal and clearly stated that "the goal of urban village renewal is to improve supporting facilities and the environment as well as focus on comprehensive remediation supplemented by demolition and reconstruction."[14] During the period of the *13th Five-Year Plan (2016-2020)*, Shenzhen city was required to provide 350,000 units of talent and affordable housing. In the context of the shortage of construction land, optimizing living conditions in urban villages and integrating brownfield land with the housing security system became a possible way to address two major problems facing the city.

The exploration of the Shuiwei model (Figure 9) began in 2015 and was completed in 2018. Shum Yip Group facilitated the negotiations, preparations and construction, while DOFFICE Architecture Urbanism Landscape was the design team for the renovation. Also, the Futian District Government and Shenzhen Housing and Construction Bureau coordinated relevant administrative departments involved in the renovation of municipal infrastructure. Shuiwei Stock-holding Corporation negotiated with more than 500 owners of 35 buildings to transfer property use rights, ultimately signing a ten-year lease agreement for 29 buildings. Shuiwei agreed to pay owners 73 yuan per square meter per month with an increase of 6% every two years. Shum Yip Group then signed a 10-year lease agreement with the stock-holding corporation to rent the 29 buildings, paying 70–80 yuan

现场指导和建议的方式把控项目质量，[15] 从而使得这项没有先例的非正式社区改造得以完成。同时，也为未来制定成熟的、非正式社区改造的技术、行政管理和机制提供了具有参考价值的实践经验。

"水围模式"的普适性亦将面临重重考验和挑战，一方面低回报率的投资模式如何能在跳脱现有市场经济逻辑的同时，依然可持续地发展；另一方面，由城中村自建房改造成人才公寓，本质上依然存在人口置换的问题。改造规模、人口置换的比例需作为前提条件确定，使得城中村在快速改造下能有效地避免因入住门槛大规模地提高，失去其原本弹性、灵活和低成本的居住条件。

6 综合整治、物业管理与商业运营三位一体

截至 2015 年末深圳常住人口为 1137.87 万人，[17] 常住居民家庭住房自有率仅为 34%[18]（全国为 91.2%），[19] 房屋租赁需求巨大。而随着城市的转型，以外来务工者为主的外来人口逐步转向以年轻白领、科技人才为核心，城中村简陋的居住条件显然无法满足年轻白领的生活需求。房地产开发企业旧改的脚步随着城中村旧改的成本越来越高逐步放缓，租金低廉的城中村成为长租公寓潜在的供给主力。此时，为解决深圳住房供应的难题，2017 年 8 月深圳市在《住房租赁试点工作方案》（征求意见稿）提出，"十三五"期间，要通过收购、租赁、改建等方式收储不低于 100 万套（间）村民自建房或村集体自有物业。9 月，《深圳市人民政府办公厅关于加快培育和发展住房租赁市场的实施意见》中提出，引导城中村通过综合整治开展规模化租赁。[7] 在政策的东风和住房租赁市场的驱动下，以"万村计划"为代表的一系列城中村改造结合长租公寓产品争相开工。

长租公寓结合城中村改造可以视为"水围模式"的市场运作模式（图 11）。模式中以政府为主导，投资基础设施和公共配套的综合整治，由开发主体配合完成，同时开发主体以略高于市场价的租金，统租 10~12 年城中村自建房物业，并针对年轻用户族群对租用物业进行改造、提升，植入文化、商业等内容，形成综合整治、物业管理和商业运营三管齐下的改造模式。2017 年 7 月万科出资 1000 万元成立深圳市万村发展有限公司，进而宣布"万村计划"。8 月，万科万村与龙岗区政府、坂田街道办紧密合作，开展岗头新围仔村的改造工作。[20] 12 月 30 日新围仔首栋长租公寓样板楼揭幕，并交

7 允许产权手续不完善、但经济利益关系已经理顺的"城中村"以及其他建筑，在符合规划、消防、安全、环保政策的前提下，用于住房租赁经营并纳入住房租赁合同备案管理。具体标准和程序尚无出台政策依据。

per square meter per month.[15] Shuiwei Stock-holding Corporation has retained the right to use all the commercial space on the first, second and part of the third floors. Shum Yip renovated the remaining residential units as rental housing for qualified talents (Figure 10), while public facilities and infrastructure, including natural gas pipelines, the water supply and drainage systems, and the power supply system were funded by the District as part of the implementation of comprehensive remediation.[16] The Futian District Government invested 100 million yuan toward the renovation project. After the renovation, Shum Yip Group rented the units at 150 yuan per square meter per month to the district government, who have allocated it to district recognized talents at 75 yuan per square meter per month. At 75 yuan per meter per month, the subsidy ratio is 1:1.[16]

The Shuiwei model, which foregrounded the experiences and results of integrating talent housing and an urban village was a milestone in the practice of urban village regeneration. This process has involved the regeneration of an informal community with an experimental spirit in terms of policy, administrative management, and building codes. The unprecedented informal community regeneration was completed through the replacement of the approval certificate with letters from government departments, with decision-making coming in the form of meeting minutes and quality control being carried out through on-site guidance and the advice of the project leaders.[15] Meanwhile, this project has also provided experience in establishing renovation techniques, administrative management and mechanisms for future informal community regenerations.

How adaptable the Shuiwei model might be will be seen in future tests and challenges. On the one hand, it needs to be determined whether the investment model is sustainable, given its low rate of return and following no existing market economy logic, while on the other, turning self-built buildings in urban villages into talent housing is a form of population displacement. To effectively avoid losing the flexibility and affordability of urban villages during rapid renovation and raised entry thresholds, the scale of renovation and proportion of population displacement must be pre-conditioned.

6 Comprehensive Remediation, Property Management and Commercial Operation Three in One

At the end of 2015, Shenzhen's permanent population was 11.3787 million,[17] and the home-ownership rate of permanent residents was only 34%,[18] compared to 91.2% nationwide.[19] Accordingly, the demand for rental housing was high. With the transformation of Shenzhen from a manufacturing to a managerial city, the city's demographic has gradually shifted from migrant workers

图 12 改造前的南头古城，2016（左）；万科改造后的南头古城，2023（右）© 都市实践
Figure 12 Nantou Old Town before renovation, 2016 (left); Nantou Old Town after the 2021 renewal project by Vanke, 2023 (right) ©URBANUS

由万科“泊寓”[8]运营，租金在 1600~2500 元 / 月，规定入住人群年龄不超过 40 岁且不能带小孩及宠物。2017 年底，深圳万村的城中村综合整治及统租运营业务已拓展至 33 个城中村，其中 10 个村已开启整租及改造运营工作。[20]

如此迅速的扩张突显其在效益平衡和人口结构置换两大问题中的矛盾。与“水围模式”中与股份公司协商的方式不同，万村直接与业主磋商签约，虽然可以避免整体统筹的空租期，但协商谈判成本相对提高。除了部分村民没有与万村合作的意愿外，万村大规模的扩张也引起现有租户对租金上涨的顾虑。2018 年 6 月，来自“无名劳工代表”的文章《13 万富士康劳工代表致万科、房东及监管部门书》引发广泛的社会关注。文中表示：“近日进驻富士康龙华工厂北门清湖新村开展‘万村计划’业务，已实质打破原有城中村租赁合同关系。随着‘万村计划’进一步拓展，后续影响未知，不确定性增大，引发

8 2015 年初，万科新城里的“万科驿”将“万汇楼”（土楼公社）纳入市场化运营，对室内进行“年轻化”装修，被视为“泊寓”的初次试水。2016 年 5 月，“万科驿”更名为“泊寓”，推出了长租公寓产品。广州万科房地产有限公司对广州棠下村 6 栋厂房进行改造，城中村厂房作为改造对象，统租价格低，改造难度相对较小，改造成以年轻人为主要族群的长租公寓，逐步形成万科泊寓这一长租公寓品牌。

to young white-collar, scientific and technological workers. The simple living conditions of urban villages no longer met the living needs of these young workers. Real estate developers have gradually slowed down the pace of urban village renewal as the cost of demolishing and constructing urban villages has increased, and urban villages with low rents have become the main source of the potential supply of long-term rentals. To solve the housing supply problem in the city, Shenzhen proposed the *Rental Housing Pilot Work Plan (Draft for Comments)* in August 2017, stating that during the 13th Five-Year Plan period, 1 million units of self-built buildings or collective properties should be acquired through purchases, leases, and reconstructions. In September of that year, the *Implementation Opinions of the General Office of the Shenzhen Municipal People's Government on Accelerating the Cultivation and Development of the Rental Market* proposed guiding urban villages in carrying out large-scale leasing through comprehensive remediation.[10] The Wancun rental apartment model is representative of the emergent projects that are driven by both these policies and the rental market.

The integration of long-term rental apartments and urban village renovation can be regarded as a market

了以富士康产业工人为主体的租客群体恐慌。”[21]

深圳市万村发展有限公司随即针对舆论做出说明，表示改造前后续的租金价格都处于相同的区间；[9] 万村在消防、管线、室内装修运营维护的投入虽然大，但是通过集约户型的经验尽可能维持租金稳定。按说明的表述，改造后租金依然维持少于 10% 的涨幅，若在没有政府补贴扶持下，[10] 作为市场运作的项目，主要依靠未来租金的涨幅回流的模式显然很难回本或是维持健康的资金流。如此一来，不仅是万村计划，越来越多以长租公寓模式参与城中村改造的可持续性将面临巨大的挑战。2018 年 11 月市场传出“万村计划”全面暂停的消息，万科予以否认。[22]

2019年9月,《万科宣布拿下南头古城改造，万村计划“旗舰店”来了？！》为题的新闻发布。万科南头古城改造计划于 2020 年 8 月完成了 299 栋房屋 9 成的统租计划（古城建筑约 1000 余栋），并率先完成了南北街 88 栋建筑的改造（图 12），同步实施雨污分流、电力、照明、消防改造等基础设施工程。[23] 2020 年 5 月 14 日，深圳万通南头城管理运营有限公司成立并投入南头古城的开发与运营，深圳市万科发展有限公司、深圳市深汇通投资控股有限公司（唯一股东为深圳市南山区国有资产监督管理局，即国有独资）及深圳市南头城实业股份有限公司分别持股 65%、30% 及 5%。改造后的南头古城的商业运营策划和规划将分为四大功能区，历史怀旧区、艺术文化体验区、品质生活区和文化创意区，植入粤港澳本土品牌、咖啡馆、米其林餐厅、文创品牌等，再以孵化器的模式，入股部分品牌店并减免店铺租金，将“万村计划”中尚未实现的收入多元化的概念引入其中，其效益和影响还有待时间的考证。南头古城的泊寓则推出多种住房租赁产品，基于楼栋条件不一，单间价格区间为 4298~5598 元 / 月，一房一厅的价格区间为 3998~6098 元 / 月，[11] 而同期南头古城其他公寓单间价格区间约 1200~2000 元 / 月，一房一厅价格区间约 1500~2500 元 / 月。[12] 显然泊寓的目标客群以消费能力更高的白领阶层为主，与南头古城原租户群体不同，也是在此次改造中约 30%“人口结构更替”的策略之一。泊寓的入驻带来周边自建房租赁价格上涨

9 “新围仔城中村未改造的单房均价在 800 元 / 间 / 月，一房一卫区间在 1100~1200 元 / 间 / 月，两房一卫均价在 1250 元 / 间 / 月，改造后泊寓的价格区间为 798~1398 元（含家私家电）；在福田玉田村，未改造的单房价格区间在 1250~2600 元 / 间 / 月，一房一厅或两房价格区间在 2600~4000 元 / 间 / 月，改造后的泊寓价格为 1398~2498 元 / 间 / 月（含家私家电）”，参见：万村复苏 . 关于万村计划参与清湖村等深圳城中村改造的情况说明 [EB/OL]. (2018-6-11) [2020-1-2]. https://mp.weixin.qq.com/s/Mw4ZIZPsIIdPUTwGPAvMVA.

10 “龙岗新围仔、福田水围新村、罗湖布吉泊寓这几个已投入运营的城中村升级项目，均不同程度享受政府的补贴。”参见：陆璐 . 破题城中村治理 深圳玉田村改造实践 [N]. 时代周报 , 2018-5-29.

11 在各户型面积的基础上，各户型价格参考各楼栋地理位置、采光条件及设施差异而设置，部分条件较好的单间价格高于条件较差的一房一厅公寓。数据来源：“南头古城泊寓”微信公众号，户型及定价，2021-8-1。

12 数据来源：深圳安居客，2021-8-1.

aspect of the Shuiwei model (Figure 11). Government-led and funded, the comprehensive remediation of infrastructure and supporting facilities has been facilitated by developers. At the same time, the developers have rented self-built buildings at a rate slightly higher than the market, for 10 to 12 years. These apartments have also been upgraded and remodeled for young people, and cultural and commercial programs have been developed to establish a renovation model that includes comprehensive remediation, property management, and commercial operations. In July 2017, Vanke invested 10 million to establish Shenzhen Wancun Development Co., Ltd. and then announced the Wancun Plan. In August 2017, Vanke Wancun worked closely with the Longgang District Government and Bantian Sub-district Office to renovate Gangtou Xinweizai Village.[20] On December 30, 2017, the first rental apartment in Xinweizai was unveiled and handed over to Vanke's Port Apartment[11] for management. The rent was from 1,600 yuan to 2,500 yuan per month, while the age of the tenants could not exceed 40 years old, and no children or pets were allowed. By the end of 2017, Wancun's comprehensive remediation of urban villages and leasing business had expanded to 33 urban villages, of which 10 had started leasing and renovation operations.[20]

Such rapid expansion emphasized the conflicts in the two major issues: cost efficiency and population replacement. As opposed to the Shuiwei Model, which was negotiated through the stock-holding corporation, Vanke Wancun negotiated directly with property owners and avoided vacant properties in the process, yet the cost of negotiation was raised. In addition to the lack of motivation of some villagers to cooperate with Wancun, the large-scale expansion of Wancun raised tenants' concerns over rent increases. In June 2018, an article from an "anonymous labor representative" entitled "Letter from 130,000 Foxconn Labor Representatives to Vanke, Landlords and Regulatory Authorities" aroused widespread social concern. The article stated that "the recent Wancun Plan in Qinghu New Village, which is located at the north gate of the Foxconn Longhua factory, has broken the current lease agreement of the urban village. With the further expansion of the Wancun Plan, future impacts are unknown and uncertainty is increasing, triggering panic among the tenants, mostly Foxconn industrial workers."[21]

10 Urban villages and other buildings without property rights confirmation, but which have demonstrated clear relationship of economic interests are allowed to be used for rental housing including in the rental housing management system under the premise of compliance with planning, fire protection, safety, and environmental protection policies. There is no policy basis for specific standards and procedures.

11 At the beginning of 2015, "Vanke Yi" in Vanke New Town incorporated "Wanhui Building (Tulou Collective Housing)" into market-oriented operation, with interior décor for young people. This was considered the first test of Port Apartment. In May 2016, "Vanke Yi" changed its name to Port Apartment and launched an eponymous rental apartment product, Port Apartment. Guangzhou Vanke Real Estate Co., Ltd. also renovated six factories in Tangxia Village, Guangzhou. As a renovation project, it is cheaper and easier to rent self-built factories than housing. The factories been transformed into rental apartments with young people as tenants, establishing the rental apartment brand.

的可能性，也使原租户群体往内城搬迁（因内城距主路远，巷道狭窄，阳光无法到达室内而租赁价格相较便宜），使城中村中原来灵活、低居住成本的租赁市场承受一定程度的考验。总体来说，涨租舆论、效益疑问以及人口结构更替过快等多种因素共同形成“长租公寓模式”的多重压力，其长期稳定的可持续发展，仍有待时间的验证。

7 结语

在城中村有机更新的命题中，基础设施的综合整治、公共空间和公共生活的重塑都在多年的探索中有了相对清晰的实施路径，而最难的是跳脱以市场作为驱动力的经济逻辑，建立引导更新的平台激发内生驱动力量。多个城中村改造示范项目在媒体和网络的传播下引发社会关注，城中村的业主和租户对空间环境改造需求比比皆是，但真正能够实现的自发更新依然是少数。结合上述经验，若实施主体在有机更新中的主要介入集中在公共设施及公共空间，进而要求实施主体或运营机构在更新中走向幕后，起到引导管理监督服务的职能，设立完善的机制和游戏规则，当地居民则可能在有管理依据的支撑下与外部资源合作，推动多主体有机更新。深圳城中村的幸运在于，不放弃的实践者并不止上述案例，而是如同星星之火，一次又一次地在极速发展的城市中尝试重建杂糅、共生的城市。

Shenzhen Wancun Development Co., Ltd. immediately issued a response, asserting that the subsequent rental prices and the pre-renovation prices were within the same range.[12] The practice of compact housing was adopted to maintain rent stability after the large investment in improving fire protection, pipelines, interior decoration and maintenance. The response emphasized that the rent was increased by less than 10% after the renovations and in the absence of government subsidies, it would be difficult to recover the cost or maintain a healthy cash flow.[13] As a result, the Wancun Plan and the long-term rental apartments in urban village renovations face sustainability challenges. In November 2018, it was reported that the "Wancun Plan" was suspended, and Vanke denied this.[22]

In September 2019, a press release article entitled "Vanke announces the renovation of Nantou Old Town. Is the 'flagship store' of the Wancun Plan coming?!" In August 2020, Vanke stated that 90% of the 299 building rental plan was complete. (There are about 1,000 buildings in Nantou Old Town). Meanwhile, the renovation of 88 buildings on the north and south streets was finished (Figure 12) with the infrastructure updated, such as the sewage system, electric system, lighting, and fire protection.[23] On May 14, 2020, Shenzhen Wantong Nantou City Management Operation Co. Ltd. was established and invested in the development of Nantou Old Town. Shenzhen Vanke Development Co. Ltd., Shenzhen Shenhuitong Investment Holding Co. Ltd. (state-owned assets of the Market Supervision and Regulation Bureau of Shenzhen Municipality) and Shenzhen Nantoucheng Stock-holding Corporation own 65%, 30% and 5% of the shares, respectively. The renovated Nantou Old Town is planned to include four programmed districts: a historical district an art and culture district a quality living district, and a creative culture district. Local brands, cafés, Michelin restaurants and cultural creative brands of Guangdong, Hong Kong and Macao have been introduced, and some were included in the incubator program with rent reductions or exemptions. The unrealized concept of income diversification in the Wancun Plan has been introduced in Nantou. The cost-efficiency and impacts have yet to be verified. Port Apartment in

12 According to Wancun Development Co., Ltd. "The average price of studio before renovation in Xinweizai Urban Village was 800 yuan/unit/month, the average price of one bedroom was 1100-1200 yuan/unit/month, and the average price of two bedrooms was 1,250 yuan/unit/month. After the renovation, the price range of a Port apartment ranged from 798 to 1,398 yuan (including furniture and appliances); in Futian Yutian Village, the price range of a studio before renovation was 1,250-2,600 yuan/unit/month, and the price range of one or two bedroom apartment was 2,600-4,000 yuan/unit/month. The price of a renovated Port apartment is 1,398-2,498 yuan/unit/month (including furniture and appliances)" See "Explanation on Wancun Plan's urban village renovation in Qinghu village and many others in Shenzhen". https://mp.weixin.qq.com/s/Mw4ZIZPsIIdPUTwGPAvMVA.

13 According to Lulu, "Longgang Xinweizai, Futian Shuiwei Xincun and Luohu Buji, all of these operating Port Apartment received government subsidies to varying degrees" See, LU L. Overcome the Issue of Urban Village Management with Shenzhen Yutian Village Renovation Practices[N]. *Time Weekly*, (2018-5-29)[2020-12-10]http://www.time-weekly.com/wap-article/251369.

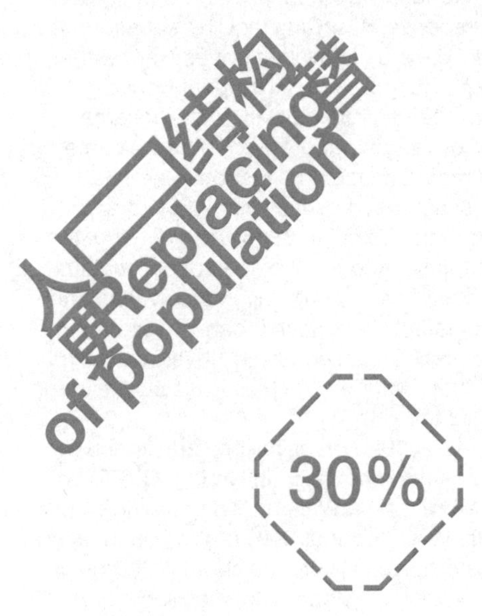

有机更新

Nantou Old Town promoted several rental products. The price of a studio ranges from 4,298 to 4,498 yuan per month, while a one-bedroom department is priced from 3,998 to 6,098 yuan per month.[14] At the same time, the price of other studio units in Nantou ranges from 1,200 to 2,000 yuan per month, and one-bedroom apartments range from 1,500 to 2,500 yuan per month.[15] The targeted group of the port apartments is members of the younger generation with higher purchasing power, which is different from the current tenants of Nantou Old Town. Nantou Old Town Port Apartments have also enabled rent increases in nearby self-built housing, causing current tenants to relocate to the inner town, away from the main streets. This challenges the flexibility and affordability of rental housing in the urban village. In general, various factors such as public opinion on rent increases, cost-efficiency, and rapid population displacement have combined to put pressure on the rental apartment model, and more time is needed to determine its sustainability.

7 Conclusion

The organic renewal of urban villages, the comprehensive remediation of infrastructure and the reshaping of public space and public life have had relatively clear implementation paths over the past years of exploration. The issue of establishing a platform to guide urban village regeneration and an active inner driving force without mobilizing market forces has yet to be solved. Several projects for urban village renovations have gained attention through exposure on media and social media. Both owners and tenants have demanded spatial improvement in urban villages, but few spontaneous renovations have been realized. Based on the above experience, when the facilitators intervene mainly in public facilities and public spaces as part of organic renewal, and they step back to become supervisors and service providers by formulating rules and mechanisms, it allows locals to cooperate with external resources with the support of regulations, thereby encouraging organic renewal by diverse parties. Urban villages in Shenzhen are fortunate that countless practitioners beyond the above precedents have repeatedly attempted to rebuild a hybrid city of coexistence in the context of rapid urban development.

14 Besides the size of the apartment, location, natural light conditions and infrusturcture of the building also affect the rents. Therefore, some of the studios with better conditions are more expensive than one bedroom units. Data Source: layout and price on Nantou Old Town Port Apartment official WeChat account, 2021-8-1.
15 Data Source: Shenzhen Anjuke, 2021-8-1.

[1] 党文婷 . 从城中村到世界油画集散地的逆袭之路 ——大芬油画村的故事（上）[N/OL]. 光明日报 , (2020-9-18) [2020-12-30]. https://wap.gmdaily.cn/article/a7e968eb416498f83891963f46276d8.
[2] 张静 . 深圳大芬村：城中村借油画再生 [N/OL]. 新京报，(2010-5-26) [2020-12-30]. https://finance.qq.com/a/20100526/001377.htm.
[3] 党文婷 . 从城中村到世界油画集散地的逆袭之路 ——大芬油画村的故事（下）[N/OL]. 光明日报 , (2020-9-18) [2020-12-30]. https://wap.gmdaily.cn/article/01f914d82bfd4b48b8daf0ca977b0499.
[4] 李振 . 大芬油画村：超集群与深圳式全球化 [N/OL]. 21 世纪经济报道 , (2018-11-5) [2020-11-18]. https://www.jiemian.com/article/2596276.html.
[5] 毛思倩 . 探访转型期的"中国油画第一村 [N/OL]. 新华网 , (2019-2-19) [2020-12-30]. http://m.xinhuanet.com/gd/2019-02/21/c_1124143647.htm.
[6] 深圳市人民政府 . 深圳市人民政府关于深圳市城中村（旧村）改造暂行规定的实施意见〔2005〕56 号 [EB/OL]. (2005-4-7) [2020-12-30]. http://www.sz.gov.cn/zfgb/2005/gb434/content/post_4994592.html.
[7] 清华同衡规划播报 . 王海红 : 深圳城市更新探索实践及城中村改造经验 [EB/OL]. (2019-11-20) [2020-12-30].https://mp.weixin.qq.com/s/uukYRlDr9hYwZ5vJxdK_VA.
[8] 深圳市人民政府 . 深圳市城市更新办法〔2009〕211 号 [EB/OL].(2009-11-12)[2020-12-29]. http://www.sz.gov.cn/zwgk/zfxxgk/zfwj/szfl/content/post_6572261.html.
[9] 王睦广，陈洁，黄志伟 . 城中村电网改造惠民生 政企合作破难题 [N]. 南方都市报 , 2016-12-15.
[10] 深圳市发展和改革委员会 . 深圳市水价改革实施方案〔2017〕897 号 [EB/OL]. (2017-7-26) [2020-12-21]. http://fgw.sz.gov.cn/fzgggz/jgzcgl/zysphfwjg/content/post_4543953.html.
[11] 孟岩 . 城市即展场，展览即实践 [J]. 时代建筑 .2018(4):174-179.
[12] 城中村爆改户 . 深圳本地智慧 城中村爆改户小房子大趣味 [Z]. (2020-5-13) [2020-12-2]. https://www.bilibili.com/video/BV1nt4y117Ni.
[13] 深圳商报 . 35 栋村民房改成人才公寓 有望今年实现政府 " 返租 " [EB/OL]. (2016-6-8) [2020-12-16]. https://sz.focus.cn/zixun/d5f1664100a40b33.html.
[14] 深圳市规划和自然资源局 . 深圳市城市更新 " 十三五 " 规划（2016—2020）[Z]. 2016-11-22.
[15] Ban Li, De Tong, Yaying Wu, Guicai Li. Government-backed "laundering of the grey" in upgrading urban village properties: Ningmeng Apartment Project in Shuiwei Village, Shenzhen, China[J]. Progress in Planning, 2021-4-5 (146).
[16] 深圳市福田区住建局 . 深圳市福田区水围柠盟人才公寓建设项目申请报告 [R]. 2019-9-9.
[17] 深圳市统计局 . 2015 年深圳市社会性别统计报告 [EB/OL]. (2017-7-14) [2021-1-2]. http://www.sz.gov.cn/cn/xxgk/zfxxgj/tjsj/tjgb/content/post_1333692.html.
[18] 房地产业处 . 深圳市住房建设规划 (2016—2020) [EB/OL]. (2016-10-25) [2021-1-2]. http://pnr.sz.gov.cn/xxgk/ghjh/content/post_5842121.html.
[19] 贺勇，谢晓萍著 . 居住在中国 1949 年以来中国家庭居住变迁实录 [M]. 南京：东南大学出版社 , 2017.
[20] 万科集团 . 万科 2017 企业社会责任报告 [R/OL]. (2018-3-26). https://www.vanke.com/responsibility/report_data?typeid=undefined&newsid=3363.
[21] 佚名 . 13 万富士康劳工代表致万科、房东及监管部门书 [EB/OL]. (2018-6-10) [2021-1-3]. https://www.sohu.com/a/235061264_100186381.
[22] 陈淑贞 . 万科"万村计划"全面暂停拿房？万科回应：不存在的！ [EB/OL]. (2018-11-8) [2021-2-2]. https://www.yicai.com/news/100055392.html.
[23] 邹曾婧，曾贤平 . 南头古城南北街示范段开街营业 [EB/OL]. 深圳晚报 , (2020-8-27) [2021-1-2]. http://www.szns.gov.cn/ztzl/nsrdzt/ntgcdb/mtbd/content/post_8232298.html.

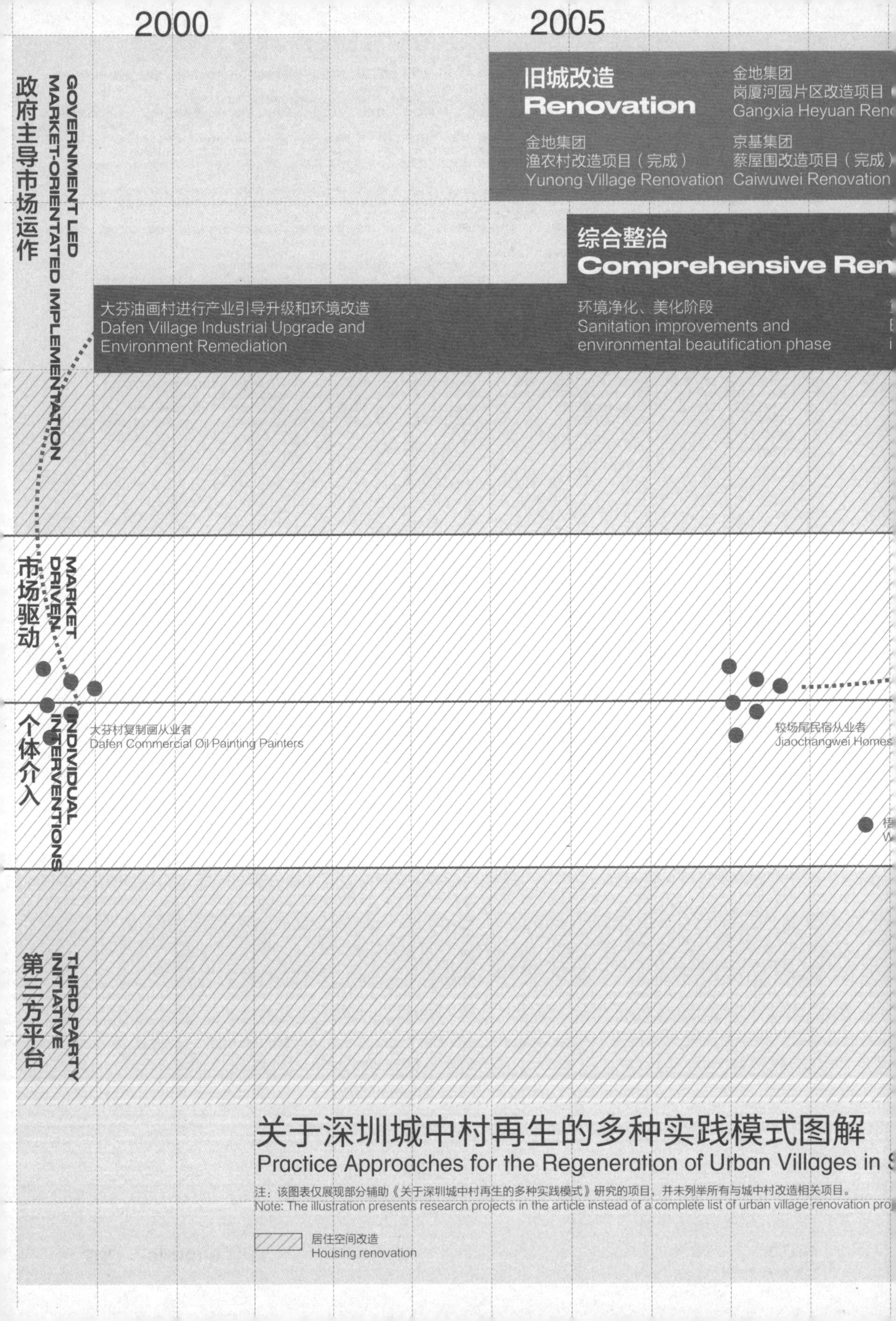

关于深圳城中村再生的多种实践模式图解

Practice Approaches for the Regeneration of Urban Villages in S

注：该图表仅展现部分辅助《关于深圳城中村再生的多种实践模式》研究的项目，并未列举所有与城中村改造相关项目。

Note: The illustration presents research projects in the article instead of a complete list of urban village renovation pro

居住空间改造
Housing renovation

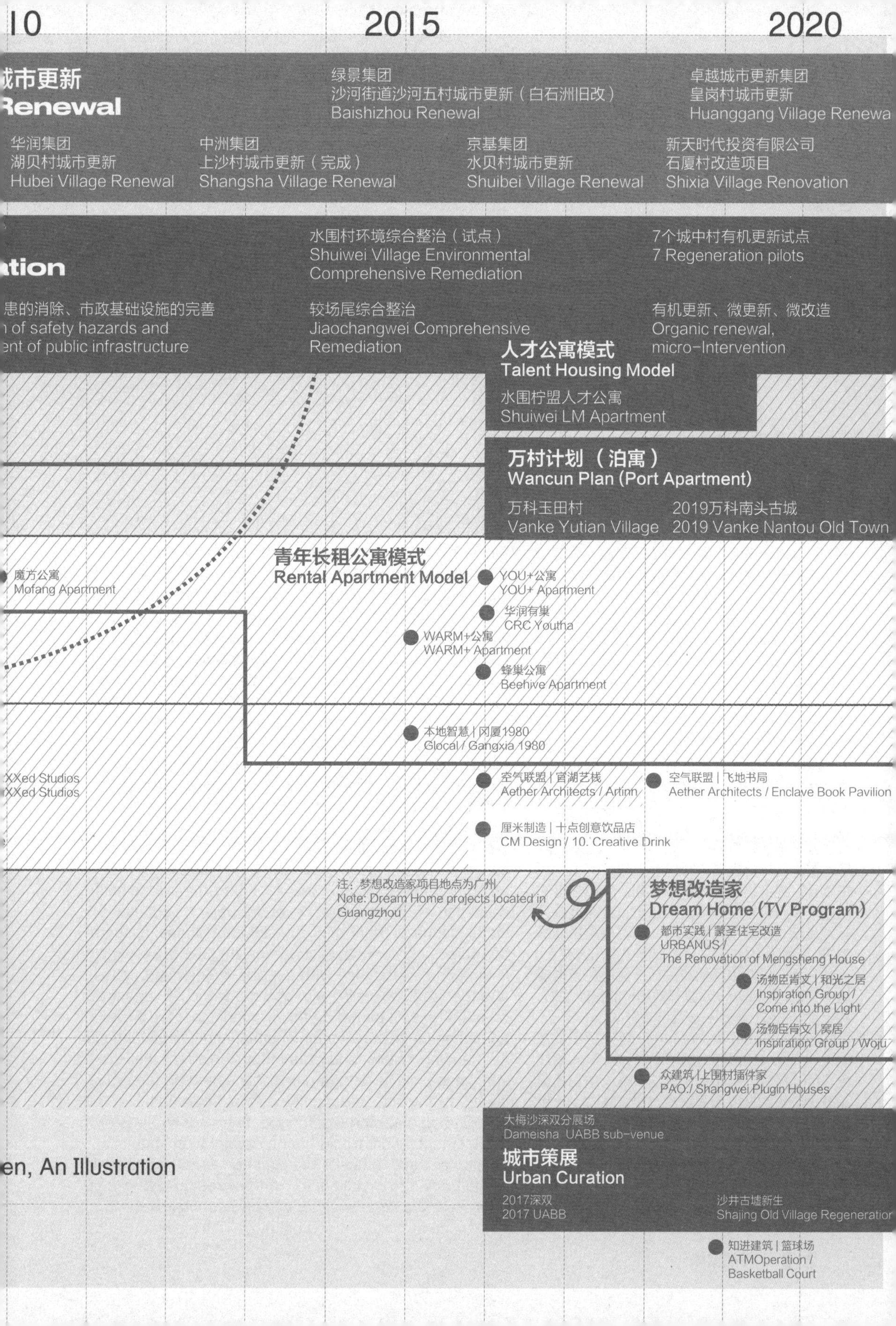

10
2015
2020
城市更新
Renewal
绿景集团
沙河街道沙河五村城市更新（白石洲旧改）
Baishizhou Renewal
卓越城市更新集团
皇岗村城市更新
Huanggang Village Renewa
华润集团
湖贝村城市更新
Hubei Village Renewal
中洲集团
上沙村城市更新（完成）
Shangsha Village Renewal
京基集团
水贝村城市更新
Shuibei Village Renewal
新天时代投资有限公司
石厦村改造项目
Shixia Village Renovation
ation
水围村环境综合整治（试点）
Shuiwei Village Environmental
Comprehensive Remediation
7个城中村有机更新试点
7 Regeneration pilots
患的消除、市政基础设施的完善
n of safety hazards and
ent of public infrastructure
较场尾综合整治
Jiaochangwei Comprehensive
Remediation
有机更新、微更新、微改造
Organic renewal,
micro-Intervention
人才公寓模式
Talent Housing Model
水围柠盟人才公寓
Shuiwei LM Apartment
万村计划（泊寓）
Wancun Plan (Port Apartment)
万科玉田村
Vanke Yutian Village
2019万科南头古城
2019 Vanke Nantou Old Town
青年长租公寓模式
Rental Apartment Model
魔方公寓
Mofang Apartment
YOU+公寓
YOU+ Apartment
华润有巢
CRC Youtha
WARM+公寓
WARM+ Apartment
蜂巢公寓
Beehive Apartment
本地智慧｜冈厦1980
Glocal / Gangxia 1980
XXed Studios
XXed Studios
空气联盟｜官湖艺栈
Aether Architects / Artinn
空气联盟｜飞地书局
Aether Architects / Enclave Book Pavilion
厘米制造｜十点创意饮品店
CM Design / 10. Creative Drink
注：梦想改造家项目地点为广州
Note: Dream Home projects located in Guangzhou
梦想改造家
Dream Home (TV Program)
都市实践｜蒙圣住宅改造
URBANUS / The Renovation of Mengsheng House
汤物臣肖文｜和光之居
Inspiration Group / Come into the Light
汤物臣肖文｜窝居
Inspiration Group / Woju
众建筑｜上围村插件家
PAO / Shangwei Plugin Houses
大梅沙深双分展场
Dameisha UABB sub-venue
城市策展
Urban Curation
2017深双
2017 UABB
沙井古墟新生
Shajing Old Village Regeneratior
知进建筑｜篮球场
ATMOperation / Basketball Court
en, An Illustration

超级乱糟糟：Mapping 南头古城

Super Messy: Mapping Nantou Old Town

何志森
图绘工作坊发起人，扉美术馆馆长，华南理工大学建筑学院副教授，墨尔本皇家理工大学建筑与城市设计学院兼职教授。主要研究领域包括图绘、城市更新、社区营造、非正规空间、微观城市、社区营造、边界、亚洲都市主义。
本文源于何志森在“超级乱糟糟：Mapping 工作坊专题研讨会”的发言，2018 年；由罗祎倩编辑和中译英。

Jason Ho
Jason Ho is the initiator of Mapping Workshop, director of FEI Arts Museum, associate professor at the School of Architecture, South China University of Technology, and adjunct professor at the School of Architecture and Urban Design at the Royal Melbourne Institute of Technology. His research focuses on mapping, urban regeneration, community empowerment, informal space, micro urbanism, border, and Asian urbanism.
The article is sourced from Jason Ho's speech at the "Symposium on Mapping Nantou Old Town - Super Messy," 2018; edited and translated from Chinese to English by Luo Yiqian.

2014 年以来，我在不知不觉中做了数十个工作坊，其间居无定所，几乎是以一种游牧的方式在教学。图绘工作坊强调透过“他者”的眼睛去理解我们今天真实生活的世界，帮助学生更好地理解“小人物们”对日常空间的使用和需求。许多人劝告说这种训练无益于今天商业化操作的建筑实践，但我认为只有当你不断地转换角色和观察范围的时候，你才能真正地拓展和延伸观看世界的视角和维度，你才能理解建筑更为宽广的意义。

图绘工作坊包括以下六个步骤：首先，在场地里面选择一个小而普通的目标对象作为切入点，可以是一个人、一只猫、一棵树，也可以是一个事件，或者一个被大家忽视的空间；然后对目标对象进行长时间的观察、记录和跟踪，挖掘目标对象在不同尺度上和其他事物的关联性；把自己转变为目标对象，以观察者的角度同步被观察者的感受和经历，进而透过目标对象的“眼睛”重新认知场地和真实生活的世界；在此认知上，在变化的尺度里揭示人和人、人和场地、场地和场地的关系，并呈现或再现这些关系；接下来一步，基于对场地的重新认知，提出设计主张；最后一步就是策展。

图绘工作坊的操作过程中，还有两个基本的要求：第一，跟踪和观察要在三个不同的尺度之间切换，比如小到身体的尺度，中到街道和社区，大到城市和国家的尺度；第二，至少扮演三个不同的角色，比如观察者、使用者、绘图者、跟踪者、设计师、管理者等。

下面我会用一个案例来具体聊一下图绘工作坊是如何开展的。

Since 2014, I have conducted dozens of workshops without realizing it. During this time, I had no fixed dwelling place and was teaching in an almost nomadic way. Mapping workshops focus on understanding the real world we are living in through the eyes of "the others" and assist students in better understanding ordinary people's use and needs of daily spaces. Many people cautioned me that the training was useless to the commercialized architectural practice of the day. However, I believe that only by constantly changing the roles and scopes of observation, can we expand and extend the perspectives and dimensions of a worldview, as well as understand the broader meaning of architecture.

Mapping workshops include the following steps: first, choose a small and ordinary object, such as a person, a cat, a tree, an incident, or a neglected space as an entry point to the site; conduct long-term observation, documentation, and tracking of the object, excavating its relevance to other things at different scales; assume the object's perspective, synchronizing the observer's feeling and experience with the observer's perspective, and then rediscover the site and the world through the object's "eyes"; based on this recognition, reveal the relationships between people, sites, and people and site at changing scales, and present or re-present these relationships; make design proposals based on the rediscovery of the site, and; finally, conduct curation.

During mapping workshops, there are two basic requirements: first, tracking and observation should change at three different scales, such as the small scale of the body, the middle scale of the street and neighborhood, and the large scale of city and country; second,

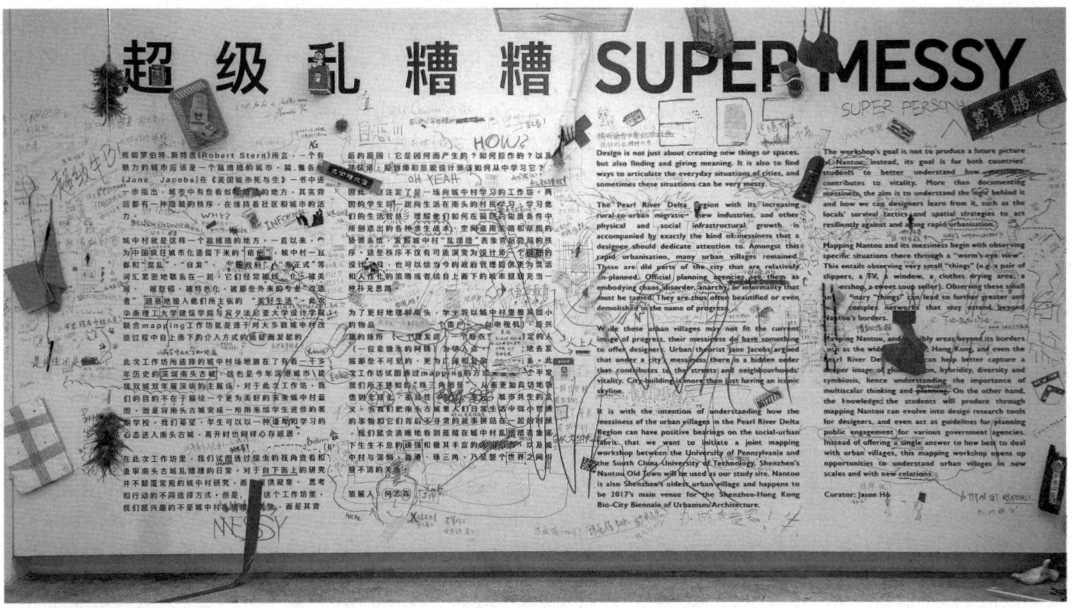

图 1 “超级乱糟糟”展览前言 ©UABB
Figure 1 *Super Messy* exhibition text ©UABB

1 都市侦探

2014 年，我在华南理工大学建筑学院开展一个名为“都市侦探”的图绘工作坊，其中一组学生选择一位在花城广场卖冰糖葫芦的小贩阿姨作为观察对象。

第一天，学生要成为“偷窥者”，他们的任务是近距离但又悄无声息地观察阿姨，理解她和广场空间的关系。例如，她是如何占用广场的空间？选择的位置在哪？和周边设施的关系是什么？还有与时间和广场使用人群的关系是什么？如果保安或城管来，她是如何躲藏的？经过一天的观察，他们发现阿姨很清楚地知道早晨 7 点小孩上学经过广场的路线，上午 10 点游客出现的位置，中午 12 点上班族在广场吃饭的地方，下午 4 点开始老人运动的地点，晚上 8 点广场舞的场地。她在广场上贩卖的位置会根据一天中时间和人群的变化、事件的上演而不断地切换。最终，学生发现了一张阿姨在花城广场的贩卖地图，这张地图完全不同于设计师、规划师自上而下的描绘。

第二、三天，学生变成“跟踪者”，跟踪阿姨一直从花城广场到她的住处，他们的任务是了解阿姨的交通工具、居住地点、居住距离、选择花城广场而不是其他地方的原因。在跟踪过程中，他们发现阿姨骑行简易的三轮车而没有使用地铁（因为冰糖葫芦的杆进不去地铁的门），在广场工作的时候三轮车寄放在附近的一个老乡家里。晚上，在接近阿姨位于城中村的家时，跟踪的三名学生被阿姨发现了，在告知用意后，被邀请到家里吃晚餐。聊天过程中，他们得知阿姨和她老公都来自山东一农村，一

each participant should assume at least three roles, such as observer, user, draftsman, tracker, designer, and administrator.

I will use the following case to elaborate on how mapping workshops unfold.

1 Urban Detective

In 2014, I launched a mapping workshop "Urban Detective" at the School of Architecture, South China University of Technology, in which a student group chose as their observation subject an auntie vendor who sold sugar-coated haws (*bingtang hulu*) in Huacheng Square, Guangzhou.

On the first day, the students needed to become "peepers." Their mission was to observe the auntie vendor closely and quietly, so as to understand the spatial relationship between the subject and the square. For example, how did she occupy the square space? Which location did she choose? What was the relationship between the chosen place and surrounding facilities? What was the relationship between time and square users? How did she hide when security guards and urban management officers came over? After a day of observation, they realized that the auntie knew very well where the children crossed the square at 7:00 a.m., where the tourists were at 10:00 a.m., where office workers ate lunch at noon, where the elderly exercised at 4:00 p.m., and where people gathered to dance at 8:00 p.m. She moved constantly, setting up her vending location based on the time, the groups using the plaza, and unexpected incidents. When the students finally mapped the auntie's

起在花城广场卖冰糖葫芦，她老公经常待在保安室附近放哨。

这次跟踪让学生意识到一个小小的活动（贩卖这个行为）需要这么多的条件和设施来满足和实现，包括阿姨的贩卖地图（时间、地点、使用者），老乡提供的放车场地，来自阿姨老公的前线情报（巡逻时间、路线）。这也让学生认识到，当在空间中设计一个物件的时候，不能只是简单粗暴地植入这个物件，还要提供各种各样在场地内外可以让这个物件运作起来的条件和设施，尤其是微观设施。

自从学生的身份暴露后，他们接下来对阿姨的观察方式变得更加亲密互动。第四天，学生目睹了阿姨的一根冰糖葫芦杆被保安没收。一根杆有 40 串冰糖葫芦，一串 10 元，一根杆共 400 元。阿姨没有了杆，就没有了一天的收入。阿姨心情非常低落，一个人坐在广场地上，半天没有说一句话。

经历这件事情后，学生开始扮演设计师的角色。他们在地图上标注花城广场里保安和城管的巡逻时间和路线、每一个摄像头的位置、每一个地铁出入口和公共汽车站的位置，甚至每一棵足够用来藏身的植物。最终，学生设计了三条花城广场的“逃跑路线”，帮助阿姨在最短的时间内消失，躲避保安和城管的追捕。

第五天，为了弥补前一天的损失，两位学生帮助阿姨卖冰糖葫芦。这个时候，学生变成了阿姨（小贩的角色）。过程中，其中一位学生发现一个很尴尬的事情：他不知道如何扛着冰糖葫芦杆上厕所。而后得知，为了避免上厕所，阿姨在凌晨 4 点起床后就不喝一滴水，直到卖完所有冰糖葫芦。

这种角色的转换又让学生回到一个真实的世界。他们最终帮助阿姨改造了三轮车，不但增加了厕所，还可以在车上售卖其他物品，比如衣服、鲜花、水果等。如果没有这种共情的能力和体验，只是坐在电脑前面画图做设计，学生永远不会做出这样的设计。我想这就是图绘工作坊的意义所在。

最后，学生成为一名策展人，把自己在不同尺度上、融入不同角色的观察和发现在一个规定的空间里有逻辑地呈现。我一直认为，未来的建筑师一定是在扮演策展人的角色，因为建筑永远不是一个人的表演，建筑师要有策划和推动各种力量合作参与的能力。

以上就是一个比较完整的工作坊实操过程，下面我来聊聊此次南头工作坊。

2 超级乱糟糟

2017 年，受美国宾夕法尼亚大学设计学院理查德 · 韦勒教授之邀，我在该院为研究生开设了一门图绘课程。正好该年深双的总策展人刘晓都老师也邀请我在双年展策划一个图绘板块。最终，课程变成宾夕法尼亚大学设计学院和

vending trajectory in Huacheng Square, it was completely different from the top-down version drawn by designers and planners.

On the second and third days, the students turned into "stalkers," following the auntie from Huacheng Square to her house. Their mission was to understand the auntie's transportation means, residence, distance between home and work, and reasons for choosing Huacheng Square instead of other places. During their reconnaissance, they discovered that the auntie went to work by basic tricycle instead of the metro (because the rod used to display sticks of sugar-coated haws were too long to fit through metro vehicle doors). She left the tricycle at a fellow villager's house while working in the square. At night, as she neared her house in the urban village, the auntie noticed the three students who were following her. After they explained what they were doing, she invited them over for dinner. As they chatted, the students learned that the auntie and her husband came from a village in Shandong, and they sold sugar-coated haws together in Huacheng Square, while her husband usually acted as a lookout man near the security office.

Tracking allowed the students to understand that even a small activity (vending) can only occur when many conditions are met and infrastructures are in place, including Auntie's vending map (time, place, user), a parking place provided by her fellow villager, and frontline information collected by her husband (patrol slot and route). It also made students aware that we cannot implement a program in an oversimplified and crude manner. Instead, we need to provide various conditions and supporting infrastructures, especially micro infrastructures, both inside and outside the site to make the program viable.

After their identity had been discovered, the students' observation of the auntie became more intimate and interactive. On the fourth day, the students witnessed a security guard confiscate the auntie's sugar-coated haws display rod. There were 40 sticks on a display rod, with each stick selling for 10 yuan, so there was 400 yuan of product on each display rod. Having lost her display rod, the auntie had no income that day. She was so upset that she sat wordlessly on the plaza ground for a while.

After this experience, the students assumed the role of designer. They marked the security guards' and urban management officers' patrol slot and route, the position of every camera, metro entrances and exits, bus stations, and even plants that could be used to hide behind or within. Finally, the students designed three "escape routes" for Huacheng Square, helping the auntie disappear in the shortest time possible to avoid being caught by the security guards and urban management officers who were chasing her.

On the fifth day, to make up for the auntie's loss

图 2 “超级乱糟糟”展览现场 ©UABB
Figure 2 *Super Messy* installation view ©UABB

华南理工大学建筑学院的联合工作坊，亦作为深双的参展作品，取名为“超级乱糟糟”（图1-图2）。

工作坊选择的场地是深双的主展场南头古城，具有一千多年历史，而今呈现为城中村面貌。我们使用蠕虫的视角察看和重新审视中国典型城中村中乱糟糟的日常。然而，我们感兴趣的不是城中村乱糟糟的表象，而是理解城中村的人们如何在简陋的物质条件中创造出各种求生战术与空间策略，发掘城中村乱糟糟背后的自治逻辑和草根的协调系统。这些秩序不仅可以演变为一个很好的设计工具，也可以给自上而下的城市规划和政府管理充当一种补充思路。

为了更好地理解城中村，学生以城中村里极其普通的日常物品、被忽视的空间和特定的人为切入点，由小及大地发掘那些不可见的、更为广阔和复杂的网络关系。当把城中村里那些细微的事物和它们背后宏大的图景拼贴在一起时，能很清晰地看到城中村里极其丰富的人间百态，以及城中村与整个世界纠缠不清的关系。

在“超级乱糟糟”工作坊里，学生在南头古城选择了拖鞋、电视、糖水铺、窗户、理发师傅作为观察对象，挖掘了五种不同的“共生”，以回应该届深双的主题。

下面我来分享其中的三组：南头的拖鞋、电视和理发师。

2.1　南头的拖鞋

初次到南头古城做田野调研时，我发现在街上行走的大多数居民都穿着拖鞋，几个流动小贩

on the previous day, two students helped her sell sugar-coated haws. At this moment, the students became the auntie (playing a vendor). Over the course of the day, one of the students found himself in an awkward situation: he didn't know how to carry the display rod and use the bathroom at the same time. He subsequently learned that to avoid using bathroom, the auntie got up at 4:00 a.m. and drank nothing until she sold all of the sugar-coated haws.

Assuming different roles brought students back to the real world. At the end of their participant-observation, they helped the auntie redesign her tricycle, not only adding on a toilet, but also a place to sell other things such as clothes, flowers, and fruits. If these students only sat in front of their computers to design without empathy and experience, they could never come up with such a design. I believe that this is where the significance of mapping workshops lays.

At the end of the mapping workshop, the students became curators, logically presenting observations of and discoveries made at different scales and through different roles in a given space. I have always believed that architects of the future will have to play the role of a curator because architecture is never a one-man show and so architects need skills such as the ability to curate and to promote collaboration.

Above, I described a relatively complete and general process of the workshop. Below, I will elaborate on the workshop in Nantou Old Town.

也都在售卖拖鞋。这种奇怪的“拖鞋现象”一下就吸引了我的注意力，前几年我住在广州城中村时很难在街上看到这么多穿拖鞋的人。学生开始在街头观察和跟踪穿拖鞋的居民，记录居民的行走路线、行走速度、行走时间、停留点，路面的铺地材质、坡度，拖鞋的大小、形状等。学生们甚至偷偷躲在居民家门口“目测”他们所住的房间大小、空间布局、空间功能以及内部的日常物件。

经过两周的观察，学生发现南头古城里的居民大部分都是外来租客，原住民寥寥无几。住宅以单身公寓居多，房间类型比较单一，面积很小，平均 10~20 平方米，一般只能放一张床。例如，一位租客租住的位于一层的房间只有约 12 平方米，她的卫生间、厨房、洗澡房、洗衣房、晾衣空间、客厅全都分散在城中村不同的地方，有的是她私人的（比如卫生间和洗澡房），有的是和其他租客共用的（比如洗衣房、晾衣空间），最远的空间（比如看电视的客厅）距离她的房间 400 多米（图 3）。

因此，很难绘制出家的边界，边界都是分离的，这迫使租客随时要通过“移动”来实现家的所有功能，而拖鞋是最方便、最节省时间的行走工具。一般来说，每个人的房间都配备了两双拖鞋：一双在睡房穿，一双在家的其他空间使用。这一组学生对家或居住空间的边界有了重新的认知，并以此探讨了城中村有关边界的共生——私密与公共，而“拖鞋”是这个共生发生的关键。

2.2 南头的电视

南头古城没有广州城中村房屋顶上的卫星接收器（广州的城中村居民很多讲粤语，都在看香港台），大部分居民来自内地，不会讲粤语。

图 3 南头的拖鞋，图绘工作坊
Figure 3 Flip flops in Nantou, Mapping Workshop

2 Super Messy

In 2017, Professor Richard Weller at the Weitzman School of Design, University of Pennsylvania, invited me to teach a graduate-level mapping course. That same year, the chief curator of the 2017 UABB, Liu Xiaodu invited me to curate a mapping section in the exhibition. In the end, the program became a joint workshop of the Weitzman School of Design, University of Pennsylvania and the School of Architecture, South China University of Technology, as well as the exhibition *Super Messy* in UABB (Figure 1-2).

We chose the workshop site in the main venue of UABB, Nantou Old Town which has carried a history of more than one thousand years and is appearing as an urban village. We used a worm's perspective to observe and re-examine the messy daily life of a typical Chinese urban village. However, we weren't interested in the chaotic external appearance of the urban village. Rather our goal was to understand the survival tactics and spatial strategies that urban villagers and inhabitants produced in basic material conditions, discovering the autonomous logic and grass-roots coordination systems behind the apparent chaos of the urban village. These systems can not only be developed into useful design tools, but also provide top-down urban planning and governance with supplemental ideas.

To better understand the urban village, the students took ordinary daily objects, neglected spaces, and specific people as their respective entry points to excavate invisible, broader, and complex social networks. By juxtaposing the minor things in urban villages to its larger context, we see clearly the rich humanity of the urban village and the twisted relationship between the urban villages and the larger world.

In the workshop "Super Messy," the student teams chose flip flops, televisions, a sweet soup store, windows, and barbers as objects through which to observe Nantou Old Town. They excavated five different kinds of "co-existence" in response to the UABB theme, "Cities, Grown in Difference."

Next, I will share work from three of the student teams: flip flops in Nantou, televisions in Nantou, and Barbers in Nantou.

2.1 Flip Flops in Nantou

When I first arrived in Nantou Old Town to conduct field

电视一般只有五个频道：深圳电视剧台、深圳公共台、深圳卫视、深圳都市和央视一台（图4）。

当一次在南头古城的小卖部看电视的时候，学生发现很多孩子妈妈的工作安排和每个电视台的节目竟然息息相关。孩子妈妈非常清楚各个电视台在各个时间段播放的节目，比如深圳卫视在早 8 点到 10 点之间有儿童节目，那么她们会在这个时间段把小孩放在家楼下的小店看电视，用这两个小时在附近餐厅做兼职，赚些外快。因此，在南头古城，很多女性住客的生活是依靠电视节目的时间来安排的。

晚上，有电视的小店会在街道上放一些小凳子，让大家可以一起看电视，这些有电视的店铺就成了租客的共享客厅和交流场所。在一些安静的小巷子里，一些固定的电视频道也成为某些特殊职业的工作信号，如性工作者，我就不展开聊了。和拖鞋一样，电视成为维系人与人关联的一个很重要的工具。因此，在“电视”这一组里，学生探讨了城中村有关维度的共生——时间与空间。

2.3 南头的理发师傅

到南头古城调研的第一天，学生就非常惊讶地发现，一个小小的城中村里面竟然藏着二十多家理发店。在这里，理发师被公认为学历较高的一群人。为了观察理发师和顾客的互动，经许可，宾夕法尼亚大学的学生在理发店理发台的正上方安装了一个摄像头。

透过摄像头，学生观察到理发师傅给本地村民理发的时间比给外地人理发的时间要长很多（平均多出 20 分钟左右）。理发师傅会和熟悉的村民交流互动，村民会经常求助有关法律、婚姻、心理、风水、疾病、身体保养的问题。因此，很多受过教育的年轻理发师身兼数职，同时也充当了律师、媒婆、心理师、风水师和赤脚医生的角色（图5）。

很多理发店会在店门外放一些椅子，方便顾客等待。学生发现这也是邻里经常聚集的地方，也是小孩子放学等待爸妈的场所。理发店不知不觉成为一个非正式的邻里瞭望所，承担了很多政府应该承担的义务和责任。试想一下，如果双年展的改造把这些理发店都改没了，城中村相互守望、互帮互助的社会结构会受到怎样的破坏？这一组学生探讨了城中村有关不同身份的共生——主业和副业。

以上就是在此次双年展中我和学生在南头做的部分观察和调查，其中有些调查是和南头的村民一起开展的，也希望这些成果可以成为未来南头古城改造的一个起点。

注：2017 深双展出五组学生在“超级乱糟糟：Mapping 南头古城”工作坊的工作成果，展览期间举办了一场专题论坛，也作为 2014—2017 年图绘工作坊的阶段性总结与回顾，嘉宾进行了延展讨论。[1]

research, I discovered that most of the inhabitants wore flip flops out in the street, while several itinerant hawkers were selling them. This strange "flip flop phenomenon" immediately attracted my attention. I had not seen as many people wearing flip flops in the Guangzhou urban village, where I lived a few years previously. The students began to observe and track flip flop wearing pedestrians, recording residents' walking routes, how fast and far they walked, where they stopped, pavement materials and slope, as well as flip flop size and shape. They even hid near the doors of residents' houses, measuring by eye room size, layout and functions as well as noting which daily essentials were present.

After two weeks of observation, the students learned that in Nantou Old Town, most inhabitants were migrant tenants, while original villagers were scarce. Most residential accommodations were efficiency apartments with a similar layout. These apartments were small, approximately 10-20 square meters on average, which was large enough for one bed. These conditions forced residents to creatively deploy other spaces in the urban village to meet their daily needs. For example, a tenant rented a ground-floor room that was approximately 12 square meters. Her bathroom, kitchen, shower, laundry room, clothesline, and living room were scattered throughout the urban village. Some of the spaces belonged to her private space (bathroom and shower), some were shared with other tenants (laundry and clothesline), while the farthest space was more than 400 meters from her apartment (the living room where she watched television) (Figure 3).

Therefore, the boundaries between a private residence and the urban village blurred, forcing tenants to opportunistically "move" in order to enjoy the amenities of a fully functioning house. In this case, flip flops were the most convenient and efficient walking tool. Generally speaking, rooms were equipped with two pairs of flip flops: one for the bedroom and one for the other rooms in the house. This student team reconceptualized the boundary limit of a house or dwelling space. They then used this insight to discuss the co-existence of private and public spaces in the urban village, as well as the critical role that "flip flops" played in producing the relevant boundaries between these spaces.

2.2 Televisions in Nantou

Unlike urban villages in Guangzhou (where most inhabitants speak Cantonese and watch Hong Kong television channels), there were no satellite dishes on the rooftops of Nantou Old Town, where most migrants hailed from China's hinterlands and didn't speak Cantonese. Televisions are usually tuned to one of five stations: Shenzhen Drama, Shenzhen Public, Shenzhen Satellite TV, Shenzhen City, and CCTV-1 (Figure 4).

Once while watching television at a convenience

图 4 南头的电视，图绘工作坊
Figure 5 Televisions in Nantou, Mapping Workshop

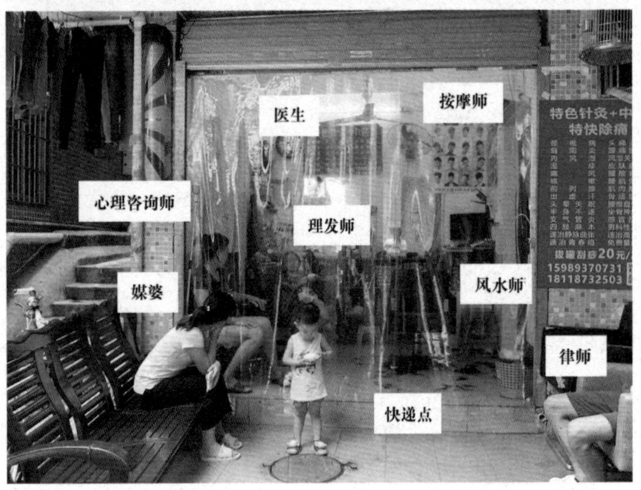

图 5 南头的理发师傅，图绘工作坊
Figure 5 Barbers in Nantou, Mapping Workshop

store in Nantou Old Town, students realized that many mothers arranged their work schedules around scheduled television programs on the different stations. The mothers were perfectly clear about when programs were broadcast, such as the children's program on Shenzhen Satellite TV from 8:00 to 10:00 a.m. During this time slot, they dropped their children off at the convince shop downstairs, where they watched television. The mothers used these two hours to earn extra income, working part-time at a nearby restaurant. Therefore, in Nantou Old Town, many female tenants organized their days around television programming.

At night, shops with televisions placed some stools in the street, allowing people to watch television together. These shops became tenants' shared living rooms and hang-outs. In some quiet alleys, dedicated channels signaled the presence of a particular occupation, such as sex work. Like flip flops, televisions had become an important tool for maintaining connections among people. Thus, the "television" team discussed how the dimensions of time and space co-existed in the urban village.

2.3 Barbers in Nantou

On the first day of conducting research in Nantou Old Town, students were surprised to find more than 20 barbershops in the small urban village. In the village, barbers were regarded as a group with a higher level of education. To observe the interactions between a barber and his customers, one of the University of Pennsylvania students installed a camera above the barber's desk.

Using the camera, the students observed that the barber spent more time (usually 20 minutes longer on average) on haircuts for local inhabitants than for outsiders. The barber interacted with familiar village residents, while the latter often sought his help in legal, marital, psychological, Fengshui, disease, and health care matters. Therefore, educated young barbers also served as lawyers, matchmakers, psychologists, Fengshui masters, and barefoot doctors (Figure 5).

There were chairs in front of many barber shops for customers to use while waiting. The students discovered that these areas were both a gathering place for people in the neighborhood and a place for children

1　超级乱糟糟：Mapping 工作坊专题研讨会，南头古城南头议事厅 B4，2018-1-15，主持：汪原 、何志森，嘉宾：刘珩、张宇星、刘晓都、朱荣远、庞伟、米笑 、李迪华、吴文媛、邱慧康、姜斌。https://urlzs.com/oBEqz.

邻里瞭望所

to wait for their parents after school. Barbershops had imperceptibly become an informal watchtower for the neighborhood, taking on some obligations and responsibilities that the government should have taken. Consider what would happen if the transformation of Nantou for the UABB were to eliminate these barbershops; would the social structure of mutual care and help in the urban village be damaged? The student team discussed the co-existence of different identities in the urban village as manifest through formal jobs and side line vocations.

I have presented some of the observations and investigations that the students and I completed during the Biennale. Some of these were shared with Nantou Old Town residents. I hope that these results can serve as a point of departure for the future transformation of Nantou Old Town.

Note: The 2017 UABB presented the results of the five student workshops as "Super Messy: Mapping Nantou Old Town," holding a themed symposium during the Biennale. This symposium also served as a retrospective of mapping workshops held from 2014 to 2017. During the symposium, invited guests held extended discussions.[1]

1 Symposium on Mapping Nantou Old Town - Super Messy, B4 Council Room, Nantou Old Town, Shenzhen, 2018-1-15. Host: Wang Yuan, Jason Ho; guests: Liu Heng, Zhang Yuxing, Liu Xiaodu, Zhu Rongyuan, Pang Wei, Mi Xiao, Li Dihua, Wu Wenyuan, Qiu Huikang, Jiang Bin. https://urlzs.com/oBEqz.

在这个工作坊里，我们感兴趣的不是南头古城乱糟糟的表象，而是其背后的原因：它是因何而产生的？如何操作的？以及设计师如何从中学习它？为了更好地理解南头古城，学生以古城里普通的电视机、拖鞋、窗户、糖水铺以及理发师为切入点，由小及大地去发掘城中村“乱糟糟”表象背后隐藏的秩序，以及那些不可见的、更为广阔和复杂的网络关系。此次工作坊试图通过图绘的方式来呈现一个平常我们所不熟知的“珠三角图景”，从而更加真切地领悟到全球化、差异性、杂糅、多元、城市共生的含义。当我们把南头古城里人们日常生活中微小普通的事物和它们背后不寻常的故事拼贴在一起的时候，就会清晰地看到古城里极其丰富的人间百态，以及古城与深圳、香港、珠三角，乃至整个世界之间纠缠不清的关系。

▷展览：Mapping 南头古城 ▷参展人：华南理工大学建筑学院及宾夕法尼亚大学设计学院

"Mapping Nantou Old Town" begins with observing specific situations there through a "worm's-eye view." This entails observing very small and super normal "objects" (e.g. a pair of filp flops, a messy window, a barbershop). Observing these small "objects" will lead to further greater and more complex networks that may extend beyond Nantou's borders. Hence, mapping Nantou, and possibly areas beyond its borders such as the wider Shenzhen, Hong Kong, and even the Pearl River Delta region can better capture a deeper image of globalization, hybridity, diversity and symbiosis. In addition, the knowledge the students produced through mapping Nantou Old Town can also evolve into design research tools for other practitioners, and even act as guidelines for renewing urban villages in China.

▷Exhibition: Mapping Nantou Old Town ▷Participant: School of Architecture, South China University of Technology and School of Design, the University of Pennsylvania

来源：深圳城市\建筑双年展组织委员会 . 城市共生：2017 深港城市\建筑双城双年展 [M]. 广州：岭南美术出版社，2021: 190-195.

南头奶茶店：一间基于社会设计的快闪店

Nantou Milk Tea Shop: A Pop-up Shop Based on Social Design

林挺、莫思飞
林挺，南头奶茶店合伙人，研进组设计事务所合伙人。
本文由董超媚编辑；莫思飞中译英。

Lin Ting, Mo Sifei
Lin Ting, a partner of Nantou Milk Tea Shop, is the co-founder of Evogma Group.
The article was edited by Dong Chaomei; translated from Chinese to English by Mo Sifei.

图 1 南头奶茶店 ©UABB
Figure 1 Nantou Milk Tea Shop ©UABB

本文作为 2017 深双中南头奶茶店快闪店（图 1）课题研究及实践的回顾，指出在城中村空间发展模式的多个利益相关方中，分离的空间所有权和使用权导致租客群体的缺位。空间设计固然是城中村更新中的重要一环，但如果建筑师一味只是以空间改造的手段来讨论城中村话题则存在局限性。建筑师要更好地履行社会责任，则需具备社会效益和经济效益双底线的思考框架，接受整合资源与建构的挑战。在“共生”主题下，南头奶茶店课题的初衷是从基本的调研中识别利益相关方的诉求，从中发现较为弱势的一方，希望通过设计有效的机制（空间体

This review of Nantou Milk Tea Shop (Figure 1), a pop-up shop and research site of 2017 UABB argues that detaching usage rights from ownership of a space leads to the absence of tenants in stakeholder conversations of urban redevelopment. Although spatial design is important in urban village redevelopment, nevertheless, if architects only discuss spatial renovations then limitations to the stakeholder model become apparent. To become a socially responsible architect, one needs to adopt a double bottom line, which denotes the importance of social and economic impacts, accepting the challenge of integrating resources and construction. Under the theme of "Cities, Grow in Difference," the Nantou Milk Tea shop aimed to recognize the demands of stakeholders, and identify the marginalized population hoping to design an effective mechanism (that integrated spatial experience, branding, and building capacity) to promote social practices to resolve problems.

From a research subject to a practice project, Nantou Milk Tea Shop spanned the two sections of the 2017 UABB, "Do Course: Store's New Image" and supporting facility bidding. Unencumbered by the content of the biennale, Nantou Milk Tea Shop enjoyed the freedom to

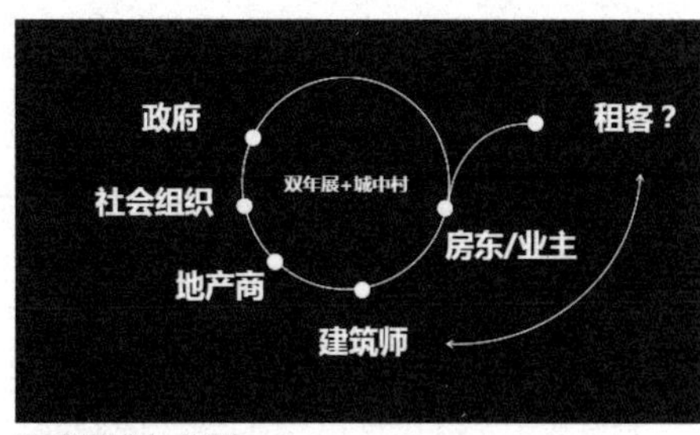

图 2 利益相关方，林挺绘
Figure 2 Stakeholders, drawing by Lin Ting

验的整合、品牌运营的建构以及能力建设），促成解决问题的社会实践。

南头奶茶店从课题到实践的过程跨越了深双展览框架中的“做课：商铺新形象”与后勤配套招投标两个板块。游离于展览内容的边缘，使南头奶茶店不必局限于空间改造，从而获得更多内容与研究的自由度。在前期调研过程中，我们在南头古城探访了皮匠、缝纫工、菜市场摊主、超市售货员等不同职业的租客。了解到租客群体最大的需求来自于生存的物质需求，简单来说是赚钱，其次是受到尊重，而自我价值的实现，不是大多数租客最迫切的需求。租客群体通过租赁合同得到使用权，是居住在城中村空间中并参与营造的群体，却因没有所有权而无法参与城中村空间开发，被视为弱势群体。我们将南头古城租客的工作分为两种类型：一是在城市中从事劳动密集型的工作，如服务业；一是在城中村开店做生意，服务城中村居住的人群。事实上这两种不同类型的人群构成一个封闭的生态或阶层，前者的成长性与竞争力不足，在城中村更新发展后势必面对淘汰的危机；后者则有通过经营与城中村共同发展的可能性。城中村的业态可分为三类：一是满足城中村生活日常刚需、已累积熟客的业态，如菜市场、士多店[1]等；二是适应城中村消费水平的外来连锁品牌；三是服务于城市需求的外卖店，以餐饮为代表，有外来连锁也有本地创办，运能能力参差不齐。对前两者进行干预无法短期见效，于是我们基于最后一类业态，以建筑师的身份与租客合作，参与到深双的课题中，提出一个以社会设计为核心，通过“能力建设”实现的共建模式（图2），回应本届深双主题——城市共生。

1 南头奶茶店的“奶茶妹”阿恋

南头奶茶店的合伙人“奶茶妹”阿恋来自广东潮汕地区，在南头古城居住已有二十余年。她与服装店老板共同租用一个店面，该店早上卖奶茶，其余时间是服装店。由于经营不善，在

1 “士多”为外来语，是英文“store”的音译。粤语中“士多店”指杂货铺。

do research without being limited to conducting spatial renovation. During our preliminary investigation of Nantou Old Town, we visited tenants who were employed in different occupations, including cobblers, sewing workers, market stall owners, and supermarket salesmen. The physiological need to earn a living was the biggest concern of the tenant population, followed by esteem, while self-actualization was not an urgent need for most tenants. Tenants acquired usage rights to a space through a lease. Consequently, they participated in the making of urban village space. However, their lack of ownership meant that they were marginalized in the redevelopment of the urban village.

We have categorized the tenant population in Nantou Old Town into two groups according to their job type. The first group of tenants conducted labor-intensive work in the city, such as food and delivery services; while the second group of tenants opened businesses that served other urban village residents. These two populations comprised a closed ecology or class. The labor-intensive working population is inevitably at risk of being excluded from the urban village renewal and redevelopment due to its lack of growth path and competitiveness. In contrast, the entrepreneurial population can develop together with urban villages through proper management. Urban village businesses can be further subdivided into three categories. First are businesses that meet the daily needs of urban village residents and have accumulated a loyal customer base, such as wet markets and grocery stores. Second are chain brands from outside the village that have adapted to the consumption level of the residents. Third are catering stores that serve the larger city, including foreign chains and local establishments, each with a different operational capacity. Short-term interventions into the first two businesses would be less effective. Therefore, in response to the theme of the 2017 UABB, Cities, Grow in Difference, we proposed a co-building model based on social design through capacity building (Figure 2), working as architects with catering store tenants.

1 Nantou Milk Tea Shop's "Milk Tea Girl,"A Lian

A Lian, the partner of Nantou Milk Tea Shop, was from the Chaozhou-Shantou (*Chaoshan*) region of Guangdong, and had lived in Nantou Old Town for more than 20 years. She and the owner of a clothing store jointly rented a storefront that sold milk tea in the morning and was a clothing store during the rest of the day. When we met A Lian during our site survey, she was about to go out of business due to poor management. Based on A Lian's previous work experience, we evaluated her demands and the possibility of collaboration. The milk tea shop was the type of business that could serve the

调研遇见她时，她正准备歇业。根据阿恋以往的工作经验，我们评估了她的经营诉求以及合作的可能性。从业态而言，奶茶店既能服务城中村居民，又能满足双年展访客的需求；从现实条件出发，可以借助阿恋已有的产品研发、采购、制作的经验，在短期内即可配合双年展开业。城中村中常提到的“熟人社会”，在前期邀请奶茶妹共同开店时成为一个埋藏的暗线。面对经营的未知与对课题的不理解，阿恋有些退缩和抗拒，但却因笔者同是潮汕人，她在沟通中逐渐放下防备，建立起相互信任的关系。为确保实践目标，我们采用团队成员现金入股、阿恋技术入股的方式，订立原则——优先保证阿恋的工资、提成和分红，即无论整体经营亏损或盈利，阿恋都会拿到一定的现金收益。于是，借助阿恋产品研发、采购、设备的经验，结合团队中品控、宣传、人事和资源调配的能力，南头奶茶店在短时间的筹备后，顺利地在 2017 深双的公共服务区中开张了（图 3）。这仅仅是一个开端，在运营的过程中，我们需要不断地设计和完善机制，以达成共同成长的目标。

在项目开始之前，合伙人将财务回报及股权架构做了清晰的测算和界定，通过制订计划和约定条款完善利益分配的问题，以财务测算和合约的方式，不仅保障了合伙人之间的信任的问题，也使得财务模型作为运营的基础为共同努力提供可参照的方向。换言之，结合运营数据，在日常运营中及时调整产品与生产流程是奶茶店顺利运转的关键。合伙人中一方（笔者，建筑师背景）具备经营及财务管理知识，另一方（阿恋）具有实践经验，双方在运营中不断地磨合、协同是必经的过程。在奶茶店经营的第一个月，阿恋出于经营本能提出引入潮汕牛肉丸等小吃等以求扩大收益；但运营团队则认为应该保证现有奶茶产品的出品效率并建立稳固的品牌效应。商量后我们选择用一星期的时间去验证，结果显示在引入小吃后使得奶茶出品效率降低，不但没有扩大收益，反倒影响了原有的效益。正是在这样相互尊重、共同决策中，我们建立了稳定的合作关系并不断地相互学习和成长。在开业一个月后奶茶店实现了盈利，也有投资者前来接洽，希望在双年展结束之后继续推动南头奶茶店的稳步发展。但在双年展作为三个月短期展览的局限性下，我们选择在利润达到高峰时结束了经营。即便经营实践时间不长，但南头奶茶店作为社会设计的一个初级模型，即与有一定技能的居民或租户合作，结合本地经验和外部资源的统筹协调的操作方式，对日后城中村的有机更新改造具有重要的借鉴意义。展览结束后几个月，阿恋筹备成立自己的新茶饮店。“南头奶茶店”的概念并没有因为展览结束而结束，我们正在寻找新的机会，将这套价值观与方法延续下去，拓展设计的边界，建构社会创新新模式。

residents of the urban village and meet the needs of Biennale visitors. In addition, A Lian's experience in product development, procurement, and production would allow the business to be set up in time for the opening of the biennale. Often mentioned in scholarly discussions of the urban village, the "acquaintance society" became a hidden thread when inviting A Lian to open a shop together. A Lian initially hesitated because the project was unknown and unfamiliar to her. However, she ultimately was convinced and established mutual trust simply because the author was also from the Chaoshan region. To ensure the goal of the project, we invested cash, while A Lian was an equity shareholder, bringing her skills to the project. Also, we agreed that A Lian's salary, commissions, and dividends were guaranteed, that is, regardless of the overall operating loss or profit, A Lian would receive a certain amount of income. Therefore, with the help of A Lian's experience in product development, procurement, and equipment, as well as the team's capabilities in quality control, publicity, personnel, and resource allocation, Nantou Milk Tea Shop successfully opened at the Public Service Area of 2017 UABB (Figure 3). This was just the beginning. Over the course of the operation, we needed to continuously redesign and improve the mechanism to achieve the goal of mutual growth.

Before the start of the project, the partners formulated plans and agreements, calculating the financial return and equity structure, and improving benefit distribution. Financial calculations and contracts not only guaranteed trust between partners, but also established a financial model as the basis for daily operations. Operational data and the timely adjustment of products and production processes in daily operations were the key to success. One of the partners (the author) had business and financial management knowledge through their work as an architect, and the other (A Lian) had practical experience. The two parties started to gel and collaborate during the operation process.

During the first month of operations, A Lian proposed introducing Chaoshan beef balls and other snacks to increase revenue. In contrast, the managing team believed that the production efficiency of milk tea products should be guaranteed and effective branding needed to be established. After discussion, we decided to use one week to test A Lian's approach. The introduction of snacks reduced the production efficiency of milk tea, did not increase revenue and affected the projected profit. We established a cooperative relationship and mutual learning based on respect and common decision-making. One month after the opening, the milk tea shop achieved profitability, and investors approached us with an eye to promoting the steady development of the Nantou Milk Tea Shop after the biennale. However, the biennale was a short-term exhibition of three months, limiting our chance to grow. We closed the shop just as

图 3 深双公共服务区中的奶茶店、单车桌、大排档花牌、你的表情就是我的符号（壁画）©UABB
Figure 3 Milk Tea Shop, Bike-Table, Dai Pai Dong with Flower Plaque, and Your Expression is My Symbol (mural) in UABB Lounge ©UABB

profits were peaking.

Despite its short lifespan, Nantou Milk Tea Shop models social design, working with skilled residents or tenants, and integrating local experience and external resources. It also serves as a significant reference for future organic regeneration of urban villages. A few months after the biennale ended, A Lian prepared to set up her new tea shop. The concept of "Nantou Milk Tea Shop" does not end with the biennale. We are looking for new opportunities to continue this system of concepts, values, and practice to expand the boundaries of design and build a model for social innovation.

2 社会设计不是慈善行为

大卫·哈维认为："要解决失控的资本螺旋上升，精英们是不会做的，因为他们是社会问题的一部分，而不是解决方案的一部分。我们需要展开一系列社会运动来解决这些问题。"[2]而社会设计就是一种行动的可能。社会设计是以解决社会问题为目标的设计方法。与传统设计方式不同的是，社会设计创造的不是物件，而是突破单一行业的局限性，针对社会系统设计新的机制，引入一种新的条件，旨在建立更深入的公众和文化参与，增加创造力、可持续性、社会平等。

社会设计并不是不求回报的慈善行为，社会效益与经济效益需要同时兼顾并相辅相成。南头奶茶店模式的目标与联合国可持续发展目标之一一致：体面工作和经济增长。[3]体面工作和经济增长，是促进持久、包容、可持续的经济增长，实现充分和生产性就业，确保人人有体面工作。创造高质量的就业岗位仍将是几乎所有经济体在 2015 年之后长期面临的主要挑战之一。可持续的经济增长要求社会创造条件，使人们得到既能刺激经济又不会危害环境的优质就业，也要求为所有达到工作年龄的人提供就业机会及良好的工作环境。

在深圳迅猛的经济发展中，为社会设计寻找适合的土壤尤为迫切。南头奶茶店的模式则是基于社会设计的一次实验，为以阿恋为代表的城中村租客群体赋能，通过社群研究、商业

2 大卫·哈维，"资本的时间与空间"，2016 年 6 月 16 日首都师范大学讲座。
3 目标 8：促进持久、包容和可持续经济增长，促进充分的生产性就业和人人获得体面工作。持续和包容的经济增长可以推动进步，为所有人创造体面的就业机会，并改善生活水平。详见：https://www.un.org/sustainabledevelopment/zh/economic-growth.

2 Social Design is not a Charitable Act

According to David Harvey, elites will not solve the issue of capital spiraling out of control because they are part of the social problem instead of a solution. A series of social movements are required to work out these problems.[1] Social design is one such possible action. Social design is a design approach that targets social problems. Different from traditional design methods, social design does not create objects, but breaks through the limitations of a single industry, designs new mechanisms for social systems, introduces a new condition, and aims to establish public and cultural engagement as well as increase creativity, sustainability and social equality.

Social design is not a charitable act. Social and economic impact needs to be taken into account and complement each other. The goals of the Nantou Milk Tea Shop model, for example, were consistent with the United Nations sustainable development goals of promoting decent work and economic growth. Decent work and economic growth are about promoting sustained, inclusive, and sustainable economic growth, achieving full and productive employment, and ensuring decent work for all.[2] Creating high-quality jobs will remain one of the main challenges for almost all economies for the foreseeable future. Sustainable economic growth requires

1 David Harvey Lecture: "Space and Time of Capital" at CNU, China, on June 16, 2016.
2 United Nations. *Sustainable Development Goals*. Goal 8 Decent work and economic growth. https://www.un.org/sustainabledevelopment/economic-growth/. Access 2020-12-12.

模式研究、能力建设、品牌运营等为更多租客在城中村发展的进程中有足够的能力“留下来”创造社会条件。

3 建筑师角色的重新定义

当建筑师在面对“城中村”话题时，多数从城市、空间、文化，甚至是政治角度去实践，少有从经济角度切入。他们往往忽略了一点——正是经济学上最为本质的供给适应需求，造就了“城中村”这种自发生长的现象。建筑师热衷谈论“城中村”现象，且试图使用专业的话语体系解决“城中村”问题，其有效性基本只局限于空间本身，这是单一行业自身的局限性。从空间设计到社会设计，建筑师的角色从建构资源整合转变到社会资源整合。在社会设计中，建筑师的角色被重新定义，行业边界得以延伸，如同跨领域合作中的履行社会责任的总导演，搭建框架、调配和整合资源。建筑不仅是功能空间、文化象征，更是一种社会行为。无论从古典主义到现代主义，再到各种分支演化，都与其时代背景、社会意识联系在一起。建筑师参与更新中承上启下的重要阶段，虽控制不了建造完成后的方方面面，但哪怕只是体现在设计中的公共空间意识和尺度，也可以触发使用日常行为的萌芽。建筑师可以选择停留在美学、材料、构造的层面去思考，也不妨将工作转向社会发展的可持续性上。由此，南头奶茶店在 2017 深双的实践提出来一个值得继续探讨的重要问题：如果将建筑专业进一步延伸到社会设计领域，除了空间之外是否能进行社会的建构呢?

society to create the conditions for high-quality employment that stimulates the economy without harming the environment. Also, it requires employment opportunities and a good working environment for all people of working age.

With Shenzhen's rapid economic development, it is a pressing issue to find fertile ground for the growth of social design. The Nantou Milk Tea Shop was an experiment based on social design, empowering the urban village tenant population as represented by A Lian. Community survey, business model research, capacity building, and branding created social conditions for her to gain the ability to stay during the subsequent redevelopment of the urban village. It is our hope that this model can be used in other urban villages.

3 Re-defining the Architect's Role

Most architects establish their urban village practice from the perspective of urban space, culture, and even politics. They rarely consider the role of the economy and often overlook the economics of supply and demand that have facilitated the spontaneous growth of urban villages. Architects are keen to talk about the phenomenon of urban villages and try to solve the issues within their professional system. This single industry focus and understanding of spatial experience limits the effectiveness of spatial transformation. From spatial design to social design, the role of the architect shifts from integrating construction resources to coordinating social resources. In social design, the professional boundaries of architects are extended and redefined. In this model, the architect becomes the socially responsible chief director of cross-disciplinary collaborations, establishing a framework, as well as allocating and integrating resources. Architecture not only refers to functional space and cultural symbols, but also to social behavior. From architectural classism to modernism and the evolution of the field's many branches, each is related to the context and social consciousness of their era. Even though architects lose control over operations upon completion of the building, nevertheless, as a connecting link between the preceding and the following forms of urban regeneration, architects can trigger daily behavior through the awareness of public space design and scale. We can choose to think at the level of aesthetics, materials, and construction, or we can turn our work to the sustainability of social development. Therefore, the practice of Nantou Milk Tea Shop during the 2017 UABB raises an important question worthy of further discussion: if the profession is further extended to the field of design, can social design be possible in addition to spatial design?

在 2017 深双筹备期间，随着展览的不断临近，置身于城中村的现实空间中并主动介入显得更为直接和必要。继 4 月 21 日公开征选“都市 | 村庄”板块中“城中村档案馆”的展览方案后，同一板块下的“城中村进行时”项目也开始公开征集，即鼓励大家进驻到城中村之中，以切身的体验、多元的方式，自发和创造性地介入城中村日常生活的再创造。

经过多个月的调研，2017 年 11 月“城中村进行时”之“做课：跟 UABB 进村做点儿什么！”项目评审结果公布，调研报告《南头古城奶茶妹创业梦想》（作者：郑婷、林挺）获得二等奖并在双年展上付诸实施，以“南头奶茶店”的形式呈现。

During the preparations for the 2017 UABB, organizers stressed interventions in keeping with the reality of the urban village. Following the open call for the "Urban Village Archives" in the "Urban|Village" section on April 21, 2017, the project of "Urban Village in Progress" under the same section called for everyone to enter the urban village, spontaneously and creatively intervening in the re-creation of daily life in urban villages through personal experience and multiple approaches.

After months of research and design, in November 2017, "UABB Do Course at Urban Village" under "Urban Village in Progress" announced the result of reviews of proposals. "Nantou Milk Tea Girl's Entrepreneurial Dream" by Zheng Ting and Lin Ting won the second prize. It was realized in the Biennale in as the "Nantou Milk Tea Shop."

单车桌与街道生活

Bike-Table and Street Life

李甫
美国建筑师协会会员 、美国绿色建筑委员会认证专家，美国德州大学奥斯汀分校建筑学硕士（城市设计方向），清华大学建筑学学士、硕士，现为深圳独特视野建筑设计有限公司执行董事、合伙人，易加设计创始合伙人。
本文首次发表于“城市中国杂志”微信公众号，2019 年；由莫思飞中译英。

Li Fu
Li Fu is an AIA member of the American Institute of Architects and LEED AP of the U.S. Green Building Council. He received his master's degree (urban design) at the University of Texas, Austin and bachelor's and master's degree (architecture) at Tsinghua University, Beijing. He is the executive director and partner of UV Architecture and a founding partner of E Plus Design.
The article was first published on the official WeChat account [城市中国杂志], 2019; translated from Chinese to English by Mo Sifei.

单车桌是我们为 2017 深双的深圳南头古城主展区公共水饮区设计的一件街道家具。设计之初，我们希望这个作品在三个月的展期结束后能持续存在，并具有不断发展下去的生命力。在设计和制造的过程中，我们就做了很多即兴的尝试。单车桌本身蕴含的灵活性在此后走进社区的多次活动中都得到应验，居民边用边玩单车桌的方式也给我们带来很多启发。

1 即兴与探险

对我们来说最有趣的是，单车桌是在制作过程中经过若干迭代逐渐成形的，并不是一设计就很完备了。我们设计的是一个不断演进的变形虫。

最初我们在电脑上草模提出了车桌的概念，并造出了第一辆样车。当时希望概念更好玩，桌腿是可以翻起来当推车把手的，测试以后觉得桌面太小，放置不稳，不够实用。而且桌腿的翻起，会给使用带来太多不便。于是调整桌面比例，改成固定桌腿，完成了雏形。后来在制作中又有更多改进，例如让全部零件可以便捷的拆装，方便运输储存。

关于桌面如何铺装，也经过若干次推敲。我们请师傅即兴测试了水泥、瓷砖和马赛克的混搭，最终确定了参展时的马赛克效果。这种即兴探索的痕迹特别宝贵，我们在参展时也展示了各种铺装的测试版本。后续每一次参展，我们都不断地做改进版本，享受这个不知道终点的探索过程。

The Bike-Table was a piece of street furniture designed for the public rest area of the main venue of the 2017 UABB in Nantou Old Town. The design aimed to last and grow beyond the three-month Biennale. Improvisational attempts were made during the design and manufacturing processes. The flexibility of the Bike-Table has been utilized in community events since then, while how residents play with it continues to inspire us.

1 Improvisation and Explorations

From our perspective, the most interesting aspect of the Bike-Table is that it was not completed all at once, instead, it gradually took shape over several interactions during the production process. What we designed was a constantly evolving amoeba.

First, we drafted a digital model of the Bike-Table and built a prototype. At the time, we hoped for a more playful concept, so the table legs were foldable cart handles. After testing it, however, we felt that the tabletop was too small, the table was unstable, and it couldn't be easily used. Moreover, the foldable table legs caused many inconveniences. Therefore, we adjusted the proportions of the tabletop and adopted fixed table legs. This became the initial prototype of the current version of the Bike-Table. During subsequent production, details were tweaked. For example, all parts can be easily disassembled for packing and shipping.

It took several attempts to determine how to finish the tabletop. We asked several craftsmen to improvise testing the mix and match of cement, ceramic tiles and

2 野生与再生

一开始我们找了几位焊工师傅，给他们看概念图，确认用角铁作为主材后，又讨论了节点的问题（主要是金属桌腿、轮子和桌面如何交接，如何实现桌腿 90° 翻转和固定，车轮 90° 旋转和固定，用什么做桌面的衬板，如何在衬板上固定马赛克）。但后来没有谈下来，原因是他们觉得这玩样儿太新奇，不熟悉，他们宁愿做自己熟悉的活儿也不愿意赚这个钱。

后来我们想这就是个自行车改造的车子，干脆找修自行车的人做得了。于是到公司附近的城中村（雷公岭村）自行车铺里找了个师傅。给他看了概念图，他说可以试试，先做个样车看看（图 1）。打样的过程其实是一边做一边发现问题的过程，改了两版之后基本定案。我们找了朋友在龙岗坂田城中村的建材铺批量生产。

准备批量生产的时候，又遇到一个难题——去哪里找这么多单车呢？那时共享单车已经普及，但是大批弃置的私人单车和残破的共享单车堆积如山。看到这种浪费资源的景象，我们找了各种牌子的残破共享单车，也回收了小区的废弃单车，用每辆车的前后轮分别制作一辆单车桌。我们也做了相应的改进，比如，车轮的固定方式本来是另外焊接螺栓，改为利用原有车闸，车把手也利用起来。

收来的二手自行车中除了三辆共享单车之外，其他都是不一样的。车轮有两种尺寸，有些车子车轮尺寸相同，但车架规格又不同。为确保桌面高度相同，每辆车子的高度都得调整，人工制作会有误差，怎么调都达不到统一的高度。我们开始也没想着把它做成那种很精致的产品，也就不那么纠结了。

经过结构和细节的若干次修改，40 辆车架安装完成。运送到南头古城的工地上之后，和公共水饮区的装修工程一起贴砖。所有的制作和材料都是村里的师傅们完成的，没有尖端的技术。桌面的村款马赛克跟南头城中村的氛围特别搭，土味十足的建筑材料看上去也很美。

原来想象的是挑选一种最佳的马赛克色彩图案，统一制作，但后来怎么挑都觉得有点单调。仔细一想，每辆回收的自行车，色彩本来就是五花八门，不如顺势而为，每辆车任意搭配一种马赛克，拼贴的方式也随意，像村子里的楼栋一样五彩斑斓。

2017 深双南头展区有一个很受欢迎的作品《城中村家具交换计划》，参与交换的有许多居民自制的凳子，令人脑洞大开。单车桌是由设计师主导设计的家具，但也是半野生的：村里研发，村里制造，村里使用。

双年展期间，位于公共水饮区的南头奶茶店大受欢迎，运营方通过能力建设、品牌运营、商业模式发展等方式，与城中村一起参与共同发展实践。南头奶茶店合伙人林挺认为，单车桌和南头奶茶店一样有很强的社区在地标签，

various mosaics, finalizing the final mosaic patterns through this process. The traces of this improvisational exploration are especially valuable and during the exhibition, we also showcased examples from the pattern tests. For each subsequent exhibition, we have continued making improved versions, enjoying the process of exploration in the absence of a final destination.

2 The Endemic and Regeneration

In the beginning, we found welders and showed them the concept drawing. After confirming that steel would be the main material, we discussed details. The main issues were: how to join the metal table legs, wheels and tabletop; how to stabilize the legs when folded 90 degrees; how could the wheels rotate 90 degrees and be stable; what material should be used to line the tabletop, and; how should mosaic tiles be affixed to the lining board? However, they eventually withdrew from the project because they would rather work familiar jobs than make some newfangled object.

Afterward, we thought, this is a bicycle remodeling project and all we need to do is find a bike mechanic. Consequently, we went to a bike shop in Leigongling, an urban village that is located near our company where we found a mechanic. We showed him the concept drawing, and he said he was willing to give it a try and make a prototype (Figure 1). The process of making the prototype was also a process of identifying issues. After two efforts, the prototype was finalized. We commissioned a friend, a building supplier who is based in Bantian, an urban village in Longgang District to mass produce the Bike-Table.

At the moment the Bike-Table was to go into production, we faced another challenge—where could so many bicycles be found? At the time, bike sharing was popular, but mountains of discarded personal bicycles and broken shared bicycles had piled up. Seeing the landscapes created by this waste of resources, we sought broken shared bicycles and recovered discarded bicycles in residential neighborhoods. The front and rear wheels of each bike were used to produce one Bike Table. We also made improvements based on the found objects themselves. For example, the original cycling brakes could be used instead of welded bolts to fix the wheel in place, and the handles were utilized.

Except for three shared bicycles, the second-hand bikes that we recovered were all different. Wheels came in two sizes, while the wheels on some bikes were the same size, but their frames were different. In order to guarantee that the height of each table would be the same, the height of every bike had to be adjusted. However, the margin of error in manual production did not allow precise adjustment for an exact height. From the start, we hadn't wanted to make pristine industrial

图 1 过程打样（上），张旭斌拍摄；在城中村里加工（下），李甫摄
Figure 1 Mock-up (above), photograph by Zhang Xubin; Processing in the urban village (below), photograph by Li Fu

单车桌易于移动，组合方式多样，非常适合客人灵活拼桌（图 2）。单车桌也引起南头本地人对艺术话题的兴趣，这也许是双年展给城中村带来的一个新鲜视角。深弘佳儿童俱乐部的张文锐老师带领小朋友们参观时，单车桌很受孩子们宠爱，奇特的结构和绚丽的桌面吸引了孩子们的注意力，孩子们还推着单车桌拼接玩耍。

单车桌是一件家具，也是一件参展作品，与展区里面大多数参展作品不同的是，它是一件观众可以触摸和使用的艺术品，很亲民。观众大多没有意识到它是一件参展作品，更多的是把它当作一个有点特别的桌子，有些好奇又细心的观众会注意到我们设计的车把手和手刹，于是解下手刹上手试着玩一把。当我们把 19 个桌子拼在一起，它变成一个大圆桌，是展区内最大的桌子。大圆桌位于 A2 和 A3 展厅之间的空地，观众到此可以稍作休息后继续看展。一到饭点，单车桌摇身一变，成为主办方工作人员的大饭桌（图 3）。

有的设计师现场与我们交流，觉得可以改进细节让它变得更精致，但我们并不是做一个高大上的东西，而是想让它像城中村的房子一样，即实用又经济，再赋予一些美感。它不是一件追求艺术性的作品，它是一个桌子，功能性会强一些。我们和鳌湖的艺术家王屹林聊到单车桌，他认为现代设计最早就认为功能即是美，功能会在不同环境下出现错位。而现代艺术有一个概念叫“新唯物论”，就是说人作为主体是不能完全穷尽物体的可能性的。以单车桌为例，它原本是水吧里的桌子，但放在村子里它又适合村民们摆摊和聚会吃饭，它总是会出现超越开始想象的功能。

products, so we just went with it.

After several structural modifications and detailing, 40 frames had been installed. They were then shipped to Nantou Old Town, where the tiling of the tabletop was done on-site, along with the renovation of the public rest area. Urban village craftsmen produced and supplied materials to make the Bike-Table, without any cutting edge technology. The tabletop mosaic patterns resonated with the mosaic facade of Nantou Urban Village, making this completely local material appear especially beautiful.

Our first instinct was to select the best mosaic pattern and produce uniform tabletops. However, no matter how we adjusted the pattern, we found it a bit banal. On second thought, we realized that as the recovered bicycles were of motley and variegated colors, it would be better to match each one with a different mosaic. These improvised patterns were as vibrant as the building facade in the urban village.

A popular work at the Nantou Old Town of the 2017 UABB was *Urban Village Furniture Exchange Program*, which showcased inspiring handmade furniture by local residents. The Bike-Table was a designer-led piece of furniture, but it was also semi-local. It was developed and made in a village for village use.

During the Biennale, one of the favorite places in the public rest area was the Nantou Milk Tea Shop. The proprietor used capacity-building, brand operation and business model to co-develop with the urban village. The founder, Lin Ting believed that like the Nantou Milk Tea Shop, the Bike-Table was a strong community emblem. The Bike-Table was movable, could be reassembled in different ways, and allowed guests to rearrange the setting (Figure 2). Nantou residents were interested in the Bike-Table as a work of art. This was perhaps a fresh perspective that the Biennale brought to the urban village. When Teacher Zhang Wenrui of the Shenhongjia Children's Club brought students to the Biennale, the children loved the Bike-Table. The peculiar structure and gorgeous tabletop attracted the children, who playfully pushed different Bike-Tables around the area.

The Bike-Table is both a piece of furniture and an exhibit. Unlike most exhibits at the Biennale, it was

3 介入街道社区

双年展是一个起点。结束后，单车桌也持续地介入各种社区活动中，也有各种好玩的再创作。

在 2017 年蛇口无车日活动中，十多辆单车桌被放到马路中间，成为街道家具。社区妈妈们用它办了一道长长的“长蛇宴”，与往来的路人分享家的味道。五颜六色的漂亮马赛克配上各色菜肴与花卉，冷冰冰的马路摇身一变，成为暖融融的蛇口会客厅。长长的桌子正好适合长长的排队人群，再多人也不会出现混乱和拥挤。

2019 年，鳌湖艺术村负责人邓春儒将单车桌引入村里举办“村民日志”艺术活动。他认为，单车桌的马赛克材质与本地社区的时代感非常契合，自行车也是村里非常熟悉的日常生活片段，它们的搭配是一种新颖的呈现。

长达三个月的时间里，单车桌被放置在鳌湖美术馆外、村头小广场上、老榕树下。平常这里有驻村艺术家的艺术作品和装置展出。后来，鳌湖的艺术家们参与了对一些破损单车桌的重新再创作，以一种更积极的姿态介入生活。

2019 年，单车桌获评红点概念奖最佳设计奖，并被新加坡红点博物馆收藏，与其他工业设计界的优秀作品一起展出，获得了国际产品设计界的认可。

4 设计平视生活

我们认为，未来的城市需要建造的建筑越来越少，但越来越多的街道和空间需要被重新介入、激活。家具是一种很重要的介入手段。华南理工大学何志森老师做了垃圾桶盖实验，对街道家具与居民行为的关联作了很有趣的阐释：在街道这个多义性的语境中，垃圾桶盖作为家具的功能性被多样化了。人的行为不是被家具规定的，而是和家具互相激发的。家具的多重内涵和街道生活的复杂性契合，引起有趣的共鸣。

有趣的家具会塑造有趣的街道。一个非常简单的家具，不同的人会从不同的角度看待它，和街道空间的互动方式之丰富更是超出了设计师的预想。我们设计制造单车桌，最初只是想做一些可移动的组合桌子，可以灵活地从室内溢出到室外空间，但是在历次活动现场发现了很多出乎意料的拼接形式。它们可以散放、接龙、摆酒、拗造型，甚至成为机动灵活的摊档利器。展览撤场时，单车桌也可以当作运输工具运送物料。

作为活跃街道气氛、增强体验的重要因素，街道家具的功能性与趣味性同样重要。我们意识到，单车桌的更多潜力还将被使用者发掘出来，为街道和生活带来更多的创意与活力。

单车桌受欢迎的过程告诉我们，美即是快乐。设计师的审美是设计的起点，更重要的是将公众的美感激发出来，才能成就一个引起强

viewer-friendly, a work of art that visitors could touch and use. Most of the visitors didn't even realize that it was an exhibit, using it as a special kind of table. Some curious and attentive visitors noticed the design of the handlebars and handbrakes, removing and playing with them. When we placed 19 Bike-Tables together, they formed the largest round table in the venue. We placed this round table at the open space between venues A2 and A3, offering people a place to rest before continuing their journey. At mealtime, Bike-Tables were rearranged to make a large dining table for the UABB staff (Figure 3).

At the venue, some designers suggested improving the details to make a more exquisite product. However, we hadn't wanted to make a high-end object, but rather to make something similar to urban village housing—useful and economical with an aesthetic touch. It wasn't an arty piece of art. It is a table, and so its functionality is primary. An artist at the New Who Art Village, Wang Qilin talked with us about the Bike-Table. He held that early modern design believed in the beauty of function, and that function would be out-of-place in different circumstances. Modern art has put forward the concept of new materialism, as subjects, human beings are unable to completely exhaust the possibilities of an object. In the case of the Bike-Table, it was just a table in a rest area, but when placed in an urban village, it could also be used as a vending cart or a dining table, its latest function always exceeding one person's imagination.

3 Street and Community Interventions

The biennale was a starting point. Ever since then, the Bike-Table has continued to intervene into every kind of community event and been subject to all sorts of creative repurposing.

In the 2017 Shekou Car-Free Day, more than a dozen Bike-Tables were placed in the middle of the road, becoming street furniture. The community of mothers served up a "serpentine potluck," sharing family flavors with pedestrians. Colorful dishes and flowers were placed on top of the vibrant mosaics, and thus a cold and unwelcoming street was transformed into the warm and welcoming Shekou Parlor. The linear configuration of tables accommodated the long line of people, preventing chaos and overcrowding.

In 2019, the director of New Who Art Village, Deng Chunru included Bike-Tables in the art event, "Villager's Log." New Who Art Village is located in Aohu, an urban village in Guanlan, Shenzhen. Deng thought that Bike-Table mosaics meshed well with the local community's lived temporality. While bicycles were a familiar part of daily life in the village, their juxtaposition was innovative.

For three months, Bike-Tables were placed outside the New Who Art Museum, at the plaza by the entrance to Aohu Village, and under the old banyan tree. These

烈共鸣的设计作品。建筑师与设计师应当以与使用者平等的视角研究生活，以开放的心态探索未知，用设计激发人和空间的互动，这样才能创造最有活力的城市场所。

图2 单车桌可拼接成各种形状，满足公共空间的场地要求和人们的使用喜好，范畴摄
Figure 2 The bike-tables can be joined in various ways, meeting the need for public space and people, photograph by Fan Chou

sites were usually dedicated to exhibiting the work of artists-in-residence. Subsequently, New Who artists fixed and creatively repurposed several damaged Bike-Tables, which re-entered daily life through this encouraging gesture.

In 2019, Bike-Table won a Red Dot Best of the Best Award for its design concept, was collected by the Red Dot Museum in Singapore and exhibited with other outstanding industrial designs, earning recognition from the international product design industry.

4 Design with Living

We believe that future cities need fewer newly constructed buildings, even as more and more streets and spaces await re-intervention and activation. Furniture is an important approach to intervention. Jason Ho from the South China University of Technology conducted research on trash can lids, establishing an interesting connection between street furniture and residents' behavior. In the heterogeneous context of the street, the functions of trash can lids as furniture have been diversified. Human behavior is not regulated by furniture, but evolves while interacting with furniture. Furniture's many connotations adapt to the complexity of street life, causing interesting resonances.

Interesting furniture makes for interesting streets. Consider a simple piece of furniture. Different people will view it from different perspectives such that its rich interaction with the street will exceed the designer's intentions. We initially designed and produced the Bike-Table to be a mobile, indoor-outdoor table that could flexibly adapt to different occasions and spaces. However, over the course of multiple events, how it was used exceeded our expectations. They were scattered, joined end to end, used as wine bars, displayed like art, and even served as flexible stalls for street vendors. Then, when the exhibition was dismantled, the Bike-Table also used to haul away material.

The functionality and playfulness of street furniture are equally important factors in activating the vitality and enhancing the experience of the street. We have come to realize that the full potential of the Bike-Table as yet waits to be realized by the next user, bringing creativity and vibrancy to the street and to life.

The popularity of the Bike-Table tells us that beauty means happiness. A designer's aesthetic may be the starting point of design, but it is more important to stimulate the public's aesthetic sensibility, because only then can one successfully design a product that resonates with users. Architects and designers should investigate life from the user's perspective, exploring the unknown with an open mind. This is the only way to create vibrant urban spaces.

图 3 大圆桌成为主办方工作人员的大饭桌，张旭斌摄
Figure 3 The big round table became the organizer's dinner table, photograph by Zhang Xubin

深双公共服务区由工厂宿舍楼改造而成，是室内外互通的半开敞式公共空间，位于 A4 东侧一层，约 160 平方米。主要功能为休息区，售卖茶饮、咖啡、啤酒，与东侧简餐区在同一空间内，无明显界限。南侧及北侧局部设卷闸，夜间封闭。内设一条贯穿南北的通道，与北侧 A5 展区连通。深双公共服务区整体设计强调对本土材料的重新审视，用“马赛克地毯”与“移动单车桌”再现城中村老式标志性材料的独特魅力。这是一次始于深双而非止于深双的“城中村制造”的永久性深度探索。

▷ 深双公共服务区
▷ 李文海、李甫

Transformed from a factory dormitory, the UABB Lounge is a 160-square-meter semi-open public space of indoor and outdoor interoperability located on the east side of A4. It serves as a place to rest with stalls selling tea, coffee, and beer that open to the dining area to the east. The roller shutters on both the north and south facades allow the place to be closed at night. A north-south passage runs through the building and connects to A5 exhibition area to the north. The overall design of the UABB public service area emphasizes the re-examination of local materials and uses "Mosaic Carpet" and "Bike-Table" to reproduce the unique charm of the village. This is a long-term exploration of "urban village manufacturing" that begins but not stops at the UABB.

▷UABB Lounge
▷Li Wenhai, Li Fu

房屋 17 号：全球化大都会的反叙事

House 17: A Counter-narrative of a Globalized City

莫思飞
2017 年深港城市\建筑双城双年展（深圳）策展助理，都市实践研究员。

Mo Sifei
Curatorial assistant of the 2017 Bi-City Biennale of Urbanism\Architecture (Shenzhen), URBANUS Researcher.

图 1 罗伯特 · 曼固彦（左三）、玛丽安 · 雷（左一）和张永和（右一），1993 年 ©B.A.S.E Beijing
Figure 1 Robert Mangurian (left third), Mary-Ann Ray (left first) with Yung Ho Chang (right first) in 1993 ©B.A.S.E Beijing

2003 年初，一直希望在中国建立一个关于建筑教育研究实验室的罗伯特 · 曼固彦和玛丽安 · 雷在偶然的机会下第二次来到北京（图 1），[1] 在草场地租下一个 7000 平方米的、由德国人设计建造的废弃厂房，作为建筑和“制造城市”的实验基地。草场地，一个居住着艺术家、出租车司机、外来务工者和当地农民的，有着强大包容力的城中村，引发两位西方学者、教育者的研究兴趣。2005 年，罗伯特 · 曼固彦、玛丽安 · 雷与密歇根大学教授罗伯特 · 亚当在草场地正式成立“建筑与城市化问题实验机构和公共论坛”[1] 以及“基地工作室”。随后密歇根大学、多伦多大学、温特沃斯理工学院、南加州建筑

1　罗伯特 · 曼固彦和玛丽安 · 雷第一次来到中国是在 1993 年，因张永和希望能在中国开设一门南加州建筑学院的研究生“平行课程”，虽然计划并未如愿实现，但他们因此和中国结下了缘分。

At the beginning of 2003, Robert Mangurian and Mary-Ann Ray, who had always hoped to establish a research laboratory on Chinese architecture, fortuitously came to Beijing for the second time (Figure 1).[1] They leased a 7,000 square meter German factory in Caochangdi as an experimental base for architecture and "city-making." A highly inclusive urban village inhabited by artists, taxi drivers, migrant workers and local farmers, Caochangdi piqued the interest of these two Western researchers and educators. In 2005, Mangurian and Ray, together with Professor Robert Adams of University of Michigan, formally established Beijing Architecture Studio Enterprise (B.A.S.E.), the "Experimental Institution and Public Forum on Architecture and Urbanization Issues"[1] in Caochangdi. Subsequently, graduate students from the Universities of Michigan and Toronto, Wentworth Institute of Technology, and the Southern California School of Architecture regularly participated in the research at B.A.S.E., forming a powerful academic wave of research on Chinese cities in the Western architectural higher education system.

Through mapping, interviews and photography, B.A.S.E has documented the ever-changing and unfinished state of Caochangdi under conditions of rapid urbanization. At the same time, they have used Caoc-

1　Robert Mangurian and Mary-Ann Ray first came to China in 1993, responding to Yung Ho Chang's initiation of setting up a "parallel course" for the graduate studies at the Southern California Institute of Architecture in China. Instead of realizing the plan, Robert Mangurian and Mary-Ann Ray became interested in Chinese architectural education and Chinese urbanism.

图 2 刘全喜（小刘）与他的 35 位老乡员工 ©B.A.S.E Beijing
Figure 2 Liu Quanxi, or Xiao Liu (Young Liu) with some of his 35 men ©B.A.S.E Beijing

学院等建筑专业的研究生不断地加入基地工作室的研究，在西方建筑高等教育体系内形成一股中国城市研究的新生学术力量。

基地工作室的研究通过映射的方式对身边居民进行访谈、摄影，形成一份草场地在快速城市化中的、一种正在变化和未完成状态的记录，并以草场地为基础研究亚洲城市化背后的推动力量。[2]

在通往机场的快速路上，自上而下地为访客建设了一些项目，[2] 种植树木形成屏障屏蔽了背后的城中村，设置尺度过大且不知能去往何处的道路，以及像现代恐龙化石般只剩躯壳的博物馆。但与此同时，草场地里曾经是农民的企业家、艺术家以及画廊老板正在以惊人的速度共同建立一个充满生命力且有经济效益的城市模式。[2]

经过在草场地多年的生活与研究，罗伯特·曼固彦和玛丽安·雷认为，城中村作为城市的一种草根形式，是被广为认知的城市模式的一剂强有力的解毒剂。[3] 作为外来者，基地工作室的研究者们通过对流动人口、外来务工者等城中村居民（图 2）的访谈，深入了解每一个个体鲜活的故事，以及他们在草场地的社会关系网络，用一份又一份的城市档案 [3] 具体而生动地反映城中村（图 3）的包容和活力。同时，这些城市研究者们又身在其中，与草场地的艺术家和居民共同过着城中村的日常生活。

hangdi as the basis for researching the driving forces behind urbanization in Asia.[2]

The top-down projects[2] plant green screens to hide the village from visitors on the Airport Expressway, install oversized roads that go from nowhere to nowhere, and build museums that sit like empty modern dinosaurs. At the same time, and even in spite of this, the former farmers turned entrepreneurs, artists, and gallery owners are building a vital and profitable urbanism in Caochangdi at an astounding rate.[2]

Having lived and studied in Caochangdi for many years, Mangurian and Ray believe that the urban village, as a physical presentation of grassroots urbanism, is a powerful antidote if not an alternative to the widely recognized urban model.[3] As outsiders, B.A.S.E. researchers have interviewed urban village residents, including members of the floating population and migrant workers (Figure 2), gaining an in-depth understanding of the vivid stories and extensive social network behind each individual. Mobilizing file after urban file, they have demonstrated the inclusiveness and vitality of the urban village (Figure 3) as well as the daily lives of Caochangdi residents.[3]

2 由国家规划的规模庞大的五环公路和机场高速，将在 2008 年 8 月带领游客们进入北京城和“同一个世界，同一个梦想”的奥运场馆，草场地这个城中村正坐落在两条公路围出的口袋形区域里。作为党中央关于城市绿化要求的一部分，北京正在种植 30 亿株树木，草场地恰好被这 30 亿中的一部分给遮掩起来——罗伯特·曼固彦，玛丽安·雷. 草场地·城乡谜题. 21 世纪共和国里远离中心的人民空间 [J]. 城市中国，2008 (8): 70-74.

2 "In a pocket defined by the intersection of the massive state-planned Fifth Ring Road and Airport Expressway, which took visitors into the city of Beijing and to the 'One World, One Dream' Olympic venues this August, sits the urban village of Caochangdi. Screened from view by swaths of some of the three billion trees now being planted in Beijing as part of a Central Party urban afforestation mandate, Caochangdi is a thriving early twenty-first-century urban space of mostly illegal structures being built by entrepreneurial farmers and contemporary art dealers and artists." – MANGURIAN R., RAY M. A. Caochangdi · Uban Rural Conundrums: Off Center People's Space in the Early 21st Century Republic of China[J]. *Urban China*, 2008(8):70-74.

图3 北京城中村分布图 ©B.A.S.E Beijing
Figure 3 Beijing Urban Village Mapping ©B.A.S.E Beijing

1　“里外不分”，从草场地到南头古城

罗伯特·曼固彦和玛丽安·雷在北美的英文出版物名为《草场地：里外不分的北京》。[4]“里外不分城市”是描述一种公共与私有边界模糊的状态。这实际上是原生村落生活状态的一种延续，与城中村原村民以及其他村落移居的务工者的生活方式相关，如在户外晾晒的被子、把一层对外的卧室加入杂货店功能、在公用走道上做饭等私人生活与公共生活的重叠。从基地工作室对城中村社会关系网络的研究中，“里外不分”也暗指一种类似农村的，但又因大量务工者的流入而形成的有别于传统的、更包容、更多元的熟人社会[5]的社会关系网络及社会资本。

基地工作室团队（图4）所描述的“里外不分”在他们的南头实践中再一次充分体现，通过南头古城在地的社会关系网络，完成展览《房屋17号：深圳迁移场所》。[4]他们修缮了一座隐匿于狭小街巷中的老房子，用作皮影戏、视频播放、现场表演等多种活动的发生地，也为居民提供休憩的公共空间。相比起我们这些说着中国话的城市人，这些语言不通的城市研究者和学生们似乎更懂得“城中村语言”。

2　发现藏于巷子深处的古董

罗伯特·亚当斯第一次来到南头古城的时候，小尺度的街巷中生机勃勃的生活方式深深地触动了他，他形容“南头古城的街巷让我们处处是惊喜”[5]。南头古城中依然存在的邻里关系密

3　B.A.S.E. Beijing的代表性研究成果有An Illustrated Lexicon of Chinese Urbanism (and Ruralism), A Tale of Two Urban Villages: Beigao and Nangao,The Beijing Houses of the Ant Tribe and the Mouse Tribe, (A) Manual (of) Urbanism: A Stop Motion Portrait of Beijing as Seen Through the Hands of It's Citizens at Work and Play.

4　在2017深双的建筑策展团队邀请下，北京基地工作室的罗伯特·亚当、罗伯特·曼固彦和玛丽安·雷带领密歇根大学和加州大学伯克利分校硕士学生参与有着“城市即展场，展览即实践”概念的“实验场”板块的介入实践项目。

1　"Inside Out": From Caochangdi to Nantou Old Town

In *Caochangdi, Beijing Inside Out: Farmers, Floaters, Taxi Drivers, Artists, and the International Art Mob Challenge and Remake the City*, Mangurian and Ray propose the concept of "inside-out urbanism," through which the boundaries between public space and private life are blurred.[4] This is actually a transplant of living practices common in traditional villages that have been redeployed by migrant workers in the city, in which the private life interacts with the public realm in different ways, such as drying quilts outdoors, adding a grocery store to a bedroom on the first floor of a residential unit, and cooking in public corridors. The concept may also be extended to include the social networks that originated in traditional villages and have evolved into a more inclusive and diverse acquaintance society[5] because of the diverse backgrounds of migrant workers in urban villages.

The adaptive reuse of House 17 in Nantou Old Town at the 2017 UABB aimed to realize the principles of "inside out" urbanism by working closely with the local community of Nantou Old Town.[4] A hundred-year-old house hidden in a narrow alley was transformed into a place where residents could gather and a venue for various activities to take place, including shadow puppet shows, video broadcasts, and live performances. During the renovation, Mary-Ann Ray and Robert Adams led groups of students from the Universities of Michigan and California, Berkeley, (Figure 4) speaking "the language of the urban village."

2　The Hidden Gem in the Alley

When Adams first came to Nantou Old Town, he was deeply touched by the lively scenes in the small-scale streets and lanes. He described how "the streets and lanes of the Nantou Old Town are full of surprises for us."[5] He also referred to the traditional daily life and cultural scenes seen in Nantou Old Town as "the counter-narrative of Shenzhen, a global city in the 21st century."[6]

The invitation from the 2017 UABB curatorial team was very open-ended, with emphasis only on the theme and concept of the biennale. The B.A.S.E. Beijing team (hereafter "the team") decided on the format of the exhibition when they discovered 17 Leping Street (hereafter "House 17") in June 2017, during their initial site research in Nantou Old Town. With the help of neighbors, the team found the owner of the house, Granny Huang

3　Examples of research by B.A.S.E.Beijing: *An Illustrated Lexicon of Chinese Urbanism (and Ruralism), A Tale of Two Urban Villages: Beigao and Nangao,The Beijing Houses of the Ant Tribe and the Mouse Tribe*, *(A) Manual (of) Urbanism: A Stop Motion Portrait of Beijing as Seen Through the Hands of It's Citizens at Work and Play.*

4　Robert Adams, Robert Mangurian and Mary-Ann Ray led students from UC Berkeley and University of Michigan to participate in 2017 UABB with invitation from the curatorial team of the architecture session.

图 4 基地工作室的展览团队与在地施工队及房东合影 ©B.A.S.E Beijing
Figure 4 Team members of B.A.S.E. Beijing with local contractors and landlord ©B.A.S.E Beijing

切的传统日常生活场景，他称之为“深圳这个21 世纪全球性城市的反叙述”[6]。

与“命题作文”式的展览不同，策展团队给基地工作室的邀请中仅强调了展览主题及策展理念，并未对展览形式和内容作预设。2017 年 6 月，基地工作室的展览团队（简称展览团队）首次到南头进行调研的时候，一座闲置的老房子——乐平街 17 号（下称房屋 17 号）引起了他们的注意。通过询问老房子周边的邻居，他们找到了房主黄笑梅奶奶并向她了解老房子的故事。

这座主体结构可以追溯至清朝的老房子是传统“三间两廊”的广府合院[6]的“廊”，即合院的厨房。后来合院的房主因经济问题，把老房子分别协议出售给黄奶奶在内的三位南头居民。[7] 房屋 17 号周边是合院的其他部分拆除后建成的多层自建房，而房屋 17 号与其南侧的小院子由于面积小，无法建成多层的楼房而得以保留。由于两廊与房间是贴合建造的，房屋 17号北侧的外墙依然保留着与其相连建筑的墙，墙上有一个当时房间内用于祭拜的神龛。老房子的门窗是 20 世纪 80 年代黄奶奶为出租给房客居住改装的，东侧朝向巷子的卷帘门则是后来理发店的店主改装的。随着时间推移，房屋结构的稳定性下降，被相关部门定性为危房，又由于历史遗留问题，[8] 城中村房屋的修缮并未有正规的实施路径，房屋 17 号在理发店搬出后就一直闲置着。虽然黄奶奶希望房屋 17 号能用作民宿，吸引前来南头游玩的游客，但当展览团队提出租用房屋 17 号进行改造和修缮时，友

5　2017 深双开幕周论坛“都市村庄：共生与杂糅”发言。
6　三间两廊式的合院是中国传统四合院的演变，“三间”指的是排成一列的三间房房，其中间为厅堂，两侧为居室，三间房屋前为天井，天井两侧的房屋即为“廊”。“两廊”一般用作厨房或门房。相较于传统北方四合院，三间两廊的空间更为开放，利于南方地区的通风。
7　由用地性质推测，协议出售并非正规房产交易，该房产是否有产权证明未明。

Xiaomei, and learned the story of the old house.

The skeleton of this old house can be traced back to the Qing Dynasty. It is the "corridor" of a traditional "three rooms and two corridors style,"[6] Cantonese courtyard house. Specifically, it was the kitchen of the courtyard house. Due to financial difficulties, the owner sold the courtyard house to three different Nantou residents, including Granny Huang.[7] Other parts of the courtyard house had been demolished to build multi-story apartment buildings, while House 17 along with a small backyard on the south side were preserved as the site was too small for a multi-story construction. The demolition of the adjacent building left one interior wall with a shrine intact and attached to the north facade of House 17. Granny Huang installed the door and window system in the 1980s, converting the house into a rental apartment, while the roller shutter on the east facade toward the alley was later installed by the owner of a barbershop. The structural stability of the house had been compromised over time, while renovations to urban village housing were unapproved due to the unclear property rights of informal land registration and a lack of regulations to follow.[8] House 17 was left vacant since an inspection of its structural safety and the relocation of the barbershop. Even though the offer from the team to renovate House 17 would have helped to realize Granny Huang's wish to transform House 17 into an Airbnb, she initially turned

5　2017 UABB opening forum, "Urban Village: Hybridity and Co-existence." Nantou Old Town. 2017-12-16.
6　It evolves from the traditional siheyan courtyard house. The "three rooms" were a living room in the middle and bedrooms on each of the two sides. The two sides of the courtyard were known as "corridors," which in practice where a kitchen and storage space. Instead of the enclosed courtyard typical of a siheyuan, in a Cantonese style compound, one side was left open for better ventilation in Guangzhou's humid and warm climate.
7　Speculating from the nature of the property, the transaction may not be recognized as formal real estate's transaction and the property rights of the house are unknown.
8　According to interviews with Nantou residents, as a result of illegal and informal construction during the 1990s, the urban management bureau now prevents any kind of construction in Nantou. Even interior renovations are stopped, while the application process is unclear, and when applications are submitted there is no response.

善热情的黄奶奶和她的家人因担心房屋的安全隐患及改造和修缮的合法性而婉拒。直到展览团队向他们解释了双年展的缘由和其半官方的性质，并得到双年展指挥部出具的信函后，才欣然同意了房屋 17 号的改造计划。

房屋 17 号的改造并非一帆风顺，在移除瓦片进行房梁修复的时候被城管制止。经过展览和策展团队多次与深双组委会和街道办事处沟通后，获取纳入展场的批复文件，才得以复工。房屋 17 号“修旧如旧”的改造保存了清代的主体结构（图 5）、20 世纪 80 年代的门窗、古老的神龛以及新加入的标识、家具和装置等等的元素。这样的改造方式恰如其分地体现了总策展人孟岩所提出的理念，“只有尊重历史原真性、且珍视各个时代的文化层积和历史印记，才能塑造一个本土文化历久弥新，永远鲜活的城市历史文化街区”。在当下以“读图思维”主导下的改造，很容易为了突显设计感而忽略原有建筑的建造逻辑和正在消失的弥足珍贵的元素。更有所谓的修缮，以外来的“传统元素”置入，使其从图面上看起来像是传统元素却缺乏对“传统”二字的理解。

3　城中村社会网络的连接

城中村小尺度、狭窄的巷道有利于社会关系的建立。“包工头”龚师傅在回家路上途经房屋 17 号，无意间听到团队和黄奶奶的对话，便自荐成为改造计划的施工人员。罗伯特 · 亚当斯在双年展的开幕论坛上介绍道：“我跟他（龚师傅）年龄相仿，相谈甚欢，刚去现场的时候我们被邀请去他家吃饭，非常慷慨大方，非常真实。他说所有的旅客都可以在南头找到一个家，这跟我们的想法很相近。”龚师傅的话被做成横幅悬挂在房屋 17 号的院子前，“每一个旅行者都能在南头城找到居所”，一方面描述了南头落脚城市的特质，另一方面也指向黄婆婆希望未来房屋 17 号成为民宿的愿望。

房屋 17 号与报德广场只有一街之隔，双年展在报德广场举行开幕式的同时，房屋 17 号的开幕式聚集了周边的街坊（图 6）。伴随着房屋 17 号热闹非凡的鞭炮声和锣鼓声，双年展的主持人在报德广场宣布“2017 深港城市 \ 建筑双城双年展正式开幕”。房屋 17 号的传统文化表演者是“包工头”龚师傅的儿子小龚先生在深圳舞狮队演出时结识的好友。在团队的邀请下，舞狮队、孙悟空模仿者、川剧变脸者、粤剧演员、传统剪纸手艺人、捏面人和糖画手艺人共同为房屋 17 号的邻里呈现了精彩的演出。

这些看似微不足道的社会网络链接的发生，

8　对南头居民进行访谈时，居民表示由于 1990 年代违章抢建问题严重，现城市管理严格，在南头内很难进行施工作业，即使是室内装修都会被执法人员制止，审批的流程没有清晰的实施路径，即使申请了也杳无音信。

down the offer due to concerns about the legitimacy and safety risks of the proposed renovation. It was not until the team fully explained to her the nature of the UABB, a public-sector-organized event, and provided an official letter from the UABB curatorial team that Granny Huang and her family agreed to the renovation projects.

The renovation of House 17 did not go smoothly. The urban management bureau stopped construction when the roof tiles and beams were removed for restoration. The construction resumed when an approval document was obtained by the curatorial team after a tremendous amount of communication between the curatorial team and the sub-district office. The renovation of House 17 restored the building's main structure (Figure 5) and the Qing-era shrine as well as the door and window system from the 1980s, while integrating a new signage system, furniture, and installations in the project. This renovation project set a great example, reflecting the concept put forward by Meng Yan, the chief curator of the 2017 UABB, who said, "Only by respecting the authenticity of the history and cherishing the cultural layers and historical traces of each era can we shape a timelessly dynamic urban community rooted in local history and culture. "Today, as visually oriented reading dominates the world, adaptive reuse projects often serve the purpose of producing stunning propaganda photos without considering the tectonics of the original architecture and undermining the protection of the disappearing architectural elements. Indeed, there is often a lack of understanding of the word "tradition" in the process of transplanting the seemingly "traditional elements" of many renovation projects.

3　The Social Network of the Urban Village

The narrow alleyways in urban villages are conducive to the establishment of social relations. The contractor, Master Gong, overheard a conversation between the team and Granny Huang on his way home while passing by House 17, so he introduced himself and became part of the team. According to Adams' introduction at the opening forum of the 2017 UABB, Master Gong generously invited the team to dinner at his living unit and had a delightful and genuine conversation. The banner hung at House 17 stated, "Every traveler can find shelter in Nantou Old Town," quoting Master Gong. The quote not only depicted the character of Nantou Old Town as a settler area, but also reflected Granny Huang's vision of House 17 as an Airbnb for travelers.

Being one street away from Baode Square, where the official opening ceremony of the 2017 UABB took place, House 17 invited neighbors to its own opening ceremony at the same time (Figure 6). The announcement of the opening of the 2017 UABB took place in

图 5 房屋 17 号修缮现场 ©B.A.S.E Beijing
Figure 5 Preservation and renovation of House 17 ©B.A.S.E Beijing

实际上是社会关系网络不断地编织和叠加新的社会关系，积累社会资本的过程。在双年展的平台下，房屋 17 号展览团队作为外来力量的介入，如同媒介，在整个改造过程中与本地社区网络紧密地结合在一起，催化社会资本的累积；同时也成为居民与城市管理机构间的一个沟通的桥梁，提供了一个可行的、具有范式性的有机更新案例。

房屋 17 号的展览介绍中提及对其未来的希望："这个温馨的小屋及南头当地友好的居民，特别是业主黄梅笑女士及建筑承包商龚师傅一行，为房屋 17 号项目带来了无限的可能：发廊、村庄档案馆、麻将室、驻地艺术家工作室、爱彼迎、供奉先人的神龛，亦或仅仅只是一个自由开放的空间，可供人们休憩并旁观乐平街上发生的点点滴滴。我们希望所有这些能带领黄笑梅女士及周边居民与房屋 17 号项目一同步向不一样的城中村未来。"[6] 房屋 17 号项目让我们看到了空间历史层积被尊重、个体价值得以彰显、生活方式和本土文化得以延续的未来图景中的一种可能性。

harmony with the sound of firecrackers and celebrations at House 17. The performers of Chinese traditional culture were the son of Master Gong, Junior Gong from the Shenzhen Lion Dance team, and his friends from different performance teams, such as imitators of the Monkey King, actors of the Sichuan opera and Cantonese opera as well as craftsmen in traditional paper cutting, dough sculpting, and sugar painting. Together, they presented a wonderful show for House 17's neighbors.

These seemingly insignificant social events were part of a process in which new relationships are continuously woven together with and superimposed on the original local social network, creating a stronger network that generates social capital. With the platform provided by the 2017 UABB, the House 17 team connected with the local community and bridged the top-down and bottom-up forces, establishing a feasible paradigm for future regeneration projects.

As has been described in the project description of House 17, "By attending to this meek building, and working with the Good People of Nantou, especially Granny Huang Meixiao the building owner, and the Gong Brothers, the contractors, HOUSE 17 is full of potential: a hair salon, a village archive, a mahjong parlor, resident artist studio, an Airbnb, ancestral shrine, or simply a free space to sit and observe life on Happy Peaceful Street. It will be left to the charismatic property manager Granny Huang and the villagers to guide HOUSE 17 into the future."[6] The success of House 17 brought one possible future to life, a future in which layers of historical spaces are respected, individuals are valued and the local way of life and culture are inherited.

[1] 罗伯特 · 曼固彦，玛丽安 · 雷 . 草场地 · 城乡谜题：21 世纪共和国里远离中心的人民空间 [J]. 城市中国，2008(8): 70-74.
[2] MANGURIN R., RAY M. A. Re-drawing Hadrian's Villa Re-writing Caochangdi Urban Village [EB/OL]. Gran Tour, 2008, 41: 120-125 [2020-11-20]. http://www.jstor.org/stable/40482320.
[3] MANGURIN R., RAY M. A. "Urbanistica Sperimentale 'Invisible': Experimental Urbanism Under the Radar," in Caochangdi and Beijing's Urban Villages[J]. Lotus International, 2010(3): 54-626.
[4] MANGURIN R., RAY M. A. Caochangdi, Beijing Inside Out: Farmers, Floaters, Taxi Drivers, Artists, and the International Art Mob Challenge and Remake the City [M]. Hong Kong: Blue Kingfisher, 2009.
[5] 吴宝红 . 城中村发展中的社区动员与青年参与 [J]. 当代青年研究，2019(2): 109-115.
[6] 房屋 17 号：深圳迁移场所 [M]// 深圳城市 \ 建筑双年展组织委员会 . 城市共生：2017 深港城市 \ 建筑双城双年展 . 广州：岭南美术出版社，2021: 379.

房屋17号的关系网络
THE SOCIAL NETWORK OF HOUSE 17

本地居民与政府之间平台的缺失，使得自下而上的更新缺少实施路径。深双的介入短期地建立起一个连接的平台，借助深双的展览机制，实现房屋17号的合法修缮，并激活本土社区网络和文化资源。
The lack of a communication platform between the local residents and the government has led to a lack of policies for self-renovation in urban villages. UABB serves as a communication platform to facilitate the legal renovation of House 17 and to revitalize the local social network and cultural resources.

探索有机更新实施路径
Explore Approaches for Sustainable Renewal

双年展办公室与南山区政府联合，形成了双年展指挥部，协调各相关职能部门，通过行政手段有效地调配力量和资源。同时，为有机更新的实验，研究和探索精细化城中村自下而上更新的审批流程。
The Biennale office and Nanshan District Government established an outpost to effectively coordinate various departments for the Biennale. At the same time, it explores the administrative process related to the renovation in the urban villages.

展览创造对话机制
Bridge the Gap via Exhibition

以房屋17号为空间载体，双年展作为平台为在地居民与城市管理者创造对话机制，联通老房、危房自我更新的实施路径。
House 17 acts as a platform that connects local residents with the city administration, enabling the potential for self-regeneration of old and dilapidated houses.

有限度的外部资源入驻
Introduce External Resources

参展人通过项目引入外部资源，如参与设计、改造的学生团队，大学研究经费支持改造以及深圳本地创客资源。
Through the project, the participants can bring resources to the urban village, such as help from architectural students, renovation funding from the University of Michigan, and support from makers.

发现、发掘本土资源
Discover Local Resources

参展人通过项目入驻到社区，从而在日常的生活和交往中，与社区建立紧密的联系，从而发掘在地资源，活化社区网络。
Participants settle in the community through the project, establishing close connections with the community through daily interactions to explore local resources and revitalize the community network.

图 6 2017 深双在报德广场举行开幕式（上）©UABB；房屋 17 号开幕式（下）©B.A.S.E Beijing
Figure 6 Opening ceremony of 2017 UABB at Baode Square (above); Opening ceremony of House 17 (below) ©B.A.S.E Beijing

房屋 17 号是清朝时建造的一栋庭院式房屋的历史遗留，坐落于南头古城聚贤街，即现今的乐平街。设计修复并改造了该历史性房屋，使之更加现代化，以此提供一个结合清朝及 21 世纪初期时代特征的空间，在双年展期间为人们带来全新的生活体验，并展现该房屋未来新的使用性。

房屋17号是一个以类似于舞台的形式、面向街道开放的家用空间，为居民和游客提供一个可以品茶、打麻将和聚会的场所。展览期间还有一系列活动、歌剧、演出着重讲述着南头村及其居民的故事。一些故事具有极高的历史意义，例如 1842 年在南头签署的《南京条约》，迫使中国在第一次鸦片战争期间将香港割让英国。虽然许多迁移场所已经转移到南头，但是村民的日常生活依然充斥着拉面、舞狮、麻将等传统项目，形成深圳这个 21 世纪全球性城市的反叙述。

▷展览：房屋 17 号：深圳迁移场所 ▷参展人：罗伯特·亚当斯、玛丽安·雷、罗伯特·曼固彦

HOUSE 17 is a small, historical fragment of a larger Qing Dynasty courtyard structure on Leping Street (Happy Peaceful Street), formerly Juxian Street (Bringing Together Talented People Street), in Nantou Urban Village. The design project restores and modernizes the house to produce a space that is simultaneously Qing Dynasty and early 21st Century to allow a new life in and around the house during the Biennale, and to open up new possibilities for its use into the future.

HOUSE 17 is a domestic space opening to the street by means of a large proscenium-like opening, staging a space for villagers and visitors to drink tea, play mahjong, or just hangout. A series of events, operettas and performances focus on telling stories of the Village and some of its Villagers. Some of these stories are of epic proportion, such as the signing of the Treaty of Nanjing in Nantou in 1842, forcing China to hand-over Hong Kong to Britain during the First Opium War. Though many migrating scenes have moved through Nantou space, its everyday practices in the life of the Village—the noodle puller and cutter, the lion dance practitioners, or the lively mahjong games—offer a counter-narrative to Shenzhen as a 21st century global city.

▷Exhibition: HOUSE 17: Migrating Shenzhen Scenes ▷Participant: Robert Adams, Mary-Ann Ray, Robert Mangurian

来源：深圳城市\建筑双年展组织委员会. 城市共生：2017 深港城市\建筑双城双年展 [M]. 广州：岭南美术出版社，2021: 379.

南头游乐场：城市体制与南头体质

Nantou Playground: The Urban System and Nantou Spirit

莫思飞
2017 年深港城市\建筑双城双年展（深圳）策展助理，都市实践研究员。

Mo Sifei
Curatorial assistant of the 2017 Bi-City Biennale of Urbanism\Architecture (Shenzhen), URBANUS Researcher.

"烂尾楼"（图 1）是常见的历史遗留问题，由于建筑的合法性、经济问题以及权属等不可调解的因素形成，却又未能寻找对应的实施路径，使其得以利用。在寸土寸金的城市里，"烂尾楼"既是对空间土地资源的浪费，也对城市环境造成负面影响，形成安全隐患。 在深圳城中村里，烂尾楼的数量绝非少数；政策对小产权违法建筑的规定及城市管理执法的日益完善，杜绝了偷建、抢建的现象，但遗留下来权属复杂、无法可依、无先例可循的烂尾楼，这些烂尾楼的再利用成了悬而未决的城市问题。除了在大规模城市更新中"一拆了之"，烂尾楼是否有再生的可能性？ 2017 深双的介入被看作是"一场自救的实验"[1]，实验之一则以南头古城中的一座烂尾楼为试点，探索再生的路径与模式，虽最终并未实现，却是具有实践意义的一次"大冒险"。如和马町评论深双时表述的："当你提出'激进的替代方案'，最有力的方式莫过

图 1 南头古城，烂尾楼现场，摄于 2016
Figure 1 The site of unfinished building in Nantou Old Town, 2016

"Unfinished buildings" (Figure 1) are a pressing urban issue. These buildings are abandoned in the process of construction due to the legality of the construction, financial difficulties, unclear property ownership and other irreconcilable factors, yet a feasible approach for their reuse cannot be found. Not only does this issue negatively impact land-use efficiency and the urban environment, it also poses huge safety risks. In Shenzhen urban villages, unfinished buildings are not uncommon. Improvements in urban management and the law enforcement of collective properties have effectively eliminated illegal construction in Shenzhen's urban villages, yet unfinished buildings, which have no regulations and precedents for their adaptive reuse, remain an unresolved urban issue. What are the possibilities of unfinished buildings surviving and regenerating in urban villages instead of being demolished during a massive urban renewal? The 2017 UABB has been regarded as an "experiment in self-help."[1] The attempt to include an intervention project for unfinished buildings in Nantou Old Town as a pilot project to explore the path toward its regeneration was a "great adventure" with practical significance despite the failure of its realization. As Martijn de Geus stated in his review of the 2017 UABB, "When advocating 'radical alternatives,' there is no stronger message than doing what you say, then to document the process, test it, use it, open it up for visitors and the professional community to assess and debate. This requires courage, strong belief in your ideals and support. It also requires to be vulnerable to possible criticism."[2]

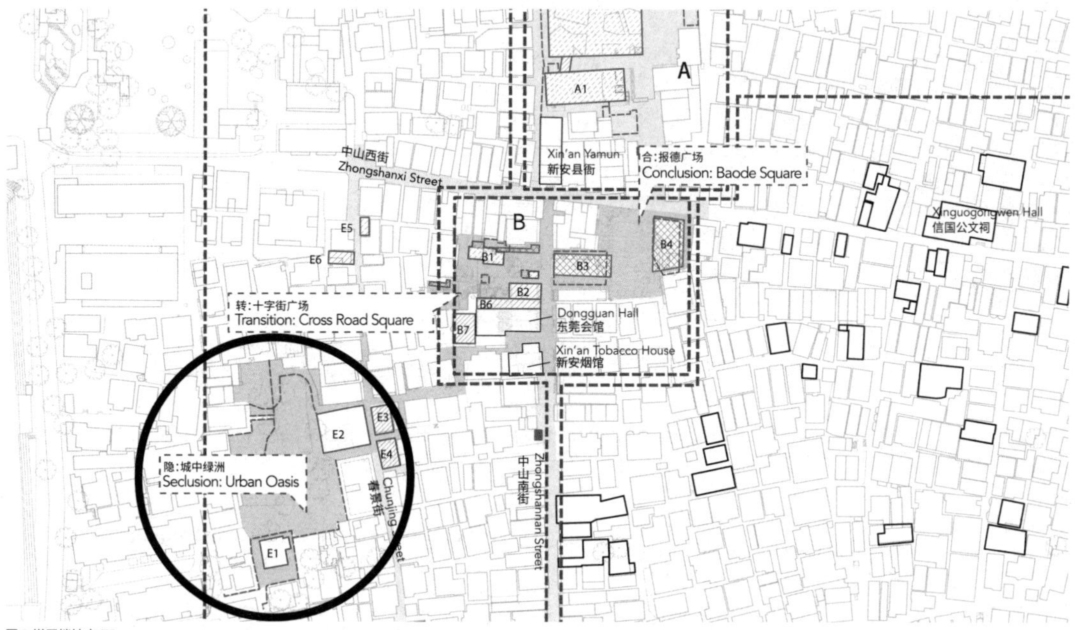

图 2 烂尾楼地点 E2
Figure 2 The location of the unfinished building, E2

于将它付诸实践，然后记录过程、测试它、使用它，向参观者和专业界开放以进行评估和辩论。这需要勇气、对理想的坚定信念和各方支持，也要承担可能面临的批评。”[2]

1 发现烂尾楼

2016 年初，都市实践城市研究团队在深入南头的场地调研中，发现了在春景街绿地入口处（图 2）的“烂尾楼”，背靠春景街绿地且紧邻绿地入口，使其西侧和南侧都有日光照耀。与街道办和房东的前期沟通中了解到，烂尾楼的用地属于房东的宅基地（或被分配用于自住的集体用地），原有占地面积约 200 平方米的两层瓦房。2000 年前后，原建筑被推倒并计划重建六到七层的住宅，但由于当时对违法建筑的认定及相关法规的出台，[3] 新建建筑在建设至二层结构框架的时候被城市管理执法部门勒令停止建设，荒废至今。此处，在堆满建材和建筑维护结构缺乏的情况下，多次成为流浪汉的居所，不仅影响春景街及春景街绿地的环境品质，其建筑质量和现状也存在着安全隐患。

“烂尾”至今是因为没有可循的先例或实施办法，对已经废弃的建筑进行重新利用。从权属上而言，建筑用地属于房东的合法宅基地（或集体用地的分配用地），并不能完全以国家所有的城市土地的开发模式开发，但土地之上的建筑依照法律法规的认定属于违法建筑，并不属于房东的合法私有财产，不能从法律上认定该建筑的权属分配办法。在这样错综复杂的情况下，若允许房东完成建设则形成对未来

1 Discovery of the Unfinished Building

In early 2016, during a site survey of Nantou Old Town, an urban research team from URBANUS discovered an unfinished building near the green space entrance to Chunjing Street (Figure 2). With the green space to its west and a gateway to its south, the building had optimal sun exposure. According to the sub-district office and the landlord, the unfinished building was located on a homestead (*zhaijidi*, which refers to collective land that has been allocated to the owner as a living unit) and had a landlord. The property registration consisted of a two-story house with a 200 square meter footprint. In 2000, the two-story house was demolished in order to build a six- to seven-story apartment building. However, the landlord did not have a building permit. Moreover, it was around that time that the laws and regulations regarding illegal building were put into effect.[3] Therefore, the city management bureau stopped construction of the new building, leaving an exposed and abandoned two-story structure framework. Over the years, the abandoned building, which had no facade and accumulated construction materials, had become an unofficial shelter for homeless people, negatively impacting Chunjing Street and the open green space, while posing safety risks due to the instability of the building structure.

Due to complications associated with the ownership of the property, the reuse of the abandoned building was put on hold. In terms of property rights, the land is legally allocated to the landlord as a homestead or collective land, and accordingly is subject to a different development

图 3 烂尾楼改造模型内部细节，卡洛斯 · M. 特谢拉
Figure 3 Unfinished building renovation model, interior details, Carlos M. Teixeira

违建行为执法的不良示范；若以政府主导的形式进行投资及重新利用，依据目前的权属界定，则属于公有资产对自有财产进行投资，可能引发社会不公。策展团队因此提出，计划把烂尾楼改造成公共空间，给予房东合法用地适当的租赁费用补偿，改造后烂尾楼结合春景街绿地的景观改造，为公共空间紧缺的南头古城提供一个“城中隐秘的绿洲”，同时也为烂尾楼这一类有权属争议性的自建房提供一种具有实践意义的范例。

2 从烂尾楼到南头游乐场

2017 年 5 月，带着“有限度地介入”和“尊重历史层积”的理念[4]，在深入研究参展人选后，策展团队找到远在南美洲的巴西建筑师卡洛斯。卡洛斯的事务所名为 Vazio S/A，Vazio 是葡萄牙语的空、空墟的意思，而他的作品则是以一系列的舞台布景设计的形式，同时结合戏剧表演艺术激活城市中的空墟，他称之“来自戏剧表演艺术的建筑”。卡洛斯与阿玛特鲁克斯街头戏剧工作室合作启动 “健忘症地形图”系列的项目。卡洛斯通过在废弃建筑中置入木制的路径，串联起四层连楼板都没有的建筑框架，巧妙地使其成为戏剧表演艺术的舞台，以节点的方式设置场景，有效地活化废弃建筑，改变城市对负面空间的消极态度，并在大规模城市更新介入城市空墟成为全球城市复兴趋势的大背景下，展示传统介入模式的局限性和临时介入的可能性。[5] 由于“健忘症地形图的场地”紧贴着另外两座居民楼，卡洛斯的设计创造了观众、舞台和城市的另一种关系。在戏剧表演的同时，附近的建筑物里上演着日常的场景，居民时而从窗户观看戏剧演出，成为戏剧表演的另一种观众。该项目同时也在批判和反思城市公共空间的缺乏，通过选取在废弃建筑中展演街头戏剧的方式，拓展“街头”（公共空间）的概念。[6]

在简单地与卡洛斯介绍双年展整体状况及南头烂尾楼项目的立意后，卡洛斯接受了策展团队的邀约。在初期的邮件沟通中，远在南美

model to state-owned urban land. However, the building was designated an illegal structure. As a result, the ownership of the unfinished building is unclear. Given this regulatory background, allowing the landlord to complete the construction might have encouraged the construction of other illegal buildings, while a government-led adaptive reuse project and investing public assets into private property might have corrupted social justice through the created inequity. In response to these complications, the 2017 UABB curatorial team proposed to transform the unfinished building into a public space with adequate rental compensation for the landlord. The renovation of the unfinished building and the adjacent open space would become one of only a few public spaces in Nantou Old Town, providing a model for the adaptive reuse project of controversial informal architecture.

2 From an Unfinished Building to the Nantou Playground

In May 2017, guided by the concepts of "limited intervention" and "respect for the layers of history"[4] and after conducting extensive research on different candidates, the curatorial team found the Brazilian architect Carlos M. Teixeira. Teixeira's firm is Vazio S/A, which means "empty" and "void" in Spanish. Vazio revitalizes urban voids through a stage design approach accompanied by performance arts, which he has referred to as the "architecture of performance art." For example, in 2004 Vazio S/A and Amatrux Street Theater Company launched, Topographical Amnesias in which a wooden path weaves through the structural framework of an unfinished building, establishing a platform with several nodes to stage performances. The project effectively reinvigorated the abandoned building and dissipated passive attitudes toward the urban void. As an alternative to the worldwide trend of mega interventions to revitalize urban areas, Topographic Amnesias showcases the possibilities of ephemeral intervention as much as the limitations of conventional mega interventions.[5] The proximity of the project to two residential buildings created a new relationship between the audience, stage and city. Scenes of everyday were displayed simultaneously with the presentation of the performance, which residents could see from their windows. The involvement of Amatrux Street Theater Company can be seen as a criticism of the privatization of the city, and a lack of public space extended the concept of public space to a new street concept.[6]

Teixeira accepted the curatorial team's invitation after a briefing on the concept of the biennale and the purpose of the unfinished building project. Located far away in Brazil, the preliminary site survey could only be conducted through photographs and communication with the curatorial team. Teixeira suggested that the best way to deal with the abandoned building would be

的卡洛斯只能通过照片评估场地现状，他提出对烂尾楼的处理要利用废弃建筑的不确定性，出乎意料的、非正规的、不寻常的元素才能让平凡的建筑充满惊喜。非事先预设是建筑不平凡的先要条件。在清理内部建筑废料时，卡洛斯主张保留现有的植物，“甚至连杂草都应该保留”。在随后提交的方案中，卡洛斯的设计以“南头游乐场”命名，为非正式聚居地补充适合儿童游玩的设施（图3），设计以金属钢筋网和木板两种材料构建游乐场中的游乐设施和栏杆格档（图4），使烂尾楼维持没有外围护结构的半室外状态，使其“身世”不被掩盖，并延伸至春景街绿地，成为一个独一无二的乐园。[7]

to take advantage of the informalities, surprises, and paradoxes that make this un-admirable work of architecture admirable. The ability to transcend the ordinary is the prerequisite to making architecture extraordinary. During the removal of the abandoned construction material, he advised keeping all the existing plants, "even the weeds." As for the concept design, the lack of a children's playground in informal settlement districts such as Nantou Old Town became the driving force for Teixeira's proposed Nantou Playground (Figure 3). The unfinished state of the building was maintained by leaving the structure open, extending the space to the adjacent green space and minimizing the interventions

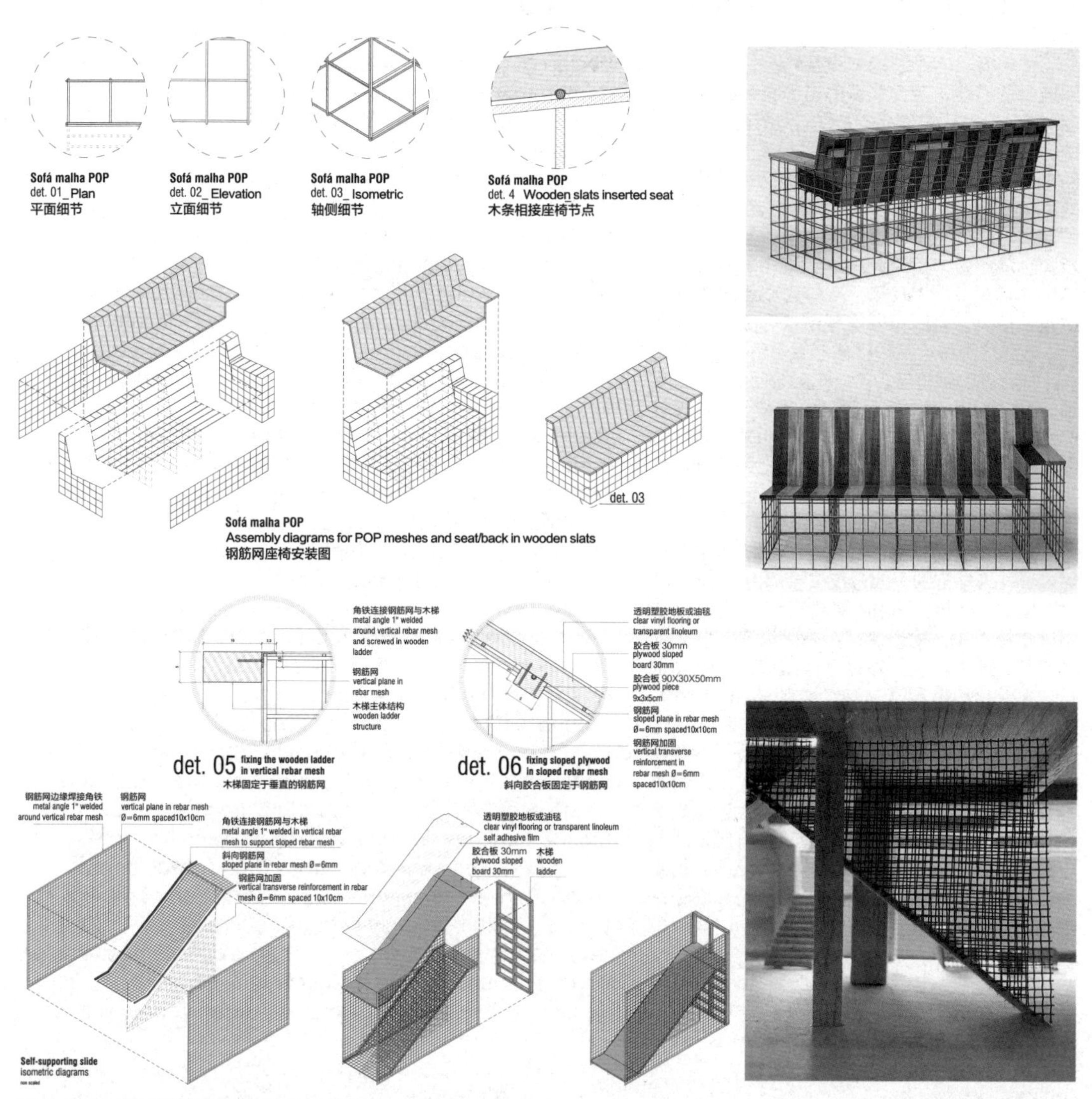

图4 烂尾楼城市家具、滑梯深化设计，卡洛斯 · M. 特谢拉
Figure 4 Urban furniture, slide of unfinished building design development, Carlos M. Teixeira

3 难以普及与应用的从自发到自觉

卡洛斯在巴西的设计常与非正规聚落相关，大部分非正式聚落与城中村有着相似的城市肌理，但都集中在权属较为清晰的公有(国有)土地上。因而有着不同用地权属的南头古城为项目增添了更多的不确定性，真正挑战的是探索一条可实施落地的路径。在与卡洛斯取得初步联系的同时，策展团队与房东的协商相对顺利，初步认可以支付三个月（双年展介入期间）的市场价租金，让烂尾楼成为展场，对于南头游乐场的设计房东也予以肯定。[1] 由于烂尾楼情况的特殊性，在与相关职能部门沟通时，策展团队强调了该试点的意义，其轻质和临时的介入方式以及项目的公共性，借此模式尝试减低引起居民不满及负面影响的可能性，从而寻找对未来有借鉴意义的改造模式。

烂尾楼试点计划的本质是希望打通从自发到自觉的通途，使无意识的本能行为转化为有意识的、经过审慎思考的行动，但保有自发的生猛、活力和多元的特质。然而“摸着石头过河”的策展团队未能通过沟通突破“缺乏政策依据”的现实局限，最终无法实施。直至2018年3月深双闭幕，烂尾楼项目在展期内依然停留在自发建造的阶段，悄然无声地完成了部分外围护结构的建设。[2] 2020年，万科启动春景街的整体改造后，卡洛斯再一次受邀作为集群设计师之一改造烂尾楼，其设计依然保留着2017年设计不加修饰的粗野感。[8]（卡洛斯的设计并未实现，2021年中旬，万科委托纵横行Plus8设计烂尾楼改造方案）卡洛斯不知道的是，烂尾楼在深双开幕前后就再一次自发地、悄然地完成了外围护结构的建设，早已不是当时只有水泥框架结构的样子。

从自发到自觉绝不是一蹴而就的，试图以烂尾楼作为展览的一部分仅仅是迈出了第一步。深双期间对其改造计划的搁浅与2021年春景街整体改造对比，揭示了在没有强大资本注入的情况下，城中村里烂尾楼乃至闲置宅基地微更新、再利用可能遇到的问题和障碍。现有的城市管理办法和政策无法适应城中村现有的问题，在为城中村量身定制的精细化“游戏”规则建立之前，自下而上的引导更新的模式难以得到普及和应用。如同南头游乐场一般，在观念上和实施路径上进行探索的实验式介入应该成为一种常态，才能有效地推动城中村精细化管理政策的形成，使城市的体制与南头的体质相辅相成。

1 信息来自尹毓俊访谈，2020-12。

2 2017深双期间，烂尾楼的外墙由业主自发建设完毕，贴上了外墙砖，只剩门窗安装及室内装修未完成。

with rebar meshes and plywood boards (Figure 4) where necessary.

3 Difficulties of Popularization and from Spontaneous to Conscious Use

Teixeira often works with informal settlements in Brazil, where the urban texture of these settlements resembles that of urban villages. However, informal settlements in Brazil are often built on state-owned land, while the different types of property ownership of urban villages contribute to the exploratory nature of the project. The conversation with Teixeria proceeded, as did the negotiation between the curatorial team and the landlord, with a deal being made for rental compensation at a market rate, allowing the unfinished building to be transformed into an exhibition venue. The landlord also agreed on the design of the Nantou Playground.[1] Due to the unprecedented intervention in an unfinished building, when communicating with government representatives the curatorial team emphasized the significance of the intervention, its ephemeral aspects and its public nature. This way, the possibility of a negative impact and negative public opinion would be minimized and a precedent for future regeneration projects could be set.

Using an unfinished building as a site for intervention was a search for a path to formalize the informal, so that unconscious, instinctive behaviors could be transformed into deliberate action while preserving the vitality and diversity of the informality. Yet, "crossing the river by feeling the stones" was insufficient for the curatorial team to break through the barrier of "a lack of policy to support," leaving the project un-realized. Until the closing ceremony of the 2017 UABB in March 2018, the unfinished building kept its informal quality by having its facade constructed undetected[2]. It was not until Vanke took over the management of Nantou Old Town and launched the overall renovation of Chunjing Street including the unfinished building in 2020, that Teixeria was invited by Vanke for a new version of Nantou Playground.[8] (The design was not realized, and Plus 8 was invited by Vanke for a proposal in 2021.) Teixeria may not have been updated with the news that the roughness of the exposed concrete structural framework and the facade that was finished in 2018 are both no more.

Formalizing the informal is by no means an overnight effort. If the failure of the intervention at the 2017 UABB and the success of Vanke's renovation in 2021 reveal anything, it has been possible difficulties that one might encounter when implementing a small-scale intervention and renewal project in an urban village without

1 According to Yin Yujun in the interview in December 2020.

2 During the 2017 UABB, the exterior of the "unfinished building" was finished with tiles by the owner. At this point, only windows and doors as well as interior were left unfinished.

mega capital investment. Yet, the attempt to make the unfinished building part of the exhibition was one baby step forward in exposing the mismatch of the existing urban management policy and the reality of urban villages. As there is a lack of defined rules for urban villages, bottom-up force-driven renovation can never be realized. There is a need to attend to experimental interventions in the search for a transformation of renovation concepts and practices. This is the way to promote the establishment of management regulations and policies for urban villages, which would allow the urban system and Nantou spirit to adapt to each other.

[1] 打边炉 DBL. 不仅是感官的爆发，也是一场自救的实验 [EB/OL]. (2017-12-20) [2020-12-12]. https://mp.weixin.qq.com/s/kPTqjgw-mNiSq9sTt50RXQ.
[2] 和马町 .2017 年深港城市 \ 建筑双城双年展评论——以全球建筑双年展的迅速增长为背景 [J]. 世界建筑，2018(1):141-144.
[3] 深圳市人民代表大会常务委员会 . 深圳市人民代表大会常务委员会关于坚决查处违法建筑的决定 [EB/OL].(1999-3-15) [2020-12-12]. http://www.sz.gov.cn/zfgb/1999/gb161/content/post_10096576.html.
[4] 孟岩，林怡琳，饶恩辰 . 村 / 城重生：城市共生下的深圳南头实践 [J]. 时代建筑，2018(3): 58-64.
[5] Carlos M. Teixeira. Entre, Architecture from Performance Arts[M]. London: Black Dog Publishing, 2012.
[6] Ana Mendez De Andes. ed. urbanaccion 07/09, 2021(2):101-115.
[7] VAZIO S/A. Nantou Playground [EB/OL]. [2020-12-12]. https://www.vazio.com.br/projetos/nantou-playground.
[8] VAZIO S/A. Nantou Playground 2 [EB/OL]. [2020-12-12]. https://www.vazio.com.br/projetos/nantou-playground-2.

本提案出于一个假定，双城双年展可能是一个绝佳的机会，可以为南头古城的烂尾楼创造新的公共用途，反之亦然。鉴于对这一地区几乎完全缺乏了解，我们从南头古城的几张照片中看到的第一印象是，许多居民都是孩子，可能这些孩子在附近没有合适的游乐场。我们希望这是一个合理的预设。现在，游乐场可以用来做想象练习，如何用不同的、不寻常的、颠覆性的方式来使用建筑。孩子们一定爱玩这个游戏，因为他们可以玩那些基本建筑元素，如楼梯、斜坡、栏杆、墙。这个游乐场的组件不是商业玩具：它们是组合的、有趣的、雕塑的集合；它们是作为正式转译其所助力的运动和游戏的装置；是一种将建筑物的不同楼层与身体的纯粹运动联系起来的策略。就像混合水和油一样，这个项目的挑战在于把废弃的混凝土框架的粗糙和原始纹理与游乐场的魔法和节日气氛融合。正是对这种粗糙的考量才使我们找到了材料：钢筋网和胶合板。在我们看来，这两种元素有足够的精美，让我们保有一种被遗弃空间的魅力、宁静和沉默。这种被废弃的结构是一种潜在的空白，等待着一种改变，它可以激发让人联想它原有状态的活动，并把时间和老化作为它的新用途的前提。与钢筋的粗糙老旧的疤痕相对应，新的暖色与混凝土的灰色相冲撞，这些滑梯将凝重的楼板转换成让人可以不用楼梯自由上下的平台。

▷展览：南头游乐园 ▷参展人：卡洛斯 · M. 特谢拉

This proposal departs from the assumption that the Bi-City Biennale can be a great opportunity to invent a fresh public use for the empty building in Nantou that was assigned to us, Vazio S/A.Given our nearly absolute lack of knowledge about that district, our first impression from the few Nantou pictures we've got is that many of its inhabitants are children, and that possibly these children don't have proper playgrounds in the neighborhood. We hope this is a reasonable forethought. Now, a playground can be taken as an exercise to imagine different, unusual, subversive ways to use a building. Children must play on it, as they can play with the general assumptions about using architectural elements such as the staircase, the ramp, the parapet, the wall. The objects of this playground do not correspond to commercial toys: they are a collection of playful sculptures that form a unit; they are devices that work as the formal translation of the motions and games they sponsor; they are a strategy to connect the building's distinct floors with the sole movement of the body. Like mixing water and oil, the challenge of this project lies in keeping something of the roughness and wild textures of the abandoned concrete frame with the magical and festive atmosphere of a playground. It was a consideration for this roughness that led us to our material palette: rebar meshes and plywood boards. Two elements that, in our view, are delicate enough to secure the enchantment, the silence and the reticence of an abandoned space. This disused structure is a latent void waiting for a change that could ignite activities reminding its previous condition, and taking time and weathering as a premise for its new use. Contrasting with the harsh materiality of the rebars and the existent scars of aging, new warm colors clash against the grayness of concrete, whereas a myriad of slides transform sober slabs into platforms for moving down without stairs.

▷Exhibition: Nantou Playground ▷Participant: Carlos M. Teixeira

来源：深圳城市 \ 建筑双年展组织委员会 . 城市共生：2017 深港城市 \ 建筑双城双年展 [M]. 广州：岭南美术出版社，2021: 368-369.

果园城记：设计介入实验

Story of Orchard City: An Intervention Experiment on Design

沈康、匡敏

沈康，华南理工大学建筑学博士、广州美术学院建筑艺术设计学院院长。

匡敏，珠海市规划设计研究院设计师。

本文首次发表于《装饰》，2019(12)：100-103；由莫思飞中译英。

Shen Kang, Kuang Min

Shen Kang has a Ph.D. in Architecture at the South China University of Technology, and he is the Dean of the School of Architecture and Applied Art, Guangzhou Academy of Fine Arts.

Kuang Min is a designer at the Zhuhai Institute of Urban Planning & Design.

The article was first published in *Art & Design*, 2019(12): 100-103; translated from Chinese to English by Mo Sifei.

中国城市在经历了近四十年的高速发展之后，政府主导的扩张型城市发展模式已经面临巨大的挑战。空间资源再分配、环境质量综合提升正成为人们日益关注的命题，增量发展主导将在今后转变为以存量提升为主。在这样的背景下，自下而上的社区营造成为新的应对选项，设计介入的方式为此提供了探索性的视角与路径。深圳南头古城春景街果园社区的设计实验便是这个背景下的一次尝试。区别于传统的规划与建筑设计实践，此次实验以日常生活的抵近观察与还原为引导，通过社会交往的设计促成社会关系的重建，激发更具自主性的创造力，营造真正为大众所有、有气息、能连接过去与未来的公共空间。

《果园城记》[1] 其实是一部短篇小说集，可读起来又像是一个可以串接完整的故事，讲述以“果园城”为背景的一系列人物和故事。《果园城记》最重要的笔意就是“果园城”，小说的作者师陀告诉我们：“我有意把这小城写成中国一切小城的代表，它在我心目中有生命、有性格、有思想、有见解、有情感、有寿命，像一个活的人。”[2] 虽然这些小说完成于 20 世纪三四十年代，但小说里的“果园城”仍然四散在我们生活的周围。由于机缘的巧合，《果园城记》的阅读也伴随了我们这次在深圳南头古城果园社区的设计实验，既是某种共鸣，也是一些线索和启发。如何从人类学的视角开展

After nearly 40 years of rapid development of Chinese cities, the government-led extensive urban development model faces huge challenges. People are increasingly paying attention to propositions to redistribute spatial resources and to comprehensively improve the environment. Future development will shift from the dominant trend of vacant land development to brownfield land improvement. In this context, bottom-up place-making and community building have become new options meeting the challenges to which design intervention provides an exploratory perspective and path. The design experiment of the Orchard Community on Chunjing Street, Nantou Old Town, Shenzhen was an attempt different from traditional planning and architectural design practices. It was guided by the close observation and representation of daily life, as well as promoting the reconstruction of social relations through the design of social interaction. The goal was to stimulate more autonomous creativity and to create a public space that would be truly owned by the public, full of life, and able to connect the past and the future.

The Story of Orchard City[1] is a collection of short stories that can be read as one story about a series of characters and stories within and against "Orchard City." The most important concept of *The Story of Orchard City* is the "Orchard City" itself. According to the book's author, Shi Tuo, "I deliberately wrote this small city as a

图 1 果园俯瞰
Figure 1 Orchard arieal view

一次设计介入的实验，跳出习以为常的建筑与规划设计思维定式，面对一处“问题”空间进行弹性介入，对日常生活的抵近观察与深度阅读，以及相关社会关系的勾连重建，成为我们这次实验最重要、也是最核心的内容。

1 南头果园

2017 深双将主展场选址在深圳的南头古城。这是一次不同寻常的突破，不仅仅是选址的突破，而且整个双年展本身亦是一场行动计划。策展人期待“这种结合城中村特殊空间、以日常生活主题为切入点的设计介入实验能够唤起人们对城中村、对城市与公共空间的重新思考，在留下视觉记录的同时，为社区营造、邻里互动、专业人士与市民的对话和交流产生长远的影响”[3]。在这样的背景下，我们获邀参与，并以联合工作坊的形式开展了一组在地艺术实验（“交织与介入”联合艺术实验，广州美术学院与台湾实践大学合作）。也是这次工作坊活动，让我们有机会接触到位于南头古城内的一处果园（图 1），并引发了持续的关注与兴趣，进行了一次新的尝试。

南头古城的这片果园位于城内的春景街。有关果园的历史源于当地街坊的描述：过去这里有一大片果园，现在周边居民的私家院子里还能看到不少幸存下来的大果树，甚至在南头古城的很多角落能意外地发现一些枝繁叶茂、树龄很长的果树。春景街的这片果园是南头古城内目前面积最大、最集中的一片果园。这片果园的发展至今经历了几个阶段的演变：1992

representative of all small cities in China. In my mind, it is full of life, characteristics, thoughts, opinions, and emotions, and has a life span just like a human being."[2] Although these short stories were completed in the 1930s and 1940s, the "Orchard City" in the book is still around us. Coincidentally, we read *The Story of Orchard City* during the design experiment conducted with the people who used an orchard on Chunjing Street, Nantou Old Town, Shenzhen. The book not only resonated with our work, but also offered clues and inspirations. The most essential content of our experiment was searching for a way to experiment with an anthropological perspective that would be free from the architectural planning and design mindset, while flexibly intervening in a "problematic" space through the observation and understanding of daily life as well as connecting and reconstructing relevant social relations.

1 Orchard in Nantou Old Town

The 2017 UABB selected Nantou Old Town as its main exhibition venue. This was an extraordinary breakthrough, not only because of its site selection, but also because the Biennale was itself an action plan. The curators hoped that "Such interventions and experiments that integrate the specific spaces in urban villages and are inspired by the daily life of ordinary people will evoke our reflections on the significance of urban villages, the city, and public spaces. While leaving visual testimonies for UABB, they will also exert far-reaching influences on community building, neighborhood interaction, as well

年，特区内进行农转非改革，南头城实业股份有限公司（简称南头股份公司）成立，将南头城村的资产转变为由村集体性质的企业统一管理，所有土地由南头股份公司管理。1994 年，南头古城内的自建房建设置换出这块空地，成为公共用地，收归南头股份公司所有，但部分村民还有这里过去私宅的地契。其后，这块土地由于长期的空置状态被周边居民逐渐分割利用，变成数片小块菜地。2008 年，政府将此处规划建设成公共花园，但由于周边的自建房切断了其与春景街、梧桐街的直接联系，被孤立成一个尴尬的岛状空地，所谓的花园被篱笆围护、隔离隐藏在内部，造成无人探访与使用的局面。刚开始还有城管过来打理，但渐渐再次陷入无人管理的境地。作为土地所有者的南头股份公司同样也无暇顾及这片土地。所有者和管理者的缺席，使得曾经的果园愈加不被人们珍惜，一直荒废，进而成为一块废弃的空地。

大约在 2010 年，附近的居民清理了这片土地上的垃圾，种上果树，这里又成了一片果园。2017 年由于双年展的入驻，展览组织方和地方政府、南头股份公司协调，将果园的一部分土地腾挪出来，作为艺术作品的展场暂时使用。正是由于双年展，把这块常

图 2 果园现状
Figure 2 Orchard status

年被人遗忘的空地带入我们的视野，引发了我们的兴趣。在密度极其夸张的城中村，应该是每一个角度都被日常生活的各种琐碎细节充斥，怎么会有这样一块颇为冷落、又权属不明的空

as dialogues and communications between professionals and the public."[3] In this context, we were invited to launch a group of on-site art experiments in the form of a joint workshop, "Interweaving and Intervention: Joint Art Experiment—A Collaboration between Guangzhou Academy of Fine Arts and Shih Chien University." During the workshop, we became aware of an orchard in Nantou Old Town (Figure 1), which led to sustained attention and interest, as well as a new intervention.

The orchard is located on Chunjing Street in Nantou Old Town. Locals narrated the history of the orchard as follows: There used to be a large orchard here in Nantou, and many surviving fruit trees can still be seen in the private yards of nearby residences, while some lush and long-lived fruit trees can be found unexpectedly in many corners of Nantou. The orchard on Chunjing Street is currently the largest and most concentrated one in Nantou Old Town. This orchard was developed in several stages. In 1992, the Shenzhen Special Economic Zone implemented rural-to-urban reform, giving rural residents urban status (*hukou*) and establishing the Nantou Corp. (Nantou Cheng Stock-holding Corporation). At that time, the assets of Nantou Village became collective assets managed by Nantou Corp., including all land resources. In 1994, the villagers exchanged the orchard land for land in Nantou Old Town to construct self-built buildings. Nantou Corp. acquired the orchard as public land, although some villagers still held land deeds to build a homestead within the orchard. The land was unoccupied for many years, and nearby residents eventually divided it up into small vegetable plots. In 2008, the government designated it a public garden. However, the self-built buildings surrounding it isolated the garden from both Chunjing and Wutong Streets. Hidden behind the fence and the buildings, the garden was neglected. At first, the city management bureau maintained the garden, but then gradually abandoned it. The owner of the land, Nantou Corp. also ignored it. Both the owner's and manager's absence made people less likely to cherish and attend to the orchard. As a result, it became a deserted open space.

Around 2010, nearby residents cleaned up the land and planted fruit trees, turning the land back into an orchard. In 2017, due to the arrival of the UABB, the exhibition organizer coordinated with the local government and Nantou Corp. to make a portion of the orchard a temporary exhibition space. Given the exaggerated population density of urban villages, every corner of Nantou should have been glutted with everyday life, begging the question: how was it possible that this neglected land with unclear ownership even existed? Why did it seem to be cared for, but the door was frequently shut? The big lock hanging on the orchard gate, the big dogs who

地呢？为什么看起来似乎有人管理而又常常大门紧闭？果园的大门挂着一把大锁，园内的大狗和墙上极其严厉的警告标语，让人不敢也无法接近（图2）。这个场景与摆放在这里的双年展作品都在各自顽强地保持着自己的姿态。

2　再识“果园”

双年展开幕两三个月后，我们持续对南头古城的在地艺术实验进行观察，这也为更好地接近和了解这片“果园”提供了机会。透过“交织与介入”工作坊期间与当地居民建立起的友好关系，我们找到了果园的两位“园主”：肖伯与叶太。

没有想到的是，与肖伯、叶太的接触和沟通竟然出乎意料的顺利，他们并不是事先想象的那样不可接近，而是两位和蔼可亲的长辈。通过多日的交谈，我们了解到：肖伯是本地的城市居民，过去是一位建筑工作者，从小在南头古城生活，对于自己生活的地方有着强烈的认同、热爱和责任感，非常希望这里能变成一个美好的地方；叶太则是本地村民，在南头有宅基地，也分到了回迁房，生活无忧，但在不久的将来会搬离南头古城，去城外的新房过年，目前主要是肖伯在打理果园。[1] 事实上，他们两位完全不需要依靠果园来获得经济上的收益，更多地是把这里当成劳动、消遣的处所。他们自愿的付出更多源于对过往田园生活的牵挂。他们并不是这片果园的所有者，当年只是因为住在附近，不能接受空地上满是垃圾的现状，想改善一下环境，才主动清理垃圾，并种上了香蕉、龙眼、黄皮果和木瓜树等。每年三月，他们甚至还会在果园里种上白菊花。但他们也苦于果树经常被外人破坏性地折枝、采摘，特别是果实成熟的季节，弄得果园一片狼藉。为了保护果树，他们不得不把果园围蔽起来。事实上，从肖伯的表述里也能感受到一种失落的情绪：“管不了了，以后怎样我也看不到了。”很显然，权属问题使管理变成一个十分敏感的话题。然而，在交谈的过程中，我们注意到一个细节——肖伯指着一棵大树说：“看！那棵树是我种的，已经十年了。”脸上抑制不住地流露出自豪的神情。这让我们突然想起《果园城记》里的描写，能让果园园主真正气愤的并不是果实被摘取，而是他们辛勤劳作的成果遭到破坏，他们不仅自己享受劳动过程中获得的成就感，其实也更期待周围邻居的认可和赞美。于是我们尝试性地问这个果园是否可以给大家使用，肖伯与叶太非常大度地表示，他们也愿意这里变成一个供大家活动的果园，他们甚至可以在开放之后继续义务打理、维护这些果树。

1　南头古城过去是宝安县的政府所在地，所以城里的本地居民有城镇人口，也有农村人口。

lived there, and the extremely stern warning signs on the wall—it all made the orchard inaccessible (Figure 2). This landscape and the Biennale exhibits placed therein tenaciously maintained their individual postures.

2　Re-cognize the "Orchard"

Two to three months after the Biennale, we were still regularly observing our local art experiment in Nantou Old Town. This allowed us to better approach and understand this "orchard." During the "Interweaving and Intervention" workshop, we established friendly relations with local residents and successfully found two of the orchard's voluntary caretakers, Uncle Xiao and Mrs. Ye.

Our conversations with Uncle. Xiao and Mrs. Ye went surprisingly well. They were not as inaccessible as we had previously imagined, instead, they were two amiable elders. Over many conversations, we learned that Uncle Xiao was a local urban resident who had lived in Nantou Old Town since childhood and that he used to be a construction worker. He was attached to and felt a strong sense of responsibility for the town, hoping that it would become a beautiful place. Mrs. Ye was a local villager who had a homestead in Nantou as well as compensation housing. She lived a worry-free life, and planned to leave Nantou Old Town before New Year, when she would move to a new home outside the town. Uncle Xiao was the orchard's primary caretaker.[1] Both regarded the orchard as a place for exercise and recreation, rather than as a financial resource. A nostalgia for past rural life motivated their voluntarism. They were not the owners of this orchard. Instead, they lived near the vacant lot, and couldn't stand the accumulated garbage. Consequently, they took the initiative to improve the land, cleaning up the refuse and planting bananas, longan, wampee and papaya trees. They even planted white chrysanthemums in the orchard every March. However, they were troubled by outsiders damaging trees and picking fruit, especially when the fruit was ripe, making a mess of the orchard. They enclosed the orchard to protect the fruit trees. Uncle Xiao said with a sense of loss, "I can no longer manage it and I won't be able to see what it will become." The ownership issue made the management of the orchard a very sensitive issue. However, during a conversation, we noticed a detail when Uncle Xiao pointed to a big tree and said, "Look! It has been ten years since I planted that tree." He looked very proud and was unable to hold it back. This reminded us of the description in *The Story of Orchard City*. What made orchard owners angry wasn't that the fruit had been picked, but that their hard work had been sabotaged. They not only

1　Nantou Old Town used to be the location of the government of Bao'an County, therefore, the local residents in the city include urban residents as well as rural residents.

故事进行到这里，所有的情境要素让我们更加萌发了行动的兴趣。公共场地的空置遗弃、环境的不断恶化、肖伯与叶太的自发打理、空地使用权的无形转移、居民的漠不关心、更加孤立和冷漠化的环境对峙，几乎就是一个以非典型方式存在的典型标本，一个失败或者消极的公共空间的标本——在南头这个土地资源极度紧张的地方，一个可以见缝插针演绎生活的地方，出现这样的失败和消极实在是令人感到匪夷所思的现实——而公共空间的缺失恰恰是我们今天的城市与社区最需要直面的问题，这也正是本次双年展以“城市共生”为题，需要着力探讨的目标之一。

3 “果园会”

只有具备公共性的空间才是真正的公共空间，这已经是一个明确的学术共识。问题是公共性如何才能抵达？在哈贝马斯的“公共领域”概念里，“公共”就是向携带各种话语的人们开放，并形成一种众生的喧哗。而众生的喧哗最直接的理解正是我们日复一日、最鲜活也最平淡的日常生活。真正实现一处具有公共性的公共空间的建构，决然不是一项常规的建筑或者景观设计，从图纸到施工的过程。在博伊斯之后，新的艺术实践提示我们，艺术、设计或者艺术行动已经发展成为社会关系建构的积极力量，同时也深刻改变了艺术与设计实践的语言形式与操作路径。

肖伯与叶太的南头果园作为一个事实上的公共储备用地，在其还没有开发之前，如果不是像开发商那样有意识地去营造良好环境、具备营销的经验，这样的弃置可能是一种常态。肖伯与叶太自觉自愿建设果园的行为创造了良好的环境，提示并促进了周围居民对此处环境的珍惜与关注。现在的困惑在于公共权属与现实管理边界的矛盾，在于个人选项（肖伯与叶太对过往田园生活的牵挂）与公共生活无法达成共振。然而，南头果园的存在其实也提示了一种可能，如果在此处建构起基于日常生活的、开放的公共性，平衡好植入日常生活的各参与方的劳作、管理与投入，是有可能创造出一处积极的公共场所的。

我们理解的社会交往方式，其实是公共形式的场地成为公共场所最为核心的要素之一。而对社会交往的关注、介入与推动等一系列行动亦是一种艺术形式与艺术实践。我们期望有一次实验，通过一组行为与活动的设计，调动参与和创造链接，有可能触发参与各方的彼此认同，进而萌生对当下共同生活状态与空间的共情，创造出较为积极的社会关系，使肖伯与叶太的果园真的变成南头的果园，如同《果园城记》里的“果园城”。

设计一次促成交往的聚会，关键在于要设计怎样的社会交往方式，如何达成交往的效果

enjoyed the sense of accomplishment they gained from their labor, but also anticipated the recognition from and praise of their neighbors. Therefore, when we tentatively asked whether this orchard could be used by the public, Uncle Xiao and Mrs. Ye agreed very generously. They also wanted it to become an orchard for everyone to enjoy, volunteering to maintain the trees after the orchard was opened up to the public.

At this point in the story, all the circumstances made us even more interested in taking action. A public space that was left abandoned and continuously deteriorated, the spontaneous management of Mrs. Ye and Uncle Xiao, the invisible transfer of use rights to the vacant land, the indifference of residents, and the more isolated and indifferent environmental confrontation—all these factors made the orchard an almost typical example of the atypical, an example of a failed or negative public space. In a place where land resources were scarce like Nantou, and where people used every bit of space just to live, this failure was an unimaginable reality. Indeed, the lack of public space is precisely the problem that our cities and communities need to face. It was also a pressing issue for the biennale, *Cities, Grow in Difference* to explore.

3 "Orchard Gathering"

Only through publicness, which is defined as the provision of open and accessible space is truly public. The question is how can this publicness be achieved? In Jürgen Habermas's theory of the public realm, publicness is defined as being open to people who engage in various discourses and form a cacophony of voices. The cacophony of voices is interpreted as our daily life, which is the vivid yet plain everyday life. The realization of a real public space does not entirely lie in conventional architectural or landscape design, nor in the process from drawing to construction. Since Joseph Beuys, new art practices remind us that art, design, or art action have developed into a positive force for the construction of social relations, and they have also profoundly changed the language form and operation path of art and design practices.

In fact, Uncle Xiao and Mrs. Ye's orchard in Nantou was reserved public land. Before its development, and in the absence of a developer to deliberately improve the environment for profit, this kind of abandonment might be normal. When Uncle Xiao and Mrs. Ye voluntarily planted an orchard, making the environment more pleasant, they also reminded nearby residents to value and cherish the environment. The current confusion lies in the contradiction between public ownership and actual management, as well as the inability to bring personal choice (Uncle Xiao and Mrs. Ye yearn for past rural life) and public life into accord. However, the existence of

图 3 果园聚会召集
Figure 3 Call for Orchard Get-together

与目标，以及以何种手段促成与放大这一过程。生活在南头古城里的居民来自不同的省份，从东到西，从南到北，是真正的五湖四海。这是深圳、也是这个时代的显著特征——空前的人口流动，空前的城市膨胀，每个新移民家庭的背后都有一个故乡，田野与果园的存在都还是不算遥远的记忆。我们希望果园以一种特别的方式被标注出来，从而显示其在日常生活的存在与意义。既标注出南头这个地方果树以及南国果园在过往的年代里曾经的普遍存在，也想让这些农业时代的印记和生活与在这里的每个人的故乡记忆与梦境联系起来。我们还希望每个人、每个家庭原生的故乡记忆与特征能够被标注出来，并且在快速流转的日常生活中保持某种清晰的、可以辨认的姿态。如果可能，也给孩子们未来一些可以回忆、思考的片段，能在他们今后的成长中有意无意地衔接故乡与此地、过去与现在，有新的认知和态度，去定义和建构属于他们新的价值与生活。

我们的计划是邀请十个家庭在果园举行一场别致的邻里聚会，时间选在“五一”长假的第一天。先是通过海报宣传、登门邀请和社区招募（图3），在候选家庭中确定十个被邀请的家庭，他们分别来自不同的地方，同时被邀请讲述一个关于自己家乡的故事或者习俗，每个家庭还需要制作一份具有自家特色的水果甜点与大家分享。我们会给予一些如何突出特色的建议。肖伯与叶太也要精心准备，分享有关南头特别是果园的种种趣事，包括南国果树的种植经验。因为是“五一”劳动节，我们还设计了一个关于劳动的仪式。甜点获得好评最多的三个家庭有机会为果园种下一棵果树，但需要孩子们和爸爸妈妈一起合作，他们或许在以后更有机会和肖伯交流果树种植的经验。活动的筹划从 2018 年的 4 月底开始，其间获得南头古城党群服务中心、村委会和南头股份公司支

Nantou Orchard suggests a possibility. If an open publicness based on daily life is constructed, and the labor, management, and investment of all participants in daily life are balanced, it is possible to create an active public place.

In our understanding, social interaction is one of the core elements necessary for a place to become public. Concomitantly, a series of actions such as paying attention to, intervening in and advancing social interactions are an art form and practice. We hope that there will be an experiment in which we, through the design of a set of behaviors and activities, mobilize participation and create links to possibly trigger the mutual recognition of all parties involved, making empathize with the current state of living together as well as the space, so as to create more positive social relations. In this way, we hoped that Uncle Xiao and Mrs. Ye's orchard could become Nantou's orchard in actuality, just like "Orchard City" in *The Story of Orchard City*.

The key to designing a get-together to promote interactions lies in three aspects, i.e. what kind of social interaction methods should be designed, how to achieve effective communication, and what methods should be adopted to facilitate and amplify this process? Many of Nantou's residents came from different provinces across the country. Unprecedented population flows and urban expansion are notable features of both Shenzhen specifically and this era more generally. Behind every migrant family, there is a hometown, and the fields and orchards are still fresh memories. We hoped that the orchard could be marked in a special way to show its existence and significance in daily life, identifying the orchard in Nantou's past and in the history of South China, and connecting the imprints of the agricultural era with the memories and dreams of everyone who came to live there. We also hoped that the original hometown memories and characteristics of each person and family could be highlighted, and a clear and recognizable posture could be maintained in the fast-moving daily life. If possible, we also wanted to provide children with reminiscences, thoughts and fragments that intentionally and not connect their hometown with this place, as well as the past and the present. These new perceptions and attitudes might allow them to define and build new values and lives that belong to them.

We planned to invite ten families to hold a special neighborhood get-together in the orchard, selecting the first day of the Labor Day holiday (May 1). We first confirmed ten invited families from the candidates who responded to our posters, door-to-door invitations, and community recruitment (Figure 3). They came from

持，驻扎在南头的共创组织“踏实玩乐”也提供了积极的建议。最重要的是得到两位园主——肖伯与叶太的认可与全力协助。肖伯负责搭建用于防晒、防雨的临时遮阳棚，包括准备活动所需的水电供给；叶太主动提出由她提供活动使用的桌椅等。果园的围墙是为保护果树而设，事实上影响了果园的公共使用，但现实中又不可能一下子拆掉。我们和肖伯商量，特别用他们种的白菊花装饰了大门，尽量让这个边界显得柔和、开放。虽然活动在下午两点半才正式开始，好些妈妈都提前到达，并在现场进行更细致的甜点加工。大家的交流出乎意料的热烈，以至于事先为活动设计的详细流程都显得多余（图 4- 图 5）。聚会和劳动让果园的午后时光变得惬意温暖，热闹非凡，有位妈妈惊喜地感慨：这里和我们老家一样呀。整个“果园会”活动其实很简单，也就是半天的聚会。虽然组织的过程比较费周折，但活动本身就是一个创作的过程，而这个过程确实唤起了居民难得的生活热情。居民们发现了他们的共同话题和经验，在深圳南头这个城中村的“异乡”重新拾起有关“故乡”的共同记忆与情绪。在这个过程中，艺术隐藏到具体的组织与行动当中，似乎没有了形式，但仍然是可以触摸的力量——“生活就是艺术，艺术就是生活”。

大约 6 个月后，我们又做了一次回访，正赶上肖伯与叶太收香蕉，剪蕉叶、晒蕉叶、烧干蕉叶做肥料，忙得不亦乐乎。活动时的临时棚架已经由肖伯拆回家，上面遮阳的蕉叶也已晒干。三个家庭种植的果树苗被肖伯用护具精心保护起来，看起来很精神。肖伯解释说这段时间没有下雨，根还不够牢，只要下雨了，这些果树苗就会越发结实。看得出肖伯对这些树苗还是非常用心也非常有信心的。果园也有了一些小小的变化，邻里的探访和对环境的共同呵护在“果园会”之后正在变成一种日常。

图 4 果园聚会
Figure 4 Orchard Get-together

different places and were invited to tell a story about or explain a custom from their hometown. Each family also needed to make a traditional family fruit dessert to share with others. We made some suggestions about how to highlight the dessert's family characteristics. Uncle Xiao and Mrs. Ye also made careful preparations. They shared various interesting stories about Nantou, especially about the orchard, including the experience of planting fruit trees in South China. Because it was Labor Day, we also designed a labor-related ritual. The three families with the most popular desserts had the opportunity to plant a fruit tree for the orchard with the stipulation that the children and their parents plant the tree together. Afterward, they might have more opportunities to exchange fruit-tree-planting experiences with Uncle Xiao. We started planning the event at the end of April 2018, and received support from the Nantou Old Town Party and Mass Service Center, the Village Committee and the Nantou Stock-holding Corporation. The Nantou-based co-creation organization "TAS" also offered positive suggestions. Most importantly, we got approval and full assistance from Uncle Xiao and Mrs. Ye. Uncle Xiao set up temporary awnings and prepared water and electrical lines for the event. Mrs. Ye volunteered to provide tables and chairs. Although the orchard walls were built to protect the fruit trees, they hindered public use of the orchard. Moreover, they couldn't be unilaterally removed. After a discussion with Uncle Xiao, we decorated the gate with the white chrysanthemums that they had planted, trying to soften and open up the interface.

Although the event officially started at 2:30 p.m., many mothers arrived early and finished their desserts onsite. Conversations were unexpectedly effortless, making the detailed schedule that was designed in advance seem unnecessary (Figures 4-5). The gathering event and working together made the afternoon in the orchard cozy, warm and lively. A mother sighed with surprise, "This is just like at home." The entire "Orchard Get-together" was a simple, half-day gathering. Despite the complicated organization process, we believe that the event itself was a process for creation, and this process piqued the residents' rare enthusiasm for life. Residents discovered common topics and experiences, reclaiming shared memories and emotions about their "hometown" in this "foreign land"—the urban village in Nantou, Shenzhen. In this process, although the art was formless, hidden in the organizational process, nevertheless it was still a tangible force—"Life is art, and art is life."

About 6 months later, we paid a return visit. When we arrived, Uncle Xiao and Mrs. Ye were busy harvesting bananas, cutting and drying banana leaves, and burning dried banana leaves for fertilizer. Uncle Xiao had removed and taken home the temporary awnings used during the event, while the banana leaves on them

4 结语

离开果园的时候，我们特意再次仔细地打量了一下周围的环境：果园周边有三栋烂尾楼，而且是烂尾了近二十年，其衰败的模样是大家可以想象得到的。四面密集的城中村建筑几乎将整个果园围得严严实实，围墙与铁网让果园的边界关系看起来还是如此杂乱不堪。但“果园会”确实让大家开始在这个地方接近彼此，尽管不是物理空间上的开放，但开放性与公共性的重建似乎可以在社会关系的层面取得突破。

图 5 果园聚会种植劳动
Figure 5 Planting at Orchard Get-together

春景街果园的局面事实上是当下中国城市中许许多多消极空间的缩影。急剧的发展对城市空间的连续性来说无疑是一种破坏，制造了大量的空间断裂，也留下许多被遗忘、被弃置的空地废墟。它们在权属明确或者被开发利用、合理使用之前，常常蜕变为城市的疮口或病灶。在这类问题面前，我们需要认识到，城建手段既不是可以立即启动的办法，也不是唯一的解决问题的方式，其路径与结果都过于单一片面，无法匹配城市的鲜活与多样性。包括艺术实践、设计介入以及人类学的角度和方式在内的各种弹性介入手段反而是可以借鉴和操作的路径。围绕存量提升的城市更新更应该跳出从图纸到施工的城建思维。

正如尼古拉斯·伯瑞奥德在《关系美学》一书中提出的那样：“艺术活动在于努力实现一些有限的串联、打通一些受阻的通道、重新让现实中被隔开的各种层次有所联系。”[4] “果园会”计划只是一场短暂的艺术实验，或许并不能彻底改变什么，但活动过程让我们发现了

were dried out. He had also carefully protected the fruit tree saplings that the three families had planted, and they looked thriving. Uncle Xiao explained that there had been no rain recently so the roots were not strong enough. However, all it would take was one good shower and the fruit trees would grow stronger. It was evident that Uncle Xiao not only took good care of the saplings, but also that he was very confident in their future. There were also some small changes in the orchard. Neighbors routinely made daily visits to the orchard, showing care for the orchard and its environment in the aftermath of the "Orchard Gathering."

4 Conclusion

Before leaving the orchard, we once again took a close look at the surrounding environment. There were three unfinished buildings near the orchard, and they had already been left unfinished for nearly two decades. The decline was just as you might imagine. On all sides, the orchard was surrounded by dense urban village buildings, and walls and iron netting made a mess of the orchard's edges. However, the "Orchard Get-together" did facilitate interactions in this place. Although it is not physically accessible, the reconstruction of openness and publicness seems to have allowed for breakthroughs in social relations.

The Chunjing Street orchard is actually a miniature of many negative spaces in current Chinese cities. Rapid development undoubtedly destroys continuities within urban space, has created a large number of spatial fractures, and left many forgotten and abandoned vacant ruins. Before property ownership has been clarified or before they have been rationally developed and used, these spaces often degenerate into urban sores or lesions. In the face of these problems, we need to realize that urban construction is neither a method that can be initiated immediately, nor is it the only way forward. Urban construction is a singular path that produces homogeneous results that are incompatible with the vitality and diversity of the city. Art practices, design interventions, anthropological perspectives and other flexible intervention methods can be adopted to clear the way forward. With respect to the renewal of extant housing stock, there is even more need to think outside the mental box of drawing-to-construction city-building.

As Nicolas Bourriaud put it in *Relational Aesthetics*, "Art activities are to strive to achieve limited connections, open up blocked channels, and reconnect various levels that are separated in reality."[4] The "Orchard Get-together" was a temporary art experiment that may or may not been able to completely change anything, but the event itself was a process that allowed us to discover the power of sharing. The moment we facilitate public sharing is precisely the process of creating publicness. Reality

分享的力量。我们在促成公众进行分享的时刻，其实正是创造公共性的过程。看起来糟糕的现实未必像其表面呈现的那样简单粗暴，各种复杂的原因造成了现实诸多的断裂与误会，只要回到日常生活本身，去发现和引导生活内在的创造性，改变就有无限的可能。

is neither as terrible nor as simple as it appears, but rather complex causes have led to ruptures and misunderstandings of seemingly terrible situations. As long as we return to daily life itself, discovering and guiding the inherent creativity of life, we believe that there are infinite possibilities for change.

[1] 师陀 . 果园城记 [M]. 北京 : 解放军文艺出版社 , 2000.
[2] 师陀 . 海上文学百家文库 · 师陀卷 [M]. 上海 : 上海文艺出版社 , 2010: 269.
[3] 侯瀚如 , 刘晓都 , 孟岩 . 策展人语 [EB/OL]. 2017 深港城市 \ 建筑双城双年展 . http://www.urbanus.com.cn/uabb/uabb2017/curatorial-statement.
[4] 伯瑞奥德 . 关系美学 [M]. 黄建宏 , 译 . 北京 : 金城出版社 , 2013: 3.

自从现代城市规划与建筑设计主导了城市治理，珠三角主要的城市化地区在现代化进程中产生各种激变。不论是在积极推动下产生，或是意外地发展出来，城中村一直作为一个复杂现象而普遍存在。这种看似复杂无序的现象背后隐藏的其实是一种长期维持社区和城市活力的运作网络系统。南头古村正是其中一个典型案例。

这个工作坊是一次期待已久的行动，有别于习惯性的宏观视野。从微观的角度入手，观察与分析生活情境与那些人们在日常生活中会遇到的复杂网络，例如片断的区块、共存的现象、求生战术、热闹的地景等等。通过仔细的辨认与深入探讨，寻求恰当的介入设计途径，依据现实条件进行具体可行的微尺度建造方案，通过五个组的足尺实验来思考南头社区空间多赢的改造方向。

院子组：一个场所如何同时满足私人摊位和公共庭院的需求？透过编织的手法，利用摊位常用材料——帆布，重新组织一种空间视觉秩序，织一面墙、一张椅子、一座灯饰、一幅画；并在入口置入一个能翻转的门牌，以正反两面的“巷子”及“院子”暗喻这个空间在不同时间段的不同属性，试图将一个原本只是“通道”的消极空间，改造成一个人们能停留、参与的积极空间，同时提升公共空间品质，进而为此处摊贩制造商机。

游戏组：南头村可以成为游乐园吗？那些具有南头典型的公共空间特征，如高低的阶梯、狭窄的建筑间隙、起伏的坡地、废弃的城市设施，都是可能创造游戏空间的组成部分。通过制作一份南头村游戏地图和几个具体的游戏介入案例，引导孩子们去发现和创造属于南头孩子们的游乐空间。

梧桐组：梧桐街为什么没有梧桐树？像城中其他街道一样，由于城市的变迁，找不到原来的印记。我们试图去城市的空间中寻找这些带有印记的角落。这角落如同一幅幅画面，而诗是建立共同认知的简明符号。于是我们用诗词加强角落的场景感和叙事性，给梧桐街种上“梧桐”，重塑街巷文化认同与社区归属。

椅子组：南头古城“充满”着被放在外面的椅子，这些椅子虽然破旧却充满活力，但它会促进空间变的更加积极，有时候它会成为城中临时的“活动中心”。于是我们用颜色和懒人夹加强椅子的意义，使其公共性得以体现，为本地居民与外来人口搭建起交流的桥梁。

东二门组：进入“城门”是进入南头古城的一个特别具有仪式感的行为。在南头古城的现状中，除了原来的东门和南门以外，还有一个因便捷而高流量进出的无名巷道，通过命名和相关视觉界面的建构，把“东二门”的概念赋予这个小而重要入口，塑造出南头古城的第三座“城门”。

▷展览：交织与介入——南头社区微观介入行动工作坊 ▷参展人：沈康、萧有志、王铬、李芃、黄全乐

Since the idea of modern urban planning and architectural design has dominated urban governance, the Pearl River Delta's major urbanization areas have been experiencing enormous changes throughout the process of modernization. Under this circumstance, the village in the city has been crucially prevalent as a complex phenomenon, actively promoted or accidentally developed.

Hidden behind this complex disorder phenomenon, there is a long-term network operation system to maintain the vitality of communities and cities. With the importance of that system, Nantou Old Town serves as an example to study.

This workshop is a long-awaited activity for Nantou Old Town. Different from the habitual macro vision, it starts from the microscopic point of view, which focuses on the observation and analysis of the living situation of those villagers in their daily lives such as segments of urban blocks, the coexistence of the phenomenon, survival tactics, lively scenery and so on. It seeks an appropriate intervention design approach through careful identification and in-depth discussion, According to the actual conditions of the specific feasible micro-scale construction program, It thinks of the Nantou Village space win-win transformation direction through full-scale experiment.

▷Exhibition: Mega-weaving & Micro-intervention—2017 GAFA-SCU Joint Workshop in Nantou Old Town ▷Participant: Shen Kang, Xiao Youzhi, Wang Ge, Li Peng, Huang Quanle

来源：深圳城市\建筑双年展组织委员会 . 城市共生：2017 深港城市\建筑双城双年展 [M]. 广州：岭南美术出版社，2021: 342.

一场展览背后：来自深双的田野笔记

Behind an Exhibition: UABB Fieldnotes

陈嘉盈
纽约州立大学宾汉姆顿分校人类学博士生，2017年深港城市\建筑双城双年展（深圳）展览助理。研究方向为城市人类学，关注当代中国政治经济，空间政治、城市更新、创意产业及设计治理。
本文首次发表于“URBAN MATTERS”网站，2018年；作者为本书补充，2021年；由莫思飞中译英；图片 ©UABB。

Chen Jiaying
Chen Jiaying is a Ph.D. student of anthropology at SUNY-Binghamton and was an exhibition assistant of the 2017 Bi-City Biennale of Urbanism\Architecture (Shenzhen). With a specialization in urban anthropology, her research interests include political economy, spatial politics, urban renewal, creative industry, and design governance in contemporary China.
The article was first published on website "URBAN MATTERS," 2018; amended by the author for this publication, 2021; translated from Chinese to English by Mo Sifei; image ©UABB.

如往常一样，我又来到南头古城，开始一天的工作。

穿过明代修建的古城门洞，迎面而来的景象熟悉又让人惊奇：紧密排列的握手楼、卖着五湖四海食物的小店、青砖瓦顶的老建筑旁散落的艺术装置、颜色缤纷的壁画、热切的游客……

我沿着中山南街走，碰见了曾在村里当会计的李伯，他告诉我自己的那栋改成民宿的高楼里今天住满了艺术家和建筑师；远远又走过来广场舞队的珍姐，刚从中山公园跳舞回来，冲着我笑说感谢我邀请她参加开幕式；路过的湛江牛杂店，经过设计师改造已经被粉刷一新，老板娘向我招手，要给我一碗今天新鲜的牛杂。

我感觉自己已经是这个社区的一分子，没有人不认识我……

1　展览：城市共生

2016年开始，我的工作场地搬到南头古城，第七届深双将在这里举行为期三个月的展览，而我负责和社区的各个群体打交道，传播双年展的信息，协助展览顺利进行。

深双素以主题的先锋性和概念性著称，这届的双年展以“城市共生”为题，批判当今中国城市空间及城市生活日趋单一和同质化的现象，呼吁“共生”的城市生态，提倡对差异性和多样性的包容。为了更好地说明这种共生的图景，“城中村”作为一个样本被推到幕前仔细端详。

除集合数百件作品，从不同的角度对城中村乃至更宏大和抽象的对象进行观察、研究和

As usual, I arrive in Nantou Old Town to start a day's work.

Passing through an ancient gate that was built in the Ming Dynasty, I come upon a scene that is simultaneously familiar and surprising: there are densely arrayed handshake buildings, small shops selling food from all over China, art installations scattered next to old buildings with blue brick walls and tile roofs, colorful murals and eager tourists...

Walking along Zhongshan South Street, I ran into Uncle Li, who used to work as an accountant in the village. He tells me that his tall building has been converted into an Airbnb. Today, it is full of artists and architects. Sister Zhen, a member of the Square Dance team walks over from a distance. She is just back from dancing in Zhongshan Park, and she smiles at me, offering thanks for inviting her to the opening ceremony. On my way, I pass the repainted Zhanjiang Beef Offal Shop, which has been remodeled by a designer. The lady boss beckons me over to give me a bowl of fresh beef offal.

I feel that I am becoming a member of this community, and everyone knows me...

1　Exhibition: Cities, Grow in Difference

Since 2017, my job has been based in Nantou Old Town. The 7th UABB will hold its three-month exhibition here, and I am responsible for communication with different groups in the community, promoting information regarding the UABB and assisting in the preparation of the exhibitions.

想象以外，策展人更把展览放置在一个真实的城中村里，使得城中村本身成为一个立体生动的展品，并把展品分布于城中村的各个角落，实际介入社区，进行“城市共生“实验。

“城市共生”的理念表达了对城中村及其所代表的草根、边缘、弱势群体的智慧、创造力和能动性的尊重及认同，并主张消除意识形态、知识和审美标准的等级秩序，这在资本和权力挂帅、城市生活及价值观日趋同质化的当今时代非常具有启发性。但是，光有理念还不足够，关键要落到实践上。

真实的城市里有各种各样复杂和多样的情况，这些理念将如何贯彻到实践中?

2 实践：城市如何共生？

正如2017深双展前册介绍的：“城中村实验场是一场城市介入实践，它涵盖了正在实施的南头更新计划以及策展人委托的户外空间装置、壁画，还有通过公开征集鼓励设计师介入古城环境改善、店面提升、民居改造的一系列实施计划，既有自上而下的策划和整体的把控，也有渐进式、自发的小介入与微更新。”

“城中村实验场”是2017深双三大板块中“都市|村庄”的一个栏目，不满足于单纯的观察、研究、议论和想象，而是通过介入社区，直接作用于人、物、空间，以获得经验和反馈。

自2016年开始，我开始参与“城中村实验场”的协调工作，深入实践的第一现场，一方面发挥了我人类学的田野功力，另一方面也为我的人类学研究提供了宝贵的素材。

通过这些实践，我们得到什么经验?

3 饕餮盛宴——观念

装置是“城中村实验场”的一种介入方式。意大利壁画师Hitnes的作品《饕餮盛宴》位于一栋临街建筑的外墙，绘有他想象中的动物和美食（图1）。作品说明中提到，他让这些绘画“超越文字的描绘，完美呈现的同时与观者进行有趣互动”。

壁画绘制中的第二天，我经过现场，看见几位老奶奶议论纷纷：“画两只这么凶的怪兽在村中央，这是要破坏风水哪！”我抬头观察，只见墙上画着两只绿色的动物，似狮子又像豹子，神情确有几分凶悍。

老奶奶们情绪有点激动。根据她们的说法，狮子是她们村的不祥之物，过去村门口曾放着两只石狮子，村里就开始噩运连连，直到把石狮子移开，村里才恢复安宁，因此村民都不喜欢狮子。

我急忙跟壁画师沟通，壁画师一脸惊讶：“我画的是貔貅，难道不是中华文化中的吉祥之物?”我哭笑不得，一番解释后，Hitnes很诚恳，也理解村民的想法：“非常抱歉带来了

Known for its avant-garde and conceptual nature, this UABB is titled, *Cities, Grow in Difference*. It takes a critical stance toward the increasingly unitary and homogeneous phenomenon of urban space and life in today's China, calling for a heterogeneous urban ecology to promote diversity and inclusiveness. To better illustrate the vision, the "Urban Village" as a paradigm has been pushed to the forefront of the exhibition to promote careful examination. In addition to collecting hundreds of works and observing, researching as well as imagining urban villages or broader abstractions from different perspectives, the curator has inserted the exhibition in a real urban village, making the urban village itself a three-dimensional vivid exhibit. Exhibits are scattered in every corner of the urban village while conducting *Cities, Grow in Difference* experiments through community intervention.

The concept of "Cities, Grow in Difference" respects and recognizes the wisdom, creativity and initiative of the grassroots, marginalized, and disadvantaged groups in urban villages, and advocates the elimination of the hierarchical order of ideology, knowledge and aesthetic standards. In this era in which capital and power are in command and urban life and values are increasingly homogenized, the exhibition is inspiring. However, ideas are not enough and the key lies in practice. The reality of an urban village is complex, begging the question: what kinds of practices will allow are concepts to be implemented?

How can cities grow in differences?

2 Practice: How can cities grow in differences?

As described in the catalogue of the 2017 UABB, "The 'Urban Village Laboratory' is the ongoing practice of urban intervention. It includes various projects, like the Nantou urban regeneration plan as well as outdoor installations and murals commissioned by the curators. It also includes UABB's open-call initiatives to improve several outdoor spaces and storefronts in the old town, and to renovate some residential buildings. The top-down strategic planning and design process is complemented by self-initiated and spontaneous micro-interventions."

The "Urban Village Laboratory" is in one of three sections comprising the "Urban Village" section of the UABB. It goes beyond observation, research, discussion and imagination, directly affecting people, things and space by intervening and working in the community to gain experience and feedback.

I have begun to take part in coordinating the "Urban Village Laboratory" and have practiced at the site since 2016. While my anthropological field study skills have been well used, the experience has also provided

图 1 饕餮盛宴 / 广场，Hitnes
Figure 1 The Feast / The Square, Hitnes

误会！不过你们请的是来自欧洲的壁画师，你懂的，我们的作品不是循规蹈矩的，要有点不合常规的冲击力。当然，村民的感受我们一定要尊重，村民会喜欢什么动物呢？鸟？猫？鱼？墨鱼？”

一番沟通后，我下午再到现场时，发现狮子已经变成了凤凰。“这个可以吗？”Hitnes站在脚手架上冲我喊。“这个很好！Very good!”听到我的回应，Hitnes露出满意的笑容，埋头继续用颜料掩盖貔貅的痕迹。

在这个介入试验中，壁画师的创意和社区使用者的观念发生了冲突，原因是壁画师不了解社区居民的宇宙观和社区史。这些地方性知识和情感结构，是由地方的文化、历史、价值观、利益以及由此形成的立场和视域塑造的。

了解及尊重地方性知识和文化，才能让社区介入的实践不是一厢情愿的好意，而是成为有效的行动。

4 商铺新形象——需求

“商铺新形象”是另一个对城中村本身已经存在的形态进行介入实践的项目，通过公开征集，鼓励设计师为城中村里的商铺进行设计改造，解决空间狭小、流线拥挤、店面招牌千篇一律，缺乏吸引力等问题。留德归来的室内设计师吴冠雄报名参加商铺改造项目并选中“锦记商店”作为改造对象。吴冠雄最开始的想法是做一个有设计感的杂货铺，经过和商店老板几次深入交流之后，却陷入反思。

对商店老板廖姨来说，最迫切的需求是翻新屋顶，老房子年久失修，瓦顶随时有坍塌的危险（图2）。而这个房子是老房东为了照顾她家庭变故，多年来低价租给她的。为了报答房东对自己的恩情，廖姨希望把房子修好，以后不做商店了把房子还给房东。而由于城中村改造的敏感性，房子的修整动工总是迟迟无法通过报批。这次遇上双年展这个“千载难逢的

valuable materials for my anthropological research.

What experiences are we gaining through these practices?

3 The Feast—Concept

Installation is one of the approaches to urban village intervention in the "Urban Village Laboratory." The Italian muralist, Hitnes has painted *The Feast* on the facade of a street-facing building, adding imagined animals and delicacies (Figure 1). In his description of the work, these paintings can "go beyond a literal representation and play with our perception."

During the second day of the painting process, I pass by and see several old grannies discussing the painting: "Drawing two fierce monsters in the middle of the village will destroy Feng Shui!" I look up and see two animals in green resembling a lion if not a leopard, and looking a bit fierce. The grannies are anxious. According to them, lions are ominous in their village. In the past, misfortune started in the village when two stone lions were placed at the village gate. It was not until the stone lions were removed that peace was restored. Therefore, the villagers dislike lions.

When I communicate with the Italian muralist, he looks surprised and says, "I am painting a Pixiu (a Chinese mythical hybrid creature). Isn't it a powerful protector in Chinese culture?" I don't know whether to laugh or cry and explain. Hitnes is very sincere and understanding of the villagers' position. He says, "I am very sorry for the misunderstanding! But you have hired a muralist from Europe. And as you know, our work does not follow the rules in order to achieve unconventional impact. Of course, we respect the feelings of the villagers. What kind of animals would they like? Birds? Cats? Fishes? Cuttlefishes?" When I arrive at the site in the afternoon after our talk, the lions have turned into phoenixes. "Is this OK?" Hitnes asks standing on the scaffolding. "This is nice! Very good!" Hearing my response, Hitnes smiles in satisfaction, painting over traces of the Pixiu.

In this experimental intervention, the creativity of the muralist and the idea of the villagers clashed with each other because the muralist didn't understand the cosmology of the residents and the history of the community. This knowledge and emotional structures have been shaped by local cultures, historical values and interests as well as the positions and perspectives established thereby. Community intervention can be effective only if local knowledge and culture are fully understood and respected.

好机会”，廖姨希望好好把握，翻新屋顶。

廖姨的诉求，与吴冠雄“做一个有设计感的杂货铺”的初衷相去甚远。他陷入思考：双年展为什么要改造商铺，他又为什么要改造商铺？改造商铺的行为到底是为了谁——是为了彰显他设计的功力，成为他个人宣传的作品？还是为了让店铺成为一个“有看点”的展品，吸引双年展观众？如果只是修葺屋顶，完工后的作品是不是毫无设计感？设计师到底是做什么，该为谁设计？

吴冠雄最后决定：为廖姨设计，不考虑美感、没有外在形式，做一个没有“设计师存在感”的案子。他做了很多看起来不是“设计”的事情：联系施工队进行瓦顶翻新，向城管局申请水表，对店内商品进行分类，摆放店内原有的街坊邻居经常来聊天坐的小凳，在门面上挂绿植，重新设计手写的招牌（新招牌被施工队无意丢掉，和廖姨一起赶到垃圾场捡回来），帮忙装卸货架，打扫卫生，开张前夜亲笔书写了一副对联祝贺店铺重新开业（图 3）。

在这个过程中，吴冠雄不是一位乙方设计师，而是和廖姨建立了感情和友谊的朋友。他抛弃设计行业对“形式”的过分追求，并认为与“创造形式”比起来，“创造条件”更重要——设计的含义应扩充为对关系的协调和整理。在改造过程中，“设计师”被称为“项目协调员”：协调施工、协调与政府机构的沟通、协调商品的摆放、协调廖姨和邻居关系的维持。这使得他的角色更像是一位社工，这是设计师的升华；用设计为弱者赋权，是一种社会性的设计。

在改造过程中，吴冠雄对自己的“审美秩序”有了觉察。例如第一次看到由廖姨相熟的施工队长挑选的厕所门，磨砂玻璃上一朵鲜红的大牡丹花，他表示不能接受，后来想，这也许就是城中村审美，“谁说我认为好看的就好看，他们觉得好看的就不行？”

吴冠雄认为，改造的过程也是个互相学习的过程，廖姨的乐观和坚韧、房东的善良（工程进行到一半，房东结算了原本廖姨打算支付的工程款）、街坊邻里的人情，让他收获良多。廖姨也从吴冠雄那里学到了，她很喜欢吴冠雄在店门上添置的绿植，显得生机勃勃，给她带来了愉悦，“小吴你真是厉害！你是怎么想到的，

图 2 改造前的锦记商店
Figure 2 Jinji Store before renovation

4 New Image of Stores—Needs

"Store's New Image" is another project that intervenes in the existing forms of the urban village. Through an open call for submissions, designers have been encouraged to design and transform urban village shops to resolve issues such as limited space, crowded circulation and unattractive store signs. Wu Guanxiong, an interior designer who studied in Germany, answered the open call, choosing the "Jinji Store" as his site for renovation design. Wu Guanxiong's first thought was to build a special designed grocery store. Yet the conversations he has had with the shop owner have made him rethink the design.

For the shop owner, Auntie Liao, the most urgent need is to replace the roof, which after years of disrepair risks collapsing (Figure 2). Auntie Liao suffered a family calamity and subsequently the old landlord rented the shop to Auntie Liao at a low price in order to take care of her family. Auntie Liao wants to repair the shop before she retires, handing back an improved shop to the landlord to repay her kindness. However, obtaining approval for a renovation keeps getting delayed due to the tightening management of retrofitting in urban villages. Auntie Liao believes that encountering the UABB is the "great opportunity of a lifetime" and she hopes to make good the opportunity to renovate the roof.

Auntie Liao's demands are nothing like Wu Guanxiong's intention of a "special designed shop." As he ponders the UABB's objective in encouraging shop renovations as well as his own motivations for participating, he ends up asking the question, "who is the shop renovation actually for?" Is it a showcase of his skill and a future part of his portfolio? Is the goal to make the shop a "spectacular" exhibit that attracts the attention of UABB visitors? Will the project show no design thinking if only the roof is replaced? What exactly does a designer do and for whom?

Wu Guanxiong decides to prioritize Auntie Liao's needs instead of his aesthetics or the external form of the shop, making a project without the "presence of a designer." In the end, he accomplishes things that are not properly considered to be "design." He contacts the construction team to renovate the tile roof, applies for water meters from the urban management bureau, sorts out the goods in the store, and places small stools for neighbors to hang out. Also, he hangs plants by the shop entrance, redesigns a handwritten store sign (which was accidentally thrown away by the contractor and he and Auntie Liao rushed to the garbage dump site to retrieve it), helps load and unload goods from the shelves, cleans up, and writes a couplet the eve of the opening of the shop to congratulate Auntie Liao (Figure 3).

Instead of being a designer, Wu Guanxiong becomes friends with Auntie Liao during the renovation

图 3 改造后的锦记商店
Figure 3 Jinji Store after renovation

我就想不到，向你学习。”

在此，我们能不能将改造实践视为一种互动、共情、学习和赋权？当以这种方式介入每个人的生活，我们是不是能在同质化的时代里回忆起，生活的本质正是异质的，是由一个又一个独一无二的、有血有肉的人和他们的故事组成。我们要了解和尊重这些个体，并赋予他们力量，让他们得以发挥创造力和价值，这才是让人之所以成为人，而城市得以共生的秘密。

5　南头会客厅——汇聚

2018 年 1 月 26 日，双年展已开幕一个多月，在南头古城朝阳北街 1 号的房子里召开了“南头会客厅”社区组织成立仪式，汇聚了南头古城社区的行政管理者和多个深圳社会组织的负责人。设计师吴鸣宣布，“南头会客厅”社区组织正式成立，将扎根社区，致力于南头古城的社区营造工作，持续开展学术研讨、调研改造、教育体验及社区活动。

吴鸣也参加了商铺改造项目，结合房东的需求，他把一个 20 平方米的麻将馆改造成青年公寓。项目落成后，他发起了一个实验：在双

process, rejecting the excessive pursuit of "formalism" in the design industry. Rather than "creating a form," "creating possibilities" is essential to extend the boundary of design to include the coordination of relationships. The designer becomes a coordinator during the process when he coordinates construction, communication with government institutions, goods displays and the relationships between Auntie Liao and her neighbors. It makes him more than a designer, rather he becomes a social worker who empowers marginalized populations through social design.

The renovation process has also made Wu Guangxiong aware of his subjective "aesthetic order." For instance, he rejects the big red peony on the bathroom door that Auntie Liao's contractor has selected. But on second thought, he now considers it part of the urban village aesthetic. "Why should my aesthetic override theirs?" he wonders.

Wu Guangxiong considers the renovation to have been a process of mutual learning. He has gained much from Auntie Liao's optimism and perseverance, the landlord's kindness (even though Auntie Liao was planning to pay for the project, the landlord paid the constructor when the project was halfway through), and the supportive neighbors. Auntie Liao has also learned from Wu Guanxiong. She likes the vibrant and pleasant vibe of the greenery by the shop entrance. "Xiao Wu, you are amazing! How did you come up with such an idea? I couldn't think of it. I shall learn from you."

Is it possible to regard the practice of renovation as a form of interaction, empathy, learning and empowerment? In such homogeneous times, is it possible to recall that the essence of life is differences formed by the unique and true stories of every individual when we intervene in everyday life? We need to understand and respect individuals, empowering their creativity and acknowledging their value. Only under these conditions do human beings become fully human and our cities grow in differences.

5　Nantou Living Room—Gathering

On January 26, 2018, UABB has been open for more than a month. The founding ceremony of the NGO, "Nantou Living Room" is being held at One Chaoyang North Street, Nantou Old Town, bringing together the administrative management of Nantou's community and

年展期间，他将开放客厅给所有人使用，实验一种社区客厅的功能，空间也命名为“南头会客厅”（图4-图5）。短短几周，许多社区的居民和来参观双年展的游客到访了这个空间，他们留在客厅里喝茶聊天，聚餐小酌。有一天，空间里来了两个从事交通规划的游客，他们想为南头古城制作一个交通梳理方案，但是不知从何入手。这启发了吴鸣和几位小伙伴：空间汇聚了这么多各具才能的人，是不是能把他们组织起来，为这个社区做些什么？于是他们准备成立一个组织，就叫“南头会客厅”。

机构还在筹备，第一个项目已经做了起来。南头会客厅隔壁是一个荒废的花园，几位小伙伴考虑，何不把花园修葺起来，变成社区的公共空间。四处寻觅，发现花园的主人已经去世了，他的子女没有办妥继承土地的手续，况且城中村禁止建设，花园也就一直荒废着。吴鸣他们找到了花园现在的管理者——物业主人的女婿叶伯，希望他同意把花园改造成公共空间。叶伯同意了，唯一要求是保留花园里他亲手种的几棵芭蕉树，而他的妻子则要求保留一堵倒塌的墙，以作为花园权属的证明。

修葺花园的想法得到很多人的支持，华南理工大学的明镜营造社为花园设计了一个改造方案，并得到两位深圳本土建筑师的点评和建议。另外，修葺花园需要经费。得知此事后，曾经来过南头会客厅的一位灯光设计师说服了自己的公司为花园免费提供户外灯具，解决了修葺花园的一部分经费。

于是围绕着这个花园，许多人和资源汇聚到一起。初试牛刀让南头会客厅的几位小伙伴有了信心：我们是不是能延续这样的模式，让

图 4 改造前的房子
Figure 4 Building before renovation

people from Shenzhen social organizations. Designer Wu Ming announces the formal establishment of the "Nantou Living Room." It will take root in the community and devote itself to building a thriving community in Nantou Old Town. Academic seminars, research and renovation, as well as experiential lessons and community events are to follow.

Wu Ming is also a participant in the shop renovation project. In accordance with the needs of the landlord, he transformed a 20-square-meter mahjong pavilion into a youth apartment. Upon the completion of the project, he initiated a social experiment. During the biennale, he opens up the living room for everyone, turning it into a community living room, which is called "Nantou Living Room" (Figure 4-5). Over a few weeks, both residents and visitors have visited the living room. They stay for tea, a chat, or a meal. One day, two visitors engaged in transportation planning came to the space. They hoped to propose a traffic planning scheme for Nantou Old Town, but they had no idea where to start. This moment inspired Wu Ming and his partners to establish a place where talents from different disciplines can meet up in a space to create different possibilities for the community. Based on this insight, they decided to set up the NGO, "Nantou Living Room."

Even as they are preparing to establish the organization, their first project is underway. The partners propose refurbishing and transforming an abandoned yard next to the Nantou Living Room into a public space. They have inquired around the community and found out that the owner of the yard passed away and his children have never completed the formalities for inheriting the property. As no new construction is allowed in urban villages, the yard has been left uncultivated. Wu Ming and his partners find the property manager of the yard, son-in-law of the owner, Uncle Ye, in the hope of gaining permission to turn the yard into an open space. Uncle Ye agrees with two conditions. First is to keep his banana trees and second, his wife requests keeping a collapsed wall in the yard as proof of ownership.

Many people support the idea of making a garden. Mingjing Society of South China University of Technology submits a renovation scheme, while two Shenzhen architects provide insights and suggestions. Also, funding is a must for making the garden. A Nantou Living Room guest is a lighting designer, he convinces his company to sponsor outdoor lighting which alleviates financial needs.

People and resources are getting together because of the yard. This emboldens the partners of Nantou Living Room to move forward. Could this be a lasting model that allows diverse community-building participants with wisdom and creativity as well as diverse perspectives and viewpoints? Could this diversified and informal organization become a new model for making urban space? If our city is produced in such a way, would cities grow

图 5 改造为南头会客厅
Figure 5 Transforming into Nantou Living Room

更多元的参与者参加社区的建设，发挥每个人的智慧和创造力，汇聚多样的视角和观点？这种多元化和自组织的方式，是不是能成为一种新的城市空间生产模式？当我们的城市是以这种方式来产生的，城市是不是能更加“共生”？

“南头会客厅”自此走上了探索的道路……

6 曙光

为期三个月的双年展将在不久结束，但是它留下了一些经验、一些感受和一个公益组织。

除了南头会客厅传承展览所倡导的对多样性的推崇及对个体智慧的拥抱，持续进行“城市共生”的实践和探索，南头古城里已经涌现了新的商业和多样化的投资开发者。未来的南头古城，是不是能孕育出一个多元化的城市发展模式？在这种新的模式下，每一个人都能参与到城市空间和城市生活的生产中，每个人的想象力、自由的心灵和创造激情都能得到最大发挥，城市的图景充满了活力和革命性！一切有待检验，我们已经看到了曙光。

7 2021 年后记

这篇文章写于 2018 年 2 月，据今已过去了 3 年半，当初的希望、憧憬、曙光，是否已实现？在完稿的 4 个月后，南头会客厅于忙乱中默认解散，曾经满怀希望地走进新的天地，理想的激情却在无数次的希望和失望间消磨殆尽了。2019 年 3 月，万科入主南头，耗时一年多的“快车道”更新，南头古城以崭新面貌重现人前。当初希冀的“城市共生”模式并未发生，各方百口莫辩。

值得反思的，是我在之前文章中曾强调的人文关怀。壁画师对本地居民信仰的尊重和体贴，设计师对杂货店老板的将心比心，这些举动固然是令人温暖的，但它们是解决社会问题

even more differently?

"Nantou Living Room" is on its way to explore...

6 Hope

The three-month biennale is almost over, but it will leave behind some experience, some thoughts and an NGO.

In addition to the concept of respecting diversity and embracing individuality advocated by the Nantou Living Room, inheriting the legacy of *Cities, Grow in Difference*, new businesses and investors have emerged in Nantou Old Town. Will the future Nantou Old Town be able to nurture a diversified urban development model? A new model in which everyone can participate in the production of urban space and life, a place where everyone's imagination, free heart and passion can be maximized, and the city is full of vitality and innovations!

Everything is about to be tested, but we have seen hope.

7 Postscript in 2021

These fieldnotes were written in February 2018. Have the hopes, vision and longing been realized over the past three and a half years? Only four months after the article was published, the Nantou Living Room tacitly closed its doors. The organization once walked into the new world with hope, yet, passion and ideals wear out in the countless moments between hope and disappointment. In March 2019, Vanke took over Nantou, starting a one-year "express" renovation, giving Nantou a fresh look. The model of *Cities, Grow in Difference* did not come to fruition, and no one remains to make its case.

It is worth rethinking the humanistic care that I emphasized in my article. The foreign muralists' respect and consideration for the beliefs of residents, as well as the designer's empathy for the shop owner, are no doubt touching. But were they the key to solving social problems? In research on the anthropology of emotion that I have studied in recent years, the close relationship between emotion and politico-economic structure is often mentioned. Do the short-term feelings that the creative class sympathizes with the lower class conceal long-standing political and economic differences, differentiation, exploitation and inequality? Has Jürgen Habermas-style negotiation and communication idealized an unshakable politico-economic structure? Has human care become a kind of emotional labor, smoothing out cracks and uneasiness in the politico-economic, so that the development train can continue running?

的万能钥匙吗？在这几年阅读的情感人类学研究中，经常提到情感和政治经济结构的紧密关系，这不禁让人思考，创意工作者对底层民众短暂的温情脉脉，是否会掩盖长期存在的政治经济差异、区隔、剥削和不平等？哈贝马斯式的协商和沟通，是否理想化了本是硬邦邦的政治经济结构？人文关怀是否成了一种情感劳动，抚平政治经济的裂缝和不安，从而让发展的列车继续运行？共情和关怀，是否成就了中产阶级的体面礼貌、自我修养和道德光辉，制造区隔和阶级再生产？

在三年半后的今天，经过艰难的阅读思考，我从自由主义的懵懂幻梦中逐渐醒来，开始面对严峻的现实。梦想的力量是巨大的，但梦想和实际之间的距离，不是想象就能消弭的。设计师和艺术家是造梦的高手，但要直面现实，才能许诺切实的梦想，而不是引人入胜却容易引入歧途的梦。

自责当初的懵懂和幻想时，慰藉我的是策展宣言上的一句话：“展览不是解决方案，不是大团圆式结局，它是一个起点，一个转折，甚至是一次探险。”

这是我此趟探险旅程得到的教训和反思，值得深思，亦不枉此行。

Have empathy and care become the decent politeness, self-cultivation, and moral brilliance of the middle class, serving to maintain class separation and reproduction?

Three and a half years later, with hard reading and thinking, I have gradually woken up from the ignorance of liberalism and begun to face a harsh reality. The power of a dream is huge, yet imagination cannot overcome the distance between dream and reality. Designers and artists are masters of dream-making, but only by facing reality can they promise true dreams instead of misguided fascinations.

When I blame myself for my ignorance and daydreams, a sentence from the curatorial manifesto comforts me, "An exhibition is neither a solution nor a production with happy ending, but a starting point, a turning point and an expedition."

These are lessons from and reflections on my expedition. It has been worth pondering and will not have been in vain.

“做课：跟深双进村做点儿什么！——南头古城居民需求调研及创新实践课程”是 2017 深双学堂的首个活动。课程招募了一批对城中村感兴趣的学员，跟随双年展进入南头古城，倾听、了解居民的需求，诊断南头古城存在的问题，评估双年展对南头的影响，鼓励学员提出创新解决方案并支持学员付诸实践。

经过两个月的调研，学员收集了不少需求，了解到城中村里的商铺面临着空间拥挤狭小、货物堆放凌乱、摆设不吸引、店铺缺乏个性等问题。在深双策展团队的号召下，做课发起了“商铺新形象”活动，征集平面设计师、室内设计师等各种专业人士和爱好者，针对位于南头古城内的商铺进行设计改造，让店铺焕发新的活力。

▷项目：做课 ▷组织人：黄伟文、陈嘉盈

"UABB Do Course at Urban Village"—Nantou Research and Creative Practice Class is the first event of the UABB School in 2017. The course invites all kinds of professionals and followers of urban village to participate in the course, following the Biennale to enter Nantou Old Town, listen and understand the needs of residents, diagnose problems of the community, assess the impact of the Biennale on Nantou Old Town, come up with solutions and put them into practices.

After two months of research, students of the course collected voices of the residents, among which those of the store owner caught our attention. Stores in urban village face problems of limited space for goods storage, bad goods display, lack of personality, etc. Under the call of the curatorial team of the 2017 UABB, Do Course initiated "Store's New Image," calling designers to redesign and reconstruct stores in Nantou Old Town, making them glow and shine!

▷Project: Do Course ▷Organizer: Huang Weiwen, Chen Jiaying

来源：深圳城市\建筑双年展组织委员会．城市共生：2017 深港城市\建筑双城双年展 [M]. 广州：岭南美术出版社，2021: 378.

2017 深双论坛现场，南头古城大家乐舞台，张超摄于 2017 ©UABB
2017 UABB Forum, Dajiale Stage at Nantou Old Town, photograph by Zhang Chao, 2017 ©UABB

论坛
FORUM

“共生实验室”回顾

编辑 莫思飞

在“城市共生”的主题之下，2017 深双是寻找新的理论、新的实验和新的实践模式的平台，是注重“发现”而非预先设定的交流空间，是一个不断生长的双年展。只有超越个人视野的的局限，强调多重身份、多元视角，才能从新观察中挖掘隐藏的机制，从根源上在文化层面、社会层面、空间层面理解和尊重城中村这样一个庞大的关系系统的复杂性、自发性以及不确定性，从而为深圳城市在“后城中村”时代提供多重的可能性。

2017 深双总策展人之一侯瀚如，曾在 2005 年第二届广州三年展策划过“三角洲实验室”的系列活动，邀请艺术家、建筑师、学者针对珠三角和中国高速城市化发展的现实进行一系列的研究、研讨和开放式交流。另外两位总策展人刘晓都和孟岩当时也曾参与其中。以“实验室”作为 2017 深双“城市共生”策展系列活动的命名，就是对这一段特殊历史的致敬。同时，此次深双因场地的特殊性变得十分复杂和不确定，且拥有不断变化的可能。在城中村做展览，不是简单的怀旧和保护，而是在讨论关乎未来的话题，因此其过程本身就如同一场实验，而“共生实验室”则是一个工作场。

- 2017 深双第一次“共生实验室”/ 城市共生
- 共生实验室（二）× 做课：城中村故事会
- 共生实验室（三）：自发的另类城市
- 共生实验室（四）：开幕式演出计划讨论 + 艺术家进村
- 共生实验室（五）× 满观城市：城市的节庆精神
- 共生实验室（六）：南头茶划会暨第一届南头街道民生论坛
- 共生实验室（七）：古城论坛——南头古城与深圳城市的再生

回顾选取共生实验室系列中的部分精彩内容进行梳理和编辑，记录对城市共生策展的思考、南头古城居民的故事以及以南头为代表的城中村未来更新实践的探讨。

1 城市共生与城市策展

（内容整理自“2017 深双第一次‘共生实验室’/ 城市共生”）

作为策展开放过程的一部分，第一次共生实验室从主题的深入阐释、内容的内在逻辑、展览的视觉呈现三个方面进行了探讨交流。活动邀请了国内具有丰富经验的策展人和专家学者共同讨论一系列命题，包括：如何为城中村现场量身定做一届双年展？如何突出展览内容的在地性，以日常生活熟悉的主题为切入点，对城中村进行深入探索和反思？如何为公众呈现一个既充满未来想象又直面生活现实的展览现场？如何从呈现方式、大众导览、公众教育及村民参与等各个角度，最大限度地调动公众对于双年展和城市议题的关注与参与？

都市 | 村庄：“自发生长”与“集体违章”

刘晓都：从社会发生机制去看待城中村，它既有世界性，也有中国特色，如四合院、工人新村、城中村都是在原本正规的体系下，在内爆式的城市化过程中向着混杂的方向演变。其背后的空间生成机制有很强的关联性，历史契机会使社会空间变化，而这种变化的共性是值得去讨论的。

史建：《落脚城市》从空间社会学的角度，重新定义类似城中村或者城市衰落区域。在高速城市化的过程中，这些区域类似一个过渡的城市空间，外来人口通过在这样的区域里生活、学习和累积，最后成为一个城市的市民。在普遍的城市化过程中，深圳面临的城中村问题，如同北京的四合院、上海的里弄，都是在不同城市所产生的新的空间环境，以及不一样的表现形式。

张宇星：城中村是在全球化新的生产消费链条上形成的一种新的空间地理学。所有跟消费相关的流动，全球的产品、信息的流动与汇聚使得城中村成为一个全新的地点。它是空间流动的聚落，是信息的聚落，是产品的聚落。这样能量集中的聚落一旦形成，人和行为都会附载在上面。传统聚落和现代城市都是通过正规体系建构起来的，没有超越体系之上的力量。城中村的机制是世界性的，是全球化新的抽象经济的基础上产生的具有超越性的地点。城中村的经验作为抵抗现代资本主义体系的参考方式值得被推广。集体化的体制和管理，加上深圳的包容，促成了集体化违章，即在集体名义之下对整个体系的违章，也是抵制私有化体制的中坚力量。集体违章的条件，也造就了客家岭南的自由生长。

孟岩：城中村是个特区，有些城市介入在城中村范围之外是难以实现的。集体违章就是制定特殊的规则。城中村看似无规则，其实是有的，我们只是把这些“潜规则”变成“显规则”，最后在这里成了城市实验。“自发”并非简单的对抗，如果我们只用“正规”和“非正规”来谈城市建筑，则会充满限制性。就深圳的城中村而言，因其各自的历史、情形、环境的不同而表现出明显的差异性，如果我们能在其中探寻一种理性的规则，能让城中村真正活下去的规则，也许这会让城市展现出不同的面貌。我们在城中村做展览，不是简单的怀旧，也不是简单地说保护这个城中村，而是在讨论关乎未来的话题。如果我们在城市里保留一些实验区，这些小的特区，未来可能是有希望的。

世界 | 南方："差异的现代性"

刘珩：2015年深双论坛，《再南方》代表了两重意义，一方面是基于南方这个地域的建筑实践，由于气候带来的可能性，形成区别于北方的明显特征；另一方面，"再"是经过多年的全球资源重新分配后，地域性问题就成了一个政治经济问题。在潜意识上，南方是一种生命力的呈现，在空间实践上应该反映南方的生猛和淡定，同时又体现出务实的一面，融入南方的生活方式。而这样的潜意识、想象和空间实践，以及生活方式和场所之间的互动应该付诸于展览，使其更贴近南方的个性。

张宇星：南方是相对的南方，它实际上是政治边缘跟中心的关系，是离权力中心距离较远的地方，受到的管控相对宽松，创造了集体违章的条件，造就了岭南客家的生长。南方允许北方与其杂交，北方是不允许南方跟它杂交的，所以南方的杂糅使其形成一个异构，出现各种可能性。

冯路：中国有两个作为文化身份的南方，一个是江南，一个是岭南。南方是远于中央的，由此滋生了大量民间文化，消弭北方带来的影响。深圳在南方，但又像个北方的飞地，作为一个偶然政治决策的结果，确实包含大量的张力。

冯原：从自觉和不自觉的关系切入，南方分为两个层面：一个是"对象层"，即这个地方、本土的；另一个是"话语层"，即南方是一种认识的结果。从这个角度来看，针对南方的话语并非属于南方。如岭南作为一个地理和地缘空间，有漫长的历史；但作为话语的历史并不长，岭南是清末民初后，由一群知识分子发起的讨论。从对象层和话语层角度，把人分为"生活者""被观察者""观察者""知道者"。后者把其扩张的世界变成一个对象世界，前者的世界则被观察和记录，但这并不代表被观察的世界没有文化，就如同今天讨论的南方，生活在这一文化中的人并没认识到自身丰富的文化，也不知道这种文化可以被讨论，他们缺乏自身的理解自觉。某种意义上，我们介入农村做乡建承担了早期西方殖民者的职能：引进先进的技术和话语，把观察和考证的结果拉到展览里，形成我们"知道世界"里的成果，这是个结构性的问题，作为身在"知道世界"的我们必须要警惕，判断观察成果与对象之间的关系。

广东省与北方中心建立的关系可以称之"皇家民间双重性"：自明清以来，整个珠三角地区的开发，都本着两重性而行：一方面要进入国家皇室的意识形态，要成为儒家统治系统的组成部分，从潮汕到广东无一不说自己是孔子传人，他们都迫切地想进入正统；另一方面他们又处在相互博弈和压制中，保有丰富的自身特征。

讨论：

展览的介入和呈现方式

张宇星：先回应刚才冯原老师说的文化反身性，我认为这种反身性在任何地方都存在。观察的对象必然受观察的影响，而且这个过程是永恒的，任何地方都会发生。这跟我们对这次展览本身的反思有关，即整个双年展放在城中村是对还是错？大家应该有一个思考的过程，因为艺术一旦介入，是把它精英化、绅士化，还是把它保存下来，这本身就是一种悖论。你批判别人，肯定也会被别人批判，与其这样不如我自己先批判自己。

第二点是站在文化和时间的角度看问题。整个深圳南方性的问题，实际上跟时间有关系，它不是传统意义上的岭南，而是通过几百年的时间慢慢形成的，可能在生成过程中很多东西消解了。深圳的南方性更多地跟时间性有关系，它通过那么短的时间积累起来，所以它的南方性特征，或者反南方性特征更都加明显。这种状态需要我们用人类学的方法剥离开，这样会更理性。但最后展览展示的是不是更理性？我觉得不一定，可以用感性的方式形成斑斓的结果。

城中村这个话题不应是我们最初定的逻辑规则，而是有其自我生长。一旦我们确定了规则，这个规则基本上已经有问题了。所以与其这样，还不如保持它的开放性，到展览最后一刻都有一个接口、有一块话题、有一个区域是反对整个话题的，允许持不同意见的观众都可以参与进来，这样一种形式也很有意思，展览不是终极的，它始终保留着接口。城中村为什么成为城中村，它允许加建和拆除，它就是自我生长的状态。

孟岩：说到展览呈现，刚才也谈到观众反应问题。在城中村做展览，我们如何健全这样一种对话机制？因为在这儿做展览的特殊性很强，观众的构成差异很大。这时候怎么动员更多人的参与，另一方面又不要出现好像我们在消费城中村的趋势。我也听到很多对双年展的批判，本来那个城中村还挺好，一进去全乱套了。怎么做到既积极介入城中村，又不要仅仅拿它只做个话题，过度消费？

张宇星：最好的方式就是让里面的村民来“消费”展览。这跟前几届不一样，前几届到晚上就关门，这次能不能允许居民随时进到展场，然后从事其他活动，彻底消费展览？我们还可以有一个展览项目，就是研究这个展览跟南头古城的共生，从第一天进入开始记录整个过程，在每天的展览过程中把结果呈现出来，包括跟踪调查一些居民。这个过程中居民慢慢对展览如何接受、磨合到最后参与，甚至展览结束以后他如何改变自己的生活状态。把展览变成一种生产的要素，让村民们来消费展览，就是最好的实验和作品。

孟岩：一个展览有它的学术规则，也有它更大的特点。搬迁的人一定是愤怒的，如果是我，我也会反抗。但这件事如果做得好，做成一个样板，使得深圳城中村不被拆除，能够生存下去，我觉得这可能造福了很多很多人。这就是冒险，但也可能有全新的经验。所以各方面都很重要，镜头的记录也很重要。不光那些人反对，其实学术界也有反对意见。作为独立策展人，我既不代表政府，也不代表民众。我有很强的价值观，就是：不拆城中村是对的，拆是不对的。

张宇星：如果村里面人都换光了呢？

孟岩：现在不动你以为它就不会换吗，一样会换。大冲都拆了，人也在换。我们可以加速也可以减速，但很多东西的方向是改变不了的。因为改造的意愿在先，双年展在后，这是另一个问题。没有双年展，改造可能会沿着另一个方向走，在两者都是重伤的情况下，宁可选择一个轻点的。建筑师天生就是要妥协的，不妥协的结果可能会把南头古城拆光变成丽江古城；但我觉得通过双年展的实验探索，有可能找到一条新的空间道路。

刘珩：我觉得这个讨论很有意思，比如广州的时代美术馆，表面上看老库（库哈斯）只是播了一粒种子，这粒种子最后也会导致士绅化，另外的方向就是维持与原居民的关系。我们没办法预测最后的结果到底朝哪个方向发展，但展览种下的种子才是我们最核心的要讨论的。把事情放在城中村里面，这已经是一个非常强的信号了。

张宇星：我同意刚才孟岩说的带价值观，但应该把反向价值观同时植入下去。不是说我的价值观多好，也不是说我的价值观多包容。在包容的同时又推销我的价值观，这样可能更好。这个过程把更多的反向价值放得越多，我认为越好，对最后生长出来的东西更有价值。

冯原：双年展放在一个生活空间里面，这个做法本身肯定会引发事态，或者引发某种矛盾。从结构上来说，村本身也是向内和外来介入之间的关系。这样一个本地和外来介入的关系，它的矛盾核心点在哪里？如果我们把双年展看成一个公共文化产品，或者称之为非排外性公共文化产品，因为双年展可以放在任何空间里，它一旦发生就不属于这个空间本身，它属于深圳甚至超越深圳，因为它成了文化产品而不能排外。但作为空间本身的主人，他们肯定有意见，因为按照企业人的说法，这个地方被他们看作一个企业：如果这个地方产出的产品归于他们所有，他们乐于支持这个项目；但如果产出的产品为所有人共享，他们就不一定会支持。他们何苦要承担公共文化的产品属性呢？所以这个矛盾是天然存在的。就像你们家本来有件旧衣服，属于你们家的，现在突然变了，规定说你只能穿这件旧衣服，还不能换新衣服，旧衣服还不能洗，还没有收益，凭什么？所以双年展的问题也很简单：你所产生的东西都是属于全人类的，所以这时候补偿给原住民的可能就是延后效应。再过些年，文化导入和保护的利益就能看出来了。

冯路：说到这个事可能就回到了主体性问题，即城中村的主人身份到底存不存在。如果是一个原始村落，从合法性来说，有原始的所有权。但我们讨论的城中村和原始村落不一样，区别在于它对城市的价值。它不再是一个原始村庄了，它承担大量外来人员中转的作用，这个作用是公共性的，而不是原生性的、本土性的。展览的价值观是不可避免的，哪怕回到人类学的身份，采用作为观察者的研究方法，尽量不触碰、不影响对象，但即便这样，无论如何小心翼翼，依然会对对象产生影响。所以进入城中村是好是坏，不可能讨论。无法考证的时候，我至少认为我们的行动是一种积极的方式，是站得住脚的。行动在这时可能比判断更为重要。

张宇星：回到最初的题目“城市共生”，差异性存在的基础就是平等，自由平等。所有东西不论一切，首先要承认它的先天平等性，这才是获得差异性的根本。所以要考虑权力之间关系的问题。展览带了价值观，这个价值观必然是一种文化权力，那么这种权力和个体发生关联的时候是如何解决的？或者你认为：因为我的价值观是正确的，所以我代表全人类，这 50 人请你走。我觉得这不叫共生，这是反共生。我们不能干反共生的事。

孟岩：我不太同意平等，我觉得是平衡。这个社会就像生物一样，在网络与各链条的发展中会达到某种平衡状态。宇宙也一样，它也有一个相对的平衡。而仅从展览的角度，解决这个问题是不可能的。但我觉得我们可以呈现它的问题，包括共生的多方诉求。我甚至觉得“城中村的未来”里面有一个板块就叫未来，但这个未来我现在在想，应该叫“共生未来”。它不是一种特定的未来，这个未来可能也包括城中村的生与死，它会有生的未来，也可能有死的未来，或许还有一种半生不死的未来，等等。所以我觉得从展览的角度，或许能呈现出这样问题。最后可能我们通过展览问了问题，而回答并非某个特定的结论。

冯原：平等和自由在西方的理解一直以来是一对产物，如果谈平等就会没有自由，如果说自由就会削弱平等。我们看到“城市共生”应该讨论的是“共同”还是“平等”的概念？英文里面说的是不同，中文叫异生，其实谈的是自由生长，即你有你的生长，我有我的生长。但共生的概念是大家要相互协调。那双年展做什么？双年展的目标主要是为了探索、批判、反思和前瞻，所有一切都表明了一个姿态。从某种意义上说，它是不需要务实的。当你要在双年展中呈现内容的时候，双年展是如何定位这些内容变成它自身的部分？城中村自身就是在一个城市发展过程中被“右派”制造出来的，右的本身是资本不断扩张，在这样过程中，才反向制造出相对的贫弱的和阶层的差异。在这种情况下，一个“左派”双年展就形成跟右派式生产之间的反向，城中村就成为必须研究和反思的东西。总策展人的使命，可能就是要把这些相互反向的东西不断地编织与编写，使得问题呈现的方式更加不同。

冯路：我们讨论这么久，是不是有点自以为是，我们都说双年展带动城中村，城中村的人就会怎么样，是不是高估自己了。因为城中村里面有方方面面的人，而双年展是相对单一的。反过来，作为一个展览，一个城市建筑类的学术型展览，今年还加入了艺术板块，如果把它从那么高的地方放下来的话，可能也没有那么宏大。

2　多面城村：九街记忆

（内容整理自“共生实验室（二）× 做课：城中村故事会”及南头古城调研访谈）

从策展之初，策展团队就启动“入村计划”，以内部视野、现实观察的方式，使得南头古城在居民的集体记忆与个人记忆、宏大叙事和个人叙事的交错中展演，从朴实的故事中瓦解偏见，构建真实的、不断生长变化的、当代城中村完整的历史链条和延续的生活方式，反思和批判单一的“地产思维”下的改造模式，发掘它已经或即将清除的多样杂糅的城市生活、文化和记忆。

“城市是一家医院，一个客栈，一个图书馆”

在 2017 深双开幕论坛上，一位居住在深圳城中村七年的居民从她一家八口的角度定义，“城市是医院、是客栈、是图书馆”。城市是医院，治好了她的“家庭病”，规范化的城市管理让她的弟弟得以戒毒；更多的可能性和工作机会让她的妹妹脱离家暴，走出抑郁症的阴霾。城市是客栈，人来人往；“富贵不归故里，锦衣夜行”。叶巧玲引用典故叙述了城中村大部分居民的“衣锦还乡”的期盼，城市只是大家暂时落脚奋斗的地方，尤其在远程办公成为主流的未来，扎根在城市不再是必须。城市是图书馆，“先进”思想扎根的地方；城市的环境和生活打开了叶妈妈的视野和对新事物的包容度，乡村与城市最根本的差异并不在于基础建设工程，而是在思想观念上的距离。“农村是抚养孩子的地方，城市才是培养天才的地方。”叶巧玲的发言引人深思，论坛上的嘉宾、策展人和观众都因她真挚而又深刻的话语对城市有了不同角度的理解。

每一个年代的深圳都有一代人的奋斗史

深圳城市发展了 40 年的今天，自媒体构建起一个故事：老深圳人等于坐收十几万元房租的包租婆，却鲜少提起上一辈，即使在改革开放前，移民到深圳的人都是深圳的“开荒牛”。黄婆婆清楚地记得，她与丈夫是在 1956 年 3 月来到南头古城的，至今扎根南头五十多年。当时 80% 的村民（原住民）都去了香港或者移民国外，还有一些年轻人最近十年也搬到小区住宅中。在黄婆婆的记忆中，南头古城的南门、东门、北门一直都是有的，但从没见过完整的西门；而城墙早几年就几乎被拆完了。“村领导把城墙拆了，建自己的房子，那时候还没有那种（文物保护）意识。现在的衙门，我来的时候也是没有的。”而现在的中山公园，则是如山林一般的杂草和小树苗，也有百多年的老树。

黄婆婆年轻时候曾参加过南头古城的生产队，后来由于四个小孩的陆续出生，就在家里做家庭主妇。生产队的工作十分的辛苦，耕地基本在距离南头古城三四公里远的地方，最近的田地也有一公里的路程。小孩子会到周边的田里面抓小动物和小鱼、小虾。“现在全都变了，什么都没了”，黄婆婆感慨道。那时候自行车很少有，早上五点钟起来打理家里养的猪后，就要挑着一百斤左右的东西往古城外走。古城周围杂草丛生，早上走一路裤子都会被露水浸湿。农忙的时候还要挑着花生、番薯这些作物回来。以前南头治安不好的时候，每家都养狗，导致路上非常多的狗跟着路人，甚至拉扯裤脚，“很害怕，晚上真的不敢出门，一出门狗就好像要追着你咬一样，我来的时候都想走了。”

以前南头附近有一个电影院，一有电影放映，在商场里就会有大喇叭广播。黄阿姨在提到看电影《阿信》的时候说：“男的也哭，女的也哭，我就不哭，他们就说你怎么不哭，那么辛苦，我说她到哪儿都有人照顾她，我从小到大比她还辛苦，怎么哭。”

黄婆婆现在与三个儿子住在东门附近的一座建于 20 世纪 80 年代的自建房里，这个房子是黄婆婆来到南头后的第三十年前后经历千辛万苦建起来的家园。前三十年，黄婆婆一家都是借住或者租住别人的房子，丈夫清廉，虽然工作努力但家庭经济困难。眼看着同期来到南头古城的邻居们都建起了自己的房子，黄婆婆也渴望有一个属于自己、安稳的家。黄婆婆说：“我们住着别人的房子是因为他们都去香港了，但人家回来之后你就要走，我住在南头三十年就搬了四次家，搬得我不想搬。”大约在 1982 年，黄婆婆向公社申请了一块地，请推土机平整了一块约 200 平方米的地，但依然没有建房的资金。黄婆婆的丈夫对此非常生气，认为自己工资一个月三十多块，养一家那么多口已经非常困难，更不可能有钱建房子。平整的地很快又长出了杂草，黄婆婆就带着十一二岁的小孩和丈夫一起去拔除茅草。慢慢地攒到建房子需要的石头后，黄婆婆请福建的工人建起一层楼。黄婆婆说：“几个铁窗很便宜的都没钱买，后来花了一年多的时间才买了那几个窗，还是我和大儿子一起搬上去，请泥水工安装。”黄婆婆的房子是那个区域第三户入驻的，因为人烟稀少，晚上睡觉连窗户都不敢开。黄婆婆的坚持和不懈的努力，才有了她现在的安乐窝。现在孙辈都已经长大，她就天天去公园唱唱歌，跳跳舞，享受着退休的日子。

苦中作乐：生活苦，助人乐

廖阿姨是广东台山人，七十年代初的时候嫁到南头，抽签到南头的第六生产队。廖阿姨刚到南头的时候住在废弃市场中一个临时搭建的棚屋里，“四面几乎连墙都没有，都漏水了。”1981 年，五队和六队位于现在中山公园以北宝安的耕地被征收，赔了每人 3000 元的果苗费。那时候建房子大约 2 万块就可以建两层，大家的房子都比较破旧，很多街坊拿这个钱去修建房子。生活艰苦的廖阿姨的第一栋房子，两层半，就是在那时向别人借钱建起来的。城里有些无主的“屋地”，业主多是在日本攻占南头的时候移居香港或移民海外，原本的老房已经倒塌废弃，街坊们得以在这些“屋地”上建起了自己的房子。

1994 年，廖阿姨的丈夫不幸中风，愈后无法继续干重活，也正是这个时候，街坊麦老先生为了帮助廖阿姨，以极低的价格把中山西街 44 号租给了廖阿姨，成为我们后来熟知的“锦记商店”。如今儿女都长大了，廖阿姨也有了自己可以外租的物业，但依然在店里忙里忙外，与来店里的街坊朋友们聊得不亦乐乎。老房东麦老先生过世后，继承房产的麦先生依然延续着三十多年前的低价租金租给廖阿姨。廖阿姨是一位非常懂得感恩的人，一直想要寻找机会，自己出资把这个据说是晚清时期的老瓦房加固修缮，在自己“退休”的时候还一个坚固不漏水的房子给房东，报答房东一家对自家的恩情。但是，由于严防违法抢建，牵涉到建筑外轮廓施工的申请修缮得以审批下来的可能性不大，一直无法动工。后来借由 2017 深双“城中村进行时”的项目，在设计师和廖阿姨自己相熟施工队的帮助下把店铺修缮好。“开始报价 10 万， 我就找了帮我和麦姨做过装修的那个年轻人帮我做，6 万元就弄好了。”

也正是这样的际遇，让廖阿姨也处处与人方便，总是说“别人帮过我，我也要帮助别人”。在与廖阿姨聊天的时候，一位来自湖南的租客刚好路过锦记商店，跟廖阿姨打招呼。廖阿姨这几年一直都以 1000 元一房一厅、比市场价低的租金租给她，后来这位大姐住得不好意思，主动让廖阿姨给她涨点租金，提了几次后，廖阿姨就涨了 300 块，当时附近的房子都已经租到 1800~2000 元了。廖阿姨说：“她带着女儿，生活也不容易。钱是需要，但不是最紧要的。”其实，这并不是廖阿姨唯一帮助的租客了，很多租客在廖阿姨的房子住了十几年都未曾涨租。在得知其中一位租客的父亲患上了癌症，家里还有上幼儿园的女儿时，廖阿姨就免除了房租，还常做好早餐叫他们一起吃。

村庄秩序下的多元共生

黄姨是义工小林的婆婆（丈夫的妈妈），今年 70 岁了。她是 1983 年从华侨城搬到南头古城的，“我自己骑单车到蛇口拉海鱼来卖，生意做到 1993 年就没做了。”而儿媳妇小林和她一样，都是潮州人，2005 年嫁入南头。虽然黄姨一家人不是本地人，但他们在这里三十多年间，邻里关系都非常好。黄姨和小林的家在朝阳南街，刚到南头的时候与当地村民、居民集资建房，村民当时居住在香港“出地不出资”，一、二层归村民所有，三层及以上分配给集资的各人。虽然没有红本，但是建房的时候找了律师公证后到政府登记信息，明确了 70 年的使用权。现在黄姨一家与集资的邻里们的关系都非常好。

陈伯，则是为数不多住在南头古城的原住民，在南头中学初中毕业后，16 岁到香港开始做生意，女儿和两个儿子都出生和生活在香港。陈伯 1989 年的时候回到南头古城建房子，在祖屋的前后分别花了 1000 块和 3000 块买下两块地，并在村委会公证、政府报批后建设。据陈伯称，政府的批示是批准拆旧建新。陈伯的哥哥则申请了靠近报德广场的一块原来家里的牛圈建房。陈伯家的门外挂着一个写着“老四居”的牌子，院子十分雅致。陈伯十分开心地表示，他现在是南头香港宗亲会的荣誉会员。

在走访南头古城居民时得知，南头古城的物业基本可分为三类，一部分是集体所有的生产队物业，第二部分是居民独资的自建房，第三部分是居民集资合作的自建房。而从用地权属上就更为复杂了，常见的一小部分是南头原住民在自家、或移居香港的亲戚的宅基地上拆老宅建新房；另一部分是生产队的居民，向生产队申请空地使用权。有个别城市居民“购买”了土地的使用权，更甚有几户是由南头古城原住民赠予城市居民的土地。从南头居民和他们各种建房故事中，可以看到上一辈的奋斗，从他们的乐善好施中也看到杂糅多元的空间背后的故事，一部南头自建房口述史。

包容，一切皆有可能

与老一辈的南头居民不同，改革开放后移居南头古城的人们是因为多元包容的深圳，让想要奋斗的人看到了机遇。小朱在 2003 年从湖南怀化来到深圳打工，没过多久就打算自己创业。当时身上只有 840 元的小朱，来到西丽的一个店里，希望在老板店里寄卖商品，没想到一下子就谈成了。2005 年小朱遇上了生意的合伙人，也是现在的丈夫，俩人找到南头古城准备开店。当时小朱资金不够，一咬牙就借了 5 万块钱，凑齐 8 万块的转让费接下这个店面，开了家药店。当时接下店面的时候，楼上三房一厅的房子也要一并承租下，每月店面 2700 元的租金以及 1500 元的房租。一开始小朱心里也没底，看到别人店里没生意，也担心自己投入过大无法回本，但幸好刚装修完就开始有生意了，一年时间小朱就连本带利地赚了回来。第一个月的时候，因为周转困难，房东老太太还帮助小朱，允许她周转后再付房租，还以当下的租金与小朱签了 5 年的约。有了稳定的做生意的门店，小朱对此至今依然铭记于心。后来，药店在东街和南街以及南油又开了三家，逐渐地把药店开成连锁店，帮助亲戚们做生意。

跌打医生是一位半盲人，他来自湖北武汉，因为眼疾在家乡连按摩的工作都找不到。因为在小说中了解到深圳，他觉得深圳是一个包容且充满可能性的地方。为了过上与原来不一样的生活，他决定扎根深圳，在南头古城里开了跌打诊所，为在古城里生活在底层的居民看病，也在南头古城有了自己的家。在故事会上，跌打医生分享着中山公园里“思源井”曾经冰凉甘甜的泉水，骄傲自豪地讲述着南头古城的往事和变化。

南头在变化，人也在变化

在居民的故事中，近年来南头古城最大的变化是治安和人群。跌打医生因为在跌打诊所帮人看病，所以感到这些年古城里人的层次在不断变化。刚来的时候看病的人以打工的为主，是工资最低的、生活在最底层的一群人，也给“黑社会”看过病；现在治安等各方面情况好了，居民大部分都是在科学园上班的白领阶层，还有大学刚毕业的小年轻。

修鞋匠大叔在南头古城算是个“公众人物”了。他 1999 年来到深圳，在古城主街附近摆摊修鞋已经有十多年了。虽然以前并没有过多地关注城市发展的问题，他说“我遗憾来了这么多年作为一个见证者，以前却没有太关注这些事情”，但深圳的变化对于他来说是一目了然的。刚到深圳的时候，深圳的城市面貌已经基本形成，高楼大厦也是随处可见，同时保有着很多的城中村。多年过去，城中村虽然规划不完整，但是基础建设、环境卫生等等经过近年的改造都已经焕然一新。大叔回忆道，“刚来的几年，小偷、扒手随处可见，我亲眼目睹的，每家每户都养狗防小偷，但最近几年有了摄像头，小偷都不敢来了。”

王大姐也表示一开始管理没有那么严格，城里还有拉帮结派的现象，“明天那个帮派来争地盘，我的锅盖、瓢天天都被他们拿跑了，我今天找这个勺子，明天找那个锅盖，因为他们一打起来什么都拿。”但是，帮派的人也不会吃霸王餐，哪怕今天没钱付，第二天也会把钱送来。王大姐说，那时候做生意也是提心吊胆的，有一次帮派报复，有人被砍倒在店门口，王大姐吓得关店一天不敢营业；但是这几年发展好了就平稳安定了。

药店小朱也经历了南头治安差的时候，被“扯耳环”耳朵受伤的受害者、打架受伤的人都跑到店里来包扎。小朱因为担心店员的安全，安排他们住在药店的附近。现在古城治安好了，但是对小朱来说也有利有弊。现在城市管理不再允许外摆，以至于小朱少了很大一部分的收入来源。“这里寸土寸金，墙面上贴广告也有钱收，门口给人家卖水果也有钱收。”

“街道小，挨来挨去就挨熟了”

把南头古城看做家的不仅是在南头古城拥有物业的资深居民，千禧年后移居南头古城来自安徽的汪大哥与王大姐、来自湖南的药店老板小朱都在南头古城找到了家乡的感觉。

汪大哥来自安徽省太湖县，小时候生活贫困，大学毕业后分配到国土资源厅。有着奋斗精神的汪大哥在 1995 年毅然决然地辞掉安稳的工作，只身来到深圳打工。工作 15 年后就慢慢开始了创业的道路。大约在 2000 年前后，汪大哥无意间路过南头古城，看到城门外有一颗大榕树，树大到四个人环抱都抱不过来，树底下还有一家人在卖凉茶。当时汪大哥就和很多街坊在树下乘凉，喝着凉茶吃着甘蔗。可惜后来，大树随着南门外的拆迁而被砍掉了。曾经在有电梯的小区住不习惯而返乡的汪大哥的父母，却非常喜欢居住在南头古城，一家人一住就十几年。汪大哥说：“因为这里的人不分外地人、不分大小，买菜也很方便，还可以讨价还价。现在我找到了目标，有自己的事业，而且能回报给父母。怎么回报？他们不需要高楼大厦，要的是一个家的感觉。让他们安安心心在这里生活，这是对父母最大的孝顺，也是古城给我最大的收获。”

王大姐来自安徽，2002 年来到南头古城的时候觉得深圳和她想象中的不同，反倒是有几分家乡的感觉。南门前有一个小土坡，一条泥巴路，下雨的时候一步一个脚印，第二年水泥路就代替了泥巴路。汪大哥说的那棵大榕树王大姐也见过，周边还有好几棵龙眼树，那时候一到夏天，爷爷奶奶们就在树下乘凉喝茶。那时候南门外的房子还没有拆，王大姐就在那里开了家包子店。王大姐做生意前也做了市场调研，“好像听说里面住了有好几万人，这么一点点地方住了这么多人，我感觉做小生意还不错。”开业后，王大姐天天凌晨一点起床做包子，熬豆浆，包子每天都供不应求，营业到晚上十点，“天天简直就是抢，因为他们有人跟我说，我们时间就是金钱。”于是王大姐就招了店员，希望能让每个顾客都吃上早餐，高高兴兴地去上班，晚上累了可以买点点心。开店两年后，南门外的房子就被拆了，王大姐就搬到城里，在城门口第一家。王大姐对南头古城有很深的感情，因为父老乡亲不会区别对待外地人，感觉大家都是一家人。

药店老板小朱在 2005 年刚到南头古城，“我很惊讶，怎么这里还有这个地方，好像回到故乡的感觉，因为这里的人很亲切。这里的街道小，也促进这些乡亲的感情。因为走路有时候都挨着，挨来挨去就挨熟了。”

九街记忆是一部深圳的口述史

南头古城里的每一个人都有一部属于自己的、在深圳的口述史。从这些故事中，我们了解到南头古城相比其他城中村，宗族关系更为淡化，也使得外来移民更容易产生归属感。大部分深圳城中村都保有以氏族、宗族血缘亲情为核心，建立在传统儒家思想忠孝礼仪和尊卑有序的宗族伦理上的宗族文化，宗族文化的延续和宗族社会生活离不开商业资本的经济支撑。古代有经济实力的宗族组织活动，巩固内部团结，增强宗族影响力，如建祠堂、组织祭礼活动、办学、赈灾等；当下城中村股份公司成为公共生活的组织者，从治理到社会调节、公益福利都起到至关重要的作用。但随着时代的变迁，传统宗族文化的“同质”特性受到现代文化“异质”的冲击和碰撞，1950 年代末到 60 年代初，村民移居香港、东南亚后，带着不同的城市观念和生活方式回来。到了 90 年代，传统只能有男性宗族成员的祭祖活动开放成为类似家族亲友聚会的活动。从空间上来说，随着宗族的包容性增强，祠堂、广场等宗族活动场地逐渐对公众开放，成为社区活动的公共空间；再到 2000 年前后，接受西方教育的“村二代”回国，有着更广阔的社交圈和生活圈，宗族关系已经不再是赖以生存的唯一选择。城中村的内部社会结构逐步地从同质单一的宗族文化，向“异质”的多元文化的社会结构发展。

与此同时，流动人口的不断涌入带动了不同文化之间的碰撞和交融。1963 年，因水利工程建设，接收来自潮州的移民；80 年代，改革开放“百万民工下珠江”的浪潮为上沙迎来数以万计的外来务工者。移民文化是一种不断自我更新、包容的多元文化，在不同的文化之间碰撞而导致移民文化的开放性和兼容性。深圳在彰显移民文化的同时，并不是采用文化熔炉式的一元主义，强制同化一种新文化，而是放大“多元文化”，使得原生文化特色得以保存的同时又互相兼容。深圳从一开始就是与“人”绑定的，深圳的故事就是每一位深圳人的故事。从南头居民的故事中，那些被遗忘的历史细节和个人梦想，每段包含时代温度的生活片段提醒着我们没有高昂生活成本驱赶下，生活本该有的样子。

3　南头古城与深圳城市的再生

（内容整理自“共生实验室（七）：古城论坛——南头古城与深圳城市的再生”）

相对于大拆大建的更新策略，针灸式的策略已成为各地城市实践的新模式。如北京的大栅栏更新计划、上海的城市微空间复兴计划等，虽不同于深圳的城市背景，但对于南头古城更新乃至深圳城中村的更新都是值得借鉴的。

2017 深双城市再生论坛作为闭幕论坛，除了回顾双年展对南头更新计划的影响，更多的是对未来可能性的探索，从实施层面对以南头为代表的城中村和其利益主体相关，包括政府、从实施角度推动的群体以及社会力量，他们之间应该怎样形成良好的机制，确保后双年展的可持续性？更重要的是什么样的模式能使得已开始的改造有力地持续下去？正如策展人在策展理念中提到的，“展览不是最终的解决方案，也不是一个有着大团圆结局的制作，而是一个起点、转折和探险。”从了解各地城市更新实践的运作机制、实施策略及后期影响中，我们或许能寻找到后双年展时期南头古城可持续发展的运营模式，以及深圳乃至珠三角地区城中村未来的更新方向。

“另类城市模本”的探索

（孟岩）

双年展在很大程度上是观念的更新，尝试在现有城市发展格局下寻找到一种新的、不同于以往的样本。2017 深双从观念上做了几个重大的挑战，同时也是南头今天变得非常有意义的原因之一。其中第一个挑战是颠覆主流媒体长期以来对深圳只有 30 年历史的描述。深圳不止 30 年，更不是一个小渔村；一千多年来，它一直持续不断地演进和重生，历史本身就充分地证明了这个城市历史文化的丰厚性，除了是一个经济发展的奇迹之外，更多是文化传承且不断演进的结果。第二个挑战是正视三十多年的城市化演进历史过程当中，深圳的两个平行现实：一个自上而下的城市设计和城市的结构，与此同时，一种平行的现代化和城市化的模式在暗流涌动地发生，成为今天所知道的城中村格局，它的活色生香在不断地延续。2017 深双所提出的主题，“城市共生”也是在大城市发展的背景下应运而生。

南头有其独特性，可以说是一个新的类型，不是城、不是村，也不是城中村或者城市化的村庄。其实，它是城村合体的一种新类型——城村。它是村中有城，城中有村，已经脱离了地域和社会学意义上的村庄，或者是简单的城市，而是一个高度混合复合的共生体，是历史古城和当下城中村完整的叠加。我们也看到南头古城里有不同时代的代表建筑以及它的空间组织方式，它非常丰厚和完整地保留了深圳城市发展的样本，形成古今共荣，新旧共生的局面。

深双的策展过程有众多的村民、政府、企业、策展人、学术机构等等多方面人员的参与，这个参与过程实际上在软性层面给未来的城中村改造提供了新的可能性。我们非常高兴地看到这个双年展是非常立体的，既有自上而下严谨的规划，也有公开招选的自愿参与者。我们作为双年展策展人，尝试搭建一个平台并努力使其得以延续，使更多不同角色的人们来参与、自发地进入城中村未来的改造。我们一直希望深双不仅是一个展览，而是一次真正的城市介入，是一次开始，更是一次行动。

双年展做不了什么事情，但是因为有了双年展就做出很多事情

（侯瀚如）

当代艺术其实做不了什么事情，就像双年展一样，实际上做不了什么。双年展是展示建筑、艺术和各种研究成果的平台，把成果带到公众的视野，除此之外它的结果一般对当地没有影响。但是，这一次我们想做成不太可能的事情，就是把这种影响不光是在文字上留下来，而且在现实里面尝试变成某一种实实在在的东西。孟岩也讲得很清楚了，南头给了我们一个机会，是实实在在的，因为有双年展就改变了很多、很现实的情况，而且还有一个可能的未来。

艺术，主要的任务是给予我们乌托邦的时刻，让艺术家“不切实际”的想法可以表达出来。我们在双年展中需要当代艺术，正是因为它虽然不一定会留下具体的痕迹，但它曾经的存在会让人们怀念，建立一些“不切实际”的讨论，形成一种新的公共领域。这种公共领域是无形的，我们称之为“飞地”。艺术板块的主题——“艺术造城”是建造在城市文化、精神或者是感情上的结构，这种结构也就是公共领域的一种，或者是公共领域里面根本的中心点。而这种无形的、延续的、让人怀念的艺术创作激发了人们对未来行动的思考。

这一次的课题既是本地化的，但同时又非常国际化，即探讨城中村的模式在国际上会被复制和扩展的可能性。同时又探讨城中村本身演进的可能性，以及城中村改造可能引发的一系列问题，如深圳城中村的改造跟社会保障房政策结合的实验，既是为群众做善事，又需要为士绅化来临做出回应和准备；再者改造对城中村中特殊的工业生产的影响又该如何应对等等，都是一系列值得深思的课题。

拓展到展览之外，艺术还具体地介入到普通人的生活里面。这种介入是短时间对普通生活方式的干预和干扰，让居民变得要么特别兴奋、要么特别难受，在一定的程度上挑衅性地干扰城中村正常的生活。短时间事件性的介入，留下来的是各种各样对于未来的想象，这一次能够尝试着去实现一下这种想象，但是以后就不一定会再有了，所以，我们会继续怀念它。

深圳 | 2015 深双西浦分展场及 2017 深双大梅沙分展场

（杨勇）

2015 深双西浦分展场位于龙岗，深圳仅存 100 多座的客家围屋中有着近百年历史的一座，大部分居住者都是老人。借由展览的机缘，仅对其中两间房子进行了微改造，希望通过微改造让居民看到改造带来的变化，同时又能融入当地的生活。除此之外，作为展览空间，也以混凝土建筑搭建了展场“共享之屋”。来自北京的青年建筑师因地制宜，把一间改造成 20 平方米的微型美术馆，另一间成为茶室，保留了很多原有的元素。整个计划参与的艺术家和建筑师多达 60 位。开始的时候，居民特别戒备，过程中给他们做了很多工作，让大家明白我们是跟他们一起来做节庆的。到了开幕那天，住在旁边村子的子女全都回来了，自发地家家户户做不一样吃的。有一位驻地艺术家生活在这，和住在这里的老年人聊天，给他们写生，把写生、交流的过程变成录像作品和版画。还有香港设计师做的公共空间，最后展览撤走的时候大家都非常不舍，居民很习惯在这聊天，而且也认为展览的介入给他们带来很多有趣的东西和愉快的事情。

2017 深双大梅沙分展场在盐田，跟我所去过的城中村不太一样，特别安逸、干净，而且密度很低，没有南头古城或者是别的城中村的紧迫感和快节奏。居民以大梅沙酒店周边的服务人群为主，他们门口如果有一平方米或者是半平方米的地方就会种上菜，如果地方更大会种更多品种的菜，生活气息极其浓厚。正是如此，这次的介入以“村是厨房”为主题，选了一条 500 米长的路径，邀请 5 位建筑师改造沿路的 10 栋房子，同时邀请 20 位艺术家展出不同媒介的作品，其中的驻留艺术家更是深入社区与居民互动。当然也会有一些艺术家的文化无法在短时间内让居民产生共鸣，但是也正因为有这样多元的碰撞，才有新的可能产生。在深圳这么长时间，我们看到了艺术家、建筑师、设计师努力对这个城市进行改造以及大家的期待，我们也看到更大的一群大众和一波一波的年轻人对艺术的期待。

香港 | 社区中的艺术空间

（梁志和）

22 年前，在香港坚尼地城开始艺术空间的时候还未出现社区艺术的概念，当时香港当代艺术的发展规模小，要做独立策展的活动是很困难的。香港艺术中心作为主要的展示场地，租金非常高昂，我们因而选择入驻社区，把租金省下的花费用于艺术制作，而没有车道的社区使艺术品运输非常困难，我们就入驻到社区做自己的作品。这个社区非常安静，街坊邻里的关系特别好，艺术家也是普通人，到了社区就成为街坊。我们在创作的时候没有想到创作与社区的关系，可能是讨论的时候想到的，当时香港本地艺术圈有给予很多鼓励。

包租婆

再过一年后，我们来到上环一个小有名气的主要卖古董的社区，除了游客会去的古董店，其他都是与生活相关的小店。我们的店原本是烧腊店，完全对外敞开的，我们还是保持了一点它过去的形态。在最初的三个月，每个月都有展览免费向街坊邻居开放，“包租婆”觉得我们是做装修的。一开始，对面洗衣店的店主也很怀疑，我们是否是真的在做艺术。但后来，时间久了他们就成了展览的常客，每当有新的展览，他们就会来一起讨论，说出自己的观点。在这个社区，街道生活的氛围非常浓，艺术介入社区，我们也成为关系非常亲密的街坊的一部分。

随着关注度的提高，策展人和收藏家的到来，就有人斥资买下店铺进行改造，致使租金开始上涨，原来的店主，像洗衣店店主和旁边七十多岁老夫妇，都因为租金上涨搬离了。社区出现了很好的餐厅，成了外国人和游客的目的地。四年前，我们也因为无法支付高昂的租金而搬离，搬离前办了一场派对，有很多人来参加。搬走时，我们是最老的街坊，这是没有想到的，很感慨但是无能为力。在正规的艺术空间，虽然空间和展览项目都很好，但是和社区艺术空间的感觉完全不一样。艺术家和社区的关系非常的复杂，艺术家的角色也很复杂，原本是想创造集体回忆，可随着原居民的搬离，原本的社区实际上什么都没留下了。

上海 | 城事设计节

（尤扬）

我们以自媒体的身份作为主体发起城事设计节，集合设计师和社会组织，自下而上地进行微空间更新的活动。作为上海市民的我们，本身既是有设计能力的团队，公司就在愚园路上，更是以在地团体的角色对上海愚园路发起微更新微改造项目。

在愚园路的生活和工作中，发现了一些有机会进行微更新的地方。于是我们作为团队向政府提出带领年轻设计师和对街区感兴趣的人进行改造，匹配政府对街区微改造的需求，促成了这次城事设计节。城事设计节分为主题论坛和街区改造活动两部分，街区改造的部分会有资深的设计师作为导师，带领青年设计师。所有的成果最终通过跟微信数据客户端的联系，比对了一年以来这个街区的人流、节点的状况，分析做的这些事情能为街区带来什么。

改造活动为互相理解带来可能性，即城市管理者对街区用户、小商贩的理解，以及居民对街区的理解，而这个思路变成组织设计节的思路。另外，通过设计师的介入，使得街区的居民意识到本地社区的价值。我们做的事情是将设计圈内积攒的资源、政府和市民的需求以及设计师和品牌方的社会责任感结合在一起，形成城事设计节。媒体出资改造是看重设计带来的内容价值，故事的传播可以为媒体带来价值。

实际上，设计落地后才是一切的起点。因此，在项目开始之初，就建立了社区营造团队，进行前期的调研，包括街区的人口状态梳理、与当地居民和居委会工作人员的座谈会等，通过磨合，逐步融入街区。为了了解社区的需求，为弱势群体发声，减少士绅化的影响，设计用社区食堂作为概念，形成一个互相交流的社区营造基地。未来的目标是完全由在地的团队接手，植根于愚园路自发地造血。虽然在日本、中国台湾等地已有专门的工作者和社会职位，社区营造的工作价值在上海还没有被广泛的认可，所以社区营造需要跟经济体绑定，能自我运营，在社区惠及更多的人。

烟台 | 广仁计划

（谭芳）

烟台市核心的区域是芝罘区，拥有美丽的山海景观资源和历史底蕴。烟台于1861 年《天津条约》（1858 年）签订后第一批开埠。2013 年，烟台被评为国家历史文化名城，广仁路历史街区也被列为历史保护的区域。目前这个街区里有法国电灯厂、青年基督教教会、公共图书馆、绣花行、东亚罐头厂等，是非常重要的见证烟台近代民族工商业崛起的区域。中国创源作为广仁艺术区的运营主体，发起“广仁计划”，对老街区进行更新，建立烟台城市品牌与形象。

整个区域由两条路构成，一是 1851 年就存在的历史悠久的广仁路，现状保护得非常好，没有做过改变；另一条路是十字街，最初的十字街很长，现在仅剩下一段。整个区域有几百栋老建筑，是本地老人家和学者们拼着性命保留下来的。作为本地运营公司接手该项目时，区域内部分建筑被用作少数富人的私人会所、高级餐厅等，成为大众无力消费的区域，业态单一使得将近 6000 万人次的旅客主要活动停留在吹海风、跳广场舞，到处拍照等。在对问题进行梳理后的设计方向是把该区域打开，还给市民和游客，让更多的人走进老建筑，感受历史的底蕴和城市的发展，感受街区的美好。

前期与欧宁老师共同组建了以北京、深圳和纽约学者、实践者组成的顾问团，进行广仁艺术街区的活化计划。通过对这些历史街区业态的更新和转换，以及文化艺术、创意力量、社会教育资源的入驻，提升整个空间的生产力，从而让它真正地成为代表烟台城市活力和城市名片的区域。计划做了三个方面的工作：一、在海边打造文化地标；二、推动传统产业升级；三、把历史街区的场景营造出来。在老建筑的改造中，以新旧共生为理念，如董功的所城里社区图书馆，以谦虚的设计姿态唤醒老建筑。众建筑打造的广仁艺术空间是艺术区的旗舰机构，是芝罘学馆，也是原创的综合书店。

这些被改造的空间都会承载一系列的文化事件和文化内容的发生。艺罘学馆有一个很核心的内容是研究者驻地，有在地研究者计划，邀请相关的专家和学者进行田野调查以及出版相关的工作，使“广仁计划”的影响力不断地发生和蔓延。

北京 | 大栅栏更新计划

（贾蓉）

大栅栏是离天安门最近的老街区，也是保留最完整的街区。首先，带着对历史的敬畏感进行介入，因为一旦做错，多年的沉淀就没有了；另外也是最重要的——尊重与这个街区相关的所有人。我们的原则是完全按照自愿的方式，不强制居民，用设计介入的角度一点点地进行微小的改造。由于没有具体的先例参考，以试点的模式进行微小的介入。微小的方式能避免因方向错误而产生的谋略性破坏。

我们相信设计的力量，因为它可以帮我们在绝境当中寻找希望。它需要顶层设计和精密的规划，如大栅栏每年设计周之前都有策展的框架，把策展的内容编制到行动计划当中，然后确保做的事能把品牌、影响力、媒体借势能量发挥出来，转变成为街区后期的价值。我们认为，每一个街区都有自己的特点和不同的资源禀赋，项目可以借鉴，但是不可以复制。老街区的魅力就在于它的与众不同，要避免因为艺术和设计的介入而变得同质化。我们要知道这个区域开放的文化精神和内涵是什么，怎样理解、传承和发扬光大，需要一点点挖掘街区独有的生活形态和邻里的文化，注重调动人的力量。于是我们加入儿童教育以及本地的手工业再生，在再生的基础上评估是不是能够反哺到社区、带动自身力量的好项目。

关于机制建立的探讨，确定了四个关键词。它一定是开放的，只有这样不同的利益主体才可以对话；同时，它们相互之间是平衡的，而且还是动态的，整体也是系统性的机制。大栅栏投资公司以国企的身份介入大栅栏街区的更新计划，并连接政府和市场，包括当地商家和居民的主体，不管是哪个阶段，都需要界定各方的责任、权力和利益归属。有一些区域里面已有实施主体，平衡多个实施主体间的关系就需要第三方平台，比如说各种各样的专家组、顾问团，甚至有一点类似管委会的第三方机制。

大栅栏是两级管理的机制，就是在政府和市场两方中间分别设立两个中间级。一方面建立大栅栏跨界中心，作为国有企业对外的开放的市场平台，涵盖专业团体和本地团体互通。另一方面，整个计划不是单方能解决的问题，而是包括文化、商业、旅游、社工各个相关部门，从发改到国土的问题。于是西城区政府成立了比各部门级别稍高半级的大栅栏指挥部，由西城区主管规划的副区长挂帅，统筹指挥各个部门。这个双层机制使得前期试点有很好的依托。永远算不平的账的问题在机制层面可以从微观和终端解决，在不同的阶段尽可能调动社会主题、社会公益项目的积极性，其次是对不同项目性质的判断，设置到“肥瘦搭配”、收益平衡。最后重要的是评估收益不应仅仅是量化的经济收益，还有社会价值等评估的标准。

在城市设计的顶层设计中，分成三个阶段：第一个阶段完善市政基础设施；第二个阶段设立平台，邀请设计师进行建筑改造等，为下一步的社区共建形成示范性的作用；第三个阶段，政府包括实施主体退回到引导管理监督服务的职能，设立完善的机制和游戏规则，慢慢地调动起当地的力量，使得当地居民可以和市场上不同的资源方、运营方，进行自由的、有依据的、有管理支撑的合作。这是从软性的不同群体共同参与，一点点活化到一个区域自我驱动的发展规划。在过程中需要不断地让新进驻的群体和本地群体进行融合，对社区生活形态产生认同和归属感，建立纽带，实现社区人口结构的有机调整，从而避免大规模结构性的人口变化。

我们认为士绅化是一个必然之路，重要的是如何平衡这个过程，能够让改造过程中不同的人各取所需。刚刚说的宝藏一般的历史和文化，当地的居民是不敏感的，不是他们更重要的需求。然而，对于历史文化敏感的，对街区文化生态、价值认同的群体（有别于富人）会先进入，当进驻的这些人把埋藏的价值活化出来，让难以理解的价值更显性，就会带动更多的人去关注文化的保护。我们只需要把握好不同群体之间的平衡和融合，再由前期种的引导种子形成内生力量。同时匹配社区活动，包括教育和社区实践，唤醒当地的社区群体，形成带动的过程，对本地的群体实现人口结构有机更新的过程。

讨论：

南头古城或深圳城中村未来有机更新实施路径的可能性

戴春：城市微空间计划给予设计师参与的机会。就案例而言，上海的介入方式与深圳既有相似之处又有不同。在上海愚园路入口的改造非常成功，这是改造团队和街道一起完成的。而通常微改造的实现方式不是一成不变的，有的是政府有意愿寻求设计师的提案，有的是建筑师自下而上的实现后政府意识到价值。虹口片区历史改造组织开发商共同开发，对标的方向是日本的开发商共同成立委员会，类似开发政协的概念，共有一个资金池。开发企业介入引导成为一个积极的状态。

徐轶婧：老旧社区的更新改造，实际上是城市空间品牌价值的挖掘。传播和媒体做的事情是一方面，深挖和梳理这个地方所承载的历史、人，以及通过设计师介入探索这个地方的价值非常重要。叫内容生产也好、空间生产也好，学术上有很多的研究，这种工作特别重要。每天具体的事情是协调各方的关系，第三方平台是一种方式，真正核心的利益相关方还是要坐在一起，成立像职委会的组织，有一个具体的目标去实现。

叶子君：其实开发商也做不了什么。每一个案例都有各自的模式和改造思路，并没有一个标准的模型。例如南头是利用双年展的契机，用展览、艺术介入迈出第一步，在短时间内呈现，完成多节点公共区域的改造与业态更新。但是对于开发商来说，需要考虑经济成本，至少不能亏本，如水围村和股份公司合作，投资四千多万元，将29栋城中村农民房改造成青年公寓。项目有政府背景，同时也有市场考量，在运营中确保项目不亏本。

培养共建意识和以居民个体作为改造主体是应对士绅化的关键

欧宁：权属问题没有厘清是导致士绅化争议的起源。介入对于居民来说可能反映了一种权力结构在里面，成为强行介入。一个社区是老百姓根据自己的生活需要不断地变化和创造出来的城市街区的风格，有很多的优点，但是某种程度上老百姓的匿名程度，街道七拐八弯错综复杂的街道格局，包括迁建，都跟所谓现代化是背离的。现代化要求清晰化、标准化和好管理。这就产生了一种冲突，才会有改造的需要，而改造是隐含权力在里面的。对城市空间权属的认知改变是第一步。比较好的是共建的角度，它不是介入，是创造、是共建。承认城中村这种街道的风格，它隐藏的地方、知识跟现代化是互补的。这个地方不单属于居民，也不单属于外来改造和参与建设的人。最重要的是在不同的阶层和民间的力量、政府的力量共同营造一个地方的时候，达成共建微改造合法性的共识。

黄伟文：权属不光是所有权的问题还是主体的问题，租户的权力需要重新认识和界定，即作为一个居住者的权力是什么。在城市更新和地区发生改变的时候，在这个地方住的人应该成为这个地区改变的重要主体。但是，这个主体现在还没有被清晰地认识和彰显出来。2017深双催生了南头会客厅，是广州的建筑师过来租下房子做实践，以租户的角色参与实践，这是特别有意思和重要的。从城市更新、城市规划、城市设计，哪怕是一个建筑设计来说，我们已经建立了经济评估、生态环境影响的评估，却缺乏对社会影响的评估。

A Summary of "Reviewing CGD Laboratory"

Mo Sifei

2017 UABB, themed "Cities, Grow in Difference," is a platform that calls for new theories, new experiments, and new practice models. It's a biennale that keeps growing, emphasizing "discovery" instead of being limited to a pre-set dialogue. Only by growing beyond one's limited vision and highlighting multiple identities and perspectives can we recognize the hidden mechanism from observations. Through understanding and respecting the comprehensive, complex, and spontaneously growing network of urban villages from aspects of culture, society, and space, opportunities and possibilities of future Shenzhen in the era of "post-urban-village" can be seen.

Among the three chief curators of 2017 UABB, Hou Hanru, curated the Delta Laboratory event for the second Guangzhou Triennial (2005), while Liu Xiaodu and Meng Yan joined the event. Artists, architects, and scholars were invited to share insights and have an open discussion of the rapid urbanization in China, especially in the Pearl River Delta Region. With "Laboratory" as the name of the 2017 UABB curatorial events, it serves as both a tribute to this shared past of the curatorial team and an embodiment of the complexity and changing uncertainty of the venue of the 2017 UABB.

Nostalgia or preservation is never the sole purpose of having Urban Village as the venue of the biennale; it is an attempt to generate discussions of future possibilities. The curatorial process is an experiment and CGD Laboratory becomes the workshop.

· The First CGD Lab for 2017 UABB
· CGD Lab 2 × Do Course: Urban Village Stories
· CGD Lab 3: Informal City
· CGD Lab 4: Opening Ceremony Agenda + Artists Mobilization
· CGD Lab 5 × Urban Full View: Urban Festive Spirit
· CGD Lab 6 Nantou Planning Symposium with Tea Reception – the First Session of Nantou Street Livelihood Symposium
· CGD Lab 7: Old Town Symposium – the Regeneration of Nantou Old Town and Shenzhen City

The selected contents (CGD Lab 1, 2, 7) review the dialogues generated during the curation of *Cities*, *Grow in Difference*, stories about Nantou residents and discussions of the regeneration of urban villages in the future.

The First CGD Lab for 2017 UABB

As part of the open curatorial process of the 2017 UABB, the first CGD Lab elaborated on the theme of the biennale, discussed the logic of the contents, and explored possible visual representation approaches. Experienced curators and scholars were invited to discuss topics such as: how to tailor-made a biennale in urban villages; how to focus on the locality of the exhibition, so as to explore and reflect on urban villages from the perspective of daily life; how to showcase an exhibition both imaginative and life-real for the public; how to draw public attention and participation into the exhibitions and topics, by optimizing means of presentation, visitors guide, public education and villagers participation?

Urban Villages Perspectives: The Memory of the "Nine Streets"

From the beginning of the curatorial process, the team launched an on-site survey to observe and learn from the perspective of the residents. It attempts to reveal Nantou Old Town through the intertwining collective and personal memories as well as the grand and individual narratives. Honest stories challenge prejudices while reflecting the inherited real-life and complete history of the ever-growing contemporary urban villages. It serves as a way to rethink and criticize the monopolistic real estate-based development model which eradicated diverse urban life, culture, and memory.

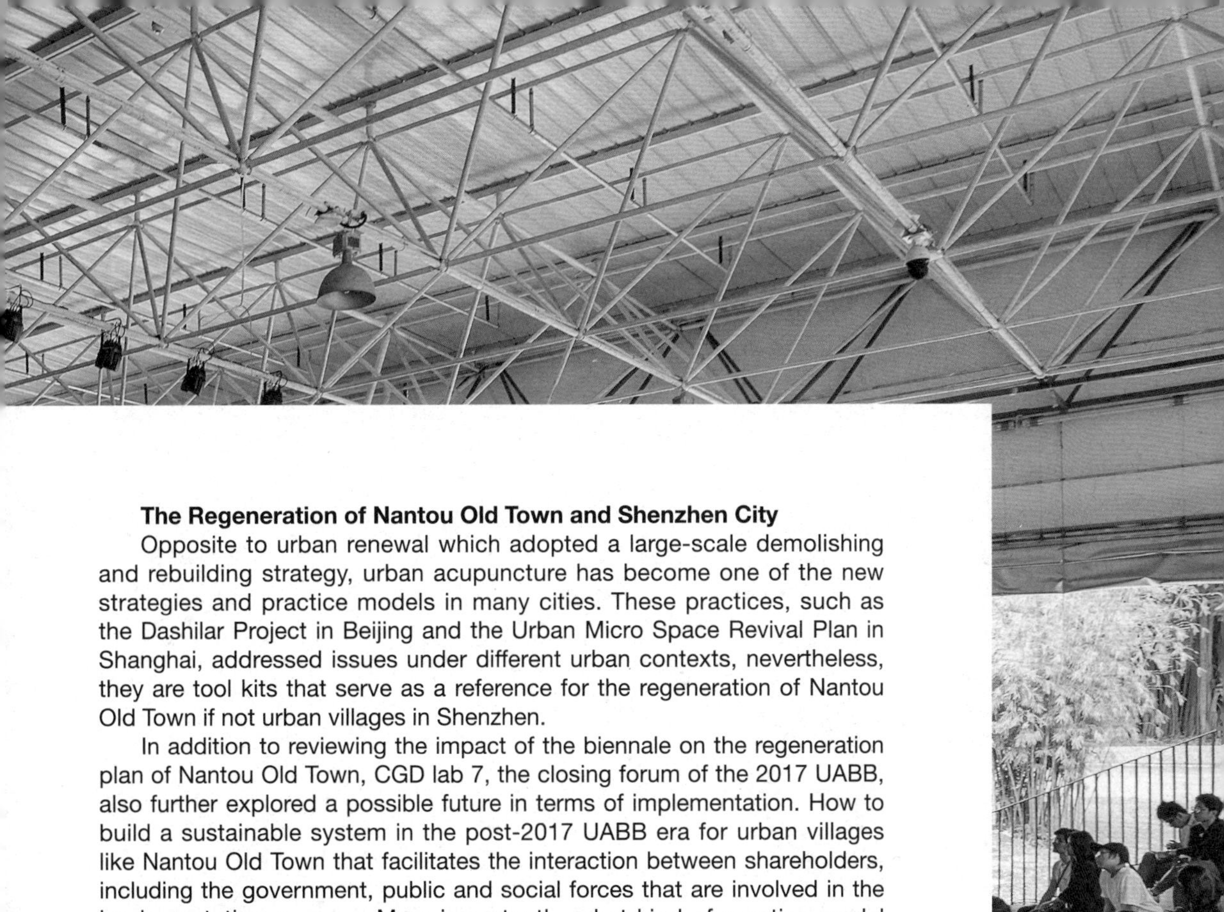

The Regeneration of Nantou Old Town and Shenzhen City

Opposite to urban renewal which adopted a large-scale demolishing and rebuilding strategy, urban acupuncture has become one of the new strategies and practice models in many cities. These practices, such as the Dashilar Project in Beijing and the Urban Micro Space Revival Plan in Shanghai, addressed issues under different urban contexts, nevertheless, they are tool kits that serve as a reference for the regeneration of Nantou Old Town if not urban villages in Shenzhen.

In addition to reviewing the impact of the biennale on the regeneration plan of Nantou Old Town, CGD lab 7, the closing forum of the 2017 UABB, also further explored a possible future in terms of implementation. How to build a sustainable system in the post-2017 UABB era for urban villages like Nantou Old Town that facilitates the interaction between shareholders, including the government, public and social forces that are involved in the implementation process. More importantly, what kind of practice model can carry on the regeneration started at the biennale? As mentioned by the curators, "the exhibition is neither a final solution nor an accomplished production, but a starting point, a turning point, an experiment, and an expedition." By learning the mechanism, strategies, and impacts of regeneration practices in different cities, we might gain insights into a way to the sustainable regeneration practices of Nantou Old Town, Shenzhen, and the Pearl River Delta area.

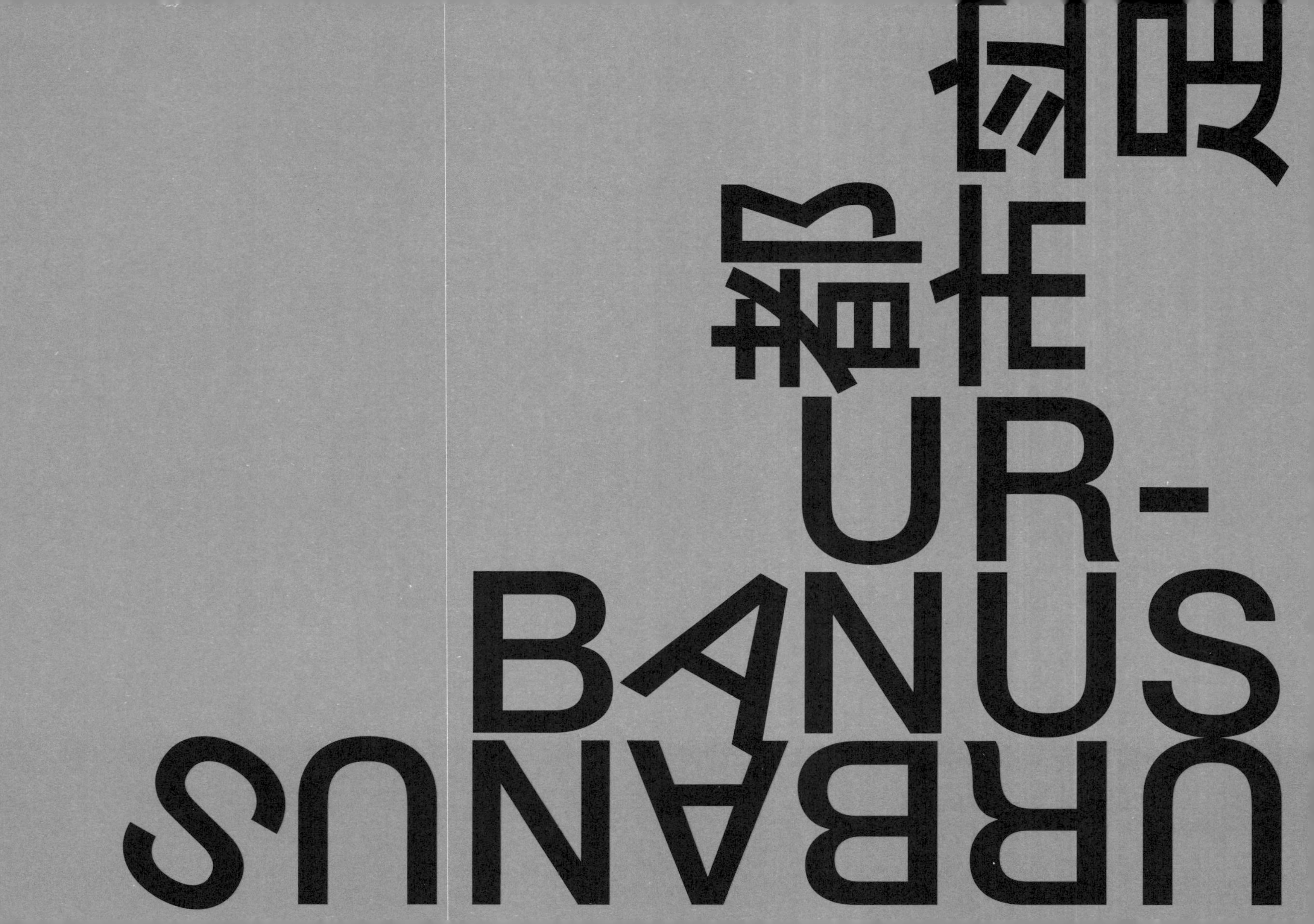
UR-
BANUS

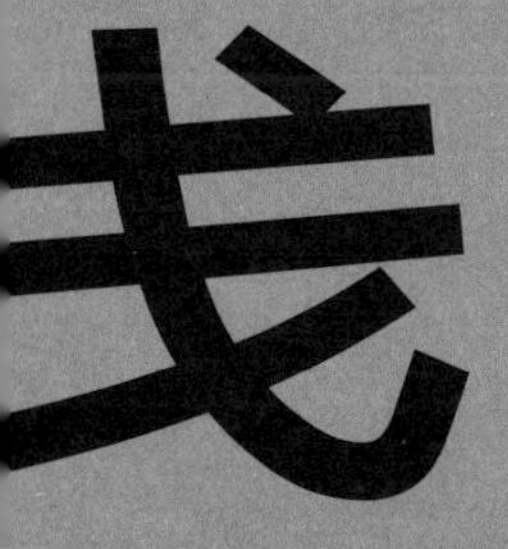

都市实践 URBANUS

肆

实践在都市
Practice for the City

“都市实践”在“深圳特区”，即是新观念的实践者游弋于新模式的实验场，在思想与行动的历险中，二者均“离经叛道”。与特区共生长，都市实践立足于建筑实践的城市性和社会性，直面都市症结，通过城市研究、建筑设计、艺术策划等多方位介入都市现场。其对城中村的研究与实践已有 22 年，随着亲历城中村的转变，对城中村的思辨和策略亦在不断演进；而未变的理念是对城市化进程中个体价值以及对未来城市中城中村多元价值的认同。在此，我们希望能够以一得之见，引众声喧哗。

《深圳，中国梦想实验场》以大芬村再生的故事讲述深圳，在 2010 世博会上呈现给全球观众。《白石洲五村城市更新研究》针对后城中村大规模开发模式提出超常规的空间策略，将空间升级与社会升级同步。《湖贝研究：城市图景与想象》复盘公共领域中关于一座城中古村去留的论辩，进而探讨城市认同感和本土文化的保育。《迈向“城市共生”》回溯南头城 / 村的改造思路和实施，在后 2017 深双的语境下再次呼吁多元共生。《这个疯狂的都市依然在不断地实践》梳理都市实践介入城中村实践的脉络和影响，以及所处的时代背景。《城 / 村计划进行时》通过城中村综合整治大背景下多方参与的龙岗六村提升计划，逐步构建出“城村共生”范式的雏形。

Within the SEZ context, URBANUS is a practitioner of new concepts, testing new models. They are unconventional and unorthodox in their ideology and action. Growing along with the SEZ, and grounded in the urbanity and sociality of architectural practice, URBANUS confronts urban issues and intervenes in the city fabric through multi-dimensional approaches, including urban research, architectural design, and public art. The experience of 22 years of research and practice in urban villages has resulted in the concomitant development of propositions and strategies. Yet the constant idea has been to recognize individual values in the urbanization process and appreciate the diversity of urban villages in future cities. We hope to initiate public discussion through these insights.

Via the EXPO 2010, "Shenzhen, Frontier for China Dreams" introduces Shenzhen to the world through the revitalization of Dafen Village. "Baishizhou Five Villages Urban Regeneration Research" proposes an unconventional spatial strategy for large-scale redevelopment in the post-urban-village era, synchronizing inevitable spatial upgrades and social improvement. "Hubei Research: Image and Imagination of the City" reviews public debates about demolishing an ancient village in the city, and discusses the concomitant cultivation of local culture and a sense of urban belonging. The essay, "Toward 'Cities, Grow in Difference'" inspects ideas about and implementation of Nantou Old Town regeneration, once again calling for coexistence through diversity in the post-2017 UABB era. "Architectural Delirium: URBABUS, Their Practice, and the City of Shenzhen" locates the context and impact of URBANUS' interventions in urban villages over time. "Ongoing City/Village Initiative" introduces the improvement project of six villages in Longgang District, highlighting multi-actor participation in comprehensive remediation and putting forward a prototype for a paradigm of City/Village Coexistence.

大芬 Dafen

大芬村，2006 © 都市实践
Dafen Village, 2006 ©URBANUS

报纸
NEWSPA-
PER
WAL-MART

⊙《南方都市报》2010年4月18日 星期日

“身尚着的蒙娜丽莎是先锋还是荒唐”

中国梦想实验场

上海世博会的倒计时牌，已越来越逼近开幕的准点。位于浦西展区的深圳馆也进入最后音效的调节与布展工作的冲刺。由深圳大芬村500多名画师集体创作的像素油画《这不是蒙娜丽莎》也已在深圳馆的外墙展现深圳的神秘微笑——欲将全球化语境下的城市秘语传递给即将到来的观众。

世博会成为世界各国，乃至各城市、地区、行业的形象展示工程，大家都铆足了劲，披挂着声光电，对自己进行宣传与讲述。比科技、比展示手法、比文化底蕴、比历史记忆、比未来空间、比想象力，也比态度的坦诚。这是一场城市营销比赛，深圳馆能否略胜一筹，能否将深圳的城市特征与城市梦想准确地传递给观众？就此本报特派记者前往世博会现场，对深圳馆进行全方位的发现与揭秘。

深圳馆以“中国梦想实验场”为主题，通过对大芬村这样一个城中村案例的声色提炼，来展现城市的变迁。这样一种以人的梦想与城市实践的跌宕讲述，是人文的，也是经济的；是城市的，也是个人的，它传递出直面城市现实的勇气。

深圳馆是一次深圳对中国当代精神的反思，它甚至带有对世博会的批判性，世博会就是技术崇拜，但每个城市最终拼的是文化，不是技术。

——周红玫，世博会深圳馆核心策展人之一、深圳市规划和国土资源委员会世博办主任

《大芬丽莎》这件貌似荒唐的作品，一旦你读懂以后，会发现它包含着诸多当前中国城市化进程中非常尖锐的话题，它敢于展示、敢于提问。

——孟岩，世博会深圳馆总策展人

世博深圳案例馆虽然是个小馆，但我希望它是个巨大的问号加感叹号，希望它能引起人们对中国30年城市化运动的反思。

——金敏华，深圳创意文化中心副主任

能够把大芬村提出作为世博案例，深

圳很勇敢。它没有偏执于展示深圳最好的一面，而是把深圳过去30年快速发展中最大的问题即城中村问题摆出来。

——廖维武，香港中文大学副教授、建筑师

在高科技、多媒体逐渐占领展览业乃至一切视觉艺术的时候，那些更真诚的讲故事的作品和态度反而有一种原生的力量。在深圳馆完成的这次行走，就像在儿童乐园。或许，这就是年轻的深圳给予参观者的最好礼物，重新找到童真、梦想、轻松等在城市中容易湮灭的东西。

——李骏，本报《城市周刊》特派上海记者

如何在80个城市案例中突围？深圳馆在这场声光电和生态化的比拼中，以馆中建馆式层次丰富的展区设计，以及实时影像和实物的结合，与诸多城市馆形象鲜明地区别开。而它以画工生态为切入点，也有别于一些城市相对保守的展示传统印象的形式

——周吟，本报《城市周刊》特派上海记者

我觉得深圳的方案是坦诚的。它坦率地承认了这座城市的当下处境，在承认现实的基础上，努力寻找自己的文化出路。比那些胡乱夸张、自我美化的假大空风格的场馆更有意思。

——张闳，文化批评家

世博会通常讲未来——科技的未来、

人类的未来、物质生活的未来，而深圳展现了一个从历史走向未来的过程，而且与这届世博会讲城市让生活更美好的主题很切合，是城市化过程当中的典型案例。

——余治平，上海社会科学院研究员

大芬丽莎这个作品放在 2010 年世界当代艺术语境里都应该算前沿的作品。

——牟森，世博会深圳馆世核心策展人之一、叙事总导演

对世博深圳案例馆，用三个关键词形容：密度、尊严、奇观。

——史建，建筑评论家、策展人

把大芬村作为主题直白地推出去，告诉观众：它是深圳城市发展造出来的现象，它足够怪异和有趣，而深圳还有很多如此怪异有趣的事，你想知道，就到深圳来看。

——于长江，北京大学深圳研究生院社会学教授

我的第一感觉是：很酷！让 500 个人来共同完成一幅作品难道不是很有创意的吗？

——萨宾娜，德国学者

深圳馆让我想到美国的一些创业历史。美好的现代的梦想正是建立在这样一些粗糙的、不那么美好的历史过程中。

——大卫，美国人，PRD 珠三角国际拓商会负责人

采写：《城市周刊》记者 谢湘南 王相明 李骏 周吟

摄影：《城市周刊》记者 胡可

Note: The *Dafen Lisa* created by more than 500 painters was featured on the facade of the Shenzhen Pavilion at the Shanghai 2010 Expo's Urban Best Practice Area, while the exhibition told narratives of individuals to represent the trait of Shenzhen City. The above news article included quotes from interviews with curators, critics, reporters, researchers, and scholars.

XIE X, WANG X, LI J, et al. "Is a lying Mona Lisa a frontier of absurdity?" [N]. Southern Metropolis Daily, 2010-4-18.

提案 PROPO-SAL

深圳大芬村：一个城中村的再生故事 2010 上海世博会城市最佳实践区城市自荐案例

周红玫
时任深圳市规划和国土资源委员会城市与建筑设计处副处长，兼深圳馆策划建设办公室主任

1　城中村——快速城市化过程中的普遍现象

在经济全球化的同时，发展中国家普遍面临着前所未有的快速城市化过程。中国改革开放 30 年，城市化水平从 20% 发展到 46% 并伴随着两个现象：一是城市边缘的农村被城市包围吞没，过程中又保留了部分的农村特征，形成“城中村”；二是大量外来人口的涌入，必然伴随着对空间的巨大需求，城中村成为容纳外来人口的重要载体。例如深圳，到 20 世纪 90 年代中期深圳共有城中村约 320 个，私宅 35 万栋，建筑面积 1.7 亿平方米，居住人口高达 600 万人，且绝大多数是城市外来人口。城中村的发展也面临着诸多问题，是“城市病”的高发地区，如经济结构单一造成抗风险能力差、人口高速流动难以形成完整有机的城市社区、人员结构复杂加大社会管理的难度，从而滋生犯罪、城市空间支离破碎等等。

2　深圳大芬村的再生过程

2.1　大芬村的概况

深圳大芬村是典型的城中村。大芬村位于深圳市城乡结合部，占地 0.4 平方公里。改革开放前村民以务农为生，原住民 300 多人，20 世纪 80 年代中后期大量外来人口涌入大芬村居住，并在周边的工厂中打工。一方面大芬村本身并没有系统的产业群和完整的产业链条，主要是传统加工制造业，居民收入很低；另一方面，没有形成良好的社区环境，社会结构松散，居民缺乏家园感和归属感。

2.2 从自发生长到主动再生

自发生长——1989 年，一个香港画商来到大芬村，租用民房进行油画的收集和转销，同时招募学生帮他完成外商订单，由此将油画复制这种特殊的产业带进大芬村。丰厚的利润和订单的增多吸引了越来越多的画工进驻大芬村，从最初的十多个人发展到鼎盛时期的上万人，且完全是一种自发的市场行为。

借文化产业政策而提升——深圳市政府在自发生长的油画产业基础上，顺势借助文化产业政策，适度投入资金，激活城市空间环境，修缮旧房屋和道路，兴建油画一条街，扶持油画交易市场，进行整体包装和推介等，开始主动促进大芬村的再生。

激发原创文化动力——随着大芬村的进一步发展，政府意识到，对产业生态的正确引导是大芬村未来发展的关键，有必要加强大芬村产业的完整产业链，并形成根植于大芬村自身的产业原创性。一个主要标志是，2007 年向大芬村植入了一颗“强有力的心脏”——大芬美术馆。大芬美术馆既是一个典型的“城市公共空间”，又是一个激发大芬村油画产业从复制向原创发展的“产业媒介”，作为一个新型城市空间复合体，其内在的高度复合性已经变成一种特殊的城市现象。某种意义上讲，大芬美术馆已经成为一个“象征性经济空间”，这种“象征性经济空间”是全球化时代的特殊景象（莎伦·佐金）。“新大芬村”已经成为深圳的一个创意产业基地、旅游景点和文化标志。

3 大芬村再生的城市实践经验

3.1 一种新的“社区活化”模式

城中村改造的普遍做法无外乎“删除”（推倒重来）和“美容”（穿衣戴帽）。推倒重来必须以开发量的成倍增加为前提，增加了城市空间、交通、市政的压力，割断城市与社会历史脉络的延续性，需要投入大量的资金、经历艰辛的谈判，且往往也难以推进；而以形象改造为主的“穿衣戴帽”也没从根本上解决城中村存在的实际问题。

大芬村再生提供了一种新的、适合于城中村发展的“社区活化”模式。其主要特点是通过市场、政府和社区的三者互动，选择一种适合于城中村的产业，通过适当的公共政策和适度的政府投入，引导原住民和外来人口加入到新的产业链之中，并借此培育良好的社会生态和社区氛围，形成城中村的“生长自循环”，从根本上解决“城市病”。正如《纽约时报》提到，“大芬村是……艺术与公共生活的结合，这正是一个有活力的城市的象征……”，通过美术馆这一公众设施将周边的城市肌理进行调整，使日常生活、艺术活动与商业设施混合成新型的文化产业基地。因此大芬村并不是一个过度补贴的样板项目，而是一个真实的社会生态和产业生态转型相结合的、为市场和大众所认同的社会功能区，是深圳众多社区活化的模式之一。

3.2 产业的培育和根植

大芬村最初以低成本优势承接国际低水平的产业分工，初具规模后又在政府的扶持下迅速壮大。历经 20 年的发展大芬村创造了奇迹，已有油画店铺 700 多家，占据了全世界 60% 的油画商品市场。2003 年大芬村的油画交易额约为 8000 万元，此后这一数字直线上升：2004 年 1.4 亿元、2005 年 2.79 亿元、2006 年 3.43 亿元、2007 年 4.3 亿元，直到 2008 年因受国际金融危机影响，交易额才降至 1.8 亿元。大芬村形成了自己的产业链，不仅出售各种风格、档次、用途的画，相关配件产品也全部能在大芬村内一站式采购，还有专门的美术培训中心、物流公司、包装公司等。政府通过主动引入原创画家、扶植创意产业等政策，特别是加大培训力度，使原住民和外来人口转变成为画工、手工艺人，建立大量的小型手工艺作坊，形成完整的生产、服务链条，大大加强了大芬村油画产业的根植性和本土性。

3.3 政府的适时、适度介入

2003 年深圳市政府确立“文化立市”战略，提出将文化创意产业打造成第四支柱产业。同年政府组织大芬村 30 多名画师前往南非、土耳其等国销售油画，大芬村的“名头”渐渐打出国门。到 2004 年底大芬村已云集全国各地的画家、艺术家、画工超过 5000 人。这一年大芬村也成为首届中国文博会唯一分会场，国家级的文博会成为推广大芬油画村和促进交易的重要平台，并又一次助推大芬村基础设施的建设。大芬村从一个密集混乱的城中村转变为与周边城市地区有机整合并具明显差异性的“典范社区”，其必要条件是：政府适时、适度、有效的投入；持续的关注、扶持油画产业发展；不断改善基础设施、完善社区服务等。这种具中国特色的公共政策扶持和引导，对于解决其他城市类似问题具有高度的示范价值。

2007 年政府投资的大芬美术馆落成。美术馆把展厅、画廊、商业、工作室等不同功能混合成一个整体，美术馆与城中村的结合，既是形式上的调和，也是让自发的社会形态能在被设计的环境中得以延续和发展的策略。大芬美术馆是一个独特的城市公共空间和产业空间的集合体。一方面它作为一个公共空间让几条步道穿越整座建筑物，人们从周边不同区域聚集于此，最大限度地提供公共交流机会；另一方面，又鼓励、推动了本土原创艺术的发展。美术馆本身也具有活力的公共空间和产业空间集合体。

4 大芬村再生的启示

4.1 城市创新文化的生长

作为独特文化奇观的大芬村，体现了都市创造性文化的社会基础及其根植于本土的创新行为。大芬村聚集了各种不同知识背景和手艺的人，他们完成了从农民向画工、画师、画商乃至艺术家的转变，跨越了阶层和文化界限。这种“非正式的、复杂的、多元的、小作坊式合作的都市模式”，反映了大芬村的开放与活力。大芬村的杂乱蕴含着无穷的创造力，在这里，有关“艺术和非艺术、艺术和技术、移民和原住民”等等传统的差异与分类并没有消失，只是变得界限模糊、相互渗透。

大芬村这一基于复杂性产生的创新空间，又成为当代文化艺术的素材和灵感，吸引着世界各地的艺术家来此创作，已形成更多元化的艺术创作领域，大芬村也成为孕育城市创新活力的重要源泉。德国艺术家克里斯蒂安·扬科夫斯基的作品《中国画家》组画的创作团队均为大芬村的画师，并在纽约、柏林、香港、深圳、广州（三年展）等地展出。他的作品突出了观念主义艺术实践中所扬弃的艺术创造力。德国著名摄影家迈克尔·沃尔夫也以大芬画师及其作品、工作环境为题材来创作。

4.2 城市“产业生态”和“社会生态”的重构

“产业生态”的重构——从农业经济到工业经济再到知识经济，快速城市化必然伴随着城市的产业转型，因此城市也面临构建新“产业生态”的机遇。城中村作为无数外来劳动力跨入城市的第一步生存空间，出现了多种多样的自我服务、自我循环的微产业体系。从自身的循环走向与城市循环紧密的融合，并进而成为城市发展的新能量机制。城中村的再生，不仅具有实践上的意义，也具有理论上的意义。大芬村再生所包含的产业发展模式，对城中村以外广泛城市地区发展也是有益启发。未来的城市空间，在适应城市产业结构升级和转型的同时，如何构筑相适应的空间经济结构，将是一个具有普遍意义的话题。

“社会生态”的重构——大芬村一方面形成了完整的产业链，另一方面也催生了各种城市活动的萌动，形成丰富的新社会生态结构。此种因地制宜的“就地城镇化”方式，使城中村以一种特别的角色重新融入城市生活，通过适度的公共政策、引导社会经济文化力量、改善环境质量与社会管理，以促进形成良性社区，这一持续的动态改良过程，最终使得大芬村成为活力、和谐、富足、安详的城市实践区。

Note: The text above, titled "Shenzhen Dafen Village: A Story of an Urban Village Regeneration—EXPO 2010 Urban Best Practice Area Case Proposal," is the approved proposal among eight submissions to represent the city of Shenzhen at the EXPO. At the time, the author Zhou Hongmei was the Deputy Director of the Design Department of Urban Planning, Land and Resources Commission of Shenzhen Municipality. The proposal elaborated on the spontaneous regeneration of Dafen Village and the insights it provided.

大芬村徐红家，孟岩摄于 2010
Home of Xu Hong, Dafen Village, photograph by Meng Yan, 2010

大芬村工作中的画师，2008 © 都市实践
A painter at work, Dafen Village, 2008 ©URBANUS

千人绘画现场，大芬村，强晋摄于 2010 © 都市实践
Thousand Painters, Dafen Village, photograph by Qiang jin, 2010 ©URBANUS

大芬美術館

深圳，中国梦想实验场

孟岩

URBANUS 都市实践建筑设计事务所创建合伙人、主持建筑师，2010 上海世博会城市最佳实践区深圳案例馆总策展人。

本文写作于 2010 世博期间，以《深圳案例馆：世博会盛宴之中的别样实验》为题发表于《新观察：建筑评论文集》，上海：同济大学出版社，2015；由孟元中译英；图片由孟岩、杨超英摄 © 都市实践。

历时 6 个月之久的史上最大世博会很快就要落幕了，随着鼎沸人声的退去和争奇斗艳、五光十色的世博场馆的拆除，城市将再次从喧嚣的大事件中回归常态。随着“后世博”时代正式来临，历史会以它一贯的方式，开始盘点并一步步过滤出光艳之后能够存留下来的世博遗产。

1　世博会看奇观

在 9 月的一天，我终于有机会再次走进世博园。与世博开展前为深圳馆的策划和落成身负重任日夜奋战那 5 个月不同，这一次，我以一个普通观众的心态，在上海的酷热稍退的季节中，几乎身不由己地跟随着或奔涌或蠕行的人流进出各大场馆。当身处由时尚的造型、新奇的材料、迷幻的灯光和影像所构筑的虚拟现实包围之中，看着身边同样鼓足勇气排队等候的人们脸上充满的渴望与好奇的神情，我不断自问：我们来世博会看什么？一个普通观众所期待的，必是奇观，必是见所未见、闻所未闻的未来奇景。

眼前世博园的宏阔场景准确印证了当下我们身处的这个充满着“大国”“盛世”豪言壮语的文化时态。30 年来经济的持续高速发展，以及从奥运会到世博会的一系列大型国际事件，把中国一次次推上国际化的大舞台，中国仿佛再次回到了世界的中心。然而这场急剧变革同时也造成了对社会心理和文化承载力的考验与挑战。今日的世博会，自然不同于几百年前的万国来朝，然而在这场历史上以规模最大、参展国最多而空前绝后的超级世博场景中，中国国家馆与各国展馆所构成的空间态势确实极具象征意义。仔细观察就会发现，这些庞大的世

图 1 大芬丽莎，杨超英摄于 2010 © 都市实践　Figure 1 Dafen Lisa, photograph by Yang Chaoying, 2010 ©URBANUS

博建筑，从国家馆和城市馆对国家 / 城市精神的表现到企业馆巨大的广告效应，它们共同交织所呈现出的正是身后不远处中国当代城市面貌的真实写照和浓缩。

1851 年首届世博会举办之后的近一个半世纪以来，它一直扮演着展示各国经济发展、科技进步和引领世界文化潮流的角色。作为世界各地共同参与的科技文化盛宴，世博会也是各地区、城市与企业的形象展示工程和吸引招商、旅游的城市营销竞技场。正是这样的语境，促进了各国展馆激烈竞争的态势：所有参展方都卯足力气，披挂着声光电加入到这场科技、文化、历史及未来想象的全方位大比拼之中。对于各个展馆的建筑来说，是否能够通过最大限度地与众不同抢夺观众的眼球，以求在这场宏大的视觉厮拼之中成功突围，成为问题的关键。然而在以往的历届世博会上，大多数争奇斗艳的展馆在这种超量信息的集体角逐之中往往相互抵消，难以形成持久的影响力以及进一步引申和讨论的学术价值。在信息爆炸的今天，在人们不遗余力地为世博会的空间集体盛宴创造技术和文化奇观的同时，也面临着在过量信息和商业竞争的裹胁之下其历史价值的自我稀释，以及这一过程的加速。而大多数场馆难逃其虽绚丽一时，但终将很快成为过眼烟云的命运。

2 深圳馆讲故事

世博会深圳馆能否超越展览瞬间的感官冲动，超越“博览”的奇观性，在追求震撼效果的同时创造更高层面的附加值，并引发更深层的文化思考和持久影响力？

一年前深圳市政府委托都市实践担任世博会深圳案例馆的策展以及总设计的工作，从展示理念的立意到展馆最终建成仅有 5 个月的时间，在策展和设计团队持续夜以继日的会战之后，深圳案例馆终于按计划落成了。

深圳案例馆的位置不在万国博览的浦东，而是坐落在浦西城市最佳实践区一栋白色张拉膜包裹的展馆内部，与韩国首尔和意大利博洛尼亚展区为邻。城市最佳实践区是上海世博会的首创特色展区，它紧密关注全球各主要城市针对其城市化过程中产生的种种问题而做出的有效和富于创造性的解决案例。不同于其他两个城市馆结合城市文化、产品和旅游宣传的开放式展示与表演，深圳案例馆可以说是一个严格意义上的案例讲述。

深圳案例馆以深圳的一个城中村——大芬村的再生故事作为叙事主线，从选题角度到展示方式都抛弃城市馆惯用的大而全的“成就展”，把对城市的宣传隐于叙事线索之间。从一个城中村的演化和再生出发，通过讲述普通人的故事，呈现底层劳动者最直白也最真实的梦想和他们为生存所激发的创造力和坚韧品德，从这里探讨深圳这座年轻城市 30 年奇迹背后的真实动力，我们相信正是这些鲜活的原生力量，汇成了一座城市的活力和尊严。

大芬村再生案例获得国际世博局的高度评价，其间也曾引发过广泛争议。在深圳人眼里，深圳有那么多高新科技产品，有那么多光鲜的地方不展示，用一个抬不上高雅文化台面的大芬村代表深圳出征世博？让一个城中村来代表深圳岂不让外人看笑话？外面也有人看不惯，有文化人甚至说：“如果我是一个深圳人，看到用大芬村的行画来演绎深圳的文化，我会像一个非洲人看到我们认为他们只会唱歌跳舞一样，内心充满遭受歧视的愤怒。”面对各种质疑，策展团队首先说服了正在全市范围征集优秀产品以在世博深圳馆展出的有关部门，让他们理解深圳案例馆不是通过在一个展馆里展示一堆优秀展品，而是让所有展品构成一次叙事，也可以说是以一个故事屋来演绎案例，而观众完整体验之后最终记住的不是深圳的哪些产品或技术，而是记住“深圳”这两个字和它所指代的城市精神气质，这是大宣传。

文化人对大芬村的不屑一顾很容易理解，在六年前我们开始深入了解大芬村时也是带着同样不屑的眼光作出评判的。问题出在只是道听途说和先入为主，这就使人们对大芬村的认识止步于“油画行画的产业村”这样一个单纯的“产业生态”。当我们把大芬村仅仅当作文化人茶余饭后的笑话谈资之际，就轻而易举地陷入了对一个异常复杂多样的文化生态简单粗暴地妄加判断的境地，成为偏见的受害者。

艺术圈对大芬村深恶痛绝事出有因，因为大芬村把艺术品的生产方式由个体的创造改为群体的大规模生产，它从根本上颠覆了艺术品传统价值的底线。对艺术家来讲，大芬村式的艺术品生产不但早已失去本雅明所称的艺术作品的“灵光”，而且嘲讽了艺术创作和艺术教育几乎所有的原则和禁忌。其实在大芬村赤裸裸地把西方经典艺术品批量生产，占领普通百姓家墙壁的同时，在艺术品生产链条的另一端，也有另一些人在有意无意间把摹仿自西方的艺术样式和观念的产品推向高端的艺术品市场。其实大芬案例关注的是人、产业和城市再生，如果从一个满足市场需求的产业发展来看，大芬村模式无可厚非，复制和标准化批量生产是产业化的基本特征。

很多新兴的中、高产阶层人士对大芬村的草根和低端文化也同样不屑一顾，他们明确指出其批量复制而非原创的艺术品生产就像盗版书一样上不得台面，而且低俗、缺乏文化品位。不过有趣的是当我们看到他们怡然自得地住在全国各地的开发商和建筑师们从欧洲、美洲、澳洲或其殖民地原封不动地复制过来的“欧陆社区”或整座“风情小镇”里时，立即使得这类文化讨论变得具有超现实的意味。

文化批评家、中山大学教授冯原先生指出了看待大芬村两种视角的不同，从现代主义和精英艺术的标准出发，大芬村的油画加工业很难称上是艺术的或原创的；但如果从后现代主义或文化多元主义的角度，在摆脱了现代主义的线性进步观和精英艺术的局限性之后，会看到它背后的另一层含义。大芬村之所以独特，是因为它成了中国进入全球化时代的一个缩影，非常形象地表达了中国深圳特区与世界的关系，即一个原来封闭的中国重新迈进全球性贸易和生产体系中的巨大力量，而其中展现的文化模仿和文化转型，都表现出强烈的中国特色。[1]

图 2 大芬画家
Figure 2 Dafen Painters

3 “大芬丽莎”的诞生

中国国家馆的“镇馆之宝”是多媒体版《清明上河图》，法国馆是卢浮宫名画，丹麦馆是原址移来的小美人鱼雕像，而深圳案例馆推出的最大艺术作品是一件 43 米长、7 米高覆盖整个展馆外墙的大型观念油画装置《大芬丽莎：这不是蒙娜丽莎》（图 1）。

达·芬奇的经典名作《蒙娜丽莎》几个世纪以来高居艺术圣殿之上，而作为一种文化符号，它早已通过不断复制的艺术品生产进入大众消费领域，成为全世界妇孺皆知的文化消费品；同时百余年间这张微笑的面孔又被无数艺术家不断改写再造，衍生出新的艺术作品和观念。

《大芬丽莎》是一个新物种，500 名大芬画家参与了这一集体绘制艺术事件。它由 999 块油画单元构成，作为原始文本的画像通过像素化的分解，画者的个性被忠实地保留在每一个单元之中。策展团队认为选用这张画非常准确地暗示了大芬村与全球产业链的关系，作为大芬村 20 年来生产最多的油画产品，它的内涵有充足的解读空间，可以作为大芬城市再生的样本。《大芬丽莎》作为深圳案例馆外墙，成了一个巨大的悬念和问号，为展览空间叙事开场。不难想象这样一个巨大问号所再次搅起的争议，使它的出世几经周折。一些人对大芬案例出征世博的疑惑更被这个巨大“山寨产品”所困扰，对于用它作为深圳案例馆的主体形象，一些文化人坦言“怎么说都看着别扭”。当然也有很多人对深圳在此所表现出的坦诚和胆量欣喜若狂，而更多的是冷静的分析解读。正如《南方都市报》的点评：“《大芬丽莎》这件看似荒唐的作品，一旦读懂之后会发现它包含着诸多当前中国城市化进程中非常尖锐的话题，深圳案例馆的价值在于敢于展示、敢于提问，而通过提问坦率地承认这座城市的当下现实并以此寻找自身的文化出路。”

《大芬丽莎》难道就能代表深圳？其实策展团队从来也没想过让它代表深圳，就像大熊猫不能代表中国一样。叙事总导演牟森解释道：选用这一形象还是为案例叙事服务，它与大芬土壤的关联性很密切，它是一个导读，是为完整叙事的第一幕设置的悬念。画背后的普通人才代表深圳，整个展馆的叙事就是要带着观众从画走到背后的人的故事和梦想。

记得《南方都市报》在报道深圳案例馆的专刊中用通栏大标题首先发问：“世界最大《蒙娜丽莎》躺着干嘛？”深具意味的是深圳案例馆的这件巨大油画装置正对面恰好是意大利博洛尼亚展区，躺着的《大芬丽莎》微笑注视着对面展出的优雅意大利古城景象和光鲜无比的红色法拉利赛车，这种超现实的空间并置是巧合，也是意味深长的一幕，以致上海一家著名报纸以惊讶的口吻感叹：意大利博洛尼亚给上海世博会带来了最隆重的

图 3 油画工厂
Figure 3 Oil Painting Workshop

合空间剧场装置，其同时承载展示和戏剧体验。说到戏剧，最初的确曾设想在剧场中用真人表演，但最后决定用空间氛围的塑造本身构成戏剧体验，这会更贴切地表现案例主题。

深圳案例馆的叙事结构采用经典的情节相贯的三部曲，也称“三联剧”。观众在展馆内几乎只能沿一条设定的线路行走，展览体验就是经历“序曲：大芬丽莎 / 深圳创世纪”“第一部曲：大芬制造”“第二部曲：大芬转型”“第三部曲：城市剧场 / ‘深圳——中国梦想试验场’”和“尾声”五个部分的串联观赏，并亲身参与这一部具有完整情节的空间戏剧作品。

礼物《蒙娜丽莎》；相对于相邻展馆超大超高清显示屏上的诱人的城市宣传片和重磅流行音乐表演，德国多媒体设计公司 THISMEDIA 用深圳 30 年前模糊不清的影像加工制作出类似 LOMO 的影效，轻松又智慧地配合了《大芬丽莎》纯手工的油画制作，深圳案例馆用更真诚地讲故事的方式反而具有一种原生的力量。

4　空间叙事三部曲

大芬村再生的参展案例以当代性、实验性和批判性介入城市现实、挖掘城市精神的策展基调得到相关领导的认同之后，如何展示成为最关键也是最困难的工作。对于一个占地 400 平方米馆中馆（42.6 米 x9.4 米，高 7.2 米），深圳可谓是重拳出击：深圳规划与国土资源委员会参照深双机制确立独立策展人，并着手组建庞大的跨界策划创意团队和学术、艺术顾问团队，引入建筑、当代艺术、实验剧场、多媒体艺术、视觉系统设计、纪录片、出版以及公共活动多个领域人员通力合作，希望各取所长，颠覆传统展示观念。更有价值的是设计同期进行了一系列跨界学术和艺术工作坊，广泛探讨有关城市精神、城市再生、城中村改造、地方性与全球化、艺术品生产与原创、民众的艺术话语权等问题。这些成果结合对大芬村以及周边地区的详细调研，出版了《特区一村》一书，并于展后结集为文献丛书。

都市实践总控展览空间设计和展示内容的完整性，先锋戏剧导演牟森勾勒出空间叙事的总体结构和拍摄纪录片，国家大剧院舞台设计总监高广建专注于城市剧场空间以及展馆整体灯效，THISMEDIA 创作大型多媒体影像，当代艺术家杨勇参与装置作品的制作，平面设计师张达利采用印刷和编织制作“深圳三十年大事件”的文献装置，80 后艺术家雷磊和由宓分别创作了多媒体动画，香港前卫音乐人龚志成用原创音乐串联起每一件作品以强化叙事结构的起承转合。在这里，每位艺术家都放弃了做自己独立作品的角色，作为设计者之一来完成一件丝丝入扣的整体作品，最后呈现出的是一个大型复

第一部曲是从一个独立的橘色空间开始，观众拾级而上，逼仄的空间飘落下百余个层叠纷乱的油画框，穿透墙壁，带出一系列大芬村史的影像档案。穿墙而入的红色集装箱下，一方迷你小剧场正在上演按大芬村史编的连环画《新山乡巨变》；接着在一个幽暗的房间，两墙相对铺满了自大芬村油画作坊采集来的风格各异的油画，以及在流水线的油画工场中分步完成的 20 幅油画样品，与此同时看到的一对影像，是大芬村两代创始人的虚拟对话；经由环绕着真人大小的大芬画工影像的空间，观众可以伸手触摸到天花板上悬挂的晒干的油画，似乎融入了影像现场；人们随即又和油画一起进入一个整装待发的集装箱内部，它是全球化产业链最恰当的空间隐喻；从《村史档案》到集装箱的现成物装置，以极高信息密度完成了第一部曲——“大芬制造”，叙事情节既有强烈的现场感，又有离奇、超现实的场景体验。

进入第二部曲的明亮开放空间，一个正在转型的大芬村被放置在 3 个活泼可爱的彩色盒子之中，成为缩微剧场，观众通过窥视感受大芬再造之中鲜活多样的文化生态。画工用过的调色板和颜料皮构成盒子的外皮，而一但走进这个区域，就可以听到大芬画工们略显羞涩地讲述他们的梦想，这里面有刚来大芬的年轻人、画家夫妇，还有残障人，他们用手艺实现梦想和生活的尊严：“我的梦想就是有房，有车，做个真正的深圳人。”这些最直白最朴素的梦想，因为真实所以感人。中国普通人的梦想其实就是一个简单的、过好日子的梦想，而正是这种梦想所催生的坚韧力量，构建了一座奇迹般的城市。

至此，对大芬村从村到城的城市再生剖面已经完整呈现，观众穿过一个通体亮绿、漂浮着几朵彩云的通道，拾级而下，转身进入一个豁然开朗的城市剧场。离村进城是这里的空间隐喻，深蓝色的数字瀑布从倾起的斜坡地面顺流而下，奔向观众，并在观众身后的墙上幻化出层叠起伏、生生不息的城市意象（图 2- 图 8）。《深圳面孔》纪录片与《深圳记忆》文献装置，一动一静相互支撑着剧场空间的叙事高潮。

当人们走出第三部曲的城市剧场，展馆尽端的大片影像墙展示着 2010 年 1 月 500 名大芬画家集体创作《大

芬丽莎》的工作场景，之后是5月底在大芬美术馆由中外艺术家及当地画家共同参与的“对流——大芬国际当代艺术展”和8月底开幕的“读村画城：大芬国际壁画邀请展”的盛况。深圳案例馆早已突破了上海世博会的空间限界，直抵深圳案例的原生地大芬村，在那里发动了一系列催生新一轮城市再生与文化生态转型的艺术和文化事件。

漫步在世博园，我会不自觉地沉浸在那些普通的中国大芬画工讲述他们直白、真实的梦想时的感动之中，在一个高科技和多媒体几乎占据一切视觉展示空间的时代，除却无限的视觉刺激之外，真实的感动似乎成了所谓的奢华。在新奇的科技盛宴之中，一点点对人的关注、对文化的敏感，都会给人留下深刻的印象和永恒的价值。几乎很偶然，在这个细雨纷飞的傍晚，当我走进芬兰馆中间的圆形露天中庭，惊喜地看到几十位来自芬兰的老人在庄重地合唱芬兰以及世界各地的古老歌曲，那韵律伴着天穹的浮云，呼唤着人们对那些清澈而古远的回忆的共鸣。这个白色的单纯得几近单调的圆形空间，在世博会科技、时尚的喧嚣盛宴之中划出一片静谧的场地，在这里人是空间的主角，人是仪式的中心，而人所创造的思想和文化，才能为城市的未来美好生活带来充满希望的一切。

图4 村史档案
Figure 4 Village Archive

Shenzhen, Frontier for China Dreams

Meng Yan
Meng Yan is the Principal Architect and Co-founder of URBANUS, and he was the chief curator of the Shenzhen Case Pivilion in the Urban Best Practice Area at the EXPO 2010 Shanghai.
The article was written during the EXPO 2010, which forms the "present moment" of text. It was published in *New Observations: A Collection of Architectural Criticism*, Shanghai: Tongji University Press, 2015, under the title "Shenzhen Case: A Pavilion beyond EXPO"; translated from Chinese to English by Meng Yuan; photograph by Meng Yan, Yang Chaoying ©URBANUS.

The largest World Expo in history lasted for six months will soon come to an end. With the receding of the crowd and the demolition of the miscellaneous pavilions, the city will once again return to normalcy from the hustle and bustle of this grand event. With the "post-Expo" era officially upon us, history will, in its usual way, begin to inventory and filter out the World Expo legacy that could last beyond the brilliance of the show.

1 Seeing the Wonders at the Expo

One day in September, I finally had the opportunity to walk into the Expo Park again. Unlike the five months before the opening, when I had worked day and night planning and completing the Shenzhen Pavilion, this time I was a mere spectator, following the flow of visitors involuntarily speeding up and slowing down, in and out of the pavilions in the midst of the receding heat of Shanghai. Surrounded by a "virtual reality" composed of stylish massing, novel materials, psychedelic lightings and illustrations, while looking around at the eager and curious visitors queuing up, I asked myself, "what am I here to see at the Expo?" Most visitors must be anticipating the spectacular and futuristic scenes that have never been seen or heard of before.

The magnificent scenes at Expo Park manifest today's cultural condition, which emphasizes the grandiloquent words "Great Nation" and "Golden Age." Ongoing rapid economic development over the past 30 years, and a series of major international events, from the 2008 Olympics to the 2010 Expo, have brought China to the international stage; China seems to have returned to its position at the center of the world. However, such drastic change poses social psychological challenges and tests our cultural carrying capacity. Today's World Expo is nothing similar to hundreds of years ago when nations paid homage to China, yet in this unprecedented huge world fair, which has attracted the most participants in history, the dynamics of the spatial configuration of China Pavilion and the rest of the country pavilions are indeed symbolic. The Expo's main structures, which range from the country and city pavilions representing national spirits to the enterprise pavilions announcing its presence, are a reflection and concentration of China's contemporary urban landscape not far behind them.

For nearly a century and a half since the first World Expo in 1851, it has played the role of showcasing the economic achievements and technological innovations of countries, while leading the world's cultural trends. As a scientific and cultural feast joined by all parts of the world, the World Expo is also a display platform and a marketing arena to attract investment and tourism for many regions, cities and enterprises. This context encourages fierce competition among the national pavilions. All exhibitors featuring lighting and sound effects are fully committed to joining the technological, cultural, historic and imaginary competitions for the future. With respect to the architecture of each pavilion, the ability to maximize eye-catching features in order to win the rivalry of visual representation is paramount. However, the flourishing pavilions of Expos past have often negated each other in the overwhelming information that such massive competition entails, making it difficult to achieve any lasting

图 5 全球产业
Figure 5 Global Industry

impact, academic value, or further elaboration and discussion. In the contemporary information bombardment, people have spared no effort to create technological and cultural spectacles for the sumptuous feast that is the Expo. Nevertheless, the historical value of each pavilion has been coerced into accelerated self-dilution by information overload and commercial competition. Most of the pavilions, despite their momentary splendor, will inevitably end up obsolete.

2 The Shenzhen Pavilion Tells a Story

Can the Shenzhen Pavilion go beyond the short-lived sensory stimulus and flamboyance of "exhibiting" to add value to a powerful experience, so as to provoke profound cultural reflection and lasting influence?

One year ago, the Shenzhen Municipal Government commissioned URBANUS to be the curator and chief designer of the Shenzhen Case Pavilion at the World Expo. From conceiving the exhibition concept to completing the construction of the pavilion took only five months. The curatorial and design teams worked around the clock to complete the Shenzhen Case Pavilion as planned.

Instead of being located at the Pudong international Expo venue, the Shenzhen Case Pavilion is located in a white, stretch-wrapped exhibition hall in the Puxi Urban Best Practice Area, next to the Seoul (Korea) and Bologna (Italy) pavilions. The Urban Best Practice Area is a first-of-its-kind exhibit initiated at the Shanghai Expo. It focuses closely on effective and creative solutions to the problems of urbanization in major cities around the world. Unlike the other two city pavilions that combine open displays and performances of city culture, product and tourism promotion, the Shenzhen Case Pavilion can be regarded as a rigorous case of storytelling.

The Shenzhen Case Pavilion takes the story of the regeneration of Dafen Village, an urban village in Shenzhen, as its main narrative thread. From the choice of topic to the style of display, it has abandoned the city pavilion model of "achievement exhibition" that unreflectively celebrates a city. Instead, the promotion of the city is immanent in narrative traces that comprise the Shenzhen Pavilion. The story starts with the evolution and regeneration of an urban village and moves through the stories of ordinary people, presenting the most straightforward and truthful dreams of the grassroots workers. Their creativity and resilience as they strive to survive is the real driving force behind the 30-year miracle of this young city of Shenzhen. We believe that these vibrant and indigenous forces that have converged to form the vitality and dignity of the city.

The case of Dafen Village regeneration has been highly praised by the International Expo Bureau, as well as sparking widespread controversy. Many Shenzheners see that their city has numerous high-tech products and wonderful places to be showcased. Why then choose the substandard Dafen Village to represent Shenzhen at the World Expo? Isn't it setting the city up for ridicule when an urban village represents Shenzhen? Some people from outside of Shenzhen are also unconvinced, including intellectuals who have said, "If I were a Shenzhener, seeing the culture of Shenzhen being interpreted by Dafen copy paintings, I would be as angry at this prejudice as an African who sees that we think they can only sing and dance." Faced with this skepticism, the curatorial team had to first convince the authorities who were collecting outstanding products citywide for display at the Expo to understand that the Shenzhen Case Pavilion would not be about displaying piles of outstanding objects, but rather about constructing a holistic narrative with all the exhibits—like a story house, so to speak—contributing to the interpretation of the case. Visitors would not remember specific products and technologies from Shenzhen, but rather the word

"Shenzhen" would be associated with the city's spirit, impressing visitors and honoring the city.

We understand intellectuals' disdain for Dafen Village because we ourselves made the same judgment six years ago before we learned more about it. Hearsay and preconceptions have limited people's imagination of Dafen Village to a "productive enterprise" like "a manufacture village for replica oil painting." When we regard Dafen Village as nothing more than the object of intellectual satire, we readily fall into the position of judging an exceptionally complex and diverse cultural ecology in a simple and brutal way, and thus become a victim of prejudice.

The art world also has valid reasons for their revulsion toward Dafen Village, which has fundamentally subverted the traditional value of artworks by changing the mode of art-making from individual creation to mass production. For artists, the Dafen Village way of producing art has not only stripped away what Walter Benjamin called the "aura" of artworks, but also made a mockery of almost all the principles and taboos of art creation and art education. In fact, while Dafen Village artists are blatantly mass-producing classical Western artworks and making art that hangs on the walls of common people's homes, at the other end of the art production chain, a number of people who, whether intentionally or unconsciously, are promoting products modeled after Western art styles and concepts to the high-end art market. People, industry and urban regeneration are the focal points of the Dafen case and there is no shame in meeting the market demand through reproduction and standardization, essential features of mass production.

Many emerging middle and upper class people are equally dismissive of Dafen Village's presumably low-end grassroots culture. They explicitly mention that the mass reproduction rather than original creation of artwork is as unseemly as pirating books, is vulgar and demonstrates a lack of taste. Ironically, it is precisely this group of people who live comfortably in "European-style communities" or "exotic towns," which throughout our country have been built by developers and architects who are replicating European, North American, Australian and colonial architecture. Once we acknowledge this truth, the cultural discussion takes on a surreal dimension.

Feng Yuan, a cultural critic and professor at Sun Yat-sen University, has pointed out the difference between the two perspectives of seeing Dafen Village. On the one hand, from the standard of modernism and elite art, the oil-painting-processing industry in Dafen Village can hardly be called artistic or original. On the other hand, however, from the standpoint of post-modernism or multiculturalism, once we break free from the linear progressive view of modernism and the limitations of elite art, we recognize the urban village's underlying significance. What makes Dafen Village unique is that it epitomizes China's entry into the era of globalization, vividly expressing the relationship between the Shenzhen Special Economic Zone and the world. Specifically, we observe the tremendous power of a formerly isolated China as it re-entering into the global trade and production system, even as the forms of cultural imitation and transformation display strong Chinese characteristics.[1]

图 6 寄往深圳的明信片
Figure 6 Postcard to Shenzhen

3 The Birth of *Dafen Lisa*

The highlight of the Chinese National Pavilion is the multimedia version of *Along the River During the Qingming Festival*. The French Pavilion has presented masterpieces of the Louvre, while the Danish Pavilion displays "The Little Mermaid" statue, which has been relocated from its original location. In contrast, the Shenzhen Case Pavilion has introduced the largest piece of art, a 43-meter-long, 7-meter-high, large-scale conceptual oil painting installation that covers the entire exterior wall of the pavilion, entitled *Dafen Lisa: This is not the Mona Lisa*.

The iconic *Mona Lisa* by Leonardo da Vinci has been considered canonical for centuries, while its long history of incessant reproductions for public consumption has made it a cultural commodity known to people all over the world. Meanwhile, relentless recreations and modifications by numerous artists over the past hundreds of years have resulted in new creations and novel concepts. The *Dafen Lisa*, a collective event of painting participated by 500 Dafen painters, is the first of its kind.

> **The value of the Shenzhen Case Pavilion lies in its audacity to show, to ask questions, and to frankly acknowledge the contemporary reality of the city through these questions so as to find its own cultural solution.**

The piece comprises 999 oil painting segments. The specifications of the original painting were pixelated in order to retain the style, individuality and mindset of the painter who is responsible for each painting unit. The curatorial team believes that the painting accurately hints at the relationship between Dafen Village and the global production chain. As the most produced oil painting product over the past 20 years in Dafen Village, its connotation allows ample room for interpretation and can serve as a paradigm of Dafen's urban regeneration. Wrapped around the exterior facade of the Shenzhen Case Pavilion, *Dafen Lisa* launches the exhibition narrative through suspense and a question mark. It is not hard to imagine the controversy stirred up by such a huge question mark, which has experienced several twists and turns since its unveiling. Some people who were doubtful about Dafen's suitability at the Expo were once again confused by this huge "shanzhai product." Some intellectuals straightforwardly opined that "it looks awkward no matter how one interprets it," when seeing it as the main image of the Shenzhen Case Pavilion. Of course, there are also many people who are enthusiastic about the honesty and boldness shown by Shenzhen here, but most of the voices are dispassionate analyses and interpretations. As the *Southern Metropolis Daily* commented, "The seemingly absurd work of *Dafen Lisa*, once understood, would reveal many critical topics in the current process of urbanization in China. The value of the Shenzhen Case Pavilion lies in its audacity to show, to ask questions, and to frankly acknowledge the contemporary reality of the city through these questions so as to find its own cultural solution."

Can *Dafen Lisa* represent Shenzhen? In fact, it was never the curatorial team's intention to have it represent Shenzhen, just as pandas cannot represent China. The chief director of the narrative, Mou Sen explained that this image was chosen to serve the narrative of the case because it closely relates to the soil of Dafen. It operates as an introduction, as a hook for the first act of a holistic narrative. The ordinary people behind the painting are the true representatives of Shenzhen, and the narrative structure of the whole pavilion takes the audience on a journey from the painting to the stories and dreams of those people behind it.

The *Southern Metropolis Daily* even published a special issue covering the Shenzhen case pavilion, which began with a banner headline asking, "Why is the world's largest Mona Lisa lying down?" The enormous oil painting installation of the Shenzhen Case Pavilion happens to be directly facing the Bologna exhibition. The implications are manifold. The prone Dafen Lisa smiles at the elegant scenes of the ancient Italian cities and a glamorous red Ferrari on display just across from her. This surreal spatial juxtaposition is a coincidence that has generated a larger picture. Indeed, a famous newspaper in Shanghai exclaimed in an astounded tone that Bologna,

图 7 深圳记忆
Figure 7 Shenzhen Archive

图 8 深圳时间
Figure 8 Shenzhen Time

Italy had gifted the Shanghai Expo with the grandest *Mona Lisa*. An adjacent pavilion showcased a tantalizing city promo and heavy pop music performance on the oversized ultra-high definition display. In contrast, the German multimedia design company THISMEDIA used blurry and indistinct images of Shenzhen from thirty years ago to create a LOMO-like shadow effect, which effortlessly and intelligently matched with the purely handmade oil painting of *Dafen Lisa*. All this to say, the Shenzhen case pavilion's use of a more sincere storytelling contains an intrinsic power.

4 A Trilogy of Spatial Narratives

Once the relevant authorities had approved the curatorial theme of the Dafen Village regeneration case—introducing a contemporary, experimental and critical intervention in the urban reality and unearthing the urban spirit—the question of how to realize this theme became the most critical and difficult task. For this "pavilion-within-a-pavilion" of 42.6 meters × 9.4 meters × 7.2 meters with a total of 400 square meters, Shenzhen has outdone itself. The Shenzhen Urban Planning, Land and Resources Commission has appointed an independent curator, following the protocols of UABB. It also set up a large cross-disciplinary planning and creative team as well as a team of academic and artistic consultants to integrate the fields of architecture, contemporary art, experimental theater, multimedia art, visual design, documentary film, publications, and public activities, hoping that together they could overthrow the conventional notion of an exhibition. More valuably, a series of cross-disciplinary academic and art workshops were conducted in tandem with the design, extensively discussing topics about city spirit, urban regeneration, urban-village transformation, locality versus globalization, art production versus originality, and the voice of ordinary people in art discourse. Integrating in-depth research on Dafen Village and the surrounding areas, the results have been published in *A Village by the SEZ* and will be consolidated into a documentary series after the exhibition.

So who did what? URBANUS has been responsible for the design of the exhibition space and the integrity of the exhibition content. The pioneering theatre director, Mou Sen outlined the overall structure of the spatial narrative and is shooting the documentary. Stage design director of the National Center for the Performing Arts, Gao Guangjian has focused on the urban theatre space and the overall lighting effect of the exhibition hall. THISMEDIA created large-scale multimedia images. Contemporary artist, Yang Yong participated in the production of the installations. Graphic designer, Zhang Dali employed printing and weaving to create an installation that depicts 30 years of major events in Shenzhen. Millennial artists, Lei Lei and You Mi created multimedia animations. Finally, Hong Kong avant-garde musician, Gong Zhicheng has linked all the pieces together through an original composition, reinforcing the narrative sequence. Here, each artist has relinquished their role of making a standalone piece, collaborating to complete a seamlessly integrated work. The result is a large composite spatial theater installation that holds both an exhibition and a theatrical experience. Speaking of theater, we did initially envision a live performance in the theater space, but eventually decided that a theatrical experience, which was shaped by the pavilion ambiance would be more pertinent to the theme of this case.

The narrative framework of the Shenzhen Case Pavilion adopts a classic trilogy of coherent episodes or a triptych. The pavilion layout practically forces visitors to walk a set route, while the exhibition experience is choreographed through a five-part series: the Prelude: Dafen Lisa/ Shenzhen Genesis; Part I: Made in Dafen; Part II: Transformation of Dafen; Part III: City Theatre/Shenzhen—The Experimental Field of China's Dreams, and; the "Epilogue. The layout enables hands-on participation in this spatial theatre piece within the context of a complete story arc.

The story starts in a standalone orange-colored space. The audience ascends the stairs, while over a hundred oil painting frames cascade within the cramped space, penetrating the walls and bringing out a series of video footage on the history of Dafen village. Under a red shipping container that crosses through the wall, a mini theater stages a picture storybook, *New Changes in the Village*, which is based on the history of Dafen. Next, in a dark room, two opposing walls are covered with oil paintings of different genres collected from the oil painting workshop of Dafen Village, along with 20 oil painting samples demonstrating the step-by-step production in the assembly line of the oil painting factory. This scene is framed by a pair of images presenting the virtual dialogue between two generations of Dafen Village entrepreneurs. While walking through a space filled with life-size videos of Dafen painters, the audience is encouraged to reach out and touch already-dried canvases that hang from the ceiling, as if they too have become integrated into the scene. One then enters the interior of a shipping container that is loaded with oil paintings This is the most appropriate spatial metaphor for the globalized production chain. From *Village Archive* to the ready-made installation of the shipping container, the first part of the exhibition is a high-density information experience—the "Made in Dafen" narrative provides both a strong sense of the site and an extraordinarily surreal experience.

Upon entering the bright open space of the second part, Dafen Village-in-transformation has been placed in three lovely and colorful boxes that operate as miniature theaters. Visitors can spy on Dafen through these panoramas, experiencing the vivid and heterogeneous culture that has characterized Dafen's reconstruction. The exteriors of the boxes are covered with used palettes and paints, and when stepping into this area, a surround sound system permits one to listen in as the Dafen painters modestly tell their dreams. Among the dreamers are young people who have just arrived in Dafen, painter couples, and people with disabilities, who use their crafts to realize their dreams and to live with dignity, "My dream is to have a house, a car, and be a real Shenzhener." The truthfulness of these straightforward and simple dreams touches visitors. Indeed, the dream of the common people in China is actually as simple as leading a decent life, and it is the resilient power spawned by this dream that has built a miraculous city.

At this point, the rural-to-urban regeneration of Dafen Village has been presented in its entirety. The audience then walks through a bright green passage, a few colored clouds floating overhead, and after descending the stairs, they turn around and enter an open urban theater. The spatial metaphor here is "leaving the village and entering the city." A digital, dark blue waterfall flows from the sloping ground toward the audience, conjuring up the image of the bustling city on the wall behind them. The documentary film *Shenzhen Faces* and the literary installation *Shenzhen Archive*, two pieces, one dynamic and one static, work together to support the narrative crescendo in the theater space.

When visitors leave the city theater in the third part, they encounter a large video wall at the end of the pavilion. Footage includes the working scene of 500 Dafen painters collectively creating *Dafen Lisa* in January 2010, the CONVECTION: Dafen International Contemporary Art Exhibition, which was held at the end of May in the Dafen Art Museum and featured Chinese and foreign artists together with local painters, as well as clips of the *Reading Village Painting City: Dafen International Mural Invitational Exhibition*, which opened at the end of August. The Shenzhen Case Pavilion has long since transcended the spatial boundaries of the Shanghai World Expo and made its way to Dafen Village, the original site of the Shenzhen Case, where it has unleashed a series of artistic and cultural events that have spawned a new round of urban regeneration and cultural transformation.

Strolling in the Expo Park, I inadvertently immerse myself in the touching moments when those ordinary Chinese painters of Dafen tell their most candid and authentic dreams. In an era when high technology and multimedia occupy almost all spheres of visual presentation, a real touching of the heart seems to have become a luxury. Amidst the feast of novel technologies, any attention to people or sensitivity to culture will leave a deep impression with enduring value. It is almost by chance that, on this drizzling evening, I enter the circular open-air atrium of the Finnish pavilion, where I am amazed to stumble upon dozens of Finnish seniors solemnly singing ancient songs from Finland and around the world. Alongside the floating clouds in the sky, these rhythms resonate with the limpid yet remote memories that each individual carries. This white, almost monotonous, circular space, in the midst of the clamorous feast of technology and fad at the Expo, demarcates a tranquil ground where people become the protagonists and the core of the ritual. Here, I believe, is where the ideas and culture created by people bring hope for a better future in every city.

[1] 勇敢的深圳，展示从中国制造到中国创造的梦想 [N]. 南方都市报 · 城市周刊 , 2010-4-17.

白石洲 Baishizhou

报纸
NEWSPA-
PER

⊙《南方都市报》2017年6月23日 星期五

“白石洲将打造350万m²城市综合体”

旧改规划草案公布，将建成住宅 1250000 ㎡，含保障性住房 50000 ㎡

6 月 15 日，深圳市规划国土委南山管理局公布了《关于南山区沙河街道沙河五村城市更新单元规划（草案）》，明确了计容面积近 350 万平方米，包括住宅、商业、办公、商务公寓等各种物业的面积，以及配套设施建设，这也意味着备受关注的白石洲旧改即将启动。

旧改规划草案出炉 旧改航母即将起航

根据《关于南山区沙河街道沙河五村城市更新单元规划（草案）》，该项目位于沙河街道的白石洲城中村，东临华夏街，南临深南大道，西临沙河街，北临香山西街，由位于深南大道北侧的塘头村、下白石村、上白石村、新塘村以及深南大道南侧的白石洲村 5 个自然村组成，又称“沙河五村”，而这片面积约为 0.6 平方公里区域也是深圳最大的城中村，居住人口约为 15 万，其中绝大多数为流动人口。

深圳市规划国土委南山管理局方面的资料显示，此次白石洲片区更新单元用地面积 480148.0 平方米，拆除用地面积 459542.1 平方米，开发建设用地面积 303793.7 平方米，计容积率建筑面积为 3479550 平方米，包含住宅、商业、办公、酒店、公寓、学校等多种业态规划。其中，有包括 5 万平方米的保障房在内的 125 万平方米住宅，104.5 万平方米的商业、办公及旅馆业建筑，以及商务公寓 112 万平方米，公共配套设施 64550 平方米。此外，社区体育活动场地用地规划有 1.8 万平方米。整个项目共由 31 栋 49~65 层住宅，21 栋公寓，3 栋 66~79 层超高层写字楼，1 栋 59 层办公楼组成。

值得一提的是，根据此次沙河五村城市更新单元规划（草案）公示，该项目共有 3 个地块规划了九年一贯制学校，办学规模分别为 27 班、36 班、27 班。此外，白石洲旧改项目还规划了 1 个 9000 平方米养老院，以及 3 个老年人日间照料中心，每个 750 平方米。

小渔村实现现代化 城市综合体的蜕变

资料显示，白石洲的历史可以追溯到 200 多年前的清代，当时这里还只是一个叫万家洲的小渔村，南面是深圳湾，因村子建在海湾的沙洲上，村后山顶上立着一块大白石，故名“白石洲”。而如今的白石洲早已成为深圳最大的城中村之一，2700余栋村民的自建楼里,住着近15万人，几乎都是外来人口，由于紧邻华侨城片区的高档住宅小区。有人用“一边是欧洲、一边是非洲”来形容这里。

早在 2005 年，深圳市和南山区两级

政府就开展了白石洲旧村改造的研究工作，2012 年 8 月，成立由南山区委区政府领导带队的沙河五村城市更新领导小组，旧改进程得到推进。

2014 年 7 月，《2014 年深圳市城市更新单元计划第二批计划》公布，明确沙河五村城市更新单元（范围调整）申报主体为深圳市白石洲投资发展股份有限公司，拟更新方向为居住、商业等功能，而白石洲旧改之所以称之为“旧改航母”，是因为拟重建的白石洲沙河五村用地面积 45.9 万平方米，重建用地面积之大，在当时 18 个城市更新单元项目中名列榜首。

2017 年 6 月，市规土委南山管理局发布沙河五村城市更新单元规划（草案）的公示，白石洲旧改将拆除用地面积近 46 万平方米，开建 348 万平方米综合体，如若项目顺利推进，将是截至目前深圳计容面积最大的旧改“航母”。

（略有删节）

《南方都市报》记者 陈文才
实习生 王凯 摄

Note: The renewal plan of Baishizhou was promulgated on June 15, 2017. The redevelopment plan proposed an urban complex of 3.5 million square meters including new condos, commercial and offices. Nearly 0.46 million square meters of land were targeted for demolition and rebuilding. The renewal project is called the aircraft carrier of urban renewal due to its large scale.

CHEN W. "Baishizhou will become an urban complex of 3.5 million square meters" [N]. Southern Metropolis Daily, 2017-6-23.

白石洲，2013 © 左氏文化
Baishizhou, 2013 ©Zeus Culture

白石洲塘头村老屋，2016 © 都市实践
Old houses in Tangtou Village, Baishizhou, 2016 ©URBANUS

新旧家私家电买卖
楼二旅
南北

白石洲塘头村古井，孟岩摄于 2010
Old Well in Tangtou Village, Baishizhou, photograph by Meng Yan, 2010

白石洲新塘村盆菜宴，孟岩摄于 2013
Poon Choi Feast in Xintang Village, Baishizhou, photograph by Meng Yan, 2013

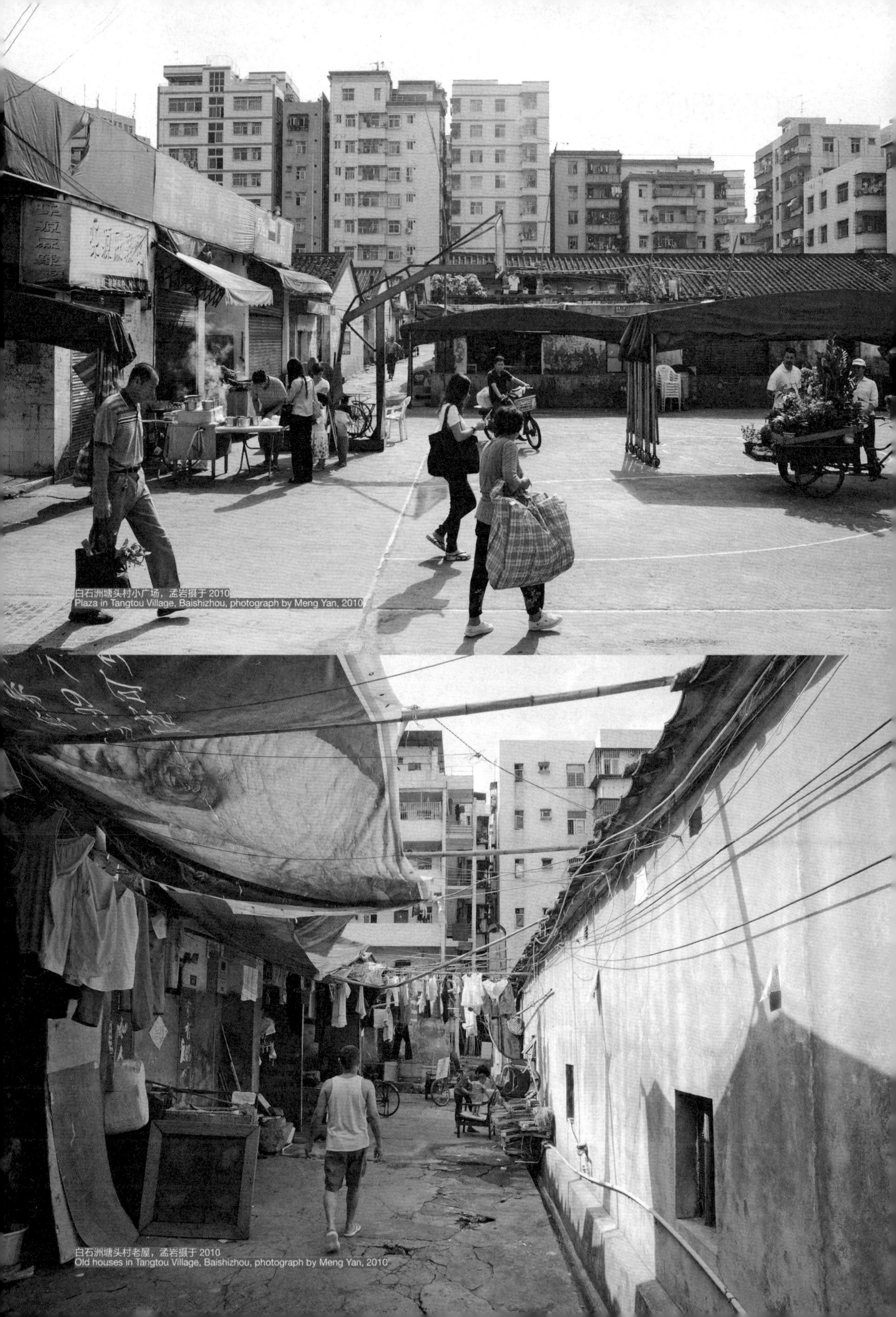

白石洲塘头村小广场，孟岩摄于 2010
Plaza in Tangtou Village, Baishizhou, photograph by Meng Yan, 2010

白石洲塘头村老屋，孟岩摄于 2010
Old houses in Tangtou Village, Baishizhou, photograph by Meng Yan, 2010

白石洲五村城市更新研究

曲韦世
ONE 建筑事务所主持建筑师，美国哥伦比亚大学建筑、规划与历史保护学院中国农村发展工作坊创立人，曾任都市实践香港团队主管（2011—2014）。
本文首次发表于《城市 · 环境 · 设计》, 2018(12): 290-295；由都市实践英译中；图片 © 都市实践。

对于世界来说，深圳几乎是一夜间建成的，这种几十年间爆炸式的城市化进程即使放到这个国家数千年的历史中也是一枝独秀。电光石火之间，珠三角的荒野和农田已然变成每年向全球经济注入万亿元的经济重镇。昨日的农民和渔夫今已身处一个拥有超过千万人口之地。劳工、商人、设计师、企业家——拥有梦想的人们从中国各地纷至沓来，在深圳追逐未来。对于这个每天都在变得更大、更密、更高的城市来说，其中的人们甚至没有时间考虑子孙后代该如何发展。在追逐现代化的旅程中，不论是城市的原貌还是发展中的点滴都在人们记录之前转瞬即逝，只在记忆中留下一道掠影——这里的人积极向前，并以外人无法企及的步伐前进。在不停运转的起重机、拔地而起的混凝土森林和无数建筑工人的身影之中，这座城市正发生着日新月异的变化。 而白石洲、湖贝、蔡屋围、大芬和其他城中村的街巷，既是对她短暂历史的记录，也是对当下快速工业化的生动注解。

在整个广东省，特别是在深圳，农业用地被蚕食，村庄被吞没，昔日的农民失去了生计。由于无地可种也无法出售房产，这些城里的农民也开始建造房子。 随着他们身边城市的发展，以及随之而来的数十万外来务工人员对廉租公寓的需求，城中村也随着城市不断发展壮大。一开始是两层小楼，接着加建到四层、六层、十层。2012 年，当都市实践第一次参与研究白石洲的城市更新时，深圳的人口已逾千万，其中 200 多个城中村就吸纳了数百万人口。 据估计，仅仅是白石洲不到 0.65 平方公里的范围内就有超过 15 万的居民。

城中村出租屋的合法性模糊不清（房屋可以出租但从未有正式的租赁协议）造成市场的极度扭曲。城中村公寓的租金也相应缩水，远低于预期的市场价值。 在一个房产价格暴涨的繁荣城市，这成了唯一的经济减压阀。在城市无数的公寓中，城中村提供了真正经济实惠的住房，给无数外来务工人员一个片瓦遮身之地。三十多年来，深圳依靠城中村悄悄地为经济注入动力。在城市发展初

图 1 白石洲愿景拼贴图，都市实践，2013 Figure 1 Baishizhou vision collage, URBANUS, 2013

期，城中村暂时安置了让这座工业大都市拔地而起的工厂职工和建筑工人。之后城中村又为第二波服务于新建的餐厅、酒店和豪华公寓的外来服务人员提供了住处。如今的城中村，特别是白石洲，又一次支撑着以科技为主导的城市转型。高昂的租赁市场制约了政府极力倡导的年轻一代应有的创业精神。对于在设计前沿团队积累经验、在刚起步的公司任职或创办自己公司的年轻人，正是城中村以相对低廉的租金消除了他们的风险。遍布“中国硅谷”的创始人和高管们越来越多地引用他们的“白石洲故事”描述他们在深圳的第一个落脚点。 从政府会议室到国际杂志，城中村所发挥的关键经济作用已成为公认的事实。

与之相对的是，除了少数活动家和学者，人们对城中村本身历史价值的认识进展缓慢。然而在缺乏共识的情况下（以及模糊的、未来不确定的现状下），各种各样的士绅化已经开始了。白石洲等位置优越的城中村中所体现的临时性和审美多样性正迅速被诸如快闪咖啡厅、精品啤酒屋和各类艺术装置同时利用。而这些城中村在不同层面上的转变也会引发这样的问题：究竟什么构成了城中村的保护？什么应当被保护？什么应当被避免？外晾的衣物、夜市、纵横交错的电缆、握手楼、狭窄的蜿蜒小巷——这些都是城中村不可或缺的关键元素吗？保留城中村在深圳经济中扮演的特殊角色是必要的吗？是应该聚焦在那些为城中村的空间肌理注入活力的人群和文化上，抑或着眼于城中村独一无二的空间肌理本身上？这些建筑有历史价值吗？应该保留建筑物本身吗？最容易的答案是“保留以上所有的东西”，但令人痛苦的悖论是，这种方法从根本上是不可能的。 虽然人们可能希望以某种方式把城中村打包并为后代整个保留起来，但不可避免地，保护本身也是一种干预行为。因为如果此时把城中村变成一件文物来收藏，就如同让时间停止一样。变化是城中村的本质，一旦城中村停止了变化，它自身也将不复存在。另一方面，如果均质化地开发城市而忽视城中村的存在也必然会使其消失。

因此我们的任务是以发展变化的视角，挖掘城中村重要的特质，使其在现代化的进程中得以保留、引导，并继续发展下去。这样城中村的历史才不会丢失，未来也不会被切断。

正如其名，城中村的特点是其与周边城市截然不同的空间结构，它以一种与深圳其他城区完全不同的尺度，按照一套完全不同的规则发展起来。同样，城中村的人际结构也完全不同，这是一种残留在大城市中的、乡村式的紧密人际关系网。城与村这两种关系可以说是交织在一起，密集且细分的城市街区使得这种乡村化社区结构得以在几十年的外部变化中生存下来。而让外界观察人士和村民自己都担心的是，按典型的深圳模式对城中村进行物理结构的推倒重来可能会危及其内部的社会结构。

由于现在的村民更愿意在城中村升级后继续依托血缘和地缘关系生活和工作，所以建筑师必须研究出适合的聚落类型策略。对于城中村区域更新，必不可少的就是维护和加强其中的邻里和社区关系，促进其成为一个整体。然而，套用网格规划或类似的表面化措施来实现保留社区的想法无疑是欠考虑的(并且可能具有破坏性)。与空间划分同等重要的是功能混合的特征。休闲和劳作、公共和私有、创业主义和集体主义空间的重叠，营造了简单均质化发展模式所不具备的动态相互关系。城中村肌理中混杂的空间结构可能和当代城市规划中明确界定的二元分区相对立，然而它却是激发社区活力的关键。白石洲和湖贝的街市上那些五花八门的商品不仅仅卖给本地居民，还能将邻近社区的居民和远道而来的游客源源不断地吸引到这里的街巷。同样，这些商业地带和周围的住宅区以及轻工业空间的相互叠加也创造了多功能的城市据点，在自给自足的同时联结了周边地区。社区

图 2 白石洲总体规划图
Figure 2 Baishizhou master plan

虽然深圳经常被认为是缺少建筑历史的城市，但在深圳仍能找到一些具有历史价值的建筑的特例。其中最重要的要数湖贝古村，其中部分建筑已有五百多年的历史。另外在白石洲也遗留着早期白石村集体农场的 11 幢公社建筑。在缺乏历史建筑的深圳，这些遗址更应该得到重视。因为它们是深圳与其建立之前的历史之间物质上仅存的联系。由于这些历史建筑处于城中村之内，开发这些区域的难度使它们得以保存到今天。而随着城中村的边界在城市发展中消失，针对这些濒临消失的遗产的保护措施势在必行。

限制城中村继续发展的极端条件（土地补偿率的天价、靠单体建筑带动大片土地的开发，以及在已经过度拥挤的社区中将城市密度翻两倍或三倍的必要财政支出）自相矛盾地形成一个这样的局面：20 世纪 60 年代欧洲的未来主义者们所描绘的巨构建筑不仅在这里找到它们第一个实践机会，而这种思想可能也是重新思考城中村下一个发展周期的最佳理论基础——在国家生产机器之中将其作为大规模居住的组件。有趣的是，正是在这种类型的土地上，中国前卫派的设计公司现在正在构思一种在以前不可能存在的建筑——那是同行前辈们没有实现的、带着怀旧色彩的未来幻想。

白石洲是（并且至今仍然是）深圳有史以来最大的城中村重建项目。提交给深圳市的“再生”提案里建议将该村的容积率从 4.5 增加到 9，并在 5 年内建成一个超过 550 万平方米建筑单体的开发项目。 这种快速而粗糙的大规模开发计划据称是解决城中村 2000 名房东极高的拆迁补偿的唯一方案。都市实践在试图完成这一不可能完成的任务时，有必要扮演一个调解人的角色，在现实的经济因素与城市设计和保护的理想目标之间寻找平衡点。这个项目恰好委托给了合适的团队，毫不夸张地说，都市实践是设计行业内城中村问题讨论的发起者。通过在湖贝村第一次螳臂当车的努力，到在大芬、岗厦、阜新、南头等地持续地在研究和实践上投入精力，都市实践一直致力于研究城中村可行的更新模式以替代原地铲平一整体出售的“焦土政策”。与许多都市实践的项目一样，白石洲工作的核心是在特定的文化和经济背景框架内探索高密度和巨型结构的空间限制

和活力对于理解城中村在城市中扮演的角色起着关键的作用。然而对于设计师来说同样重要的是，发现不同城中村各自独特的空间特征，并了解这些物质特征是如何运作并服务居民的。

事实上，如今矗立在深圳的几乎所有建筑本身并不具有历史意义。它们之中能追溯到特区成立初期的少之又少，更不用说成立之前的。大多数城中村建筑物都是在过去十年内建造的。如果只从建筑的角度出发，特区初期遗留下来的可能只有藏在下面的基地网格了。那么这似乎引发了另一个保护困境——比起建筑本身，其下的基地网格甚至更有价值。也许正相反，每个城中村的独特之处恰恰在于打破地块固有界线的限制。随着个别建筑物的兴建、倒塌和更替，或是贯穿肌理的步道，抑或是在其中蜿蜒的小路和台阶，都在整个城中村的历史中留下痕迹。聚落中的水井和高大的遮阴树依然能让人回忆起过去农村的景象，并用它们的历史说明了现有建筑保留的优先次序。所有这些元素交织形成的网络共同构成这一场所独特的基因。而当务之急便是让这种代代相传的基因继续传承下去，否则这里将变成一具空壳。

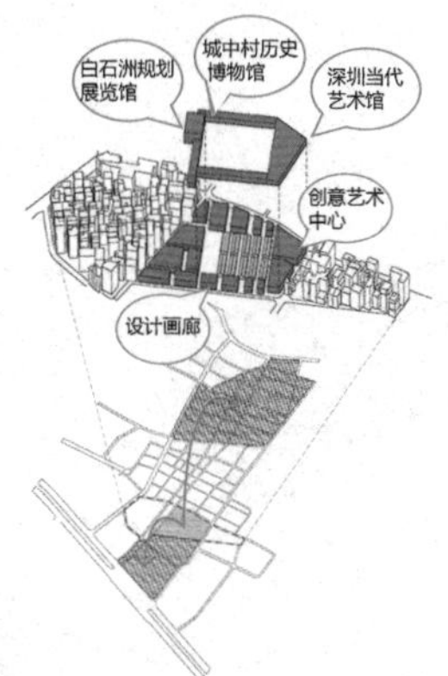

图 3 分期实施：准备期与第一期
Figure 3 Scheme phases: preperation and the first phrase

（图 1、图 4）。与 10 年前的湖贝非常相似，该团队开始对需要保护的最关键因素进行深入（并且最大限度地保持务实）评估，并判断哪些特征和结构对于城中村的重新使用具有最大价值，哪些是可以放弃的。

不同于传统的激进主义者的方法论，孟岩主张，通过牺牲一些城中村土地来做开发并以极高的密度（远高于通常允许的容积率）集中建设这些区域，保留大片城中村以求得有机发展（图 2、3、5、6）。这便是都市实践对于白石洲城中村改造提出的最终方案。如果开发商、政府和居民都能够同意并接受这种权衡的提案，那么城中村的难题将会变成一个设计问题（也就是说有解决的可能），而不再是一个难以解决的社会学问题。

图 4 白石洲愿景拼贴图，都市实践，2013
Figure 4 Baishizhou Vision Collage, URBANUS, 2013

COFFEE
COFFEE

Baishizhou Five Villages Urban Regeneration Research

Travis Bunt
Travis Bunt is the Principal of One Architecture & Urbanism, founder of the Rural China Lab at the Columbia University Graduate School of Architecture, Planning and Preservation, and he was the director of URBANUS in Hong Kong (2011—2014).
The article was first published in *Urban Environment Design*, 2018(12): 290-295; translated from English to Chinese by URBANUS; image ©URBANUS.

The incendiary urbanization of Shenzhen might well have occurred in secret, so far as it progressed before the rest of the world realized what was happening in the swamps of Canton. In a nation that dates its history in millennia, there exists little framework to evaluate how a mega-metropolis appeared from nothing in the span of a few decades. In a flash of light and fury, the fields and farms of the PRD were transformed into a sprawling, economic powerhouse pumping over a trillion yuan a year into the global economy. The farmers and fishermen of yesterday's generation have been, quite literally, surrounded by a populace of over 10 million. Laborers, traders, designers, entrepreneurs—dreamers, all; they are drawn from the furthest reaches of China to chase the future in Shenzhen.

For a city growing ever-bigger, ever-denser, ever-taller every day, there has been scant little time to note the stages of that development for posterity. The origins and the way stations are quickly wiped from memory in pursuit of modernization, having appeared and disappeared before any chronicler could take proper note. The local populace looks forward, and the outsider can hardly keep up with the pace.

Amongst the cranes and concrete and hard-hatted teams of construction workers, there is one ledger of the city's incredible development. Logged in the plots and alleyways of Baishizhou, Hubei, Caiwuwei, Dafen, and Shenzhen's other remaining urban villages is the record of that fleeting history, a living palimpsest of rapid industrialization.

The villages themselves are the legacy of those left out of modern China's initial spurt of wealth creation, due to the long-lasting ramifications of Mao's 1958 legal division of "rural" from "urban" lands. Although these collectives cum cooperatives did come to "own" their homes (and a limited portion of the immediate farm lands surrounding the villages), theirs was a joint-ownership – much like a modern co-op, where all members must decide together on any financial transaction. In practice, this meant that, unlike their urban brethren, rural citizens were not legally able to sell their homes, effectively locking them out of the burgeoning real estate market. At the same time, the value of undeveloped land drew the cities ever closer, as local governments snapped up the surrounding farmland and put it to auction. Throughout Guangdong province, and especially in Shenzhen, agricultural lands were swallowed up and the villages engulfed, relieving the now-former farmers of their livelihoods.

Unable to farm, and unable to sell, these villagers began also to build. As the cities around them grew, and with them the demand for affordable apartments to house hundreds of thousands of migrant workers, the village houses too grew. First to two stories, then four, six, ten. As tall as they could go without elevators and without drawing the ire of the government. By 2012, when URBANUS was first enlisted to study the

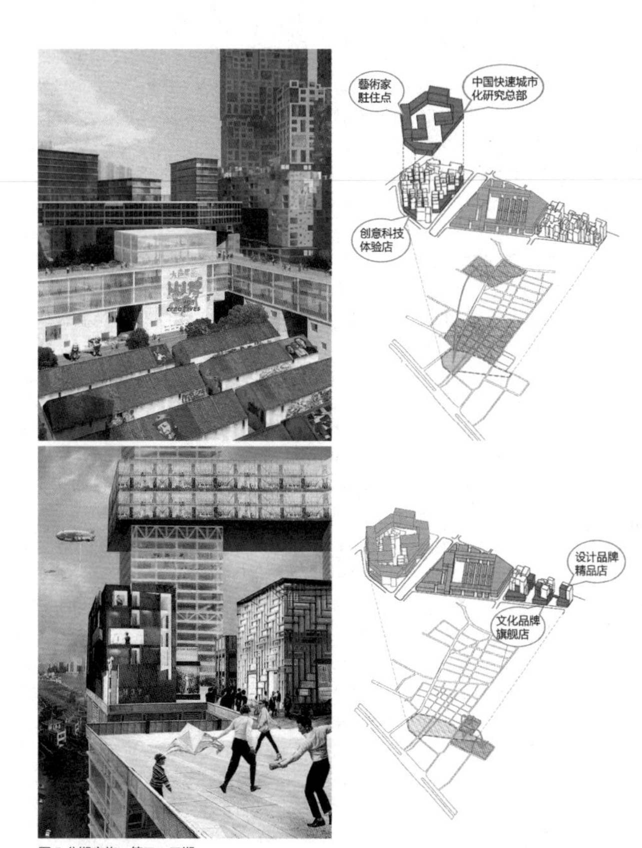

图 5 分期实施：第二、三期
Figure 5 Scheme phases: the second and the thrid phrases

regeneration of Baishizhou, Shenzhen's population was racing past the 10 million mark and its two hundred plus urban villages had absorbed many millions of that total. Baishizhou alone is estimated to hold over 150,000 residents on less than 0.65 square kilometers.

The ambiguous legality of urban village tenements (where you can rent a room but never have a formal lease agreement) creates an incredible market distortion – and apartments are discounted accordingly, renting for a fraction of expected market value. In a booming city with skyrocketing real estate prices, this has been the only relief valve, *de facto* affordable housing for uncounted workers within clusters of uncounted apartments.

Shenzhen's leaders have relied on the urban villages to secretly power the economy for more than three decades, first to temporarily house armies of factory and construction workers to construct an industrial megalopolis from scratch, and later, to domicile a second wave of migrants to staff, service, and clean the newly built restaurants, hotels, and luxury apartments.

Today, as the city is once more reinventing itself, this time as a tech hub, the urban villages are again being called upon to support another revolution—Baishizhou in particular. The relative cheapness of the village has allowed for the type of risk-taking the government is at pains to encourage for its current crop of young graduates—whether apprenticing at a design vanguard, working for a start-up, or starting their own company—entrepreneurial decisions that would be stifled by the sheer expense of market rents. Founders and executives across "China's Silicon Valley" increasingly cite their "Baishizhou origin stories," as their first foothold in Shenzhen. From government boardrooms to international magazines, the critical economic role played by urban village clusters has become an accepted fact.

Conversely, the recognition of the physical Urban Village itself as something historically valuable has been slow coming, outside of scattered activists and academics. In the absence of formal recognitions (and in the ambiguous, future-not-assured present), a gentrification of sorts is already underway, with pop-up coffee shops, craft beer houses, and art installations leveraging at once the temporality and alt-aesthetic of primely located villages like Baishizhou. These layered paradigm shifts necessarily raise the question of what, exactly, constitutes preservation; and, accordingly, what is it that is meant to be saved, and from what?

Hanging laundry, night markets, criss-crossing powerlines, kissing buildings, narrow winding alleys—are these the critical elements that cannot be lost? Is it the peculiar economic function Urban Village plays within Shenzhen that is necessary to maintain? Should the focus be on the people and the culture that enliven the fabric—or on the incomparable fabric itself? Is it the architecture that is valuable to history, the buildings themselves that should be saved? Although the ostensible answer is "all of these things, together," the bitter paradox is that such an approach is fundamentally impossible. While one may hope to somehow shrink-wrap the Urban Village and preserve it in whole for future generations, unavoidably, the act of preservation is itself an act of change.

An attempt to stop time dead, at this particular moment, is to turn the village into an artifact. Change is the essence of Urban Village, in removing it, the village ceases to exist. On the other hand, doing nothing in the face of generic redevelopment likewise ensures its extinction. Our task, then, is to envision an evolution of the Urban Village—to identify the key features that should be maintained, cultivated, and built upon during the modernization process—such that its history is not lost, and its future not cut-off.

As evident in the name, the Urban Village is characterized by a spatial fabric distinct from the surrounding city, having developed at a fundamentally different scale from greater Shenzhen and according to a different set of rules. Likewise, the human fabric of Urban Village is markedly different, a remnant of close-knit rural life preserved inside the larger city.

These two fabrics are arguably intertwined, the dense, subdivided blocks allowing the village community to survive over decades of external change. It is the expressed worry of both outside observers and villagers themselves that a *tabula rasa* reorganization of physical structure toward the typical Shenzhen model might jeopardize the underlying social structure.

Insomuch as the current residents prefer to continue living and working together with their kin in the next generation of the village, the Architect is compelled to investigate typological strategies for clustering. Maintaining and developing adjacencies and common, community areas is integral for the success of the regeneration effort.

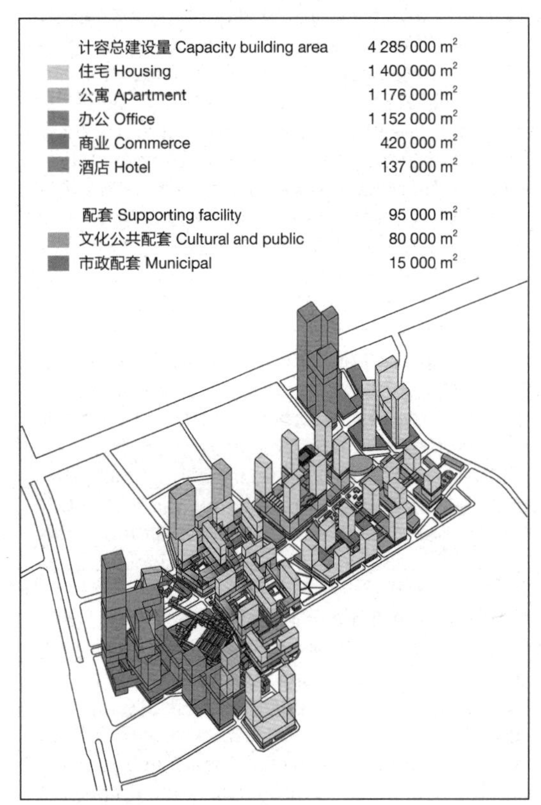

图 6 功能分布
Figure 6 Scheme program

However, it would be overly simplistic (and potentially damaging) to assume that preservation of the current community could be accomplished through a regurgitation of the grid plan, or similar superficial carryover. The characteristic programmatic mixing is at least as important as spatial allocation. The overlapping of spaces for leisure and labor, of public and private, of entrepreneurialism and collectivism, create dynamic interrelationships not readily produced by generic development patterns.

The mish-mash of uses within the fabric of the Urban Village may be antithetical to contemporary city-planning—with its clearly delineated two-dimensional zoning—but it is absolutely key to the vibrancy of the community. The panoply of offerings along a market street in Baishizhou or Hubei do not merely serve the residents, but draw a regular stream of visitors from adjacent communities, and tourists from further afar, into their alleyways. Likewise, the layering of residential and light-industrial spaces above and around these commercial strips create multifunction urban anchors that are slmultaneously self-sufficient and acting as key nodes within the surrounding city.

While the concepts of community and vibrancy are vital to understanding the role of Urban Village within the city, it is also important for the designer to recognize key spatial features specific to a given village, and how these physical characteristics shape the way the village is used and experienced by its residents.

In truth, nearly all of the buildings standing today across Shenzhen's urban villages are not themselves historic. Few date back to the incorporation of the SEZ, much less to prior generations of farmers. Most structures have been constructed within the last decade. When one considers only the buildings within the village, more often than not, the only remnant of pre-Shenzhen is the underlying lot line. At first glance, this itself raises another preservation quandary, if it is not even the buildings themselves that matter, but the grid beneath.

Perhaps instead, the uniqueness of each village lies in the violations of that grid. The sharp cut of one upward sloping footpath through the fabric, or the stepping, meandering route of another act as special thumbprints, creases and lines that persist throughout a village's history, as individual buildings rise and fall and are replaced. Community wells and massive shade trees hearken back to the rural past, and in their age, demonstrate an existing hierarchy of preservation.

The network of these elements, read together, constitute the spatial DNA of a place. It is imperative that they be maintained into the coming generations, as they have been through the past generations, or it will be as if nothing ever existed here.

Although it is generally true that the city has no architectural past, there are a few special examples of legitimately valuable historical buildings in Shenzhen. Of primary note are the original village houses of Hubei, some of which date back five centuries, and the eleven remaining commune houses in Baishizhou, from the original collective farm at Baishi Village.

Given the dearth of historically relevant architecture in Shenzhen, special attention must be given to these sites, as they represent the last physical ties to important periods in the city's pre-history. Their respective locations within urban village boundaries—and the difficulties of developing such areas—have been the central reason for their preservation up until today. As those boundaries disappear into the continually growing city, it is imperative that specific protections be planned and implemented for these newly endangered relics.

Market forces and the will of the government virtually guarantee full-scale modernization of Shenzhen's urban village sites within the decade. If concerned urbanists demand only full-scale preservation and nothing less, failure is ensured. The task of the urban designer is to recognize the fleeting opportunity to not only preserve, but also to innovate.

The extreme conditions constraining urban village redevelopment (sky-high land-compensation rates, single-entity led development of massive tracts of land, and the financial necessity to double and triple urban density in already overcrowded neighborhoods) paradoxically create situations where the futurist paper mega-structures of 1960's Europe not only find their first realistic opportunity to become manifest, but may be the best basis for rethinking the next iteration of Urban Village—a component machine for mass-living within a national machine of production. Interestingly enough, it is on these types of sites that the avant-garde design firms of China are now sketching out visions of such formerly impossible architecture—nostalgic fantasies of futures not attained by the profession's forebears.

Baishizhou was (and remains still) the largest urban village redevelopment project ever attempted in Shenzhen. The "regeneration" proposal as submitted to the city would double the FAR of the village from 4.5 to 9, and construct a single development of over 5.5 million square meters within five years. Such a rapid and brutal mega-scale development was claimed to be the only solution to fund the extremely high financial loss from compensating the 2,000 landlords of the urban village. URBANUS, in attempting to tackle this impossible task, is necessarily a mediator—the intermediary that must seek to balance the real-world economic parameters with the idealistic goals of urban design and preservation.

The commission was apt. It would be no great overstatement to say that URBANUS was the initiator of Urban Village discussion within the design world. From its first quixotic efforts in Huibei, through extensive efforts and studies in Dafen, Gangxia, Fuxin, Nantou, and others, the firm has been at the forefront of conceptualizing alternative (re)development models for urban villages—that is to say, viable and compelling alternatives to wholesale, scorched earth destruction.

As with many URBANUS projects, the core of the Baishizhou work was an exploration of the spatial limits of densification and megastructure, within a specific cultural and economic contextual framework (Figure 1, 4). Much like Hubei a decade before, the team commenced a deep (and largely pragmatic) evaluation of the most critical elements deserving preservation, which features and structures had the most value for re-appropriation, and which parts of the village could be let go.

Unlike a traditional activist approach, Meng Yan's working theory held that by sacrificing some land to development and intensely focusing density there (in far higher proportions than typically allowed), large swaths of Urban Village could be retained to evolve organically (Figure 2, 3, 5, 6).

This is ultimately the proposition laid forth in the URBANUS vision for Baishizhou. If the developer, government, and residents can agree to and accept such a trade-off, the Urban Village conundrum then becomes instead a design problem (that is, presumably solvable) and ceases to be an intractable sociological one.

湖贝 Hubei

报纸
NEWSPA-
PER

⊙《深圳商报》2016 年 6 月 23 日 星期四

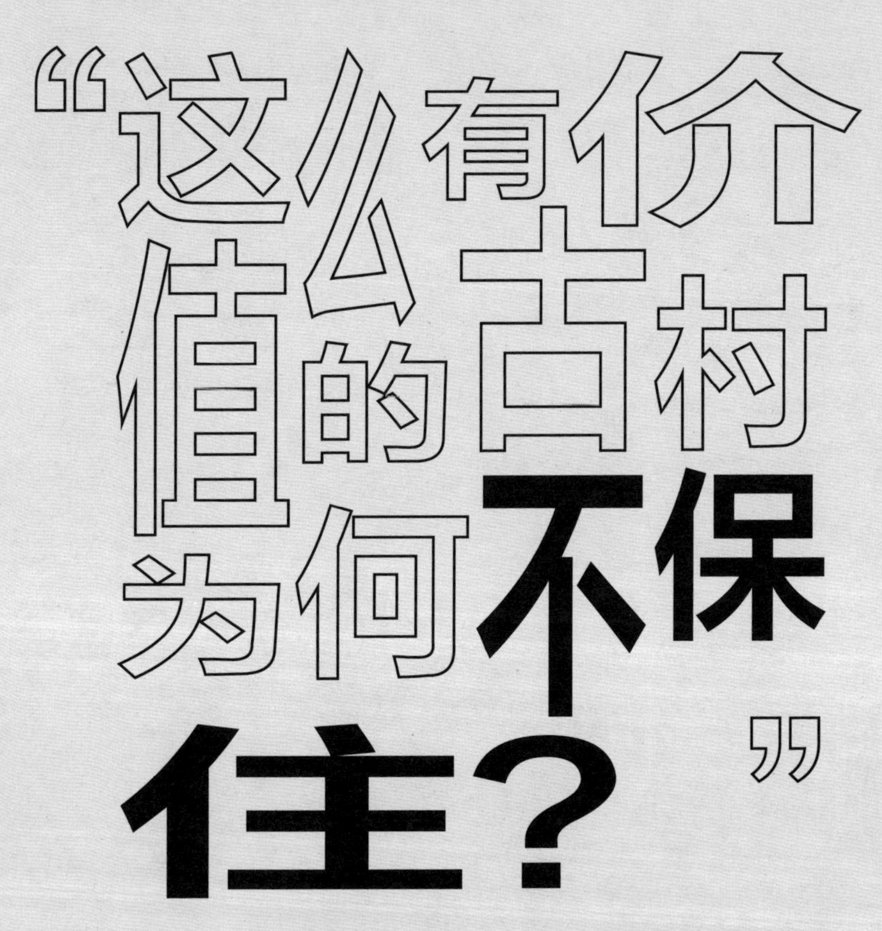

300 亿元改造湖贝古村，引发深圳建筑界的“拯救行动”

“湖贝这么有价值的古村为何不保住？”

周红玫2012年10月开始多次进入湖贝古村调查，之后邀请孟岩、张之扬、刘珩、郭湘闽、马立安等人进行了一场讨论，那是专业界人士举办的第一次民间讨论。

周红玫给记者的微信列出了近年来专业界人士对湖贝古村保护的努力。那次讨论，周红玫邀请了一位非常有意思的嘉宾——香港著名导演张经纬，他是湖贝张氏的后代，其创作的《天水围》后来被许鞍华拍成《天水围的夜与雾》。他对古村、旧社区和城市变迁有着浓厚兴趣。

之后，都市实践著名建筑师孟岩牵头做了关于湖贝古村的独立研究，这些年他一直在奔走呼吁。“后来听说他进入开发商的决策会议，甚至努力以经济测算数据说服，但效果并不理想。”周红玫说。

和周红玫一样的多名专家在微信群里没日没夜讨论，刷屏的速度和热度令人迷惑：这群专家如此关心湖贝的命运，到底为什么？记者电话访问了黄伟文。

“在很多地方，深圳被描述为从小渔村发展成国际化大都市。在对深圳发展奇迹夸赞的同时，深圳或者昔日的宝安、新安地区的历史被遮蔽了。其实深圳地区一千多年的历史是连续的，没有断裂。湖贝恰好和深圳的历史吻合。比如，深圳地名就是来源于深圳河边的深圳墟，深圳墟则是由包括湖贝在内的张氏一族建设。湖贝古村的基本格局很完整，所以它和深圳的历史是零距离的。我们理解湖贝古村，不能把它简单看成几百间老房子，应该把它看成深圳的历史标识，湖贝可以说是深圳人所共有的文化资产。”黄伟文说。

据文献记载，湖贝历史上为深圳墟发源地之一，村内湖贝路北侧的怀月张公祠为区级文物保护单位。湖贝立村已有500年历史，明朝时期张氏爱月、思月、怀月、念月四兄弟分别在向西、水贝、湖贝立村，

张爱月的叔叔张靖轩则创建了黄贝岭村。“深圳墟”的建立、维护与发展，就是张氏村落共同经营的结果。

“还有，我们认识湖贝的价值，不能简单看房子建筑本身。深圳不能和吉安、徽州去比古建筑资源。我们需要评估的是，这是深圳数量不多的古村落，而且是处于城市商业中心、处于最繁华地段的古村落，是与深圳有着历史渊源的古村，凭着这些，我们自然应该重新考虑它的价值。”

作为本土人且有过长期在东门生活经历的深圳民俗专家廖虹雷近年一直为保护湖贝古村奔走呼号。“湖贝古村保有三纵八横的村落结构，保存有清嘉庆九年（1804）重建的宗祠，另有门楼、水井和200多间民居，是典型的广府系坊巷式排屋村。1984年，深圳进行过一次全市地名普查，在后来出版的地名志中，深圳古村落数量为1500个；到1992年剩下1200个；到2012年只有200个。现在，深圳市区只有笋岗和湖贝两个古村。湖贝这么有价值的古村为何不保住？我百思不得其解。”廖虹雷告诉记者。

“保护古村和城市发展没有矛盾”

近日记者再次探访湖贝旧村。黄昏时分，

⊙《深圳商报》2016年6月23日 星期四

走进狭窄的巷道，古村入口处香火散发的气息扑面而来。湖贝的张氏族人大多已经不住在旧村，如今的居民多数是在东门市场做生意和打工的潮汕人。小孩子在巷道玩耍，几名摆摊的小贩慵懒地靠在小店的柜台上。往里走，海鲜的腥味让人有些不适，地上到处有污水。一间一楼的小房子开着门，里面有两个人正在开着生蚝……

环境恶劣，交通、消防不配套，正是主张拆除湖贝旧村人士的重要理由。确实，对于绝大部分深圳人来说，很难想象在东门这样的商业繁华地段，还有湖贝古村存在。

对此，黄伟文表示："从专业角度看，保护古村和城市发展没有矛盾，保护它也能带来价值。比如上海的新天地，现在是一个具有上海历史文化风貌、中西融合的都市旅游景点。这里是石库门建筑旧区，当时卢湾区的政府官员认为，这些房子太破败了，没有办法改造，希望推倒重来。可是后来专家说服了开发商，也说服了政府，抱着试一试的态度由美国建筑师本杰明·伍德主持，改变了石库门原有的居住功能，创新地赋予其商业经营功能，现在，新天地已经成为上海的名片。上海田子坊由一个街道工厂和废弃仓库改造成居民和艺术家共同生活的特色里弄，也是中西文化交融的地方。这两个地方保护都很成功。"

廖虹雷告诉记者，新华社香港分社原副社长、广东省政协原副主席祁烽曾经是深圳解放时最早成立的沙深宝边区委员会书记，后来到深圳，看到东门以前的6座碉楼一座也没有了，非常痛心。"改造过后就留下一堆假东西，老领导怎么会不痛心呢？看看近几年来，北京重视胡同文化，上海重视里弄文化，广州则重视街巷文化。广州的街巷文化，给我留下非常深刻的印象。到了我这样的年纪就明白，一个地方能不能吸引人，说到底是你的文化。文化是内核，虽然看不见却很重要。并不是拆了以后建高楼大厦，就会有发展。"

黄伟文认为，在中国的一线城市中心地带，只有深圳有这样的村落，这是它的独特性，这个价值是不可估量的。"我认为，保留它是对抗'千城一面'的最好的办法，能够使得一个农业文明聚落在一个创造经济奇迹的城市核心区焕发生命力，融入现代生活。如果湖贝古村保护下来，将来一定是深圳的骄傲，值得夸耀。否则，将来的深圳拿什么和广州、北京、上海这些城市平起平坐一起对话？"

"我害怕推土机一来古村就完了"

在"湖贝古村在行动"微信群里，大家讨论最多的是如何保护，这里既有观点的碰撞，也有智慧的火花。

有人提出"三个湖贝"的概念，一是"深圳人的湖贝"，就是强调湖贝是深圳人共有的记忆和文化资产；"五万人的湖贝"，指出必须从湖贝现有五万居民的角度评估，包括现在居住在旧村里的租户，如何顾及他们的利益；"每个人的湖贝"，就是所有人，包括游客、外地教授学者、路人甲、路人乙等，要听听他们对湖贝的命运有什么想法。黄伟文说，他在这个基础上还提出了"湖贝人的湖贝"，古村保护也要顾及湖贝村民的利益。"深圳在中国城市建设中是勇于创新和有足够包容度的。我觉得这是一个很好的机会，我们可以把一个500年前的农业村落保留下来，通过改造，融入现代商业世界，最后让各方获得利益。如果能做到这一点，深圳就有故事可讲，值得自豪。"

饶小军2015年开始受华润的委托，对古村进行保护性测绘，最后形成一个专业报告。"结果这个报告被解读成我是支持拆除古村的。其实我的态度非常明确，湖贝有着丰富的文化沉淀，同时也有着深圳城市化、工业化快速发展的印记，这种双重的特色非常值得保存。所以我不仅反对拆，而且主张对古村进行完整的保留。”

黄伟文反对在保护古村实践中所谓的“复古”。他认为，复古只是模仿，保护性迁移也有问题。湖贝不是某一栋房子有价值，而是整个村落，它哺育了张氏家族、老深圳墟、深圳人。黄伟文举了北京大栅栏改造的例子。大栅栏做了很多技术上的探索，主要就是对舒适度比较低的民宅通过改造，在不改房屋架构的基础上提升到现代生活水准。“当然，这对于发展商来说，需要社会责任感，需要信心。现在的开发商，一是观念和眼界有局限性，在城市改造和建设中比较喜欢从众，为避免风险，就只会采用保险的旧有模式；然后就是对历史文化带来的综合价值认识不到位。当然最重要的一点是，发展商不愿意保护古村，还是目前商业开发模式甚至是城市开发建设都采用成片改造、大面积开挖的僵化模式和路径依赖造成的，但实际上那样是不可持续的。”黄伟文说。

廖虹雷在采访最后告诉记者，“历史上，湖贝村的商业文化基因很发达。整个张氏家族很团结，所以才能保证深圳墟长盛不衰。当年对深圳墟的改造，就应该遵循对文化遗存保护的五个原则：原址、原材料、原结构、原功能、原生态。我曾经批评一些房地产开发商，看中哪个山头，看中哪个人气旺的村落，哪个地方就会遭殃。现在动不动就几百亿砸一个古村，我就很害怕。我很害怕推土机一来，古村就完了。”

《深圳商报》记者 蒋荣耀

Note: Through the lens of architects, urban planners, film directors, scholars, etc., and their efforts to protect Hubei from urban redevelopment, the article advocates the preservation of the 500-year-old heritage in the city center.

JIANG R. "Why not preserve the valuable ancient village?" [N]. Shenzhen Economic Daily, 2016-7-5.

◉《南方都市报》2016 年 7 月 5 日 星期二

“湖贝旧村非文保單位年底将启动一期拆”

注
消

政府部门和开发商分别对近日的湖贝争议作出回应，开发商呼吁专家“理性和客观”

曾向省里申请文物保护未获批

湖贝旧村整体保护派指出，湖贝旧村南坊最早可追溯到明成化年间，它的空间与建筑形态是从明清到改革开放500年的深圳历史与文化变迁、甚至是“深圳”名称来源和最后的物质证据。

对这一说法，深圳市文物局并没有给出直接回应。但在该局发回给南都记者的书面回函中明确表示：到目前为止，罗湖区湖贝旧村未纳入法定“不可移动文物”中。

深圳市文物局称，目前深圳市有区保文物76处、市保37处、省保13处、国保1处。在该局提供的《深圳市各级文物保护单位名单》中，罗湖区湖贝旧村并不属于区保、市保、省保、国保中的任何一项。而在该局提供的另一份深圳市《非物质文化遗产项目（名录）保护单位和代表性传承人统计表》中，湖贝旧村也没有任何民俗被记录在列。

此前，曾有本地媒体报道称，据文献记载，湖贝历史上为深圳墟发源地之一，村内湖贝路北侧的怀月张公祠为区级文物保护单位。但是湖贝村实业股份有限公司董事长张齐心证实，早年曾向省里提出对湖贝古村进行保护的申请，但没有批下来。

事实上，在前日由专家学者组成的保育团体主办的研讨会上，曾经活化过乌镇、周庄等古建筑的古建保护专家阮仪三也坦言，湖贝旧村的建筑物大多都建于清末民初，从古建筑价值来看意义也许并不大，阮老主张要保护湖贝旧村现存的民俗，他前日赴旧村实地踏勘就对不少住户的焚烧祈祷之举很感兴趣。但南都记者实地采访后却发现，这些住户实际大部分都是潮汕人，前日是潮汕地区“谢神”的日子。

城市更新对文物有制度保障

市文物局介绍，目前深圳市不可移动文物的申报及公布流程：首先是由所在辖区文物部门公布为未定级不可移动文物；第二步再遵照逐级申报的原则，区政府挑选部分价值较高的未定级文物公布为区级文物保护单位；接着市政府再挑选部分价值较高的区保单位公布为市级文保单位；以上均包括申报、专家评审、公示及正式公布等程序。

至于非物质文化遗产代表性项目的申报和公布流程则可以首先由公民、法人和其他组织进行申报，再由所在辖区区政府公布为区级非遗代表性项目，此后再遵照逐级申报的原则，市政府挑选部分价值较高的区级非遗公布为市级非遗项目；以上也均包括申报、专家评审、公示及正式公布等程序。

深圳城市更新系统一名不愿具名的内部人士称，实际上深圳市城市更新办法与实施细则对于项目内的文化保护已经有制度上的安排，比如更新办法第四十四条规定，城市更新项目应当遵守保护历史文化遗存的法律、法规，依法保护城市更新范围内的历史文化遗存。实施细则第十五条则规定，城市更新单元规划的编制应当推进文化遗产融入城市发展，保护城市肌理和特色风貌，改善生态环境和人文环境。因此，在这位内部人士看来，政府在前期规划编制中对于湖贝旧村是否保留如何保留会有考虑，并非如外界学者所担忧的会完全拆除。

2015年，龙岗区还发布《龙岗区未定级不可移动文物管理办法（试行）》，其中规定，城市更新单元在制定计划申报过程中，区城市更新部门需征求区文物行政部门的意见，区文物行政部门对城市更新范围是否存在未定级不可移动文物及其保护要求提出意见。区文物行政部门对文物进行等级分类及制定相应的保护及利用导则，明确周边需预留的保护范围。

市委领导调研要求“双赢”

不久前，省委副书记、市委书记马兴瑞调研罗湖城市更新工作时，也实地考察了湖贝旧村的情况。马兴瑞指出，要通过城市更新改造改善旧村面貌，进一步消除安全隐患，提升市民群众生活质量。在这一过程中尤其要坚持开发与保护并重，不断完善规划方案，切实做到文化传承和城市更新双赢。昨日，南都记者就此也向罗湖区询问湖贝旧村城市更新与文化保护的情况，目前是否有相关方案，不过截至发稿前暂未获得回应。

统筹：《南方都市报》郭锐川
采写：《南方都市报》晏婵婵 崔欣 陈博 郭锐川 谢湘南 实习生 王宜嘉

Note: Through the lens of municipal administration and developers, the initiative to protect Hubei is irrational without objective evidence. The preservation of Hubei was not supported by the Shenzhen Municipal Cultural Heritage Bureau as Hubei was not officially listed as a heritage site that is protected against demolition.

GUO R. Old Hubei Village is not a cultural heritage site and the first phase demolition will begin by the end of this year [N]. Southern Metropolis Daily, 2016-7-5.

湖贝旧村，陈冠宏摄于 2014 © 都市实践
Old Hubei Village, photograph by Alex Chan, 2014 ©URBANUS

湖贝古村怀月张公祠建筑细部，孟岩摄于 2012
Architectural Details of the Zhang Huaiyue Ancestral Hall, Ancient Hubei Village, photograph by Meng Yan, 2012

湖贝古村巷道与青砖瓦房，孟岩摄于 2012
Alleyway and houses with grey brick and tiled roof, Ancient Hubei Village, photograph by Meng Yan, 2012

湖贝古村巷道，孟岩摄于 2012
Alleyway, Ancient Hubei Village, photograph by Meng Yan, 2012

湖贝古村怀月张公祠前，孟岩摄于 2012
In front of the Zhang Huaiyue Ancestral Hall, Ancient Hubei Village, photograph by Meng Yan, 2012

湖贝古村巷道，王大勇摄于 2016
Alleyway, Ancient Hubei Village, photograph by Wang Dayong, 2016

湖贝古村巷道，孟岩摄于 2012
Alleyway, Ancient Hubei Village, photograph by Meng Yan, 2012

湖贝研究：城市图景与想象

罗祎倩

都市实践研究员

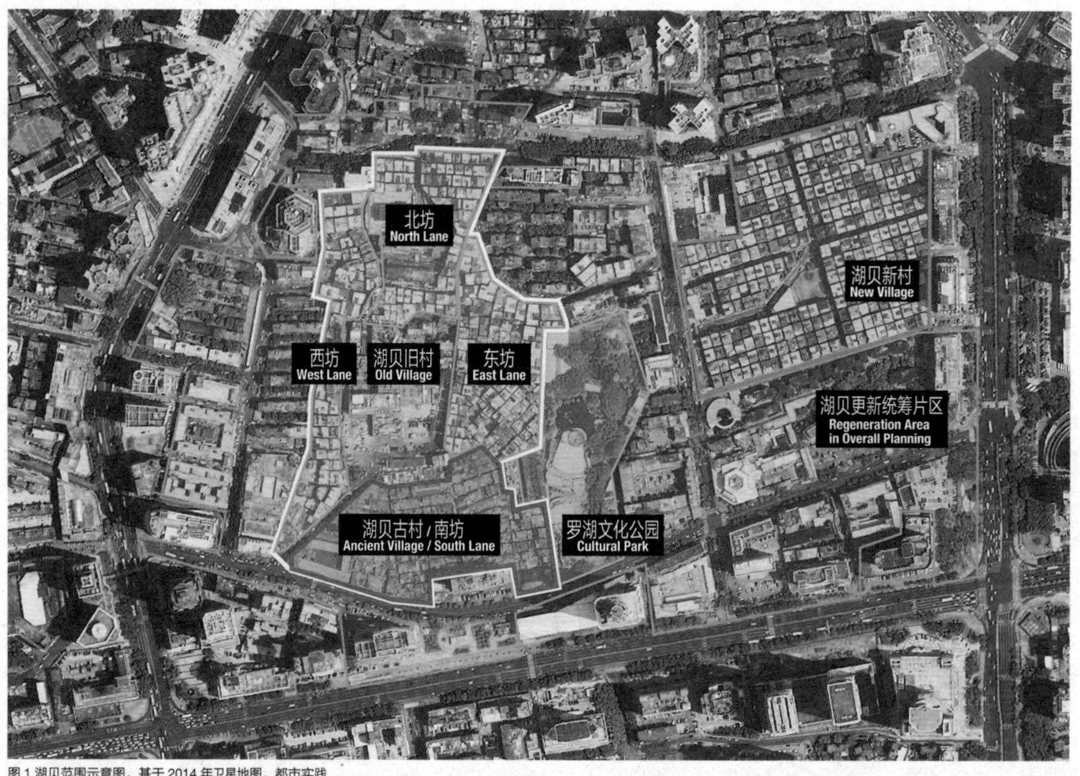

图 1 湖贝范围示意图，基于 2014 年卫星地图，都市实践
Figure 1 Hubei Area based on the satillite map in 2014, URBANUS

湖贝，似乎由于太老，而无法存活在现代城市图景中；又似乎由于太新，而无法立足于主流历史想象中。

1　空间扩张的反向

在漫长的深圳历史中，湖贝村因对居住空间的需求不断增长逐步向外扩张。1466 年（明代成化年间），它立村于山陵和堰塘之间；[1] 随着村民增多，村外建围，围外再建围墙，以抵御袭扰；1 而后村庄扩大，拆除围，东西两面建房；最终形成保留至今的三纵八横肌理，东西宽约 180 米，南北深约 120 米，有着窄巷、门楼、水井、青砖灰塑的瓦房。2 它的外部景观从山陵变成农田，又从农田变成楼房。历时五百余年，湖贝村成为湖贝古村，也成为高速城市化席卷过后的城市中心仅存的历史空间。1986 年，村民迁入不远处的新村，古村转变为数千名潮汕租户的聚居地，3 历史空间成为当代城市生活的发生地。同年，罗湖文化公园紧挨古村东部落地，在随后的三十年间既上演着传统粤剧，又萌生了当代诗篇。4

在深圳短促的城市建设史中，外部城市不断地反向吞噬位于罗湖区的湖贝古村。1992 年，古村被纳入旧改范围，多次讨论，皆因建筑密集、权属复杂等种种原因而搁置。2011 年，区政府与华润置地（简称华润）签订战略合作框架协议，5 湖贝城市更新统筹片区项目（简称湖贝项目）正式启动（图 1）。6 2012 年，深圳市湖贝实业股份有限公司（简称股份公司）投票表决，7 同意华润作为改造实施主体。华润委托深圳市城市规划设计研究院（简称深规院）和华阳国际设计（简称华阳）进行规划设计，2010—2014 年间的数版方案都将古村和公园替换为超大型购物中心和超高层办公楼。彼时，城市管理者将罗湖瞄向国际消费中心，[2] 实施主体预估湖贝项目的税后利润可达 500.83 亿元。[3] 城市面貌和经济效益直接推动空间扩张的彻底反向，利益相关方在此划定共赢的阵营，怀抱历史上前所未有的雄心。

2　城市空间中的公共生活

2.1　复合空间

特区成立初期，湖贝东、西、北三坊围山而建，与湖贝古村（又称南坊，图 2）共同组成湖贝旧村。20 世纪 80 年代，村民迁入不远处湖贝新村中统一规划建起的两三层高带小院的楼房，出租旧村瓦房。90 年代，村民将旧村三坊和新村均改建成七八层高的握手楼出租，自己则搬至村外的商品房（部分定居港澳和海外地区）。之后，他们已鲜少回到湖贝旧村，仅在重阳节秋祭时到旧村祠堂上香，再到新村空场吃盆菜。8

2016 年，湖贝旧村共有租户 20028 人，其中潮汕租户占比 66.1%。[4] 大部分租户从事水产生意和服务业，工作地点位于湖贝村、东门步行街和东门海鲜市场，9 距离住所步行可达。湖贝片区有近千家固定商铺和摊贩，既有辐射市域范围的水产批发和海鲜食街，也有服务于社区的菜市场和便利店。他们带来 24 小时不间断的商业活动——批发从凌晨到清早，生鲜零售从清早到傍晚，餐饮从早上到晚上，夜市又从傍晚到凌晨，还有全天营业的生意，商铺既是工作场所也是住所。灵活混杂的业态、可负担的生活成本为新移民进入城市提供了途径，也为城市带来一个具有活力的街区。有着数十家海鲜餐厅、吸引无数食客的乐园路也成为一种深圳特色。

在湖贝古村内，三间三进的怀月张公祠是张氏宗族的情感依托，而街巷和南围门则成为潮汕租户的祭祀空间（图 3）。潮汕人信奉多位神明，拜神的日期和习惯众多。他们在房屋外墙壁设简易香炉，在街巷中设临时供案，以做日常礼拜；在南围门两侧门跺设置神像供奉台和烧香处，作为公共祭祀空间。每逢初一、十五、神明寿诞时，家家户户到南围门拜神。围门原本作为村庄的交通节点，随着祭祀活动产生了精神依托和集聚功能。

2.2　多元社群

大部分租户在湖贝旧村生活多年，甚至长达 30 年，因亲缘、地缘、业缘生成紧密的社群网络。他们或独自落脚在这里，或和家人一同搬到这里，又或在这里成立自己的家庭。正如一位在湖贝出生、租住时长 15 年的学生所说，“我们所有的时光都是在这度过的，就算是我们的故乡了。”[5] 这种平实真挚的归属感来源于经年累月的日常生活。农历节庆，很多租户选择留在当地庆祝而非返乡，在湖贝旧村营造了热闹的氛围，这与深圳其他移民为主的社区全然不同。

在湖贝项目中，租户作为真正使用者，和所在地有着紧密的物理和情感联系，但因为不拥有产权，所以不被定义为利益相关方。长期以来，村民和租户形成互利关系，前者出租房屋以获得经济回报，后者租赁房屋以获得生活（甚至是生存）空间。而在湖贝古村去留的议题上，两者转向对立面，前者希望拆迁以获得经济补偿，后者希望保留以维系低成本的生活。仅以产权和户籍作为标准，村民拥有房屋，租户仅是使用房屋；村民拥有名正言顺的乡愁，租户不属于这个地方。村民有权利留下但选择离开，租户选择留下但被要求离开。

3　城市驱动力

1978 年改革开放以来，风靡全球的现代城市图景被投射到中国上空，推动各地昂首迈入城市化进程，一往无前。这幅图景处处簇新和规整，以效率为导向，有着更大宽度的道路和更高密度的建筑。城市生活以商品为媒介，方方面面围绕着消费展开。这幅图景也成为深圳这座新城的驱动力，驱使它不加思索地建造，又驱使它毫不犹豫地重新建造。

1　故湖贝古村又称湖贝大围。
2　古村现存房屋多建于清代末年。空间变迁及建筑现状，见深圳大学建筑与城市规划学院 . 深圳市罗湖片区城市更新项目湖贝旧村南坊调研报告 [R]. 2014.
3　以潮汕话为母语的汉族民系，主要来自广东潮汕地区，包含汕头、潮州、揭阳、汕尾市。
4　园内曲艺台是粤港两地民间粤乐社进行艺术交流的重要平台，持续二十多年每周举办两次粤剧粤曲大家唱活动；剧场西附楼是创刊于 2012 年的《飞地》诗刊的编辑部，也是深圳诗人的聚点。
5　2012 年，两方签署《政府所属物业拆迁补偿框架协议》。
6　该片区位于罗湖区东门街道，东南西北分别临近文锦中路、深南大道、东门中路、中兴路，用地面积 400701.8 平方米。湖贝古村和罗湖文化公园位于片区中部。
7　2017 年，两方签署《集体物业拆迁安置补偿协议》。
8　张氏族人的秋祭传统曾中断 40 年，1989 年恢复。
9　东门海鲜市场由村民集资，1989 年建成，2018 年 12 月 31 日结业，曾是全国十大肉菜市场之一，日客流量近 10 万人次，年交易总额达 15 亿元。

图 2 湖贝古村，又称湖贝南坊，王大勇摄于 2016
Figure 2 Ancient Hubei Village, also known as the South Lane, photograph by Wang Dayong, 2016

用以争夺现代城市图景所描绘的城市中心，高度无疑是最具有视觉冲击力的符号。罗湖曾在30年里独占“深圳第一高楼”的符号。1985 年，160 米的国贸大厦一跃成为深圳和全国第一高楼；1996 年，383.95 米的地王大厦成为深圳第一高楼时更是刷新了亚洲高度; 2011 年，441.8 米的京基 100 大厦仍将深圳第一高楼锁定在罗湖区，直至 2016 年才被位于福田区的平安金融中心取代。在 2017 年一版湖贝项目方案中，紧挨湖贝古村东部的湖贝塔高达 830 米，超过世界第一高楼哈利法塔，对高度这一符号的追求毫不掩饰。

与此同时，罗湖创造了并长期拥有商业用地开发的符号。1984 年，全国第一个商品房小区东湖丽苑落地罗湖。1987 年，全国第一次以公开拍卖的方式有偿转让一幅罗湖地块的使用权，次年建成商品房小区东晓花园。[10] 这直接促使全国土地利用方式的转变，[11] 之后随着住房制度改革和商品房普及，“房屋是商品”的观念从深圳走向全国。[12]2004 年，全市建筑面积最大的高端购物中心万象城落地罗湖，这也是华润置地在深的第一个项目，随后，其不断升级购物中心的开发模式。

2012 年，根据开发商介绍，“按照规划，湖贝片区改造更新后，总计容建筑面积约为 192 万平方米，其中零售、娱乐部分约为 50 万平方米、办公部分约为 45 万平方米、公寓及居住部分约为 97 万平方米。计划投资 300 亿元。……将对其中近 10 万平方米用地建筑物进行综合整治，另外还需拆除旧建筑物面积约 80 万平方米，涉及 87 块宗地，共有业主近 4000 户。”[6]

4 公共领域中的城市空间

4.1 拆或留？（2012—2014）

在这一幅崭新的现代城市图景中，遍布使用痕迹的湖贝古村显得无所适从，它的巷道狭窄、房屋低矮，在尺度上无法抗衡。然而在真实的城市生活中，古村和公园是市民日常和本土文化的空间载体。面对图景和现实的矛盾，业界和学界从专业视角对现有城市观念提出质疑：

10 一幅 8588 平方米地块的 50 年使用权由深圳经济特区房地产公司以 525 万元拍下。东晓花园当时的售价为每平方米 1600 元，整个项目净利润近 400 万元。

11 1988 年，《中华人民共和国宪法修正案》修改条款为“土地的使用权可以依照法律的规定转让”；《土地管理法》也作了相应调整。

12 深圳在 1988 年 1 月 1 日发布《深圳经济特区住房制度改革方案》，成为全国第一个地方性房改文件。

湖贝古村和罗湖文化公园的价值如何定义？它们之于罗湖、之于深圳的价值如何定义？

2011 年，都市实践合伙人获悉湖贝古村的拆除重建方案，即刻带领团队开展独立研究，指出存活了五百余年的自然村之于深圳城市中心的稀缺性。湖贝古村的格局和建筑、居民的习俗和生活既栖息在城市中，又因自身的特殊性游离于逐渐趋同的城市。基于研究，都市实践提出完整保留古村三纵八横的肌理，通过允许片区局部高密度建设，满足经济测算的提案（图 4）；次年形成研究报告递交区政府。这种同时满足古村保护和开发指标的城市更新思路获得认可。接下来，都市实践持续动员身边资源关注古村保护，举办工作坊“古村保护与更新”，邀请城市管理者、实施主体和社会各界共同探讨。[13]2013 年“两会”期间，9 名人大代表呼吁整体保留湖贝古村古民居，作为深圳“最直观古老的发源地标本”[7]。

项目初期，尽管公众参与引起城市管理者的关注，但未能影响实施主体对于湖贝更新方向的判断。2013 年，华润表示拆除古村后，将通过新建一组岭南建筑群来提现文化；与华润多次协商后，股份公司向市规土委申请异地重建怀月张公祠，以配合片区的整体规划。[7] 湖贝项目第一轮规划方案递交市规土委审议，因未原址保护紫线范围内的怀月张公祠和未保护古村特有的空间和文化，未能通过。[14]

与此同时，华润委托都市实践进行古村保护及活化专项研究、深圳大学建规学院（简称深大团队）进行古村保护测绘，并组织两方与华阳参与总体规划及旧村改造设计工作坊。至此，具有专业能力和素养的第三方正式进入对话平台。2014 年，都市实践进一步研究罗湖的新定位，强调湖贝古村作为独一无二的文化街区的重要意义；在完整保留古村的基础上，提出社区升级、商业分区、单体改造的策略，形成第二轮提案，探索三维立体、多元文化共融的城市空间模式（图 5）；[8] 同时呼吁各界关注，举办工作坊“城中村——湖贝老村对新型文化商业的启示”[15]。深大团队对湖贝古村进行详尽调研，通过分析空间变迁和现状，建立历史风貌、改加建空间质量和建筑结构质量的评价指标体系。根据调研报告，古村现存的 200 多间房屋多建于清代末年，具有广府民居的布局和传统样式的细部，又因城市化带来生活方式的改变而产生空间变形（多数为改建和扩建小院以增加空间）。[9]

4.2 留多少？（2016）

2014 年，因种种原因，湖贝项目停滞。2015 年，罗湖成为城市更新工作“强区放权”试点，[16] 罗湖区委将 2016 年定为“城市更新突破年、城市管理治理年”，湖贝项目再次提上日程。同年 5 月，区更新局召开湖贝项目专家评审会，规划方案未得到多数专家赞同。[17] 委托第三方进行古村保护研究后，实施主体虽然在说法上认可古村价值，但是在做法上却提出“迁建、仿建、创建”，仅划定 6000 平方米的古村保护范围，西部保留怀月张公祠，中部打造风情街，东部拆除以让位给购物中心。这给深圳呈现了一个去真存伪、存表去里、支离破碎的“古村”。这一名为保护、实为全盘毁坏的规划方案触发了一场在深圳城市历史上重要的公众参与事件。

在这个急迫的关口，深圳社会、文化、规划、建筑、艺术各界的知识分子自发地搭建平台，共同出谋划策，希望能够为深圳和深圳人留下仍存活着的古村。主要推动者包括长期研究深圳城市化进程的文化学者、人类学家、大学教授，以及深耕深圳近二十年且自 2011 年持续关注湖贝古村的都市实践。

6 月，六位院士联名致信市委市政府领导，指出“（湖贝）古村落继续留存的价值是不可以用经济利益来估量的”，建议启动未定级文物的先予保护的机制。[10] 城市设计促进中心举办酷茶会“城中古村的活下、活化和活计”，深规院的湖贝专项规划负责人指出更新机制存在的问题，直言“开发商站在前沿位置，去主导和实施规划，从而导致政府缺位的问题。”[11]《深圳晚报》《深圳特区报》《南方都市报》《羊城晚报》和自媒体等发出相关报导，《深圳商报》更是进行持续半个月的追踪——“这么有价值的古村为何不保住？”[18] 大众媒体为公众提供了知情渠道和与利益相关方对话的平台。

图 3 湖贝古村，王大勇摄于 2016
Figure 3 Ancient Hubei Village, photograph by Wang Dayong, 2016

13　E6 空间，2012-10-29，参与人员包括湖贝村民、建筑规划专家、社会人文学者、媒体和政府相关部门。
14　怀月张公祠始建于明代中期，1804 年（清代嘉庆年间）重建，1935 年重修，2008 年被划入《深圳市城市紫线规划》保护规划范围，核心保护范围 3600 平方米，建设控制范围 6918 平方米。
15　E6 空间，2014-3-14，参与人员包括湖贝村民、建筑规划专家、政府相关部门、文化品牌运营、项目开发商、规划和建筑设计团队和商业策划团队。
16　罗湖区城市更新项目的部分事权由市规土委调整至区更新局行使，试点期 2 年，见深圳市人民政府关于在罗湖区开展城市更新工作改革试点的决定 . 深圳市人民政府令（第 279 号）. 2015-8-29. 次年，城市更新工作改革推行至全市，见深圳市人民政府关于施行城市更新工作改革的决定 . 深圳市人民政府令（第 288 号）. 2016-10-15.

7月初，深圳知识分子共同发起“湖贝古村120城市公共计划”（简称湖贝120），发表《湖贝呼吁共识：拯救我们的历史记忆》（简称《共识》）[12]61-64，得到近千名社会人士联署。此时，关于湖贝古村和罗湖文化公园的价值和未来发展的讨论正式在公共领域展开。《共识》阐述湖贝古村之于深圳历史的唯一性，对于城市认同感的意义，认为它既是原住民的资产，也是所有深圳人的共同社会资源。

综合深大团队的测绘报告，《深圳市历史风貌区和历史建筑评估标准》研究团队对湖贝古村进行价值评估，提出古村核心保护范围15961.5平方米、协调保护范围18696.1平方米。基于此，湖贝120提出湖贝项目的两条底线：核心保护区先予保留、新罗湖公园完全落地；[12]22-25 7月2日，举办“共赢的可能：湖贝古村保护与罗湖复兴设计工作坊”[19]，数十名规划师和建筑师在优先保护公共利益的前提下，针对局促的建设用地提出8条出路，寻求切实可行的方案（后文展开）；7月3日，举办“对话湖贝”论坛[20]，古城保护专家强调湖贝古村能为后辈留下乡愁，应作为文化遗产保护，而非旅游开发（图6-图7）。

基于工作坊和论坛的讨论，湖贝120对湖贝古村保护的实施路径提出了五点建议：①对历史文化资产划线立法；②通过公众参与听证会，实行未定级文物的先予保护机制；③通过制度与设计创新，实现老旧建筑的历史文化与经济价值；④完善城市更新中的第三方综合评估和公共参与制度；⑤增加城市历史空间的保护和研究的资金投入。[12]54-59 这场公共计划形成了三份正式报告——《致深圳市建环艺委的意见书》《致罗湖区城市更新局的意见书》《致深圳市规土委的“湖贝预保护议案”》，递交相关部门。

在密集的公共讨论中，城市管理者迅速地作出回应。6月25日，市委书记调研湖贝，强调在城市更新过程中“尤其要坚持保护与开发并重”。[13] 6月30日，区城市更新领导小组会议议定在湖贝项目坚持开发与保护并重的思路。7月6日，区更新局召开“城市更新媒体发布会”，首次公开回应古村保护问题，提出从较大范围保留古村的三纵八横肌理，协调城市开发与旧村保护的关系，但未明确保护的面积和策略。

尽管未出席湖贝120的活动，华润和华阳迅速地对规划方案作出调整。7月12日，区更新局再次召开湖贝项目专家研讨会。[21] 这轮规划方案原址保留古村三纵八横肌理，保留范围扩大至10000平方米，同时规划在古村东面的地标塔楼仍需要拆除29栋保留完好的历史建筑。7月22日，湖贝120对此发出优化建议，在完整保留15961.5平方米核心区域的前提下，通过设计手法提升公共空间和商业空间（后文展开）。

7月26日，区领导带队到北京拜访院士，在这轮湖贝项目的规划方案文本上得到“积极保护、整体创造”的评语。[14] 7月29日，区更新局召开见面会，邀请华润、华阳、湖贝村民和湖贝120代表参加，但仍未明确古村保护范围、未来产权、管理运营等公众关注的问题。[22] 8月，省、市领导调研湖贝，对区更新局、华润、华阳汇报的方案作出肯定的评价，大众媒体就此发表相关报道。[15] 至此，这场集结各方观点和专业力量的公众参与事件的舆论渐熄。

此间，通过展览和论坛，湖贝古村和罗湖文化公园保护的讨论扩展至更大范围和更多层面。都市实践以“湖贝古村城市更新研究”项目参展“村镇-城市，再创造”。[23] 湖贝120参与2016中国古村与新乡村主题展。[24] 通过活动，更多的人真正地走近古村和公园，通过自己的方式与这个地方发生连接。在湖贝120发起的“每个人的湖贝”公共艺术计划中，有“深二代”离开“万象城-书城-购物公园套路”的探访，建筑师从南头古城跑到湖贝古村的“圈地”，策展人互相喂食的饭局，艺术家在祠堂的砸琴等等。[25] 土木再生城乡营造研究所和未来+学院发起“湖贝请留门！”独立研究计划，最后形成开放的公益课程。[26]

4.3 “保护与开发并重”（2017—2019）

2017年2月，区更新局进行次湖贝项目规划（草案）公示，初步认定的古村保护面积为10350平方米。[27] 针对仍未完整保留古村核心区域的方案，土木再生发表《湖贝少拆一半增值10,000,000,000的秘诀！》，湖贝120发表《深圳人的深圳缺的就是一个折衷开发方案吗？》。

2018年，政协委员提出紧急抢救古村的提案，[16] 区政府回复项目规划研究体系正在完善。[17] 8—10月，区更新局就湖贝项目召开两次城市设计国际咨询会、交通专题评审会、公共配套及市政专题评审会，得到专家通过的意见。[28] 12月，区更新局进行湖贝项目规划（草案优化）公示，初步认定的古村保护面积调整为10016平方米。[29] 湖贝120针对公示方案发表意见，再次呼吁真正地完整保留古村的三纵八横肌理，质疑步行街和地下开发的影响，要求公开详细的保护范围、原则、拆除的建筑位置和面积。

17 专家包括朱荣远、司马晓、王富海、覃力、孟岩、万众、杨晓春、黄伟文、张之扬、王晓峰、唐志华。

18 系列报道包括“这么有价值的古村为何不保住？”“古村落保护将有‘规’可依”“先保护起来再讨论更新”“古村保护需要相应政策支持”“湖贝古村改造有‘第三条路’”“专家工作坊为湖贝旧改献策”“‘湖贝可以为后辈留下乡愁’”“旧村核心区域保留并活化利用”“新版方案强化旧村保护被肯定”。

19 有方空间，2016-7-2，主持人：饶小军、孟岩，邀请专家：文化/社会学者：马立安、史建、廖虹雷、杨阡、王大勇、滕斐、张星、张琴，规划/建筑专家：饶小军、孟岩、朱荣远、曾群、杨晓春、费晓华、汤桦、钟兵、张之杨。拟邀请机构：罗湖区政府、罗湖区更新局、深圳市规土委城市与建筑设计处、湖贝股份公司、华润置地、深规院、华阳国际/RTKL、深圳市城市交通规划设计研究中心、深圳市城市规划学会、深圳市勘察设计行业协会、深圳市城市设计促进中心、深双组委会办公室、深圳大学、未来+学院、都市实践。大部分机构未派人出席。

20 有方空间，2016-7-3，主持人：史建，对话嘉宾：阮仪三、饶小军、廖虹雷、马立安、杨阡、张之扬、孟岩。论坛记录，见湖贝120. 致深圳市规土委的湖贝预保护案 . 2016-7-21: 36-42.

21 专家包括许安之、费晓华、万众、朱荣远、左肖思、司马晓、汤桦、何昉。

22 与会人员包括区更新局领导，华润置地赵荣、祖基翔，华阳国际薛升伟，湖贝股份公司张齐心，湖贝村民代表张炜良，湖贝120代表杨阡、廖虹雷、吴然、温洲冰、王大勇。会议记录，见杨阡 . 城市客厅的吝啬与豪奢——湖贝观察（一）.“湖贝古村120城市公共计划”微信公众号 . 2016-8-11. https://urlzs.com/2mMMG.

23 雕塑家园，2016-7-29—8-31。

24 深圳会展中心，2016-8-23—25。

25 发起于2016年7月8日，详见“湖贝古村120城市公共计划”微信公众号及“湖贝生死进行时”网站：http://hubei120.mysxl.cn/3。

26 发起于2016年8月31日，详见“土木再生”微信公众号。黄伟文发表《新遗产与多地面：一种包容现状的开发模式》，并在2016中国城市规划年会自由论坛“城市非保护街区有机更新”发言；同济大学408研究小组发表《湖贝村调研报告》。

27 2017年2月24日起，公示期30个自然日，收到来自华润、湖贝原住民、丽苑酒店、中兴路两侧居民、保障房潜在使用者、黎明大院的七份意见。2018年11月27日，区领导小组审议并原则同意公示意见处理。

28 专家包括朱荣远、许安之、左肖思、杨晓春、倪有为、顾新、庄葵、陆轶晨、丘建金、邱维扬、张俊杰、谷茂、南凌、费晓华、汤桦。

29 2018年12月3日起，公示期7个自然日，收到湖贝原住民、社会人士、中物大厦、湖贝120的四份意见。2018年12月19日，区领导小组审议通过第二轮

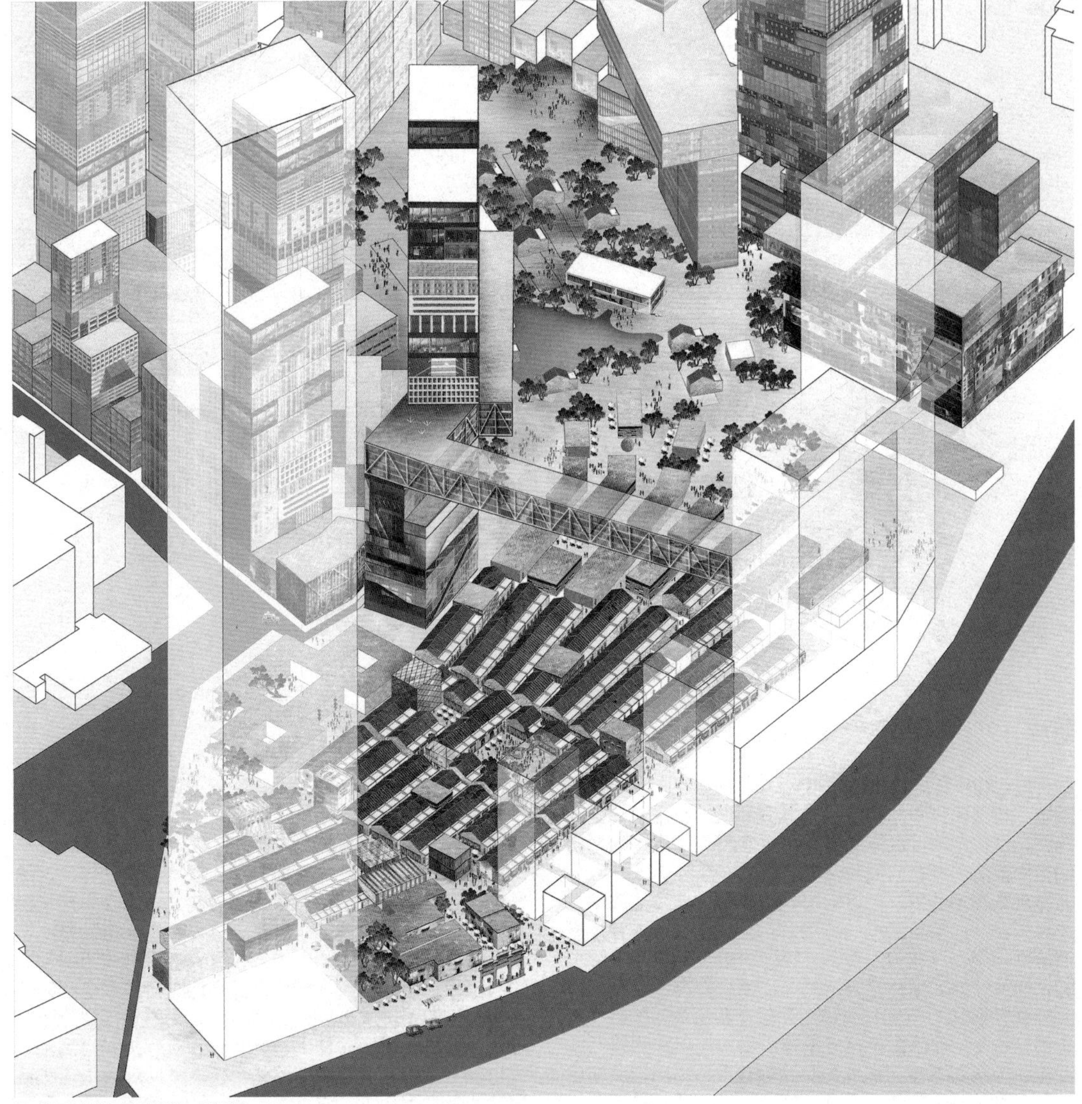

图 4 湖贝古村保护与更新提案，都市实践，2012
Figure 4 Scheme of the preservation and regeneration of Ancient Hubei Village, URBANUS, 2012

2019 年，人大代表再次提出紧急抢救的建议，建议政府推动古村保护立法和公众参与机制。[18] 区政府在回函中阐述湖贝项目审批情况，并肯定了目前规划的工作成果。[19]2 月，市人大常委会召开代表建议办理座谈会 30。3 月，区检察院对湖贝项目进行公益诉讼立案调查，并召开座谈会，约谈相关责任单位。31 9 月，市政府通过《湖贝城市更新统筹片区专项规划》，方案中的古村保留范围为 10016 平方米。

后续几次公众参与并未能进一步影响实施主体扩大湖贝古村的保留范围。诚然，城市更新涉及方方面面，关系错综复杂，古村保留面积这项单一指标无法概括全貌，但它在很大程度上反映了各方对古村价值的判断和片区的定位。在湖贝古村保护议题上，从经济视角出发，0、6000、10000 平方米代表的是不同的土地利用率和开发回报率。而从公共利益视角出发，15961.5 平方米包含的是一个可进入的城市空间，一段可触摸的历史文脉。在两组数字背后，两种视角形成对峙局面，最终得到折衷的结果。

尽管各方达到“保护与开发并重”的共识，但并没有得出具体的设计方案和实施路径。华润表示“一个具有历史文化的建筑和地域更具备商业开发的潜力，而商

公示意见处理，12 月 23 日举办第二次公示意见座谈会。
30　2019-2-26，参会记录及反馈，见湖贝 120. 人大开会关注湖贝保护，专家担心掏空地下会让保护落空．“湖贝古村 120 城市公共计划”微信公众号．2019-3-3. https://urlzs.com/U3Dqk.
31　这是罗湖区检察院第一起涉及古文化保护的公益诉讼案。由区检察院副检察长黄海波主持，约谈市城市更新和土地整备局、区更新局、区文体局、东门街道办、华润置地等 8 家相关责任单位，邀请市人大代表陈锦花、王建锋，文物保护专家黄伟文、杨晓春、张一兵、廖虹雷，及市检察院公益诉讼部部长刘汉俊，2019-3-4。

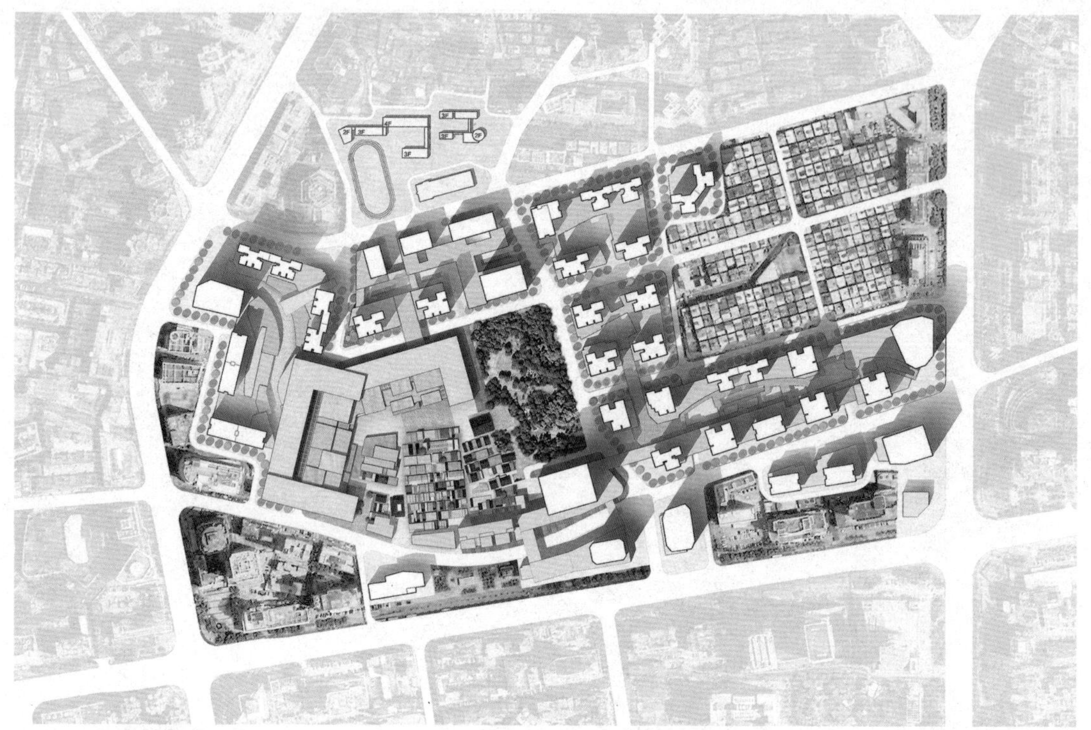

图 5 湖贝古村保护与更新提案，都市实践，2014
Figure 5 Scheme of the preservation and regeneration of Ancient Hubei Village, URBANUS, 2014

业开发也作用于历史文化，对文化起到一个宣传、保护、发扬、传承和升华的作用。"[20] 罗湖区"十三五"规划提出，"在湖贝片区探索城区古村落与现代商业相融合的更新改造新模式，推动湖贝旧村保护性改造。"[21] 当消费成为基本参数，"保护与开发并重"实质指向"开发再开发"。实证在前，1996 年的东门老街改造将建于清康熙年间、已被列为市级文物保护单位的思月书院拆除，而又"为了弥补这一遗憾"，特意将其迁址至老街风貌街区重建。[22] 那些有着生动起源和日常的街巷（上大街、鱼街、鸭仔街、猪仔街等）和广府民居则是彻底消失。[23] 人们常说"深圳先有墟再有市"，旧称深圳墟的东门是湖贝、向西、黄贝岭等几个村落建起的集市，但因失去了存续的空间而逐渐沦为布景。"以商业开发升华文化"的论调实际是有名无实。

5 关于城市的想象

5.1 "共赢的可能"

毫无争议地，深圳的城市化速度比任何时期的任何地方都要迅猛。它的建设者对开发熟稔无比，无论在荒地、农田、村庄、城中村还是已建城市等各类土地上都能够快速地覆盖现代城市图景，而对有待商榷的，保护是否应该服务于开发，甚至等同于开发，就难以为继了。

近二十年来，地产开发商大举介入城市更新项目，以资本影响城市发展方向和决策，他们应用的经济算法压倒性地超过其他评价标准。地产开发逻辑主导的城市更新编写了"拆迁造富"的神话，村民对家乡和家宅的情感链接被巨额的经济补偿所淹没。对于湖贝村的拆迁补偿，股份公司董事长称"村民普遍满意，又有一批千万富翁诞生。剩下不同意的村民基本是利益问题，比如自家有门面房，嫌赔偿的价格不够。"[24]

然而，城市系统复杂而开放，以湖贝为载体，这场发生在公共领域的讨论正是试图脱离"城市作为增长机器"的框架，将久违的"城市公共性"重新带回城市发展决策的价值判断体系中。产权所有者拥有对所持物业的决策权，与此同时，全体市民也拥有对城市空间的参与权。正如湖贝 120 强调，"'活化'不是商业盘活的意思，而是让资源总体上有利社会……保护和'活化'古村唯一重要的意义是：重新创造一个深圳人生活和认同的社会空间。"[25] 在各方的博弈和求同下，原本面临拆除的怀月张公祠入选于 2017 年《深圳市历史建筑（第一批）保护名录》，湖贝南坊（即古村）于 2019 年入选《深圳市历史风貌区（第一批）保护名录》，这可以说是中国城市发展史上样本级的事件。

在新的发展阶段，深圳仍受现代城市图景的惯性驱动，各区追求高度和商业的符号，争夺城市中心的地位。然而，历史上的罗湖拥有的不仅仅是符号，还有符号生成所需要的、通过时间积淀的本土文化——五百余年的湖贝古村所承载的岭南文化、深圳墟所代表的市场文化、数千万来深建设者所创造的移民文化，与近当代的深圳身份紧密相关。

面对罗湖重新定位和深圳培育本土文化的需求，具有专业背景的知识分子并非以怀旧的情绪呼吁不拆古村，而是以面向未来的态度呼吁保护古村。在 2016 年"共赢的可能：湖贝古村保护与罗湖复兴设计工作坊"中，[32] 规划师和建筑师综合完整保留古村的诉求、项目愿景和现实情况，对于局部保留 6000 平方米的规划方案提出以下八条出路，为实施主体出谋划策，向城市管理者建言（图 8）。[12]17-36

（1）在场地上建立一个 Y 形的空间结构，向西连接东门老街，向东连通湖贝新村，使整个旧改项目有机融入罗湖整体城市生态的格局。
（2）新罗湖公园以带状姿态向东北连接湖贝新村，成为新老两村的空间纽带，创造该片区最具活力的大众共享空间系统，而不仅仅是高楼中的孤岛花园。
（3）“三位一体”或“三体合一”，古村、公园和商业综合体三种功能无界融合，让古村、公园和商业综合体在立体空间上咬合、渗透，在平面上叠合。
（4）万象新城购物中心尽量向南侧深南大道寻求发展空间，上跨湖贝路，与现有的“大金牙”高层建筑裙房融合，可借鉴香港 IFC 和深圳南山宝能城的做法，确保满足现有建设规模。
（5）围绕现有地铁站营造一个尺度更为宜人的城市广场，同时在项目用地西南角增加一栋超高层标志性塔楼，释放项目建设用地的紧张压力。
（6）重新设计住宅塔楼户型，在外廓尺寸不变的情况下，将原来不规则的 X 形户型改为口形平面，每层面积可由 700 平方米增加至 900 平方米，在日照条件基本不变的情况下，可增加约 25% 的住宅面积。这样可以消化原来需要拆除古村的新建面积，从而保住古村又不牺牲太多开发商利益。
（7）通过调整更新改造范围，腾挪土地，调整功能搭配，为保护旧村创造有利条件，并研究设立因古村保护而给予容积率奖励的政策。
（8）在目前拆掉南侧两栋高楼的假定下，万象新城完全可能沿深南路展开并跨越湖贝路。其首层和二层可成为开放商业街区，打通深南路与古村的联系；建筑体量南高北低，与古村尺度衔接。兴湖东路仍有很大余地向西转弯，移出更大空间给古村和公园。

虽然开发商和其委托的设计方均未出席工作坊，对八条出路的提案未有明确回应，但迅速地调整了规划方案，将古村保留面积从 6000 平方米扩大至 10000 平方米，地下开发也去除部分古村对应的范围。对于调整方案，湖贝 120 跟进提出以下六条城市设计层面的优化建议，希望进一步落“保护与开发并重”的观念（图 9）。[26]

（1）现有古村的核心区域（15961.5 平方米）全部完整保留，包括东侧重点区域，为此 400 米地标塔楼需要向东稍移，并适当减小商业综合体目前吞掉古村东半部的部分体量。
（2）为弥补综合体面积之不足，建议环绕古村的东南侧建造开放的商业街区，并进而向南延展整合现存金色大楼裙房，设置大体量、跨湖贝路的垂直购物中心——万象新城。
（3）取消湖贝路下穿，参考罗湖万象城以及南山海岸城商业中心跨路的经验，上部跨越湖贝路，使万象新城可以临深南东路，加大商业展示面和标识性，极大地提高商业价值。
（4）兴湖东路不必直接联通深南路，规划西部商业步行街区以及住宅塔楼的布局仍有很大优化空间。塔楼可整体西移，使内部商业街尺度更加宜人，土地使用更加紧凑（参考现规划方案东部街区与住宅塔楼的空间尺度），这样可给公园留有更大的空间，同时减少对湖贝古村西侧的吞噬。
（5）目前规划方案东部临文锦中路处现存有精心设计、绿化良好的街头公园应予以保留，可将塔楼向西移动，现规划方案中正对十字路口设置花园广场的方式只有视觉功能，并不聚集人气，在城市设计方法上是不可取的。
（6）应保留罗湖文化公园内部的大树，结合新建社区室外公共空间布局，并强化中心公园这一绿化及公共步行系统与湖贝新村步行系统衔接，也能借此打破现有规划方案中住宅塔楼过于均质化分布的简单呆板格局。我们希望规划方案应更强调组团式格局，与罗湖该片区现有城市肌理更好融合，避免“大冲式改造”的塔楼堆积形成的非人性化、单调乏味的城市空间。[33]

这六条建议也未得到正面回应，在 2017 年 2 月和 2018 年 12 月公示的两版规划草案中，大部分未被采纳。虽然湖贝 120 提出的方案不尽全面，但切实地从公共利益视角出发进行专业探讨，在一定程度上填补自上而下的城市规划体系和地产开发逻辑的盲区。更重要地，对湖贝项目规划方案的“修正”直接反映了专业人士对于当代城市的想象——既有高效的物质生活，亦有人文精神，它的人民能够对其进行创造而非被动依附。他们批判单一的城市更新模式，是以反抗这种模式必然带来的同质化的城市未来。

5.2 “大刀阔斧展未来”

2019 年 9 月，湖贝城市更新统筹片区一期项目（不包括古村范围）奠基，项目共四期，预计 2030 年完成。根据开发商介绍，湖贝项目“更新范围约 34 万平方米，公共利益用地占 54%，总建筑面积 300 万平方米，总投资 700 亿元，以新型文商旅产融合发展的都市综合体为定位。”[27] 对比 2012 年项目初期的 192 万总建筑面积和 300 亿元总投资，这两项指标增加幅度巨大。正如在工地围挡上，鸟瞰效果图配合着“新貌可期：大刀阔斧展未来”的口号（图 10）。在这幅未来图景中，湖贝古村在中心熠熠发光，被绿地环绕，东北角是高耸的湖贝塔，周边是景观屋顶覆盖的购物中心和玻璃幕墙包裹的超高层住宅楼和办公楼。

无论在垂直还是水平的尺度，无论在空间还是时间的维度，这幅图景都传递了湖贝焕新的决心与雄心。这里的一切都将是崭新的，农耕时期的山陵、堰塘和农田，工业化时期的工厂和自建房，40 年高速城市化下的规划

32 参与设计师包括汤桦、曾群、吴文媛、张星、朱荣远、钟兵、张之扬、习哲森、冯果川、费晓华、孟岩。

33 大冲改造项目将约 1500 栋建筑物拆除，重建为数十栋超高层塔楼，总建筑面积约 380 万平方米。

图 6 湖贝 120 海报，黄扬设计，2016
Figure 6 Hubei 120 Poster, huangyangdesign, 2016

和自发建造交织而成的复杂面貌都未能留下痕迹。古村也是崭新的，它的内外都被抽空，甚至净化，准备被高度控制、一次成形的肌理覆盖。真空式保留的湖贝古村成为一个象征历史的符号，通过新造布景为现代城市图景增添时间厚度。

经过 2016 年的公众参与事件，城市管理者和实施主体明确表示将完整保留湖贝古村的三纵八横肌理，坚持“保护与开发并重”原则。然而，2017 年与 2018 年的两版规划草案公示均未深化古村保护和活化方案。2019 年与 2020 年的现场效果图中，古村东、西两侧仍有数十间民居被拆除，村内的东、西两纵变成内外边界，三纵八横有名无实。

2021 年 1 月，华润发布湖贝塔建筑设计和湖贝古村规划设计的国际竞赛，划定大部分古村范围为公共管理与服务设施用地（GIC）、右部为公园绿地（G1）。任务书明确古村核心保护范围 10016 平方米，建设控制地 4462 平方米；要求“按照三纵八横肌理进行综合整治、保护和活化利用，地下空间不开发，历史建筑怀月张公祠保留，现状建筑经评价后分类施行保护措施。”[28] 相较于初始的整体拆除重建方案，这一设计要求体现了实施主体对古村价值判断的深刻转变。在各方关于城市发展的讨论中，湖贝古村的历史价值得到明确认可。但它之于罗湖、之于深圳的当代价值仍需厘清。目前，规划方案与“保护与开发并重”口号之间存在一定差距，关于古村的未来发展仍不明晰。

在 2021 年夏本文写作时，湖贝古村已被腾空，四周竖起铁皮围挡，仅留怀月张公祠一处隐蔽入口，曾经作为祭祀空间的南围门附近地面堆放着破碎的神像和香炉（图 11）。除了古村，湖贝片区几近被夷为平地，包括东、西、北坊的握手楼和罗湖文化公园。如此做法与“保护与开发并重”这一原则存在一定差距。最新公告的规划图和现场展示的效果图也与“完整保留湖贝古村的三纵八横肌理”这一策略也有所出入。[29]

5.3 谁的城市

深圳的先锋性与特殊性体现在它在接纳意外事物的大胆试验中，生成既有规划的中心区，亦有自发的城中村（包括难能保留至今的城中古村）的城村并置肌理；也体现在市民话语的发声，对城市的未来有着深远影响。长期以来，城市更新项目中的公众参与“点到即止”地停留在咨询阶段，“呈现出参与权利的封闭化和参与过程的形式化特点”。[30] 而在湖贝项目中导向城市发展决策改变的公众参与可谓是公共参与程度加深的一个重要开端。

湖贝 120 的知识分子致力于拓宽深圳城市形态的公共讨论，而不是阿附政府和开发商的声音。在城市总体的规划层面，尤其是湖贝古村的开发方面，他们组成一个包含政府、开发商、各专业人士的公共氛围。这三类人在对话中讨论什么是公共产品，即便大家各自代表不同的侧重：政府对社会秩序负责，主要体现在总体规划上；开发商为实现特定方案的物质资源负责——一个包含商场、办公、住宅塔楼的巨构综合体；湖贝 120 成了“监察者”，提醒政府和开发商，关于规划方案和实施上的弱点以寻求一个更好的深圳城市空间。在这个意义上，湖贝 120 的知识分子不是把自己视作政府和开发商的反对者，而是和他们一起雕琢更好的方案和城市。[31]

特区伊始，深圳赤裸而无畏地走进了前所未有的城市化时空隧道。它在限定的情境里生猛生长，每个人都在想象它的未来，而它也一直在超越每个人的想象。随着特区的成功，“深圳奇迹”的叙述反而将它置于成为一种确定范式的危险位置。尽管研究者和实践者力图对深圳的成长和成功祛魅，更多人发问：“其他地方能否复制‘深圳奇迹’？”而深圳之所以成为如今的深圳，正是源于“一种拒绝僵化和确定性的精神。当你想要复制这种精神的时候，就已经赋予它一种强烈的确定性，这种精神就已经死了。”[32]

Hubei Research: Image and Imagination of the City

Luo Yiqian
URBANUS Researcher

Hubei Village seems too old to survive in the modern city but too young to be located in the mainstream imagination of history.

1 The Inversion of Spatial Expansion

In the long history of Shenzhen, Hubei Village expanded slowly to fulfill the increasing demand for residential space. It was established amid hills and ponds in 1466 (Chenghua Reign, Ming Dynasty).[1] As its population grew, it was enclosed by walls (*wei*) to protect itself from harassment.[1] When the village expanded, it demolished the walls, building houses on its western and eastern sides. Eventually, it assumed the three-vertical-and-eight-horizontal alleys structure that it has retained to this day. Its width is approximately 180 meters east to west and its length is approximately 120 meters north to south, containing narrow alleys, a gate tower, water wells, and houses with tiled roof, grey brick and plaster sculptures.[2] Over the years, its surrounding landscape has shifted from hills to farmland, and then to buildings. Having endured more than five hundred years, Hubei Village has become Ancient Hubei Village, which is the only remaining historical space in the center of a city swept up by rapid urbanization. In 1986, its villagers moved into Hubei New Village, which was located nearby. Thousands of Chaoshan tenants settled in the Ancient Village,[3] where contemporary urban life took place in historical space. In the same year, Luohu Cultural Park was built next to the east side of the Ancient Village. The Park not only staged traditional Cantonese operas, but also became a breeding ground for contemporary poems.[4]

Over the short history of Shenzhen Municipality, Ancient Hubei Village has been devoured. The village is located in Luohu District and has been subject to the District's urban planning. In 1992, Hubei was included in Luohu renewal areas, but after several rounds of discussion, the renewal plan was held up because of dense buildings and complex property rights. In 2011, the Luohu Government and CR Land (China Resources Land Limited) signed the *Outline Agreement on Strategic Cooperation*,[5] which is

1 Ancient Hubei Village is also called Hubei Dawei (large enclosure).

2 Most existing houses in Ancient Hubei Village were built in the late Qing Dynasty. On spatial changes and architectural conditions, see School of Architecture & Urban Planning, Shenzhen University. *Survey Report on the South Lane in Old Hubei Village of Urban Regeneration Project in Luohu, Shenzhen*. 2014.

3 A branch of Han Chinese people who speak Chaoshan dialect, mainly from the Chaozhou-Shantou area in Guangdong Province, including Shantou, Chaozhou, Jieyang and Shanwei Cities.

4 In the Park, the Folk Arts Stage was an important platform for the communication of non-governmental Cantonese music societies between Guangdong and Hong Kong, holding the twice-a-week event *Sing Cantonese Opera and Music* for more than 20 years. Founded in 2012, the editorial office of *Enclave* poetry magazine settled in the west annex building of the theater, becoming a gathering spot for Shenzhen poets.

5 In 2012, they signed the *Outline Agreement on the Demolition Compensation for Government-owned Property*.

regarded as the kick-off of the Project of Urban Renewal of Hubei Coordinated Area (hereinafter referred to as Hubei Project) (Figure 1).[6] In 2012, through a vote, Hubei Company (Shenzhen Hubei Joint-stock Company) agreed that CR Land would be the implementation body.[7] CR Land commissioned UPDIS (Urban Planning & Design Institute of Shenzhen) and CAPOL (CAPOL International & Associates Group) to formulate the urban plan. From 2010 to 2014, various planning schemes replaced the Ancient Village and Cultural Park with a mega shopping mall and high-rise office buildings. At that time, the city administrator aimed to produce an international consuming center in Luohu District,[2] and the implementation body estimated a net profit of 5.0083 million from Hubei Project.[3] The city image and economic interest converged, directly pushing to completely invert the spatial layout. The alliance of these stakeholders formed was based on mutual benefit, the scale of their ambition unprecedented in history.

2 Public Life in Urban Space

2.1 Hybrid Space

At the establishment of the SEZ, the East, West, and North Lanes (*fang*) were built around a nearby hill, comprising Old Hubei Village with Ancient Hubei Village (also known as the South Lane) (Figure 2). In the 1980s, villagers moved into the two-to-three-story houses with courtyards in New Hubei Village, and leased the brick-and-tile houses of the Ancient Village. In the 1990s, except for the ancient part of the village, villagers re-constructed buildings in the Old and New Villages into seven-to-eight-story "handshake" rental tenements (similar to "kissing buildings" in English). At the same time, many villagers moved into commercial housing in the city, while others settled in Hong Kong, Macau, or overseas. Since then, they have rarely returned to Hubei Village apart from the Autumn, when they attend the Double-Ninth Festival, offering incense in the Ancestral Hall in the Old Village and eating Poon Choi (*pencai*, literally means basin cuisine) in the open plaza of the New Village.[8]

In 2016, there were 20,028 tenants in Old Hubei Village, of which 66.1% were from the Chaoshan area.[4] Most of the tenants worked in the aquatic business and service industry in Hubei Village, Dongmen Commercial Street and Dongmen Seafood Market, which were within walking distance of their residences.[9] There were nearly a thousand shops and vendors in the Hubei area, including aquatic product wholesalers, seafood street restaurants that attracted customers from the entire city, and food markets and convenience stores that served nearby communities. Commercial activities ran 24 hours a day. Wholesale markets from dawn to early morning, fresh food retail sales from early morning to dusk, catering from morning to night, street markets from evening to late night, and all-day shops with both working and sleeping functions. Flexible and mixed business forms and affordable living costs provided a path for new migrants to land in the city, creating a vibrant neighborhood for the city. Innumerable diners with dozens of seafood restaurants gathered along Leyuan Road, which became a feature of Shenzhen.

Located in Ancient Hubei Village, the Zhang Huaiyue Ancestral Hall has three courtyards and a three-bay main room. This building is the emotional heart of the Zhang clan. Nearby alleys and the village's South Gate became ritual spaces for Chaoshan tenants, who continued to worship their gods and observe traditional holidays and customs even after migrating to Shenzhen (Figure 3). They attached simple incense burners to the outer walls of their houses and set up temporary alters in the alleys for daily worship. On the opposing walls of the inner passageway of South Gate, there were prayer tables that were available for public use, along with numerous icons. On the 1st and 15th of every lunar month as well as on gods' birthdays, Chaoshan families worshiped around the South Gate. Indeed, South Gate, once the traffic node of the village, generated spiritual support and provided migrants a place to gather through situating ritual activities.

2.2 Diverse Communities

Most tenants stayed in Old Hubei Village for years, some for as long as three decades, during which time they formed tight social networks due to kinship, geographical connections, and working relationships. Some arrived here alone, some

6 The area is located in Dongmen Sub-District in Luohu, with Shenzhen Boulevard to the south, Zhongxing Road to the north, Wenjin Middle Road to the east, Dongmen Middle Road to the west, and covers 400,701.8 square meters. Ancient Hubei Village and Luohu Cultural Park were located in the area center.

7 In 2017, they signed the *Agreement on the Demolition and Relocation Compensation for Collective-owned Property*.

8 The Zhang clan's tradition of the autumn ritual was interrupted for 40 years and resumed in 1989.

9 Dongmen Seafood Market was built in 1989 with funds raised by villagers. It was closed on December 31, 2018. It used to be one of the ten biggest food markets in the country, with nearly 100,000 visits per day and an annual total transaction amount of 1.5 billion yuan.

moved here with family, and some started a family here. A student who was born in Hubei and stayed as a tenant for 15 years said, "We spent all our time here, hence this place can be regarded as our hometown."[5] This kind of natural and sincere sense of belonging derived from years of daily life. During lunar festivals, many tenants chose to celebrate in Old Hubei Village, creating a joyful atmosphere, rather than returning to their native place. This practice differed from other immigrant communities in Shenzhen.

As the actual users of Hubei, tenants had close physical and emotional ties to the place, but they were not defined as stakeholders in Hubei Project because they did not have property rights. For a long time, villagers and tenants had a mutually beneficial relationship. The former leased houses for economic returns, while the latter rented houses for living (even surviving) spaces. However, in the issue of Ancient Hubei Village preservation, they became opponents. The villages wanted the houses demolished to secure economic compensation, while the tenants hoped that the houses could be preserved to maintain their low-cost urban life. When property rights and registered residence are the only legal standard for belonging, then villagers who owned houses were the only ones to claim a legitimate nostalgia for Hubei. In contrast, the tenants did not have any legal standing in the village. In the end, villagers chose to leave Hubei even though have the right to stay, while tenants were forced to leave even though wished to stay.

To demolish 拆 or 或 留 to preserve

3 Driving Force for the City

Since the reform and opening up in 1978, the image of the global modern city has been projected over China, forcing different places to enthusiastically urbanize without looking back. According to this image, the built environment is everywhere brand-new and ordered, oriented by efficiency, paved with wide roads and filled with denser

图 7 湖贝 120 工作坊“共赢的可能”和论坛“对话湖贝”，2016 © 都市实践
Figure 7 Hubei 120 workshop "An All-Win Possibility" and forum "Conversation with Hubei," 2016 ©URBANUS

buildings. Commodities are the medium through which consumption becomes the focus of urban life. This image has also become the driving force for Shenzhen, driving it to build without consideration and rebuild without hesitation.

To successfully occupy the center status depicted in the image of the modern city, height is undoubtedly a symbol with the strongest visual effect. This pursuit to be the tallest building has been an important part of both Luohu District's identity and Hubei Project. For thirty years, for example, Shenzhen's tallest building was the Guomao Building, which is located in Luohu District. In 1985, Guomao Building measured 160 meters, not only becoming the tallest building in the city, but also the nation. In 1996, Diwang Building, which is also in Luohu measured 383.95

meters when it took over the title of Asia's tallest building. In 2011, Kingkey 100 Tower measured 441.8 meters, keeping the title of Shenzhen's tallest building in Luohu until 2016, when the Ping An International Finance Center in Futian District became the city's tallest building. The following year, a 2017 proposal for Hubei Project, Hubei Tower was planned right next to the east side of Ancient Hubei Village and designed to be as high as 830 meters, surpassing the tallest building in the world, Khalifa Tower.

At the same time, Luohu District created and was long held to be the symbol of commercial land development in Shenzhen. In 1984, for example, the first commercial housing community in China, Donghu Liyuan, was built in Luohu. In 1987, China's first auction of use rights to state-owned land applied to a plot in Luohu, where a commercial housing community, Dongxiao Huayuan, was built the following year.[10] These actions directly changed land use nationwide.[11] Moreover, the concept of "housing is a commodity" spread from Shenzhen to the rest of the country through the reform of the housing system and the popularization of commercial housing.[12] In 2004, the largest high-end shopping mall in Shenzhen, MixC, was opened in Luohu, which was also CR Land's first project in Shenzhen. Since then, CR Land has continuously upgraded its development model of shopping malls.

In 2012, for example, the developer introduced its redevelopment model for Ancient Hubei Village, "According to the planning, after the renewal of the Hubei area, the GFA included into FAR calculation is approximately 1.92 million square meters, including 500,000 square meters for retail and entertainment, 450,000 square meters for offices, and 970,000 square meters for apartments and residences. The estimated investment is 30 billion yuan. [...] The project includes comprehensive remediation of buildings on nearly 100,000 square meters of land, and the demolition of old buildings of approximately 800,000 square meters, involving 87 plots of land and nearly 4,000 owners." [6]

10 The 50-year use right of the 8,588-square-meter plot was obtained by Shenzhen Special Economic Zone Real Estate and Properties for 5.25 million yuan in the auction. Dongxiao Huayuan was sold for 1,600 yuan/square meter at the time, and the net profit of the project reached nearly 4 million yuan.

11 The 1988 *Amendment to the Constitution of the People's Republic of China* amended the article as "The right to the use of land may be transferred according to law." The *Land Administration* amended accordingly.

12 Shenzhen introduced the *Reform Plan for the Housing System of Shenzhen Special Economic Zone* on January 1, 1988, which became the first local document of housing reform in the country.

4 Urban Space in the Public Sphere

4.1 To demolish or to preserve? (2012-2014)

In comparison to the image of the modern city, Ancient Hubei Village failed on all counts; it was too small scale, its alleys were too narrow, and its houses were too short. However, in real urban life, the Ancient Hubei Village and Luohu Cultural Park spatially mediated residents' daily activities and local culture. Facing the contradiction between the image and reality, professionals and academics questioned the prevalent view of urbanity. How should we define the value of Ancient Hubei Village and Luohu Cultural Park as well as their value to Luohu District and Shenzhen City?

Upon learning about the *tabula rasa* development scheme for Ancient Hubei Village in 2011, the partner of URBANUS launched independent research on the village, drawing attention to the rarity of a 500-year-old village in the city center of Shenzhen. The fabric, architecture, customs and daily life of Ancient Hubei Village differentiated it from the homogenizing city. Based on this research, URBANUS proposed a scheme that matched the economic projects of the raze and rebuild plan and preserved the complete three-vertical-and-eight-horizontal alleys fabric of Ancient Hubei Village by allowing high density in some areas (Figure 4). The following year, URBANUS compiled a research report and presented it to the Luohu Government. The report was praised for its idea of achieving both village preservation and development indexes in urban regeneration. The firm also mobilized its resources to pay attention to the redevelopment of the ancient village, holding the workshop "Preservation and Regeneration of Ancient Hubei Village." During the workshop, city administrators, representatives from the implementation body, and citizens were invited to discuss the future of Hubei Ancient Village.[13] During the 2013 Two Sessions (NPC and CPPCC) Meetings, nine people's representatives raised an appeal for the overall preservation of Ancient Hubei Village and ancient domestic architecture as the "most intuitive example of ancient Shenzhen's homeland."[7]

Despite public participation drawing city administrators' attention to the issue, nevertheless,

13 E6 Space, October 29, 2012, participants included Hubei villagers, architecture and urban planning experts, social and humanities scholars, media, and relevant government departments.

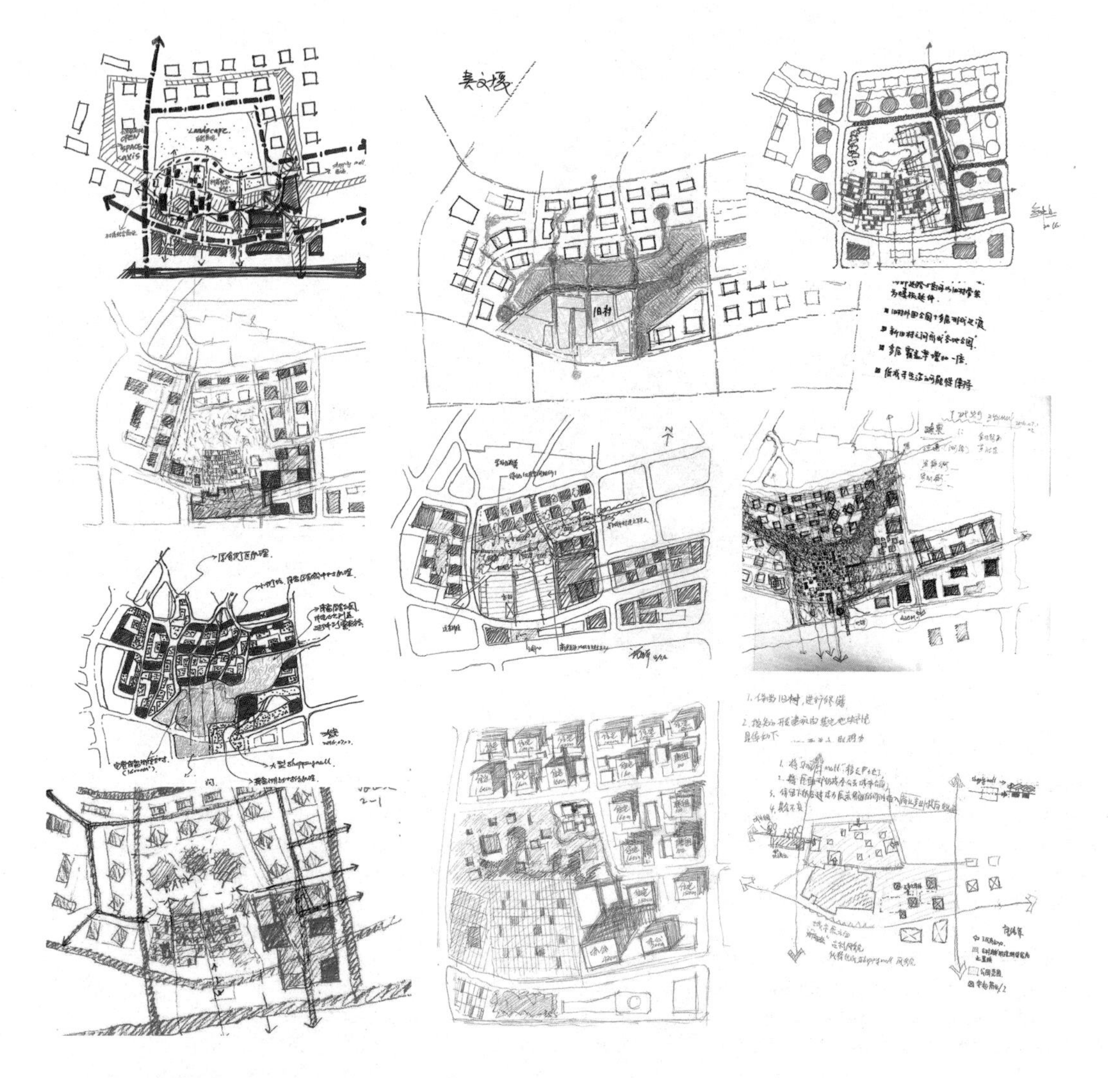

图 8 “共赢的可能：湖贝古村保护与罗湖复兴设计工作坊”提案，2016-7-2
Figure 8 Proposals of "An All-Win Possibility: Design Workshop of Ancient Hubei Village Preservation and Luohu Revival," 2016-7-2

at the early stage of Hubei Project, public opinion did not sway the implementation body's decision on how the Hubei area should be redeveloped. In 2013, CR Land claimed that after demolishing Ancient Hubei Village, it could reproduce local culture by building a new group of Lingnan architecture. Hubei Company, after several discussions with CR Land, submitted an application to the Planning and Natural Resources Bureau of Shenzhen Municipality (Municipal Planning Bureau) to relocate and reconstruct the Zhang Huaiyue Ancestral Hall in cooperation with the overall planning.[7] However, the Municipal Planning Bureau rejected the first-edition planning scheme of Hubei Project because it neglected the unique space and culture of Ancient Hubei Village and did not take consideration in-situ preservation of the Ancestral Hall, which had been included inside the city's designated conservation spaces.[14]

At this time, CR Land commissioned URBANUS and the SZU Team (School of Architecture & Urban Planning, Shenzhen University) to conduct research on the preservation and activation of Ancient Hubei Village, prepare a preservation

14 The Zhang Huaiyue Ancestral Hall was first built in the middle of the Ming Dynasty, rebuilt in 1804 (Jiaqing Reign, Qing Dynasty), and refurbished in 1935. It was included in the conservation area of the *Purple Line Planning of Shenzhen*, with the core conservation area of 3,600 square meters and controlled construction area of 6,918 square meters.

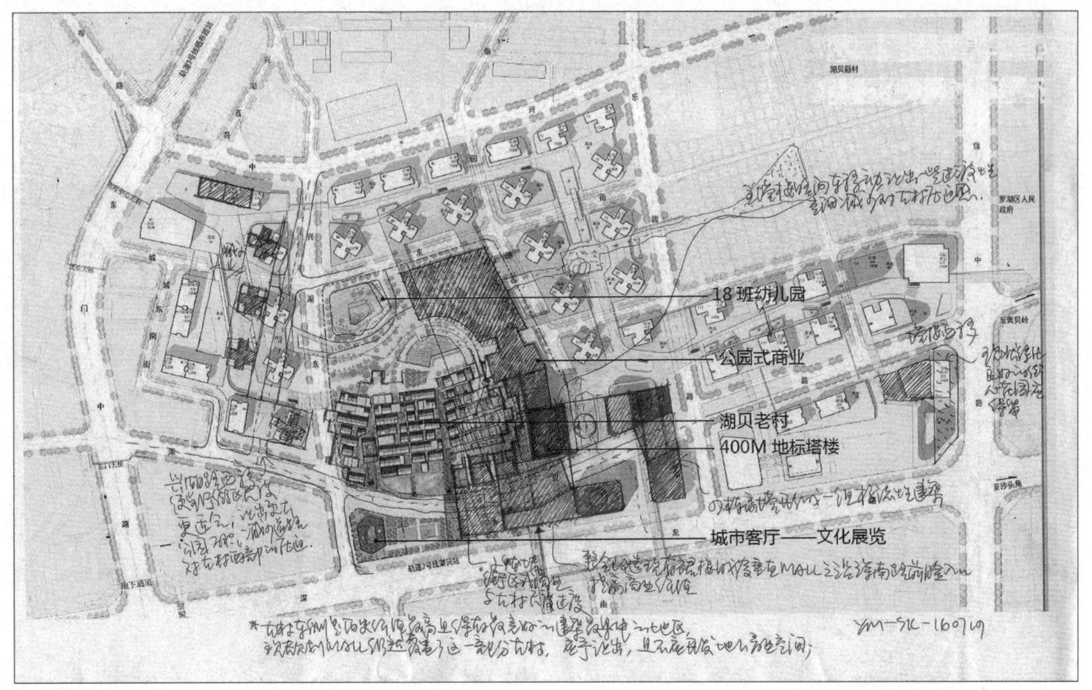

图 9 孟岩手稿，2016-7-19
Figure 9 Drawing by Meng Yan, 2016-7-19

survey of Ancient Hubei Village, and to organize workshops, inviting the two research teams and CAPOL to discuss the overall planning and renewal design of Old Hubei Village. Third party professionals also came to the table to participate in the conversation. In 2014, URBANUS studied the overall layout of Luohu District, emphasizing the significance of Ancient Hubei Village as a unique cultural block. URBANUS proposed strategies of community upgrading, commercial zoning, and architectural transformation based on the complete preservation of Ancient Hubei Village. The result was a second scheme, which explored a three-dimensional urban spatial model with multi-cultural integration (Figure 5).[8] The firm again appealed for social attention, holding the workshop, "Inspiration of Old Hubei Village on New Cultural Commerce."[15] Based on the survey analysis of spatial changes and status quo, the SZU team established an evaluation index system for Ancient Hubei Village, including historical features, reconstructed and added spatial qualities, and building structure quality. According to the survey report, the more than 200 existing houses were primarily built during the late Qing Dynasty, exhibited Cantonese domestic architecture layout and traditional style details, and had spatial modifications (mainly reconstructed and expanded courtyards) caused by lifestyle changes under urbanization.[9]

4.2 How much to preserve? (2016)

In 2014, Hubei Project was stalled for various reasons. In 2015, Luohu District was designated the pilot of "strengthening district power" in urban renewal.[16] The Luohu District Committee of CPC set 2016 as "the year of making breakthroughs in urban renewal and making improvements in urban administration," putting Hubei Project back on the agenda. In May 2016, the Urban Renewal and Land Development Bureau of Luohu District (Luohu Renewal Bureau) organized an expert review for Hubei Project, in which the planning scheme failed to receive the endorsement of most experts.[17] Although commissioned third parties had conducted research on preservation, nevertheless, the implementation body merely

15 E6 Space, March 14, 2014, participants including Hubei villagers, architecture and urban planning experts, relevant government departments, cultural brand operators, and Hubei Project developer, planning and architectural design teams and commercial teams.

16 Some of the administrative rights of urban renewal in Luohu District was transformed from the Municipal Planning Bureau to the Luohu Renewal Bureau for a two-year pilot period, see *Decision of Shenzhen Municipal People's Government on Luohu District as a Pilot for Urban Renewal Reform*, Order of Shenzhen Municipal People's Government (no. 279), August 29, 2015. In the following year, urban renewal reform was implemented over the city, see *Decision of Shenzhen Municipal People's Government on Implement Urban Renewal Reform*, Order of Shenzhen Municipal People's Government (no. 288), October 15, 2016.

issued a statement on the value of Ancient Hubei Village putting forward renewal methods that included "building through relocation, imitation, and creation." It designated 6,000 square meters of Ancient Hubei Village for preservation, retaining the Zhang Huaiyue Ancestral Hall in the western section, constructing a traditional-style street in the middle section, and demolishing buildings in the eastern section to build a mega shopping mall. This plan offered Shenzhen a fragmented and broken "ancient village," replacing the original with a fake and leaving the facade without a core. This planning scheme that comprised a statement of preservation and a fact of destruction triggered one of the important public responses in the history of Shenzhen.

At this urgent juncture, Shenzhen intellectuals from different sectors including society, culture, urban planning, architecture and art spontaneously set up discussion platforms, hoping to preserve a living ancient village for Shenzhen and its citizens. Main promoters included cultural scholars, anthropologists and university professors who had studied Shenzhen urbanization and architects, such as URBANUS who had practiced in Shenzhen for nearly 20 years and studied Ancient Hubei Village since 2011.

In June 2016, six national academicians jointly wrote a letter to the leaders of the Shenzhen City Committee of the CPC and City Government, pointing out that "the value of retaining Ancient Hubei Village cannot be measured by economic interests," and suggested "using the mechanism of in-advance protection for unclassified cultural relics."[10] Shenzhen Center for Design held a seminar "Survival, Activation, and Livelihood of the Ancient Village in the City" in its Cool Chat series. During this seminar, the Hubei Project planning director at UPDIS pointed out one of the main problems with the renewal process, mainly that "the developer is leading and implementing the planning at the front row, resulting in an absence of the government."[11] The *Shenzhen Evening News*, *Shenzhen Special Zone Daily*, *Southern Metropolis Daily*, *Yangcheng Evening News*, and we-media reported on the issue, and the *Shenzhen Economic Daily* followed up, publishing a half-month series, "Why Can't We Preserve Such a Valuable Ancient Village?"[18] Mass media provided the public with a channel to gain information and a platform to have conversations with stakeholders.

In early July of that year, Shenzhen intellectuals initiated the "Hubei 120 City Public Plan" (Hubei 120) and published the "Consensus of Hubei Appeal: Rescuing Our Historic Memory (the "Consensus")"[12]61-64 which was later signed by nearly a thousand people. At this time, the discussion on the value and future of Ancient Hubei Village and Luohu Cultural Park formally entered the public sphere. The "Consensus" describes the uniqueness of Ancient Hubei Village in Shenzhen history and its significance to urban identity, regarding Ancient Hubei Village as not only being the villagers' private property, but also a social resource for all Shenzheners.

Integrating SZU Team's survey report, the *Evaluation Criteria for Historical Areas and Historical Buildings in Shenzhen* assessed the value of Ancient Hubei Village, suggesting a core preservation area of 15,961.5 square meters and a coordinated preservation area of 18,696.1 square meters. Based on this assessment, Hubei 120 put forward two limits to the Hubei Project. First, to retain the core preservation area of 15,961.5 square meters in advance and second, to make a new Luohu Cultural Park fully on the ground.[12]22-25 On July 2, 2016, Hubei 120 held "An All-Win Possibility: Design Workshop of Ancient Hubei Village Preservation and Luohu Revival,"[19] in which dozens of planners and architects provided eight solutions for the Hubei area, basing their suggestions on the priority of protecting the public interest protection and seeking a feasible scheme (to be discussed below). On July 3, 2016, Hubei 120 held a forum "Conversation with Hubei,"[20] during which an expert in ancient cities/towns protection stressed that Ancient Hubei Village could be a homeland for future generations and should be preserved as cultural heritage rather than developed as a tourist site (Figure 6-7).

Based on the workshop and forum results, Hubei 120 put forward five suggestions on how

17 Experts included Zhu Rongyuan, Sima Xiao, Wang Fuhai, Qin Li, Meng Yan, Wan Zhong, Yang Xiaochun, Huang Weiwen, Zhang Zhiyang, Wang Xiaofeng, and Tang Zhihua.

18 *Shenzhen Economic Daily*, June 23 - July 13, 2016.

19 POSITION Space, July 2, 2016. Forum hosts were Rao Xiaojun and Meng Yan. Invited experts included cultural and social scholars Mary Ann O'Donnell, Shi Jian, Liao Hongle, Yang Qian, Wang Dayong, Teng Fei, and Zhang Qin as well as planning and architectural professionals Rao Xiaojun, Meng Yan, Zhu Rongyuan, Zeng Qun, Yang Xiaochun, Fei Xiaohua, Tang Hua, Zhong Bing, and Zhang Zhiyang. The invited institutes were Luohu Government, Luohu Renewal Bureau, Shenzhen City Planning Bureau, Hubei Company, CR Land, UPDIS, CAPOL/RTKL, Shenzhen Urban Transport Planning Center, Urban Planning Society of Shenzhen, Shenzhen Exploration & Design Association, Shenzhen Center for Design, UABB Committee Office, Shenzhen University, Future Plus Academy, and URBANUS. Most of the institutes did not send a representative.

20 POSITION Space, July 3, 2016. The forum host was Shi Jian in conversation with Ruan Yisan, Rao Xiaojun, Liao Honglei, Mary Ann O'Donnell, Yang Qian, Zhang Zhiyang, and Meng Yan. Forum record, see Hubei 120, *Prior-Preservation Proposal of Hubei*. 2016: 36-42.

Ancient Hubei Village should be preserved: First, legislate a definition of the scope of historical cultural assets; Second, use the mechanism of in-advance protection for unclassified cultural relics through public participation and/ or hearings; Third, realize the historical, cultural and economic value of old buildings through system and design innovation; Fourth, improve the system of third-party holistic assessment and public participation in urban regeneration, and; Fifth, increase capital investment in the preservation and research of historical spaces in the city.[12]54-59 This public plan was expressed through three formal reports, *Opinions to Committee of Architectural and Environmental Arts*, *Opinions to Urban Renewal and Land Development Bureau of Luohu District*, and *"Prior-Preservation Proposal of Hubei" to Planning and Natural Resources Bureau of Shenzhen Municipality*. The reports were submitted to relevant government sectors.

With intensive discussions in the public sphere, city administrators responded quickly. On June 25, 2016, the then Secretary of the City Committee of the CPC investigated Hubei, emphasizing to "insist on laying equal stress on protection and development" in urban renewal.[13] On June 30, 2016, the leading group meeting of the Luohu Renewal Bureau agreed to the idea of laying equal stress on protection and development in Hubei Project. On July 6, 2016, the Luohu Renewal Bureau held an Urban Renewal Press Conference, in which for the first time made a public response on the issue of Ancient Hubei Village preservation, stating to preserve its fabric of three-vertical-and-eight-horizontal alleys on a larger scale and coordinate the relationship between urban development and old village protection. The Bureau did not, however, specify preservation area and strategy.

Although CR Land and CAPOL did not show up to Hubei 120 activities, they nevertheless made a quick adjustment to the planning scheme. On July 12, 2016, the Luohu Renewal Bureau once again organized an expert review for Hubei Project.[21] The adjusted scheme proposed the in-situ preservation of the three-vertical-and-eight-horizontal alleys fabric of Ancient Hubei Village, expanding the preservation area to 10,000 square meters, and demolishing 29 historical buildings in good condition in the east part to build the landmark tower (as per the previous planning scheme). On July 22, 2016, to optimize the new scheme and preserve 15,961.5 square meters of the core area, Hubei 120 provided advice to improve public space and commercial space by design means (to be discussed below).

On July 26, 2016, Luohu Government officials led a team to visit a national academician in Beijing to discuss the issue, returning with his comment "active protection, integrated creation" written on the booklet of the latest Hubei Project planning scheme.[14] On July 29, 2016, the Luohu Renewal Bureau held a meeting with CR Land, CAPOL, Hubei villagers and Hubei 120 representatives. This meeting did not address public concerns, including the preservation area, future property rights, and management and operation.[22] In August 2016 provincial and municipal government leaders investigated Hubei and gave positive comments on the scheme presented by the Luohu Renewal Bureau, CR Land and CAPOL. This official recognition was reported by the mass media.[15] Subsequently, the diversity of voices included in the Hubei Project through public participation faded.

During this period, discussions on Ancient Hubei Village and Luohu Cultural Park preservation expanded to a larger area and included more layers through exhibitions and forums. URBANUS presented "Urban Regeneration Research on Old Hubei Village" in the exhibition, Rural-Urban Re-Inventions.[23] Hubei 120 participated in the 2016 Chinese Ancient Villages and New Countries Theme Exhibition."[24] Through these activities, more people became aware of both the ancient village and the park, making a connection with these places in unique ways. Hubei 120 launched the public art plan *Everybody's Hubei*. Art projects included a second-generation Shenzhener who had a visit beyond the "MixC-Book City-Shopping Park routine," an architect "enclosed land" by running from Nantou Old Town to Ancient Hubei Village, a curator hosted a meal with an audience member during which they fed each other, and an artist performed and smashed his guqin (Chinese zither) in the Zhang Huaiyue Ancestral Hall.[25] In addition, Retumu Urban/Rural Institute and Future Plus Academy launched

21 Experts included Xu Anzhi, Fei Xiaohua, Wan Zhong, Zhu Rongyuan, Zuo Xiaosi, Sima Xiao, Tang Hua, and He Fang.

22 Participants included Luohu Renewal Bureau leaders; Zhao Rong and Zu Jixiang from CR Land; Xue Shengwei from CAPOL; Zhang Qixin from Hubei Company; Zhang Weiliang representing Hubei villagers, and; Yang Qian, Liao Honglei, Wu Ran, Wen Zhoubing and Wang Dayong representing Hubei 120. For meeting minutes, see Yang Qian. "The Extravagance and Stinginess of City Living Room, Hubei Observation (1)." Hubei 120 official WeChat account [湖贝古村 120 城市公共计划]. August 11, 2016. https://urlzs.com/2mMMG.

23 Sculpture Homeland, July 29 - August 31, 2016.

24 Shenzhen International Convention & Exhibition Center, August 23-25, 2016.

an independent research plan, entitled "Please Leave a Door Opened for Hubei!" Which was also an open course.[26]

4.3 Laying Equal Stress on Protection and Development (2017-2019)

In February 2017, the Luohu Renewal Bureau publicized the "Hubei Project Planning Scheme (Draft)," which included preserving 10,350 square meters of the ancient village.[27] This scheme did not preserve the complete core area of Ancient Hubei Village. In response, Retumu issued "The Secret to a 10,000,000,000 Increase in Value by Halving the Demolition Area of Hubei!" and Hubei 120 issued "Does the Shenzheners' Shenzhen Lack a Developing Scheme Compromise?"

In 2018, members of the Shenzhen People's Political Consultative Conference raised a proposal for the emergent rescue of Ancient Hubei Village,[16] to which the Luohu Government replied that the research system of the project planning was under construction.[17] From August to October 2018 the Luohu Renewal Bureau held two international consultations for urban design, transportation review, and public facilities and infrastructure review, which were approved by experts.[28] In December 2018 the Luohu Renewal Bureau released the "Hubei Project Planning Scheme (Optimized Draft)," which had adjusted the preservation area to 10,016 square meters.[29] Based on this scheme, Hubei 120 again appealed for the complete preservation of the three-vertical-and-eight-horizontal alleys fabric of Ancient Hubei Village, questioned the influence of the pedestrian street and underground development, and demanded the disclosure of the preservation area, design principles, and detailed locations of buildings to be demolished.

In 2019, Deputies to the Shenzhen People's

图 10 湖贝项目工地围挡，摄于 2020
Figure 10 Construction fences of the Hubei Project, 2020

Conference again raised a proposal to rescue Ancient Hubei Village, suggesting that the government promote preservation legislation and public participation.[18] In response, the Luohu Government explained the approval process of Hubei Project, confirming the planning work to date.[19] In February 2019, Shenzhen Municipal People's Congress held a symposium on deputies' proposal.[30] In March 2019, Luohu People's Court filed and investigated the public interest litigation of Hubei Project, and held a meeting with responsible sectors.[31] In September, the Shenzhen City Government approved the *Special Planning of Urban Renewal of Hubei Coordinated Area*, in which the preservation area of Ancient Hubei Village was designated to be 10,016 square meters.

Subsequent public participation after 2016 did not further affect the implementation body in terms of expanding the preservation area of Ancient Hubei Village.

Urban regeneration involves all levels of society, and the relationships between and among these groups are complex. Although the preservation area of Ancient Hubei Village is only one indicator and cannot summarize the whole picture of urban renewal in Shenzhen, nevertheless, it does reflect different parties' judgment of the value of the ancient village as well as the position of Hubei Area within larger urban configurations.

25 Initiated on July 8, 2016, see the Hubei 120 official WeChat account [湖贝古村 120 城市公共计划] and the website, "Life and Death of Hubei," http://hubei120.mysxl.cn/3.

26 Initiated on August 31, 2016, see Retumu official WeChat account [土木再生]. Huang Weiwen presented the lecture, "New Heritage and More Ground: A Development Model with Inclusive Existing Conditions," which was included in the free forum "Organic Regeneration for Non-Protective Urban Blocks," the 2016 Annual National Planning Conference. 408 Team of Tongji University issued *Hubei Investigation Report*.

27 The publicity period lasted for thirty days from February 24, 2017. The Lou Renewal Bureau received seven opinions from CR Land, Hubei villagers, Liyuan Hotel, residents living in Zhongxing Road, potential users of affordable housing, and Liming Dayuan. On November 27, 2018, the Luohu District leading group reviewed and agreed in principle on the answers for publicity opinions.

28 Experts included Zhu Rongyuan, Xu Anzhi, Zuo Xiaosi, Yang Xiaochun, Ni Youwei, Gu Xin, Zhuang Kui, Lu Yichen, Qiu Jianjin, Qiu Weiyang, Zhang Junjie, Gu Mao, Nan Ling, Fei Xiaohua, and Tang Hua.

29 The publicity period lasted seven days from December 3, 2018, and received four opinions from Hubei villagers, concerned citizens, Zhongwu Building, and Hubei 120. On December 19, 2018, the leading group of Luohu District reviewed and agreed in principle to the answers for the second publicity opinions, and held a symposium on December 23.

30 February 26, 2019, minutes and feedback, see Hubei 120. "Focusing Hubei Preservation on People's Congress, Experts Worrying Underground Development to Fail the Preservation." Hubei 120 official WeChat Account [湖贝古村 120 城市公共计划], March 3, 2019. https://urlzs.com/U3Dqk.

31 This was the first public interest litigation involving ancient culture protection heard in the Luohu People's Court, March 4, 2019. Presiding official was Deputy Chief Procurator Huang Haibo and eight relevant offices were invited, including Shenzhen and Luohu Renewal Bureaus, Luohu District Culture and Sports Bureau, Dongmen Sub-District Office and CR Land. People's representatives were Chen Jinhua and Wang Jianfeng. Experts in cultural relics protection included Huang Weiwen, Yang Xiaochun, Zhang Yibing and Liao Honglei. The director of the Shenzhen People's Court Public Interest Litigation Department, Liu Hanjun also participated.

图 11 湖贝古村南围门，摄于 2020
Figure 11 South Gate, Ancient Hubei Village, 2020

From an economic perspective, for example, 0, 6,000 and 10,000 square meters represent different rates of land use and development return. From the perspective of public interest, however, 15,961.5 square meters contain an accessible urban space and tangible historical context. Behind these two sets of numbers, two perspectives formed a confrontation and eventually compromised with each other.

Although different parties reached a consensus on "laying equal stress on protection and development," they did not settle on a practical design scheme and implementation path. Said CR Land, "A building and an area with history and culture have more potential for commercial development, and commercial development also can publicize, protect, develop, inherit, and sublimate history and culture."[20] The *13th Five-Year Plan of Luohu District* stressed that it was important to "explore a new renewal model that integrates the Ancient Village and modern commerce in Hubei Area, promoting the protective transformation of Old Hubei Village."[21] When consumption became a basic index, "laying equal stress on protection and development" in practice continued to be "develop and also develop." For example, the 1996 Dongmen Old Street transformation project demolished the Siyue Academy, which had been built during the Kangxi reign (Qing Dynasty 1662-1722) and listed as a historical and cultural unit protected at the municipal level. It was subsequently relocated to the traditional style block to rebuild "in order to make up for the regret."[22] The demolition of local streets and alleys, including Shangda, Fish Street, Duck Street, and Pig Street as well as Cantonese domestic architecture was also a demolition of vivid roots and the context of daily life.[23] Shenzhen has been referred to as "a city that emerged from a market *(xu)*." Dongmen, formerly known as Shenzhen Market, was a market established by several villages including Hubei, Xiangxi and Huangbeiling. However, it became a stage set alienated from its history. "To sublimate culture by commercial development" is, in fact, a package without meat.

5 The Imagination of the city

5.1 An All-Win Possibility

Undoubtedly, the urbanization of Shenzhen has been faster than anywhere else at any other time in history. The city's constructors have been familiar with the development and are capable of imposing the image of the modern city on all types of land, including empty fields, farms, villages, urban villages, and built urban environments. However, it remains to be discussed whether protection/preservation should serve the purpose of development or even be equated with development?

Over the past two decades, real estate developers have taken on urban renewal projects, with capital outlays greatly influencing the future imaginary of the city. These developers have applied an economic algorithm the importance of which has overwhelmingly exceeded many other evaluation criteria. This real-estate-driven urban renewal model has furthered the myth of "getting rich by demolition," drowning villagers' emotional links to their homeland and homes in huge economic compensation packages. The chairman of Hubei Company has said that in terms of the demolition compensation, "Most villagers are satisfied. There will emerge another group of multimillionaires. The villagers who disagree do so basically out of private interest. For example, those who have houses with shop fronts demand higher compensation."[24]

A city system, however, is complex and open. The discussion about Hubei and its redevelopment occurred in the public sphere with an eye to breaking an organizing framework of the "city as a growth machine" and returning the "city's publicness," to the value system of the decision-making in urban development. Owners have the right to deal with their properties, and at the same time, all citizens should have the right to participate in urban spaces. As Hubei 120 stressed, "'Activation' does not mean revitalizing with commercial development, but making resources beneficial to the society on the whole... The only significance of preserving and activating the ancient village is to re-create a social space where Shenzheners live with a concomitant identity."[25] Through negotiation and the consensus of different parties, the Zhang Huaiyue Ancestral Hall, which had faced the threat of demolition was included in the Protection Catalogue of Historical Buildings in Shenzhen (Batch I) in 2017, and Hubei South Lane (i.e. the ancient village) was included in the Protection Catalogue of Historical Areas in Shenzhen (Batch I) in 2019. This process can be regarded as a sample event in the history of Chinese urban development.

In its latest stage of development, Shenzhen has still been developing within the inertia of the modern city image. Various city districts have been chasing the symbols of height and commercialization, competing for the status of the city center. However, Luohu District had not only achieved these symbols, but was also home to the local culture required to form and inherit these symbols. This local culture and heritage included Lingnan culture that was embodied by the over-500-year-old Ancient Hubei Village, a market culture that was represented by Shenzhen Market, and the immigrant culture that was forged by the tens of millions of migrants who built Shenzhen. Thus, the built environment of Luohu is intimately related to the modern and contemporary identity of Shenzhen.

Faced with the demands to both redirect Luohu and cultivate local culture, intellectuals with professional backgrounds did not oppose the demolition of the ancient village based on nostalgia, but appealed for the preservation of the ancient village with an eye toward the future. In the 2016 All-Win Possibility Workshop,[32] planners and architects took comprehensive consideration of the complete preservation appeal, project goals, and practical conditions, and then put forward the following eight suggestions for the planning scheme, which at the time only included a partial preservation area of 6,000 square meters. This advice was compiled for the implementation body and city administrators (Figure 8):[12]17-36

> *(1) Establish a Y-shape spatial structure, connecting Dongmen Old Street to the west and New Hubei Village to the east, so as to organically integrate the renewal project into the overall urban ecology of Luohu;*
> *(2) Take New Luohu Park as a spatial link between the New and Old Villages, connecting New Hubei Village to the northeast in a belt shape, so as to create the most dynamic space system shared by people in the area instead of creating an isolated garden among high-rise buildings;*
> *(3) Three-in-one design to functionally integrate the ancient village, park and commercial complex, interlocking and interpenetrating in three-dimensional space and overlapping on a two-dimensional plane;*
> *(4) Place the New MixC shopping mall on Shennan Boulevard as far as possible to the south, over-crossing Hubei Road, integrating with the existing podium of the so-called "Big Gold Teeth Towers" to meet the required construction scale, with a reference to IFC in Hong Kong and All City Mall in Nanshan;*
> *(5) Build a city plaza around the existing metro station to a proper scale, and add a landmark skyscraper tower in the southwest corner of the area, reducing the pressure of limited construction land;*
> *(6) Re-design the layout of residential towers, adjusting the irregular X-shape plan into a square-shape plan and maintaining the outline dimension unchanged, whereby the area of each floor could be increased from 700 to 900 square meters and the residential area increased by 25%, even as sunlight exposure remained unchanged. This approach covered the new-built area that required the ancient village to be demolished, preserving the Ancient Village without sacrificing the developer's economic interests;*
> *(7) Provide advantages for ancient village preservation by adjusting renewal scope, relocating land and adjusting function collocation. Research and set up a policy to reward FAR for ancient village preservation strategies, and;*
> *(8) With the plan to demolish two high-rise buildings in the south, expand the New MixC along Shennan Boulevard and across Hubei Road. Make its ground and second floors an open commercial block to channel Shenzhen Boulevard and the ancient village. Decrease the architectural volume in height from south to north, matching the scale of the ancient village where it conjoins the new*

32 Participating designers included Tang Hua, Zeng Qun, Wu Wenyuan, Zhang Xing, Zhu Rongyuan, Zhong Bing, Zhang Zhiyang, Jason Hilgefort, Feng, Guochuan, Fei Xiaohua, and Meng Yan.

development. Turn Xinghu East Road to the west to make more space for the ancient village and park.

The developer and commissioned design companies neither attended the workshop, nor made any direct response to the eight suggestions. Nevertheless, they quickly adjusted the planning scheme, expanding the preservation area from 6,000 to 10,000 square meters and excluding some corresponding area of the ancient village from the scope of underground development. Following up on the adjusted scheme, Hubei 120 put forward the following six suggestions for optimization at the level of urban design, hoping to find a path for the idea of "laying equal stress on protection and development" (Figure 9).[26]

(1) Preserve the complete core area of the ancient village (15,961.5 square meters), including the eastern section. This would mean moving the 400-meter landmark tower slightly eastward and appropriately reducing the volume of the commercial complex that is presently planned to occupy the eastern section of the ancient village;
(2) To make up the lost area, build an open commercial block around the southeastern section of the ancient village, extending it to the south in order to integrate it into the existing podium of the gold tower cluster and situating the New MixC, a large-volume vertical shopping center on the other side of Hubei Road;
(3) Cancel the under-crossing design of Hubei Road, referring to the crossing-road examples of MixC in Luohu and Coastal City in Nanshan. Crossing Hubei Road, the New MixC is situated closer to Shennan East Road, enlarges its commercial display and strengthens its identity, which will greatly improve its commercial value;
(4) There is no need to connect Xinghu East Road directly to Shennan Boulevard. Instead, improve the layout of the commercial pedestrian block and residential towers in the western section, by moving the towers westward to make the scale of the inside commercial streets more appropriate and to densify land use (here, refer to the spatial scale of the block and residential towers in the eastern section), which will provide more space for the park and supplement land occupied by the western section of the ancient village;
(5) Retain the well-designed greening pocket park near Wenjin Middle Road in the eastern section of the area, and move the tower westward. The planned square garden facing an intersection only works visually, but will not encourage people to gather. This is an inappropriate urban design;
(6) Preserve the big trees in Luohu Cultural Park, incorporating them into the layout of outdoor public space in the new development and enhancing the connection of the pedestrian system between the central park and New Hubei Village. This will break up the simple and stiff pattern of homogeneously placed residential towers. The planning scheme should emphasize a cluster pattern, integrating better with the extant urban fabric of Luohu and thereby avoiding the inhuman and monotonous urban space of stacked towers, which is exemplified in the transformation of Dachong.[33]

There was no direct response to these six suggestions, and most of them were not adopted into the two planning schemes that were released to the public in February 2017 and December 2018. Although intellectuals did not provide a perfect proposal, nevertheless, they did—within limits—engage in a professional debate from the perspective of public interest, supplementing the blind spots of the top-down urban planning system and logic of real estate development. More importantly, the revisions to the Hubei Project planning scheme directly reflected their understanding of the contemporary city, in which a sufficient material life co-existed with humanistic needs and where citizens actively participate in the construction of the city, rather than being passively incorporated. They made the critical point that to resist the single-minded model of urban regeneration is also to resist the inevitably homogeneous urban future that is brought into being by such a model.

5.2 Leap forward in Big Jumps

In September 2019, the foundation stone for the first phase of the Project of Urban Renewal of Hubei Coordinated Area was laid (Ancient Hubei Village was not included in this phase). The project is expected to be completed by 2030, over four phases. According to the developer, Hubei Project "covers 0.34 million square meters of renewal area, of which land for public use accounts for 54%. The development will have 3 million square meters of GFA and a total investment of 70 billion yuan. The goal is to develop an urban complex that integrates culture, commerce and tourism."[27] Compared with the 2012 plan, which called for 1.92 million square meters of GFA and a 30 billion yuan investment these two indexes have surged. Aerial photos of the area show the slogan, "leap forward in big jumps" on construction fences (Figure 10). In this image of the imagined future, Ancient Hubei Village glitters in the center stage, encircled by greenspace. Hubei Tower rises in the northeast corner. It is surrounded by a shopping center that includes landscaped and residential and office skyscrapers that have been wrapped in glass curtain walls.

In both its vertical or horizontal scales, in its temporal or spatial dimensions, this image conveys the determination and ambition of the redevelopment of Hubei. Everything is brand-new. There are no traces of the past, even those complex memories of hills, ponds and farmland of Hubei's agricultural past, the factories and self-built buildings of Shenzhen's early industrialization, and the formal and informal construction of 40 years of high-speed urbanization. The ancient village is also new. Within and without, it has been evacuated and sanitized, prepared for highly controlled and instantly molded texture to cover its bones. Ancient Hubei Village will be preserved in a vacuum as a symbol of history, flavoring the thickness of time on the image of the modern city with new artificial settings.

In January 2021, CR Land launched an international competition for the architectural design of Plot C2 (i.e. Hubei Tower) in Hubei Project, including the conceptual planning design of the Central Area (including the ancient village). The design brief shows that most areas of the ancient village has been classified for public administration and service facilities (GIC). Its eastern section is classified as a park of green space (G1). As specified in the brief, "Comprehensive treatment, protection and active utilization will be carried out based on three-longitudinal eight-transverse texture, and keep the historic building Zhang Huaiyue Ancestral Hall. The total protection scope of Ancient Hubei Village is 14,478 square meters (including the core protection area of 10,016 square meters and construction control zone of 4,462 square meters). The underground space will not be developed, and the existing buildings will be put into different classes for protection after evaluation."[28] Compared with the initial scheme of overall demolition and reconstruction, this design requirement reflects a profound change of the implementation body's view toward the value of Ancient Hubei Village. Through different parties' discussions on urban development, the historical value of Ancient Hubei Village has been recognized, but its contemporary value to Luohu District and Shenzhen City remain unclear.

In the summer of 2021, when this essay was written, Ancient Hubei Village has been vacated, enclosed by construction fences, with a hidden opening near the Zhang Huaiyue Ancestral Hall. The South Gate, once a ritual space, is now buried by broken icons and incense burners (Figure 11). Except for the ancient village, nearly all of the Hubei Area has been razed to the ground, including the handshake buildings of East, West and North Lanes and Luohu Cultural Park. There are certain gaps between this approach and the

33 The transformation of Dachong demolished about 1,500 buildings and built dozens of high-rise towers with a total GFA of approximately 3.8 million square meters.

principle of "laying equal stress on protection and development." The latest planning scheme announced and the renderings displayed on site are also different from the strategy of "completely preserving the three vertical and eight horizontal texture of Ancient Hubei Village."[29]

5.3 Whose city is it?

The pioneering spirit and special character of Shenzhen are revealed in bold experiments, which allowed unexpected things to happen, giving rise to the mosaic fabric of its urban center, where planning and urban villages collided (including the surviving ancient village in the city). The voices of public discourse, which have a profound influence on the city's future are also part of what makes Shenzhen special. For too long, public participation in urban regeneration projects has merely stayed in the consultation stage, "featured with the isolation of participation right and formalization in the participation process."[30] The public participation in Hubei Project that led to a change in urban development decisions can be regarded as an important beginning of deepening public participation.

Hubei 120 intellectuals aimed to widen public discourse about Shenzhen's urban form beyond the voices of the government and developers. With respect to urban planning in general and the redevelopment of Old Hubei specifically, their actions constituted a public sphere comprised of the government, the developer, and intellectuals. These three actors were in conversation to determine the public good, even as each actor represented a different aspect of it: The government was responsible for social order, realized as the Master Plan. The developer was responsible for material resources to realize the plan as a specific place – a mega multiplex comprising a shopping mall, office buildings, and residential towers. Hubei 120 asserted a "watchdog" responsibility, reminding both the government and the developer about the weaknesses of the plan and its implementation in order to make Shenzhen even better. In this sense, it is notable that the Hubei 120 intellectuals did not see themselves as opposing the government and developer, but rather working with them to craft better plans and build a better city.[31]

At the establishment of the SEZ, Shenzhen freely and fearlessly entered the temporal-and-spatial tunnel for unprecedented urbanization. It grew vigorously in the expected contexts, and molded an evolving context in turn, in which everyone was imagining its future while it was keeping surpassing everyone's imagination. With the success of the SEZ, a narrative about the "Shenzhen Miracle" faces the danger of becoming a definite paradigm. Although researchers and practitioners have tried to moderate stories about Shenzhen's growth and success, more and more people are asking, "Can other places copy Shenzhen Miracle?" However, the reason why Shenzhen has become today's Shenzhen lies in "a spirit that rejects rigidity and certainty—whenever you want to copy this spirit, you have already given it a strong certainty, and then the spirit is dead."[32] The progress in public participation is another critical evolution in its history of urban development.

深圳大奇迹

[1] 胡爱民，主编．湖贝村村史 [M]. 深圳：金岛出版事务中心，2012.
[2] 深圳市人民政府．罗湖先行先试建设国际消费中心的行动计划 [R]. 2010.
[3] 戴德梁行．深圳华润湖贝项目投资可行性研究报告 [R]. 2012-12: 103.
[4] 李景磊．深圳城中村空间价值及更新研究 [D]. 广州：华南理工大学，2018: 44.
[5] 齐珉华．深圳市湖贝古村城市更新中的第三方参与研究 [D]. 哈尔滨：哈尔滨工业大学，2017: 88.
[6] 华润置地．深圳旧改巨幅地块引房企“争食”[EB/OL]. 2012-11-25. https://www.crcsz.com/detail.aspx?cid=397.
[7] 湖贝旧改将建岭南文化风情街 [N]. 羊城晚报，2013-1-25.
[8] 都市实践．湖贝古村保护研究与“湖贝 120”公共计划 [J]. 城市环境设计，2018(12): 296-300.
[9] 深圳大学建筑与城市规划学院．深圳市罗湖片区城市更新项目湖贝旧村南坊调研报告 [R]. 2014.
[10] 吴良镛，何镜堂，崔恺，孟建民，王建国，常青．关于呼吁在湖贝旧改项目中保留湖贝古村事宜致深圳市领导的函 [R]. 2016-6.
[11] 深圳市城市设计促进中心．城中古村的活下、活化和活计 [EB/OL]. 2016-6-24. https://www.szdesigncenter.com/Busj653YUlygyA.
[12] 湖贝 120. 致深圳市规土委的湖贝预保护案 [R]. 2016-7-21.
[13] 书记调研湖贝旧村城市更新 [N]. 深圳商报，2016-6-26.
[14] 湖贝片区规划方案问计北京专家 [N]. 深圳特区报，2016-8-1.
[15] 省住建厅领导调研湖贝旧村保护开发工作并予以充分肯定，湖贝旧改规划方案是旧村保护典范 [N]. 深圳晚报，2016-8-22.
[16] 黄伟文，王雪，李毅等．关于紧急抢救特区内唯一幸存古村完整命运格局的提案．第 20180471 号 [R]. https://www.szzx.gov.cn/content/2018-04/09/content_18838734.htm.
[17] 深圳市罗湖区人民政府．深圳市罗湖区人民政府关于深圳市政协六届四次会议委员提案第 20180471 号办理情况的函 [EB/OL]. 2018-7-3. http://www.szlh.gov.cn/zwgk/jggk/jytabl/jytabljg/content/post_10116470.html.
[18] 陈锦花，王建锋．关于紧急抢救深圳原特区内唯一幸存古村完整格局的建议．第 20180883 号 [R]. https://mp.weixin.qq.com/s?__biz=MzI5ODMzMjY3NA==&mid=2247484492&idx=1&sn=8b440b643ec8cfe49772adaa11ec96cf.
[19] 深圳市罗湖区人民政府．深圳市罗湖区人民政府关于深圳市第六届人民代表大会第六次会议代表建议第 20180883 号办理情况的函．2019-1-22.
[20] 湖贝旧村非文保单位，年底将启动一期清拆 [N]. 南方都市报，2016-7-5.
[21] 罗湖区人民政府．罗湖区国际消费中心建设“十三五”规划 [R]. 2016-12.
[22] 深圳市规划和国土资源委员会，深圳市城市设计促进中心．趣城・深圳建筑地图 [M]. 深圳：海天出版社，2014: 78.
[23] 廖虹雷．井，街，市：从深圳墟到东门商业区 [J]. 世界建筑导报，2013: 18-20.
[24] 丘濂．深圳湖贝村：“城中村”的另一种选择 [J]. 三联生活周刊．2016-8-1: 104.
[25] 古村保护需要相应政策支持 [N]. 深圳商报，2016-6-27.
[26] 湖贝 120. 针对深规院 / 华阳湖贝旧改规划方案（最新版）的调整建议 [R]. 2016-7-22.
[27] 华润置地华南大区．罗湖焕新——华润置地又一城市更新巨著悄然启航 [EB/OL]. 2020-12-1. https://urlzs.com/oMMU6.
[28] 华润置地．湖贝统筹片区城市更新单元 C2 地块（含中央区）项目建筑专业国际竞赛阶段设计任务书．2021-1-4: 11.
[29] 深圳市罗湖区城市更新和土地整备局．关于罗湖区东门街道湖贝统筹片区城市更新单元的公告．2021-3-12.
[30] 深圳市城市规划设计研究院．深圳城市更新探索与实践 [M]. 北京：中国建筑工业出版社，2019: 494.
[31] 马立安．深圳之心的再发现：保护湖贝古村运动．棱镜，译．杨阡，编辑．原文见 O'DONNELL M A. Heart of Shenzhen: The Movement to Preserve "Ancient" Hubei Village [M]// BANERJEE T, LOUKAITOU-SIDERIS A, eds. The New Companion to Urban Design. London: Routledge, 2019: 480-493.
[32] 孟岩访谈．在深 21 年：都市实践 [J]. 建筑实践，2020(11): 95.

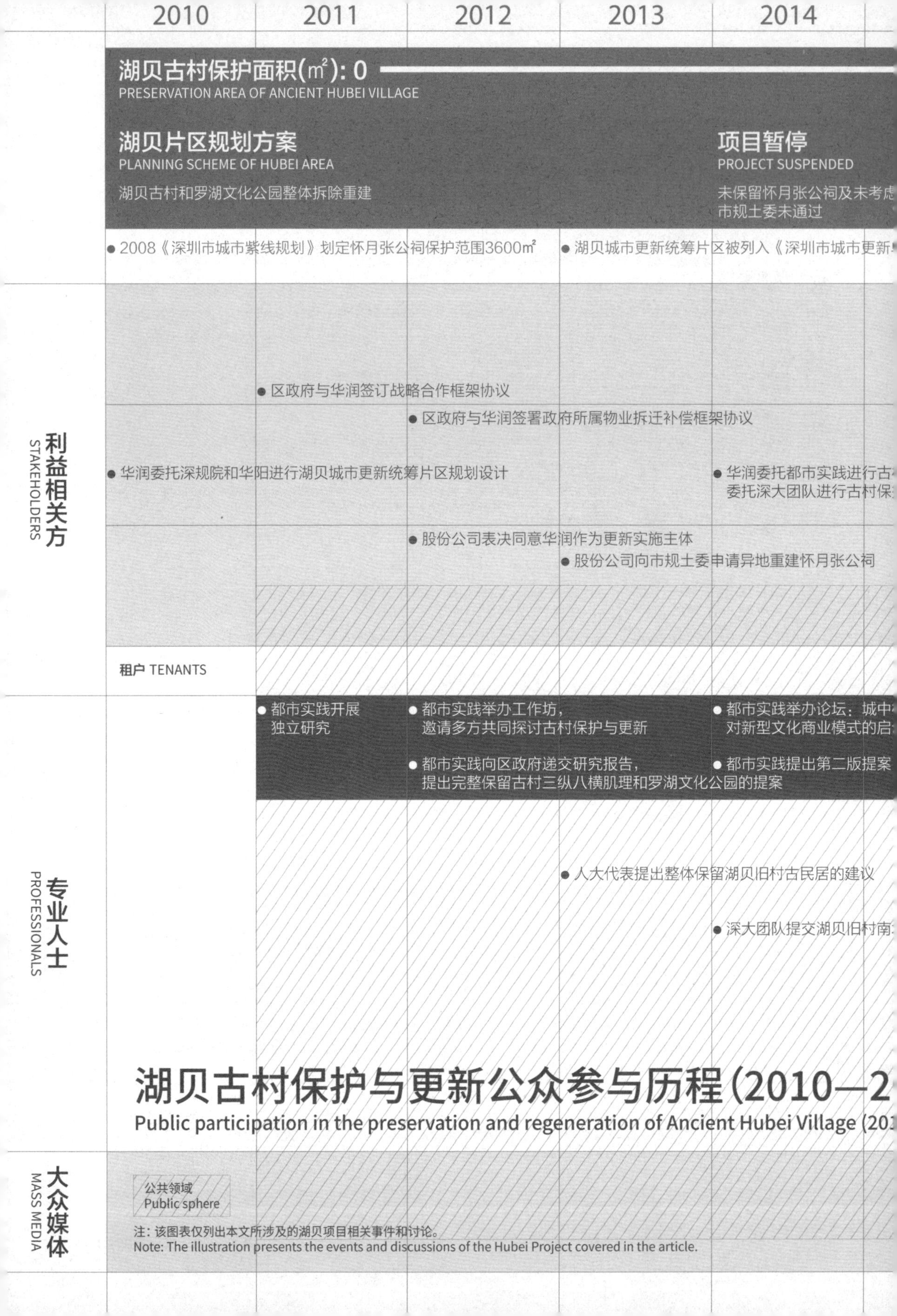

2010
2011
2012
2013
2014
湖贝古村保护面积(m²): 0
PRESERVATION AREA OF ANCIENT HUBEI VILLAGE
湖贝片区规划方案
PLANNING SCHEME OF HUBEI AREA
湖贝古村和罗湖文化公园整体拆除重建
项目暂停
PROJECT SUSPENDED
未保留怀月张公祠及未考虑
市规土委未通过
2008《深圳市城市紫线规划》划定怀月张公祠保护范围3600㎡
湖贝城市更新统筹片区被列入《深圳市城市更新
利益相关方
STAKEHOLDERS
区政府与华润签订战略合作框架协议
区政府与华润签署政府所属物业拆迁补偿框架协议
华润委托深规院和华阳进行湖贝城市更新统筹片区规划设计
华润委托都市实践进行古
委托深大团队进行古村保
股份公司表决同意华润作为更新实施主体
股份公司向市规土委申请异地重建怀月张公祠
租户 TENANTS
专业人士
PROFESSIONALS
都市实践开展
独立研究
都市实践举办工作坊，
邀请多方共同探讨古村保护与更新
都市实践举办论坛：城中
对新型文化商业模式的启
都市实践向区政府递交研究报告，
提出完整保留古村三纵八横肌理和罗湖文化公园的提案
都市实践提出第二版提案
人大代表提出整体保留湖贝旧村古民居的建议
深大团队提交湖贝旧村南
湖贝古村保护与更新公众参与历程(2010—2
Public participation in the preservation and regeneration of Ancient Hubei Village (20
大众媒体
MASS MEDIA
公共领域
Public sphere
注：该图表仅列出本文所涉及的湖贝项目相关事件和讨论。
Note: The illustration presents the events and discussions of the Hubei Project covered in the article.

2016	2017	2018	2019

6000 — 10000 — 10350 — 10016 — 10016

项目重启
PROJECT RESTARTED
未保留古村肌理，专家评审未赞同；
区更新局再次召开专家研讨会，明确保留古村的三纵八横肌理，未划红线

规划公示
PLANNING SCHEME PUBLICIZED

方案通过
PLANNING SCHEME APPROVED
古村划为公共管理与服务设施用地
未保留公园，市政府通过

一批计划》

- 怀月张公祠入选深圳市历史建筑保护名录
- 湖贝南坊（旧村）入选
  深圳市历史风貌区保护名录

- 区更新局召开见面会
- 区领导带队拜访院士
- 区更新局召开城市设计国际咨询会
- 市人大常委会召开代表建议办理
  座谈会

化专项研究，

- 华润对公示方案发表意见

- 股份公司与华润签署集体物业拆迁安置补偿协议

- 湖贝原住民、保障房潜在使用者对公示方案发表意见
- 湖贝原住民对公示方案发表意见

老村

15961.5m²
PRESERVATION AREA SUGGESTED
《深圳市历史风貌区和历史建筑评估标准》
研究团队提出湖贝古村核心保护范围

- 政协委员、人大代表分别提出
  紧急抢救湖贝古村的提案和建议
- 区检察院立案调查涉古村保护
  公益诉讼，
  召开座谈会，约谈相关责任方

湖贝古村120城市公共计划

- 六位院士联名致信市政府领导，呼吁保留湖贝古村
- 上千名知识分子签署《湖贝共识》
- 湖贝120举办工作坊：共赢的可能：湖贝古村保护与罗湖复兴，
  举办座谈会：对话湖贝，
  发起公共艺术项目：每个人的湖贝
- 湖贝120对公示方案发表意见
- 湖贝120针对公示方案发表文章：
  湖贝更新：深圳人的深圳缺的就是一个折衷方案吗？
- 形成《深圳市建环艺委的意见书》《致罗湖区城市更新局的意见书》
  《致深圳市规土委的“湖贝预保护议案”》，并递交相关部门

- 土木再生和未来+学院发起独立研究计划：湖贝请留门
- 城市设计促进中心举办酷茶会：城中古村的活下、活计和活化

- 《深圳晚报》《深圳特区报》《南方都市报》《羊城晚报》报导
- 《深圳商报》系列报导：这么有价值的古村为何不保住？
- 区更新局答记者问

南头 Nantou

报纸
NEWSPA-
PER

⊙《南方日报》2016 年 2 月 29 日 星期一

“明年深港双年展或在城中村举办”

700多万市民生活的大家庭，可以把深圳的近现代历史和改革开放历史、市民现代生活历史都丰富地展现在大家面前

2月28日，第六届深港城市\建筑双城双年展（以下简称“深双”）闭幕式在本届展览的主展场大成面粉厂落下帷幕。据统计，自展览于2015年12月4日开幕起，“深双”共举办了209场活动，展出展品160件，超过25万人次到访参观。

与往届相比，2015年深双极具创新，第一次出现分展场，25个外围展参与度更大，也实现了覆盖深圳的深双地图。从策展思路方面，本届展览着眼于“重新利用”、“自下而上”等世界各地的城市化案例，拓展了人们对于身边的城市和城市化的固有观念。同时，本届展览首次深度聚焦珠三角，策展人提出珠三角2.0时代需要城市的人、土地、精神、历史、建筑多维的流动平衡。

闭幕式现场为本届深双参展作品颁发了组委会奖、学术委员会奖、公众奖和独立评委奖四项大奖。其中，“蛇口浮田：蛇口大成·浮城桑田”获得组委会大奖。由五位国际知名建筑师、评论家组成的独立评审团也选出了金奖“西撒哈拉国家馆”，银奖为“深圳未来高密度环境中多元混杂栖居方式的推测与设想”；“此时、彼地：后地理时代的城市场景”与关注本地设计的“无名的设计行为”共同获得铜奖。

深圳市委常委、党组成员杨洪表示：“本届双年展倡导建筑对城市的现状、再利用、再思考和再想象，高度契合深圳‘十三五’期间要打破空间资源的约束，为城市可持续发展提供保障的战略。本届双年展让更多的人认识到要注重利用现有的城市和建筑空间。”

展期内，组委会还召开了“启动2017”头脑风暴会。在展场选址方面，有专家建议从工业遗址的改造向城中村等其他空间形式转向。杨洪非常赞赏2017深双将在城中村中举办的建议，他认为城中村是承载着现在700多万市民生活的大家庭，可以把深圳的近现代历史和改革开放历史、市民现代生活历史都丰富地展现在大家面前。

《南方日报》记者 苏妮

Note: At the end of the news article regarding the closing ceremony of the 2015 UABB, it revealed the venue of the next biennale would be in urban villages. "Yang Hong, member of the Shenzhen Municipal Standing Committee, highly appreciated the suggestion. He believes that urban villages form a big family that houses the daily life of more than 7 million citizens. It can bring stories about Shenzhen's modern history, the history of reform and opening up, and the everyday life of ordinary citizens to a larger audience."

SU N. "Next year's UABB may be held in urban villages" [N]. Southern Metropolis Daily. 2016-2-29.

南头古城，2016 © 都市实践
Nantou Old Town, 2016 ©URBANUS

南头古城南城门，2017 © 都市实践
South Gate, Nantou Old Town, 2017 ©URBANUS

南头古城关帝庙，2016 © 都市实践
Temple of Guan Yu, Nantou Old Town, 2016 ©URBANUS

南头古城报德广场，2017 © 都市实践
Baode Square, Nantou Old Town, 2017 ©URBANUS

南头古城制衣厂，2016 © 都市实践
Garment Factory, Nantou Old Town, 2016 ©URBANUS

南头古城大家乐舞台，2017 © 都市实践
Dajiale Stage, Nantou Old Town, 2017 ©URBANUS

南头古城工厂区入口，2016 © 都市实践
Entrance of the Factory Zone, Nantou Old Town, 2016 ©URBANUS

迈向“城市共生”

罗祎倩
都市实践研究员
本文首次发表于《建筑实践》, 2020(11): 154-163；由 Killiana Liu 中译英。图片 © 都市实践 ©UABB。

2004 年，深圳成为全国首个 100% 城市化的城市，在名义上，没有农村行政建制、没有农村社会体制。[1] 自 1980 年特区成立以来，深圳的城市化进程先由政府和国有企业主导，后以规划为依托，由开发商大规模介入实施。然而，与自上而下的建设并行，在农村集体用地建起的厂房和住房以及其他农村和城市输出的劳动力人口同样对深圳的高速城市化给予巨大的推动。在城乡土地二元化的历史遗留和日益增长的市场化需求的双重裹挟下，个体的非正规建设形成了今天所见的城中村。

同在 2004 年，深圳出台城中村改造政策，正式启动城中村改造工作。[2] 彼时，非正规状态的城中村被看作是“完美城市化”的瑕疵，一度被称为“城市毒瘤”。2012 年，深圳的存量用地首次超过新增用地，土地开发利用模式出现拐点。[3] 2019 年以来，二次城市化呼声不断。[4] 未按规划建造、未按政策改造而又位于城市中心的城中村无法避免地成为新的博弈现场。

城中村与外部城市的关系愈发紧张。从时间的维度上讲，两者并不冲突，分别从宏观和个体的视角叙述深圳的城市发展史；实际上，两者无法切分，深圳无法走向一个不包含自身历史的未来。应该说，城中村的更新模式预示着深圳城市发展的未来方向。

1　主流的城中村更新模式

近二十年来，深圳数百个城中村都被笼罩在即将拆迁的狂喜和担忧中。大型城中村蔡屋围、岗厦、大冲被整体拆除重建。[5] 由政府主导，开发商建设的高层办公楼、购物商场、高容积率住宅楼盘一次性、大规模地替代“握手楼”，成就一种现代城市的形象。通过这些城中村更新，开发商获得企业利润，原村民获得拆迁赔偿。

然而，深圳需要多少座京基 100，多少个万象城？城中村拆迁的巨额补偿款又是由谁承担？

在这种全然替代的更新模式下，城中村原本的生活瞬时消失殆尽，仿佛不曾存在。失去可负担的生活区，大量租户搬至尚存的城中村，甚至搬离深圳。随着这些人群的离开，城市运作成本提高，城市记忆消失，城市认同感难以积累。这种更新模式以牺牲深圳自身的特质、一段历史、一种甚至多种未来的可能性作为代价。

与其他城中村相同，南头也在土地和市场的双重裹挟下形成城中村肌理（图 1- 图 3），也经历着一次次更新浪潮的冲击。但与其他城中村不同，南头具有深圳最悠长的历史，肩负着“源头”的荣光与使命。在 1300 多栋建筑中，违章违法建筑占九成多，[1] 仅存九街肌理、部分古城垣遗址、南城门、东莞会馆、信国公文氏祠等少量古迹。[6] 因而，南头深陷于既定现实与历史想象的矛盾中。

在初期规划中，1998 年的《南头古城文物保护规划》提出“规划南门广场，凸显标志性；改造整治中山南街、东街，恢复古城风貌”。2000 年的《南头古城南门广场改造设计方案》获批，提出“修一个广场，恢复一段城墙，改造一条街”。同年，南门广场的 65 户拆迁工作启动，历时 16 年完成。[7]

1997 年香港回归之前，由于急需确认与巩固南头“港澳源头”的地位，[8] 城市管理者和企业联合对南头进行修

1　深圳在 1992 年完成“统征”，在 2004 年完成“统转”，详见《关于深圳经济特区农村城市化的暂行规定》（深发［1992］12 号）和《深圳市宝安龙岗两区城市化土地管理办法》（深府［2004］102 号）。
2　深圳在 2004 年发布《深圳市城中村（旧村）改造暂行规定》（深府［2004］177 号），成立深圳市查处违法建筑和城中村改造工作领导小组。
3　2012 年，深圳市存量用地供应占供地总量的 56%；2013 年，达到土地供应计划的 70%。
4　2019 年，国家发改委发布《2019 年新型城镇化建设重点任务》。
5　蔡屋围、岗厦、大冲旧改的运作主体分别为京基、金地、华润。
6　南头古城的建城史长达 1700 余年，可追溯至晋代。南城门始建于 1394 年（明代洪武年间），是古城现存最古老的建筑物。
7　南门广场改造工程于 2000 年立项，同年启动拆迁工作启动，2004 年暂停，2009 年重启，2016 年完成。

图 1 南头古城南门广场改造方案，都市实践，2017
Figure 1 Transformation scheme of the South Gate Square, Nantouu Old Town, URBANUS, 2017

复，新建县衙、烟馆、聚秀楼、海防公署等 8 处仿古建筑，吸引了大批游客。[9] 为兴建县衙，南头城街市于 1997 年被拆除，之后南头未有集中的肉菜市场。2005 年，南山区政府斥资重修清代遗址东莞会馆和信国公文氏祠。[10] 为迎接 2011 年大运会，城内中山南街和东街首层统一绘制仿青砖外墙，安装仿古屋檐、仿古店招。[11]

在这些规划和实施中，行政决策占据主导，侧重恢复古城风貌，但未对城中村的空间形态及相应的业态和人群提出可行的改善方案。出于对速生属性的抗拒，深圳被安排以“美好的历史场景”自证来源。因此，每逢深圳大事件发生之际，“源头”南头都会经历一番以溯源为名的行动，试图将其推回到某一个时空。各个发展阶段的历史层积因为不符合主流宏大历史的想象而被漫不经心地刻意抹除。随着时间的积累，历史层积却在消逝。每一次“溯源行动”都是为了覆盖过往而产生，然而又被下一次“溯源行动”覆盖而消失。这种单线的历史观导致深圳愈是溯源，愈发速生。

2 共生实验

城中村现状与古城定位的矛盾使南头的当代价值和历史价值含糊不清，以至于互为掣肘。但在某种程度上，这种矛盾也为南头在高速变化的深圳中争取了生存空间。

随着城市观念的进步，城中村之于城市的价值被逐渐认可，由“城市病症”转变为“城市器官”。2016 年，城中村更新政策转向“综合整治为主，拆除重建为辅”。[2] 就实施而言，城中村综合整治的决策和实施仍然由具有支配性力量的单方主导。2017 年，万科启动针对城中村改造的“万村计划”，计划统租 10 万间以上住宅，改造后作为品牌长租公寓出租。[3]

图 2 南头古城中山东街，2017
Figure 2 Zhongshan East Street, Nantou Old Town, 2017

2016 年，受设计联合会委托，都市实践开始进行南头研究，提出强调本土文化复兴的更新策略。都市实践合伙人孟岩和刘晓都作为 2017 深双总策展人，基于城市研究和可行性建议主展场选址南头，提议通过。总策展人批判性地解读世界和中国城市化现实，提出另一种城市未来图景的尝试——“城市共生”。此届深双不仅是一个展览，同时也是一场介入计划，一次直接的造城行动（图 4- 图 9）。

正如当时孟岩起草的策展宣言所揭示，“当代中国的造城运动在权利和资本裹挟下经历了三十多年的高速发展，城市面貌渐趋同质化和单一化。近年来旨在提高生活品质的‘城市更新’往往把层积丰富的历史街区和多样杂糅的城市生活进一步清除，代之以全球化、商业化的标准配置。”面对这样的现实，“共生”愈发必要。

“城市共生”的核心在于对不同的认同与包容，强调多重身份、多重视角。城中村之于外部城市，因其自发地应对现实而有着不同的进化，这种自发性由人的生活需求和城市的发展需求激发，甚至比规划来得更为直接。此届深双在策展、设计、实施、展览环节均试图囊括各方参与和交流，寻求自上而下和自下而上的平衡。南头不仅承载着过去，它还活在当下和指向未来。这一行动并不试图补写南头的历史，而是基于真实的历史层积，尝试为南头和深圳的未来发展开启另一种可能性。此届深双开展历史上首次公开征集，汇聚关于城中村议题的深入研究和实践，拓展展览边界。其中，“城中村实验场”配对设计师与有意愿改造房屋的南头居民，在改造住宅、改造商铺、提升公共空间、社区营造四个方面的 5 个合作项目得以实现。[12]

因土地制度的历史遗留问题和城市政策的疏漏，村民可对宅基地进行建设；但又由于突破面积，甚至突破红线等建设行为的合法性，村民对自建房的改造受到限制。行政部门和村民对城中村自建房的合法改造路径都面临重重阻力。在“实验场”改造项目中，屋顶修复工程因涉及建筑外轮廓而被社区工作站喊停，策展团队进行多轮沟通后才得以完成。深双的介入为这种困境提供一个切入点，通过文化事件沟通行政决策和居民需求，提升城中村环境。

贯穿展期的“共生实验室”系列论坛则探讨更广泛的社会、文化、空间议题，包括城中村居民对反映深圳特质的生活、经营、创业故事的分享，艺术家对城中村日常生活的挖掘和创作，规划师、建筑师对非正规城市的研究与借鉴。其中，在“南头茶划会暨第一届南头街

8　香港曾是新安县属，而县治就位于现在的南头古城。
9　“原南头古城仿古建筑系采用电影制景手法，柱体、屋檐等大量采用塑料泡沫、涂料喷绘等工艺。这些仿古建筑既非另地复建亦非原址重建。”来源：仿古建筑将升级为小型博物馆 [N]. 晶报 , 2011-4-5.
10　东莞会馆建于 1907 年，信国公文氏祠建于 1804 年。
11　“投资 380 余万元对 178 家商铺招牌进行改造，首层外墙墙面批绘画。”来源：迎大运惠民生，激情南头喜迎八方宾朋 [N]. 南方日报 , 2011-8-2.
12　南头会客厅（朝阳北街 1 号）、湛江牛杂店（中南南街 72 号）、锦记商铺（中山东街 44 号）、房屋 17 号、铁皮房。

图 3 南头古城中山南街，2016
Figure 3 Zhongshan South Street, Nantou Old Town, 2016

道民生论坛”，居民、工作站代表、社会组织成员、行政部门负责人、业界专家从多个维度探讨南头在“后双年展时期”的发展方向和实施路径。

基于现实需求并结合展览叙事，都市实践对展场进行设计和改造，希望藉由深双的契机梳理南头公共空间脉络，为南头的未来提升提供一个结构性框架。工厂区三栋厂房被改造为展场，展后可成为开放的创意集聚区和古城内年轻人的新生活区；报德广场的两栋临时铁皮房被改造为“信息中心 + 书店”和“南头议事厅”，展后可保持公共属性，服务于社区；大家乐舞台的改造保留社区中心的功能，并容纳专业论坛，展后可作为多功能表演空间。这些公共空间节点在展期作为文化事件的发生点，在展后将继续作为社区生活的载体。

都市实践通过城市研究和策展实践介入南头更新。此次介入力争与居民的日常生活并行不悖，由轻度改造和深度讨论激发社区的文化自觉和自信。更新强调多方参与、有限改造，不同于主流模式的单方主导；同时引入合法合规及专业化建造，有别于城中村无序的自开发模式。展期中，各方从不同角度对南头更新提出解读，在同一平台上协商，最终根据实际需求和资源确定方向和路径。

3 南头模式

2019 年，《深圳市城中村（旧村）综合整治总体规划（2019—2025）》发布，明确划入分区的城中村居住用地在七年规划期内不得纳入拆除重建类城市更新单元计划、土地整备计划及棚户区改造计划，这为部分城中村留下喘息空间。[4] 但另一方面，经过十几年的多方博弈，湖贝和白石洲的拆除重建已成定局。[13]

在这一关键的时刻，都市实践以深双介入为基础，以“城市共生”理念为核心，提出寻找“南头模式”的必要性，在新的语境下再次探讨城中村的社会、文化和空间议题，旨在为城市发展模式提供一种新的选择以应对当下紧迫的现实。

在“南头模式”下，每一个环节都有多元主体参与，不同的主体因能力、资源、诉求、目标不同，能够导向多样化的开发、运营、使用模式。这不同于主流的城中村更新模式下，垄断性的支配力量以自身利益为导向，大规模占领空间，排斥其他各方的合理存在和发声，其结果必然导向单一的城市图景。

城中村的空间生产不仅是空间产品的创造过程，而且是社会关系的再创造过程。由于村民谋生需求和低收入住房缺口，城中村自建房常见的租赁模式为业主自住其中一层、面向大量的外来务工人员出租其他楼层或将整栋出租。在保留这种选择的前提下，小型开发商可与业主协商，统租部分自建房，适当改造再面向进入城中村的初级白领出租；行政部门也可沟通将一部分自建房纳入住房保障体系。[3] 原有的社区业态既为居民提供工作，又为社区提供便利的服务和可负担的商品。随着居民的能力提升和不同文化圈层的进入，新的业态因社区新的需求而产生，提供更多的生活选择，出现肉菜铺、快餐店、杂货店和网红店并置的可能。每一个个体都可以做出自己的生活选择，不被设定，不被强制。

不同的参与方在南头或并行、或交融、或冲突，在时间的磨合下逐渐达成一种动态平衡。旧的东西在积累，新的东西在介入，这种未完成的状态为创造提供了孵化条件。具有文化自信和文化自觉的社区逐渐形成城市认同感，逐渐形成更加独特、动人的本土文化。直面现实问题，这种以日常生活出发的改变在本质上区别于推倒重来的“溯源行动”或现成置入却难以维系的热闹图景。

4 南头现状

深双闭幕后，南头改造历经多方讨论，最终确定其公共部分由政府投资建设，由深圳万通南头城管理运营公司（由万科、区属国企深汇通、村股份公司合资成立）执行与运营。2019 年 10 月，南头古城保护与利用项目动工。2020 年 8 月，南北街示范区开街。短短几个月，行政部门和代建企业统租约 300 栋建筑，其中，示范段 88 栋建筑的统租在 10 天内完成。[14]

该项目将南头定位为“深圳最有特色的城市传统风貌展示区、最有底蕴的人文旅游目的地和最有内涵的文化创新产业集聚区”。[5] 传统风貌被定格为清代与民国时期的广府民居建筑，两栋仿古建筑紧挨南城门，由精品咖啡店和香港品牌酒楼运营。南北街的大部分建筑外观都经重新设计，配置粤港澳特色餐饮和文化创意商业。除首层设置商业外，部分建筑楼上改造为品牌公寓出租。

主街上的新商铺有着响亮的名号，而看似多样的选择背后却是相同的模式、相同的客群。原有的平价餐饮、便利店、诊所均被一次性清退，更新后的几十家商铺中仅保留一家南头原有的糖水店。原本在主街上进行日常

13 湖贝旧改由华润运作，2019 年 9 月，一期工程动工；白石洲旧改由绿景运作，2019 年已进行清租。

14 “据悉，今年 3 月 5 日，区政府专门就南头古城保护与利用发布了 1 号任务令，短短 10 天时间，南头街道办完成了与 1324 人的谈判工作，一举攻克示范段 88 栋约 4 万平方米建筑统租任务。”南山区政府 1 号任务令助推南北街示范段 88 栋建筑统租完成 [EB/OL]. 深圳政府在线，2020-4-9. http://www.sz.gov.cn/cn/xxgk/zfxxgj/gqdt/content/post_7132111.html.

生活消费和休闲的居民将主街空间让渡给打卡游客。统一改造的品牌公寓也替代了原有的出租房，面向年轻的、更有经济买实力的租户，力图呈现一个有活力的社区。这种“焕然一新”正是该项目的目标——“人口结构改变，产业结构升级”。[6]

图 4 中山南街上 2017 深双壁画作品
Figure 4 Mural work of the 2017 UABB on Zhongshan South Street

此次南头更新指向一个节选的过往片断，而不是一种未被构建但有着各种可能的未来。特区 40 年的真实历史被清代和民国风情涂抹，自 1949 年以来形成的移民文化被符号化的粤港澳氛围覆盖。在“源头”的重担上，南头还背上创意文化的指标，外来案例疾速填充，南头自身生长的本土文化反而被忽略。

文化没有速成法，但是此次南头更新对时间的压缩、对效率的追求更甚以往。在一次性更新的建筑空间里，繁荣的生活是现成的——从空间提供者、内容到消费者都是现成的。这种模式在短期内快速见效，但以置换大量的日常性空间、更替原有业态和常住社群为代价；在远期需要不断地更换刺激点，吸引游客，而越是频繁更替，失去基础的本土文化越是难以生成。

5 结语

特区成立之初，深圳的造城与深圳人的造梦一致，得以成为“中国梦想实验场”；[15]40 年后的今天，地产思维强势主导的城市发展模式却在一定限度上以牺牲深圳人的梦想为代价。丰富的城市风貌、多元的社群加速消失，城市生活似乎已被设置成单一选项：现成的、快速的、消费的、娱乐的。

同质化的发展模式不断吞噬城中村和一切与现有方式不同的异质体，城市观念亟须进步。城中村之于城市，不是病症、不是阻碍、不是土地耗尽时的储备粮。城中村就是城市，它承担了容纳大量新移民（深圳人）的城市职能，它承载了各个阶段的城市记忆，它随着时间的流淌滋养了本土文化。城中村是疏解同质化城市问题的另类样本，它无法复制、无法仿制，且一旦消失将无法复原。

面对当下紧迫的现实，城中村更新不应只有一种模式。“南头模式”旨在为城市发展提供另一种思路、另一种工具。在城市演变中，当城中村已与外部城市发展不同步、外部力量必然进入时，“南头模式”是一次力图寻找各种力量动态平衡的尝试。向南头学习，向城中村学习，也许深圳可以迈向一个更加共生的未来。

对于作为先行示范区的深圳，中国和世界正待其再度以“敢为人先”的精神不断地冲破藩篱。

租租租
租租租
租租租

15　2010 上海世博会深圳案例馆主题。

Toward "Cities, Grow in Difference"

Luo Yiqian
URBANUS Researcher
The article was first published in *Architectural Practice*, 2020(11): 154-163; translated from Chinese to English by Killiana Liu; image ©URBANUS ©UABB.

In 2004, Shenzhen became the first city in China with 100% urbanization, officially free of rural administrative divisions and the rural social system.[1] Since the establishment of the SEZ in 1980, the urbanization process of Shenzhen has been primarily dominated by the government and state-owned enterprises, followed by extensive implementation, which was carried out by real-estate developers that were guided by urban planning. In parallel with the top-down approach of development, factories and housing built on rural collective land as well as an influx of labor from other rural and urban areas have also contributed to the rapid urbanization of Shenzhen. Driven by the residual dual system of land tenure and the surging demand for marketization, self-led informal construction activities developed today's urban villages.

What also happened in 2004 was that the Shenzhen government promulgated the renovation policy on urban villages, officially starting the work of transforming urban villages.[2] At the time, urban villages as a phenomenon of informality were regarded as flaws in the city's "perfect urbanization" and for a time referred to as the "urban tumor." In 2012, for the first time in the city's history, the area of land being used exceeded that of land available for development, which marked a turning point in land development.[3] Since 2019, there have been continuous calls for re-urbanization.[4] The urban villages, which were neither built according to the urban plan, nor transformed according to policy but which are situated in the city center, have inevitably become new frontiers in the game.

The tension between urban villages and the city proper has become more pronounced. From a temporal perspective, the two are not in conflict as both narrate Shenzhen's history of urban development, albeit one from an individual perspective and one from a macro perspective, respectively. In fact, the two are inseparable. Shenzhen cannot move toward the future by discarding its own history. To a great extent, the regeneration model for urban villages indicates the future pathway of the city's urban development.

1 The Mainstream Regeneration Model for Urban Villages

Over the past two decades, the prospect of demolition has caused hundreds of Shenzhen

1 Shenzhen completed the "unified acquisition" (of land) in 1992 and the "unified transformation" (of residence) in 2004. See *Interim Provisions of Urbanization of Villages of Shenzhen Special Economic Zone* (1992) and *Management Measures on Urbanized Land in Bao'an and Longgang Districts of Shenzhen* (2004).

2 In 2004, Shenzhen issued the *Interim Provision of Shenzhen Urban Village (Old Village) Redevelopment*, establishing the Shenzhen Municipal Leading Group on the Investigation and Handling of Illegal Buildings and Urban Village Renovation.

3 In 2012, land under use in Shenzhen accounted for 56% of the total land supply, reaching 70% in 2013.

4 In 2019, the National Development and Reform Commission issued the *2019 Priority Tasks for New Urbanization*.

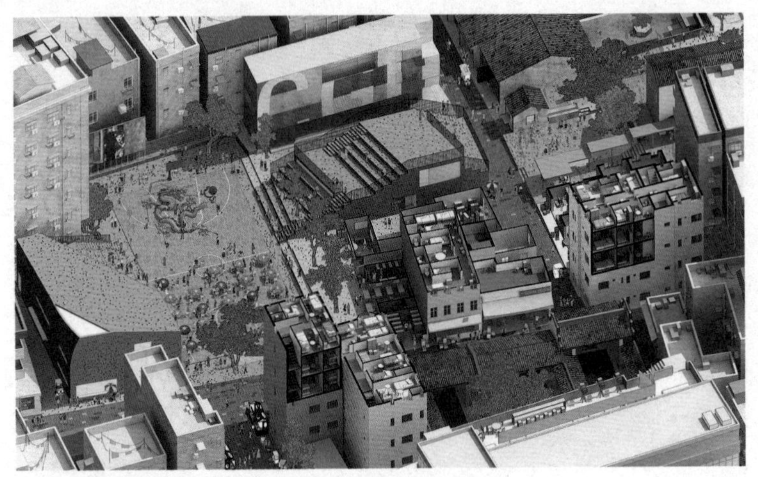

图5 南头古城报德广场改造方案，都市实践，2017
Figure 5 Transformation scheme of the Baode Square, Nantou Old Town, URBANUS, 2017

urban villages to experience both ecstasy and anxiety.[5] Large urban villages, such as Caiwuwei, Gangxia and Dachong were thoroughly demolished and rebuilt. Led by the government and constructed by developers, high-rise office buildings, shopping malls, and high floor-area-ratio residential buildings in one fell swoop replaced "handshake buildings" (similar to "kissing buildings" in English), thereby creating an image of the modern city. The regeneration of these villages meant profits for developers and compensation packages for villagers.

Nevertheless, how many KK100s or MixCs does Shenzhen need? Who is paying the huge cost of compensation for urban village demolition?

In the aftermath of regeneration by completely replacing an urban village with another urban model, the original lifestyle of the urban village disappears instantly and thoroughly, as if it never existed. Losing affordable housing, a large number of tenants have had to move to surviving urban villages or leave the city for good. As these people move away, the city's operational cost rises, its memory fades, and its urban identity becomes difficult to accumulate. This regeneration model comes at the cost of Shenzhen's distinctive features, its history and future possibilities.

Similar to other urban villages, the fabric of urban village life in Nantou (Figure 1-3) formed due to the combined effects of the land system and market demands, as well as the experience and challenges of regeneration. However, what makes this village unique is that Nantou has the longest history in Shenzhen and thus bears the glory and mission of being the city's origin. Among the nearly 1,300 buildings in Nantou old town, illegal ones accounted for more than 90%,[1] while only few historic sites remain, including the nine streets fabric, sections of the ancient city walls, the South Gate, the Dongguan Assembly Hall, and the Ancestral Hall of Wen Tianxiang Duke of Xinguo.[6] Therefore, Nantou finds itself caught in a contradiction between extant reality and historical imagination.

In terms of early planning, the 1998 *Planning for Cultural Relics Protection of Nantou Old Town* proposed the idea of "building the South Gate Square into a landmark; transforming Zhongshan South Street and Zhongshan East Street to restore features of the old town." Approved in 2000, the "Renovation Design Scheme of the South Gate Square in Nantou Old Town" proposed "building a square, restoring a part of the city walls, and transforming a street." In the same year, the demolition of the housing of 65 households for the South Gate Square started, taking 16 years to complete.[7]

Before the Return of Hong Kong in 1997, due to the urgency in defining and consolidating the role of Nantou as the "origin of Hong Kong and Macao areas,"[8] the government and enterprises worked together on the restoration of Nantou, producing eight new pseudo-classic architectural works, including the Nantou County Office, the Opium Den, Juxiu House (a pleasure quarter in ancient times), and the Coastal Defense Commission, which attracted a large number of tourists.[9] To build the new County Office, the street market of Nantou Town was demolished in 1997. As a result, there has been

5 The operation parties of the Caiwuwei, Gangxia, and Dachong projects are Kingkey, Gemdale, and CR Land respectively.

6 Nantou Old Town has a history of more than 1,700 years, dating back to the Jin Dynasty. The South Gate was built in 1394 during the Hongwu Reign, Ming Dynasty and is the oldest existing built structure in Nantou.

7 The South Gate Square renovation project was approved in 2000. The demolition work was initiated in the same year. It was then interrupted in 2004 and restarted in 2009. In 2016, the demolition was finally completed.

8 Hong Kong was once under the jurisdiction of Xin'an County, which had its county seat at Nantou Old Town.

9 The original pseudo-classic style buildings in Nantou used film-making techniques, where the pillars and eaves were largely made of plastic foam and used paint spray. These buildings are neither reproductions in another location nor rebuilt on the original site. See *Daily Sunshine*, April 5, 2011.

no fresh product market in Nantou since then. In 2005, the Nanshan District Government funded the reconstruction of the Dongguan Assembly Hall and the Ancestral Hall of Wen Tianxiang Duke of Xinguo, which were historic sites built during the Qing Dynasty.[10] To welcome the FISU World University Games in 2011, on housing facing Zhongshan South and Zhongshan East Streets, grey bricks were painted over the ground-level facades, ancient-style eaves were installed, and older signage was replaced with uniform signboards.[11]

The above mentioned planning and implementations featured government-led decision-making and a focus on restoring the old city image. However, there was no feasible plan to improve the spatial forms and corresponding businesses and communities in the urban village. The speed of urban growth led to Shenzhen being arranged in a series of "beautiful historical scenes" to prove its origin. Subsequently, during each major event in Shenzhen, Nantou stood as the city's "origin." In the name of tracking the city's source, Nantou would be forced back into a certain time and space in history. The evidence accumulated throughout different development stages was carelessly and deliberately erased, as these traces were inconsistent with the grand historic imaginary held by mainstream society. As time passed, living historical strata disintegrated. Moreover, every time an "origin-tracing movement" was launched, it took shape by burying the previous effort to track the city's origin. In the aftermath of these processes, the harder people have tried to trace Shenzhen's origins, the faster the unilateral narrative of Shenzhen's history has grown.

2 An Experiment in Growing in Difference

The contradiction between the present state of an urban village and its positioning as an old town has led to such ambiguity in the contemporary and historical values of Nantou that the two hinder the development of each other. However, this contradiction has also partially created a survival space for Nantou within and despite Shenzhen's rapid development.

As the city's understanding of urbanity has progressed, society has gradually recognized the value of urban villages for a city, turning away from the metaphor of "urban disease" toward "urban organs." In 2016, the policy of urban village regeneration instructed an approach that focused on integrated renovation, with the demolition and reconstruction as supplementary methods.[2] On a practical level, however, powerful entities still unilaterally controlled the decision-making and implementation of the integrated renovation of urban villages. In 2017, Vanke launched its "Wancun (Ten-thousand Villages) Plan," a project targeting the transformation of urban villages in order to source over 100,000 units and rent them out as brand long-term apartments after standardized renovation.[3]

In 2016, commissioned by the Shenzhen Culture Creative & Design Association, URBANUS began its research on Nantou, proposing regeneration strategies with a focus on the revival of local culture. As chief curators of the 2017 UABB, Liu Xiaodu and Meng Yan, co-founders of URBANUS, proposed Nantou as the main venue based on urban research and feasibility. Their proposal was approved. The chief curators provided a critical interpretation of the current urbanization in the world and China, attempting to envision an alternative future for cities through the theme, "Cities, Grow in Difference." The 2017 UABB was not only an exhibition but also an intervention program and a city-making action (Figure 4-9).

The curatorial statement drafted by Meng Yan argued that "mobilized by power and capital, China's city-making has gone through rapid development for over 30 years, and the city tends to be homogeneous and generic. In recent years, urban renewal, in a bid to improve quality of life, has further swept away time-honored historical areas and a hybrid urban life, which have been replaced by a globalized and commercialized standard configuration." Faced with this reality, "growing in difference" had become increasingly critical.

Cities, Grow in Difference was fundamentally the recognition and inclusion of differences, emphasizing plural identities and perspectives. Compared to the city at large, urban villages experienced a different path of evolution, which was a result of its spontaneous response to realities. This spontaneity was inspired by the needs of both everyday life and urban development, and was even more direct than planning. The 2017 UABB attempted to engage various

10 The Dongguan Assembly Hall was built in 1907, and the Ancestral Hall of Wen Tianxiang Duke of Xinguo was built in 1804.

11 An investment of more than 3.8 million yuan was given to renovate the signboards of 178 shops and paint on the exterior walls of the first floor of the buildings. See *Nanfang Daily*, August 2, 2011.

图 6 在报德广场举办的 2017 深双开幕式
Figue 6 2017 UABB opening ceremony at the Baode Square

parties in conversations at curation, design, implementation and exhibition levels in search of a balance between top-down planning and bottom-up growth. Not only has Nantou carried the past within itself, but it also lives in the present and moves toward the future. Instead of reclaiming the history of Nantou, this action tried to introduce an alternative possibility for the future development of Nantou and Shenzhen, which was based on authentic historical layers. The first open call of the 2017 UABB brought together in-depth research and practice on the topic of urban villages, expanding the exhibition boundaries. The "Urban Village Laboratory" paired designers and residents who intended to have their houses renovated. It achieved five cooperation projects, including residence renovation, shop renovation, public space improvement, and the completion of a community building.[12]

Due to problems left-over from the land system and the lack of policy, villagers had been allowed to build on their homesteads. Concurrently, the vague legal status of buildings that exceeded size requirements or extended beyond the red line meant that villagers were limited as to how much they could renovate their self-built buildings was restricted. Government bodies and villagers faced multiple barriers in finding legal paths to renovate self-built buildings in urban villages. In the "Urban Village Laboratory" project, for example, roof restoration work was suspended by the community workstation as it has the potential to alter building outlines. The curatorial team went through many rounds of negotiation before finally receiving the green light for the work to be finished. The intervention of the UABB provided a starting point to address these challenges. Here, a cultural event facilitated communication on government decision-making and residents' needs, effectively improving the conditions of urban villages.

Throughout the exhibition, the "CGD Lab" forums were held to explore wider issues with respect to social, cultural and spatial spheres, including story sharing by urban village residents about daily living, businesses and entrepreneurship, studies and works performed by artists on daily life in the urban village, and research and applications by urban planners and architects on informal cities. Among them, the "Nantou Planning Symposium and the First Nantou Sub-district Livelihood Forum" brought together residents, representatives from the sub-district office, social organizations, directors of government bodies, and industry professionals to explore, along multiple dimensions, development directions and a path for Nantou in the post-2017 UABB era.

图 7 信息中心 + 书店，2017
Figure 7 Infomation Center + Book Store, 2017

Based on real-life needs and facilitated by the exhibition narrative, URBANUS designed and transformed the venue. Taking advantage of the opportunity offered by the UABB, the firm aimed to clarify the skeleton of public spaces and

12 These projects included Nantou Meeting Room (1 Chaoyang North Street), Zhanjiang Beef Offal Restaurant (72 Zhongnan South Street), Jinji Shop (44 Zhongshan East Street), House 17, and the temporary sheds.

provide a structural framework for the future improvement of Nantou. The three factories in the industrial area were transformed into exhibition venues, which could be later used as an open creative cluster and a new community space for young people. The two temporary sheds at the Baode Square were transformed into an information center + book store and the Nantou Conference Hall. They would later remain as public spaces, continuing to serve the community. The renovation of Dajiale stage retained its function as a community center with added capacity to hold professional conferences. It could later become a multi-functional space for various kinds of performance. These public space nodes held cultural events during the biennale and were expected to accommodate community life thereafter.

URBANUS intervened in the regeneration of Nantou through urban research and curatorial practice. It strives to respect residents' daily life and to stimulate cultural awareness and community confidence through minimal transformation and in-depth discussion. The intervention emphasized the engagement of various parties and moderate transformation, which was different from the traditional model of unilateral decision-making. At the same time, the intervention also introduced a construction approach that was legal and professional, unlike the unordered self-led development of urban villages. During the exhibition, stakeholders shared interpretations of the regeneration of Nantou from different perspectives, negotiated on the same platform, and determined the direction and path based on actual needs and available resources.

3 The Nantou Model

In 2019, the *Shenzhen Urban Village (Old Village) Comprehensive Remediation Plan (2019-2025)* was released, clarifying that within the subsequent seven-year planning period, residential land in urban villages that lay within specified zones should not be included in the demolition and reconstruction category of urban regeneration unit plans, land rearrangement plans or shantytown renovation plans. This has left breathing space for some urban villages.[3] Nevertheless, after a decade-long multilateral game, the demolition and reconstruction of Hubei and Baishizhou Villages were *fait accompli*.[13]

At this critical juncture, URBANUS has pointed out the necessity of seeking the "Nantou Model," leveraging its experience of the 2017

图 8 工厂区展场入口，2017
Figure 8 Entry to the venue in the Factory Zone, 2017

UABB intervention and the "Cities, Grow in Difference" concept. In a new context, it is once again exploring the social, cultural and spatial issues of urban villages, with an eye to providing alternative development models for the city in response to the pressures of contemporary reality.

In the "Nantou Model," multiple stakeholders participate in each stage, where their differences in the capacity, resources, demands, and goals bring in diverse models of development, operations, and uses. This is different from the mainstream regeneration model for urban villages, where a monopolistic and dominant power, oriented by self-interest, occupies extensive space, and rejects the rational existence and rights of other parties, inevitably leading to a unitary urban image.

The production of space in urban villages is not only a process of creating spatial products, but also a process of recreating social relations. The demand for villages' livelihood and the shortage of low-income housing has led to common tenancy in the self-built housing in urban villages, where the whole building is rented out to many migrant workers, with the owner possibly living on one floor. Retaining this option, small developers can negotiate with property owners to source a certain number of self-built buildings and rent them out, after proper renovation, to junior

13 The redevelopment of the old town of Hubei was undertaken by China Resources with phase-one construction launched in September 2019. The redevelopment of Baishizhou was undertaken by LVGEM with all rentals terminated in 2019.

图 9 南头古城工厂区改造方案，都市实践，2017
Figure 9 Transformation scheme of the Factory Zone, Nantou Old Town , URBANUS, 2017

white-collar employees moving into urban villages. Government bodies can also be involved, including some self-built housing in its housing security system.[4] Local businesses provide work opportunities for residents and offer convenient services and affordable commodities to the community. With the improvement of residents' capabilities and the expansion of cultural circles, new businesses will emerge as a response to new needs, thus providing more lifestyle options. It is likely that butcher and farmer shops, fast food restaurants, groceries, and internet-famous shops will open next to one another. Every individual is allowed to make their own lifestyle choices in the absence of coercion.

As different members of society co-exist, converge, or collide with one another in Nantou, over time, a dynamic balance can develop. The old accumulates, while the new intervenes, an ongoing state that provides favorable conditions for incubating creativity. A self-aware community with confidence in its culture can gradually generate a sense of urban identity and a specifically unique and charming local culture. Changes predicated on daily life are a response to real problems and fundamentally different from changes based on "origin-tracking," in which everything is overthrown or becomes a lively and ready-made, but ultimately short-lived landscape.

4 Current Nantou

After the UABB, the transformation of Nantou underwent multi-party discussions. It settled in a collaboration in which the government invested in the public space and Shenzhen Wantong Nantou City Management and Operation Co., Ltd (a joint company by Vanke, the district state-owned Shenzhenhuitong, and the village joint-stock company) execute and operate. In October 2019, the Nantou Old Town Conservation and Utilization Project was launched. In August 2020, the demonstration zone covering Zhongshan South and North Streets went into operation. In just a few months, the responsible government department and the agent developer managed to rent approximately 300 buildings, among which the sourcing of the 88 buildings in the demonstration zone was finished in ten days.[14]

The project positioned Nantou as "the most distinctive display area of traditional style, the most profound tourism destination with cultural significance, and the most well-grounded industry cluster of culture and innovation in Shenzhen."[5] The traditional style was defined as Guangzhou residential architecture *(Guangfu minju)* that was popular between the late Qing Dynasty and the Republic of China. It was represented by two pseudo-classic buildings adjacent to the South Gate, which are currently operated by a premium coffee shop and a famous Hong Kong restaurant. The exteriors of most buildings on Zhongshan South and North Streets were redesigned and then occupied by Guangzhou, Hong Kong and Macao-style restaurants as well as cultural and creative businesses. While the ground floor is used for commerce, the upper floors of some buildings have been renovated as branded rental apartments.

While the new shops on the main streets have introduced famous brands and appear to offer a variety of options, the business model behind each shop is the same, targeting the same customers. The original affordable restaurants, convenience stores, and clinics were all removed at the same time. Among the dozens of newly opened shops, only one local dessert shop remains. Residents who used to perform their daily con-

14 It is reported that on 5 March, 2020, the district government issued the "no. 1 special task" order on the conservation and utilization of Nantou Old Town. In just 10 days, the Nantou Sub-district Office completed negotiations with 1,324 people and accomplished the rental sourcing work of 88 buildings, approximately 40,000 square meters, within the demonstration section zone. See, Shenzhen Government Online, April 9, 2020. http://www.sz.gov.cn/cn/xxgk/zfxxgj/gqdt/content/post_7132111.html.

sumption and leisure activities on the main streets have given up the space to one-day tourists. The standardized, renovated and branded apartments, which target younger tenants have replaced the former rental housing, and they strive to present a vibrant community. Such a "fresh look" was precisely the goal of the project, which is "changing the demographic structure and upgrading the industrial structure [of Nantou]."[6]

The renovations keep close to a chosen clip from the past, instead of orienting toward an open future that has yet to be built. Forty years of actual history in the SEZ has been casually graffitied over with the styles from the Qing Dynasty and the Republic of China, while the immigration culture formed since the founding of the People's Republic of China has been covered by the symbolized atmospheres of Guangzhou, Hong Kong and Macao. On top of the heavy burden which Nantou bears as the city's "origin," the border of another layer of requirements of "creative culture" has been added. External cases are quickly filling up the space, while the local culture of Nantou is ignored.

There is no shortcut for the generation of culture. The most recent instance of Nantou regeneration demands greater efficiency, compressing time. In spaces that have been renovated all at once, the good life is readily available and providers, content and consumers are there to be had. This model produces short-term results, but they are achieved by replacing a huge amount of daily spaces, local businesses and communities. In the long term, this model requires the constant provision of new stimuli to attract tourists. However, the more frequent the replacement is, the more difficult it becomes for a local culture, which has already lost its roots to grow.

5 Epilogue

In the early days of the SEZ, the city-making of Shenzhen was consistent with the dream-chasing of Shenzheners, making the city a "frontier for China dreams."[15] Forty years after the establishment of the SEZ in 1980, the urban development model has become real-estate driven aggressively, functioning at the cost of Shenzheners' dreams. As the richness of the urban landscape and the diversity of its communities have disappeared at an increasing pace, urban life seems to be programmed with one single option, which is ready-made, rapid, consumer-based, and entertaining.

The homogeneous development model has consistently cannibalized urban villages nor the heterogeneous bodies that are different from the mainstream. There has been an urgent call to improve our understanding of urbanity. Urban villages are neither diseases and barriers to urbanization, nor are they reserves when developable land has been exhausted. The nature of urban villages is their urbanity. They have undertaken the urban function of accommodating large numbers of new immigrants (Shenzheners), carried the city's memory of various development stages, and nourished local culture. Urban villages are precisely the alternative sample that can solve the city's homogenization. They cannot be copied or mimicked, nor can they be restored once they are gone.

Given the challenges of the contemporary realities, there should be more than one regeneration model for the urban villages. The "Nantou Model" aims to provide an alternative way of thinking and an enabling tool in urban development. During urban evolution, when there is a gap between the development of urban villages and that of the city at large, and an inevitable intervention by external forces, the "Nantou Model" is also an attempt to find a dynamic balance of these various forces. By learning from Nantou and learning from urban villages, we may be able to move toward a more diverse and inclusive future Shenzhen.

Insofar as Shenzhen is a pilot demonstration zone, China and the world are expecting it—once again—to make breakthroughs with the "dare-to-be-the-first" (*gan wei tianxia xian*)[16] courage.

15 Theme of the Shenzhen Case Pavilion at the EXPO 2010.

16 A phrase used to describe the Shenzhen spirit as Shenzhen set many first-time records in China.

[1] 千年古城承载深港文化 [N]. 人民日报 , 2016-6-1.

[2] 深圳市城市更新“十三五”规划（2016—2020）. 深规土［2016］824 号 [R]. http://www.sz.gov.cn/szcsgxtdz/gkmlpt/content/7/7019/mpost_7019458.html#19171.

[3] 万科“围猎”深圳城中村 [N]. 中国经营报 , 2018-4-24.

[4] 深圳市城中村（旧村）综合整治总体规划（2019—2025）. 深规划资源［2019］104 号 [R]. http://www.sz.gov.cn/cn/xxgk/zfxxgj/ghjh/csgh/zt/content/post_1344686.html.

[5] 深圳南头古城蝶变，将打造成特色文化街区 [N]. 中国改革报 , 2019-10-18.

[6] 万科 36 载：深耕厚植，做城乡建设与生活服务商 [N]. 南方都市报 , 2020-6-30.

2017 深双南头主展场，都市实践，2020
2017 UABB main venue Nantou , URBANUS, 2020

南头模式，都市实践，2020
Nantou Model, URBANUS, 2020

城市共生
金皇冠印刷有限公司
城市共生
金皇冠印刷有限公司

你的表情是我的符号（原稿节选），刘庆元，作品于 2017 深双展出
You Expression is My Symbol (excerpt), Liu Qingyuan, work on display at 2017 UABB

A3
城市共生
A3
南頭奶茶店
城市共生

这个疯狂的都市依然在不断地实践

莫思飞
2017 年深港城市\建筑双城双年展（深圳）策展助理，都市实践研究员。

深圳速度

从 2002 年都市实践正式介入城中村的城市研究和实践至今已有 20 年之久。20 年间，深圳的城市发展被形容为“弯道超车”，没有先例的“深圳速度”也导致了一系列以城中村为代表的新的矛盾。在城市不同的时代背景下，不同条件的城中村都需要量身定制的策略。都市实践以城市建设积极参与者的角色，通过对当前城市问题的敏锐观察，进而提出前瞻的、有效的应对策略。总体来说，都市实践在城中村的介入，从保护和提升两个层面双管齐下：一方面是整合、动员文化资源，拒绝简单拆除承载着市民生活的城中村；另一方面是积极地探索通过植入公共空间、公共功能，挖掘城中村本土文化资源和提升民生条件，让“城市化的村庄”融入城市中（图 1）。

1 在快速城市化中如何推动单一城市建设观念的转向？

1980 年深圳经济特区正式成立，在新的建置下深圳开始了任重而道远的探索式的城市发展，同时也是深圳经济特区和原村落二元并置发展轨迹的原点。随着改革开放的进程，二元土地政策和城市管理制度使得城、村二者虽然都以“深圳速度”发展，却形成截然不同的城市肌理。“九二南方讲话”后，深圳迈入深化改革开放的阶段，深圳常住人口持续暴增（在册常住人口数：1979 年 31 万人，1992 年 268 万人，2004 年 800 万人）[1]，使得深圳的正规住房短缺。在“法不责众”的心理预期下，非正规的自建房在高额房租的利益驱动下迅速生长，进而形成容纳深圳约 60% 人口的高密度的城中村。[2] 由于大部分城中村缺乏城市基础设施的建设和管理，形成“脏乱差”“违章抢建”等积重难返的城市问题，成了有着“超一线国际城市”野心的深圳不得不面对的“包袱”。

2005 年 10 月，蔡屋围金融中心改造项目的实施单位（简称“拆迁人”）确定为深圳市京基房地产开发公司和蔡屋围股份公司。两公司同年 11 月 2 日取得《房屋拆迁许可证》。经过近一年的协商，到 2006 年 10 月 23 日，蔡屋围有 95% 以上的业主与拆迁人签订了拆迁补偿安置协议。至 2007 年 4 月市国土房管局作出书面裁决，仍有 6 户未与拆迁人达成一致拆迁补偿意见。其中态度最为强烈的拆迁户是蔡珠祥、张莲好夫妇。
昨日，记者获悉，深圳蔡屋围“钉子户”蔡珠祥的拆迁补偿问题得到协商解决。开发商与蔡家经过一年多漫长协调，最终双方达成一致意见。蔡珠祥得到超过千万元的“天价”补偿同意搬家。[3]
——梁永健《南方都市报》，2007-9-30

2004 年，在建市二十余年之际，城市化率达到百分之百的深圳出台城中村改造政策《深圳市城中村（旧村）改造暂行规定》（简称《暂行规定》）。《暂行规定》通过“政府主导 + 市场运作”的改造模式，由政府监管控制风险，并通过减免地价等一系列优惠政策吸引开发商负责项目、资金的运转，从而提高改造效率。此举开启了开发商在城中村“跑马圈地”的新局面。[4] 2005 年，城中村旧改试点项目渔农村的第一爆，更是把特区内的

图 1 与城市并置发展中的城中村，蔡屋围，孟岩摄于 1995
Figure 1 Urban village and the city in development, Caiwuwei, photograph by Meng Yan, 1995

城中村推向了旧改的风口浪尖，村民和各大开发商都纷纷前赴后继地想从中分一杯羹。根据《暂行规定》中明确规定的补偿细则，[1] 渔农村 76 户原村民（平均 1700 平方米 / 户）中诞生了不少的千万富翁。赔付巨额拆迁款的房产商则是以《暂行规定》中的地价优惠政策，和高容积率的开发强度获利。2005 年在土地紧缩政策下，特区内不再供应新增土地，[2] 开发商面临着无地开发的困局（图 2）。在新的游戏规则下，虽然开发商在其中的收益依然未公之于众，但被视为"价值洼地"、地处城市中心的城中村，成为了各大地产开发商趋之若鹜的破局机会。

据《深圳商报》报道，地产研究机构的监测数据显示，去年深圳房价涨幅创历年之最，成交均价达每平方米 6952 元，比 2004 年上涨 16%。[5]

——董超文 . 中新网，2006-1-30

"拆临城下"的局面引起包括都市实践在内的一些学者和实践者的关注，并将该议题带到 2005 年首届深双中。都市实践展出了与黄伟文规划师 [3] 合作的城中村研究"城 / 村 · 剖面"，并于次年出版了《村 · 城 城 · 村》（图 3）。书中对未来的旧改发展进行了假设性的推演和预判，提出了未来城中村更新模式和城市政策导向的另一种可能性。这是首次国内外参与深双的实践者、学者和学术机构通过展览的平台共同发声，把城中村旧改的话题推到城市决策者的面前。在《村 · 城 城 · 村》中，以现有简单的地产思维逻辑和开发商的盈利模式为假设条件，推演可实施的推倒重建模式：即使忽略村民对拆迁巨额获益的心理预期一次次上涨的因素，至少需要提高原容积率的 3 倍，以支付巨额的拆迁款并获取开发的利润。推倒重建模式看似甩掉了城中村这个包袱，却引发一系列复杂的城市问题。城市公共空间、配套、设施和交通是否能承载推倒重建后突如其来的密度增长？巨额的拆迁款由谁买单？是否值得牺牲城市脉络和日常生活换取所谓的国际形象？由政府倡导，村民通过违法、低质资产翻倍增值获利如何体现城市利益分配的公平？大面积拆除城中村后，廉租房的替代品是什么？[6] 面对推倒重来的尴尬，都市实践以岗厦村（与黄伟文规划师合作）、福新村、新洲村和大芬村为例分别提出了有机更新的提案。提案试图从空间整改的层面着手，以局部拆除、缝补、插建、挖填和加层等设计策略，提出有机整改的可能性，倒逼城中村有机更新机制和相关城市管理政策的形成。学者的发声加上蔡屋围和岗厦拆迁中一轮一轮的拆与被拆方的博弈提升了城市更新的门槛，深圳城中村旧改政策导向从粗犷的一刀切推倒重建的更新模式转为针对不同城中村以不同办法应对的更新策略，但"推进城市更新工作刻不容缓"[4] 的主旋律依然盘旋着。

《村 · 城 城 · 村》中的大芬案例在都市实践持续的研究和介入中，走进了国际的视野。大芬村这个位于深圳边缘的传统客家小村落，跟随着城市的脚步，逐步发展成典型的城中村，同时也成为全球商业油画产业链不可或缺的一环，彼时 60% 的商业油画都出自大芬村的画工们。大芬村把个体的艺术创作转为一个群体的"山

1　补偿给居民的房地产面积原则上不超过每户 480 平方米，超过的合法住宅面积实行货币补偿。

2　2002 年深圳市政府实行土地紧缩政策，土地供应量日益减少；2005 年起，特区内不再供应新增土地。

3　黄伟文时任深圳市规划和国土资源委员会副总规划师。

4　深圳市城市更新办法 .（深府 2009〔211 号〕）2009 年 12 月 1 日，深圳市政府出台了《深圳市城市更新办法》，明确原权利人可作为更新改造实施主体，改造项目无需由"发展商"实施，同时政府鼓励权利人自行改造；突破更新改造土地必须招拍挂出让的政策限制，规定权利人自行改造的项目可协议出让土地。

图 2 深圳建设用地逐渐饱和，刘焯摄于 2018
Figure 2 Development land in Shenzhen getting saturated, photograph by Liu Zhuo, 2018

寨”的复制及标准化生产行为引起不同观点的文化讨论。对于很多新兴的中高产阶层人士来说，批量复制非原创的艺术品是上不得台面的草根和低端文化，低俗、且缺乏文化品位。基于研究，都市实践回应“其实大芬案例关注的是人、产业和城市再生，如果从一个满足市场需求的产业发展来看，大芬模式无可厚非，复制和标准化批量生产是产业化的基本特征。”[7] 大芬村是自下而上与自上而下结合的范例，为因缺乏城市管理而落下“脏乱差”污名的城中村提供了一剂良药。自发的油画产业发展，推动了政府基础设施建设和环境卫生管理；人人都从事着与油画相关的生产和创作，使其从小型工作室、小型油画厂，逐步生长变化，到书店、咖啡店、年轻人的各种设计小店等的出现。

2007 年由都市实践设计，全国第一个城中村美术馆，大芬美术馆的落成广受国际媒体的赞誉，这个美术馆所在的大芬村也逐渐地受到国际关注。2010 年都市实践合伙人孟岩受命为上海世博会深圳案例馆的总策展人，与提案人周红玫[5]和总叙事导演牟森作为核心策划人，把以大芬村为代表的城中村多元共生发展模式首次在世博会“城市最佳实践区”向世界展示（图 4）。国内外参展城市都迫切地向世界展示科技文化最新成就，百花齐放的时候，三十周岁的深圳直面讲述“城市伤疤”的愈合史，关注城市化进程中底层普通人的个体境遇，以回应城市的本质问题。500 名画师共同创作的“大芬丽莎”本身并不代表深圳，而是在其背后 500 个普通人的故事和梦想代表着深圳这座城市。这也是中国当代第一次把这类话题置于官方大事件的国际舞台上，获得了国内外各界的广泛赞誉。深圳案例馆清晰地展现了深圳的创造力、思想力、开放度和“敢为天下先”的深圳精神，也包含了深圳馆的参与者以深圳城市作为实验场，引领中国城市发展的新观念、创造新思想和理想的决心。[8]

“大芬案例”指向着另一种颠覆性的城市建设发展观念的树立。著名独立学者秋风在解读 1982 年宪法时提出，“城市的土地属于国家所有”是对既成城市土地权属的认定，而非对未来农村集体土地发展的规定；他认为“城市化可以在任意一种所有权的土地上展开，不管公有还是私有”。[9] 大芬自发的城市化正指向这样一个未来，消灭并不是城中村城市化的唯一途径。当政府承担起辖区内公共服务和基础设施的供应，使城中村的居民享有城市居民同样的城市权利之时，城市将会迎来真正全面的城市化。

2　在全面城市化后如何找到经济逻辑和文化保育的平衡点？

2012 年，深圳存量建设用地首次超过新增建设用地，面临着二次城市化的浪潮，以湖贝、白石洲为代表的城中村处在了新一轮博弈的前沿。在大芬村获得国际广泛关注之际，深圳城市中完整保留古村空间格局和潮汕民俗文化的湖贝古村在 2011 年已悄然纳入湖贝旧村改造规划方案中完成了立项。眼看着这个拥有着 500 年历史和

5　周红玫时任深圳市规划和国土资源委员会设计处副处长。

多样杂糅的日常生活的“活古村”即将被全球化、标准化、布景化的商业配置所取代。都市实践研究部由合伙人孟岩带领，自发开启了湖贝古村的研究，利用自身资源，邀请相关利益方和建筑规划专家、社会人文学者、政府工作人员、社会学者、媒体人士等在同一个平台上探讨和挖掘湖贝古村的可能性。持续的研究和协调工作，使得湖贝古村的价值和其升级的建议引起了罗湖区领导、实施主体高层和湖贝股份公司董事长的关注。2013 年，都市实践将“湖贝古村保护研究”（图 5）第一轮提案递交罗湖区政府，提出完整保留三纵八横的古村肌理和局部高密度的策略；[10] 古村保护与满足开发指标的城市更新思路引起了政府的重视。后于 2014 年，受实施主体委托，都市实践提交第二轮提案，探索更全面的指向多元文化的保护与开发模式。

在湖贝古村研究的同时，规模最大的白石洲也面临着几乎不可避免的拆除的命运。2013 年，在《白石洲五村城市更新研究》中，都市实践再次充当起城市的调解人，开始探索是否有可能化解重建与保育的悖论：有没有一个新的方法能在后城中村时代满足建设量需求和促成社会升级的同时，激发城市功能多样化并延续城中村的文化肌理？在《白石洲五村城市更新研究》的城市设计提案中，权衡了经济发展的必答题和保育的理想，以远远高于通常政策允许的容积率的高密度巨构，探索城中村生活方式和社会肌理在巨构中延续的可能性，同时，以集中建设的方式，使得大部分城中村区域得以保留和发展。白石洲的更新提案看似是有机更新的另一个极端，却是都市实践再一次以设计为手段，试图用另一种可能性，推动城市政策和城市管理制度的创新，使深圳依然能保有其自发和原生的城市文化和社会生态系统。[11]

白石洲项目位于深圳市南山区深南大道，毗邻科技园及华侨城片区，是深圳目前航母级的城市更新项目。根据已经获批的专项规划，项目计容面积约358 万平方米，将会分三期进行开发，总体开发周期预计为八至十年。目前，白石洲项目正在积极推进前期工作。白石洲项目于 2018 年底取得专项规划批准，2019 年 7 月启动私人物业拆赔签约 ......[12]

——绿景中国新闻中心，2020-8-25

经过一轮又一轮的博弈，2016 年华润公布湖贝片区城市更新方案（简称更新方案），正式启动旧村全面改造。更新方案与都市实践的古村保护及活化专项研究背道而驰，打破了四年来看似达成共识的局面，暴露了开发商的利益与城市公共利益的矛盾。更新方案以“迁建、拆建、创建”[6] 等名义上保护，实则全盘拆毁的方式，试图抚平公众舆论，却引发了关于湖贝古村命运广泛的公共讨论。在第三方专家的质疑下和经历多轮沟通无果后，湖贝村的保护与活化被推向城市公共领域，成了一场话语权的争夺。据《南方都市报》报道，文保部门的回应疑似质疑专家们出师无名，而实施主体的回应更是质疑专家对文化价值判断的理性程度和客观程度。[13] 正是在城市更新力量步步紧逼的时候，文艺界、社会研究与规划建筑界人士共同发起“湖贝古村 120 城市公共计划”，在深圳有方空间举办了第一场设计工作坊。而后在一系列的发声、展览和斡旋下，罗湖区城市更新局在 2017 年公示的更新方案中 10350 平方米的湖贝古村暂时转危为安。但保留面积远远小于湖贝 120 建议的保护面积，而且古村的活化实际上是依然存在很多悬而未决的问题。为此湖贝 120 作出三个呼吁：建立信息和方案分享平台，建立公开的、定期的沟通渠道，以及邀请和参加由任何一方举办的事关更新的学术性活动。[14]

虽然城中村旧改的步伐依然急促，公共展览、各界发声持续地驱动着城市观念的转变。早在 2013 年，深圳市福田区就支持城中村市政基础设施和公共配套设施建设，提出了资金管理实施细则，随后各区政府也逐步把城中村纳入城市管理系统中。城中村环境的提升使得“脏乱差”“城市毒瘤”这些标签逐渐淡去，也意味着从观念层面重新定位城中村和再定义城中村更新的条件逐步成熟。但是，就本质而言城市发展模式并未另辟执行路径，试图用地产开发思维解决地产开发思维所带来的问题，使得没有地可开发依然成为掣肘深圳城市发展的客观条件。

3 城市策展如何探索和先行示范多元的未来城市？

城中村改造的微观研究已经非常深入全面，微观层面城中村改造的规划建设技术流程，包括拆迁、补偿、土地出让、建设回迁房与商品房等等全过程，也都非常成熟和规范。但是对于微观改造引发的宏观问题——非户籍常住人口被驱赶、缺乏公共空间、逆向垄断或暴力拆迁导致的不可持续等等——认识和研究不够。[15]

——叶裕民 . 2015-8

2016 年深圳的 GDP 几乎与全国排名第三的广州持平，2017 年更是超越广州成为全国 GDP 位居第三的城市。同年，明确了粤港澳大湾区的建立。深圳面临着从经济资本发展转向社会资本累积的拐点，从单纯的“时间就是金钱，效率就是生命”的空间建设迈向与人文精神生活建设同步的时代。在这一时代背景下，以拆除重建为主要手段的城中村旧改已初现慢下来的趋势。2016 年，《深圳市城市更新“十三五”规划（2016—2020）》提出规划期内力争通过城市更新，实现违法建筑存量减少 1000 万 ~1200 万平方米，但同时提倡有机更新，以综合整治为主，拆除重建为辅。《关于加强和改进城市更新实施工作的暂行措施》明确在城市更新中要注重保留城市记忆，提倡有机更新。在《关于完善人才住房制度的若干措施》中更是提出，充分挖掘存量住房资源，支持原农村集体经济组织继受单位依法将已建成、审批手续不完备的住房改造为租赁型人才住房，所在区政府给予适当的改造和运营资金补贴。至此，与城市文化和生活方式的延续相关的城中村的更新改造不再仅仅是一个技术问题，而是在深圳城市发展模式转型的拐点上，有着深刻社会意义的，关乎未来的城市问题。

6 详细内容参见 2016 年由深圳市城市规划设计研究院提出的《罗湖区东门街道湖贝城市更新统筹片区规划》。

图 3《村 · 城 城 · 村》，都市实践
Figure 3 *Village/City City/Village*, URBANUS

杨洪[7] 非常赞赏 2017 深双将在城中村中举办的建议，他认为城中村是承载着现在 700 多万市民生活的大家庭，可以把深圳的近现代历史和改革开放历史、市民现代生活历史都丰富地展现在大家面前。[16]
——苏妮《南方日报》，2016-2-29

南山区在 2010 年出台了《南头古城保护规划（2010—2012）》，提出保护与发展相结合的规划保护概念和原则，突出历史文化主题。在审议会上，区领导表示南山区要成立领导小组，推进南头古城的整体整治工作，加强文物保护，拆除违法建筑。但就拆除违法建筑而言，拆迁量极其巨大，南头将近 90% 的建筑都被定性为违法建筑，承载着 3 万多外来人口的生活与居住；若按计划打造主题化的历史街区，约 800 多位内城居住的原居民将移居新村，却忽略了关于 3 万多外来者的安顿计划。[17]“许你一个真正的古城”[18] 一时之间成为主流媒体的论调。继而 2012 年的《南头古城保护规划实施方案研究》延续原规划“就地保护，局部整治”的思路继续探讨实施的策略与路径，并明确提出“空间置换，文化重构”的主张。推荐方案的改造思路在观念层面和实施层面都具有很大的局限性，仅是风貌控制区就需要拆除厂房 27000 平方米，住宅 16000 平方米，[19] 根据当下周边房价的平均价，预估赔迁款将近 4 亿元。事实上在过去的十几年里，由于种种复杂的历史遗留问题，与《南头古城保护规划》一样，大量研究、规划和设计提案都难以实施。多个提案都试图重新开发古城的文化和旅游资源，侧重南头作为深港之源千年古城的历史意义，却忽略了南头作为将近 4 万居民生活方式和文化的空间载体的当代价值。

2016 年，面对城中村、古城发展互相掣肘的困境，都市实践受托对南头进行新一轮研究与设计，以期识别出南头古城破局的关键点。都市实践研究团队在研究初期提出一个关乎城市未来观念走向的问题：如何准确定位同时是历史遗迹和当代遗产的南头？与一般城中村不同，南头古城是当代城中村与千年古城的共生体，城中村作为“活着的”当代遗产的价值与其千年古城的定位相互牵制。南头古城与湖贝古村的“古”并不相同，南头古城经历了历史的起伏后，真正存在的历史古迹寥寥无几，八成以上的建筑都是改革开放后的自建房。作为城中村，南头多样杂糅的城市生活，面临着被清除殆尽的命运；另一方面作为千年古城的遗址，南头被一次又一次地推回到某种历史的场景之中。

在对南头的历史文脉、空间肌理进行梳理和研究，以及对国内外大量城市历史片区改造实例和历史保护与改造观念不断迭代分析后，都市实践提出南头古城重生的城市定位：“一部南头古城的发展史就是一部完整浓缩的深圳城市发展史，相比空间改造而言，古城重生更需要的是居民生活品质的提升以及它所承载的本土文化的复兴……只有尊重历史原真性、且珍视各个时代的文化层积和历史印记，才能塑造一个本土文化历久弥新，永远鲜活的城市历史文化街区。今日的南头不再是传统意义上的‘古城’，而是承载着千年古城文化、且沉淀了深圳各个发展时期空间、社会、和文化遗产的‘南头故城’。她是深圳仅有的能将千年文化传承谱系与近三十余年中国高速城市化的过程全光谱式并置呈现的珍贵城市文化样本。”[20] 都市实践在《南头古城保护与更新》提案中，主动建议搭接历史与深双。深双作为介入型的文化事件与都市实践的城市设计不谋而合，以介入实施为导向、由点及面渐进式激活、以文化活动促进南头复兴。[21] 都市实践合伙人孟岩和刘晓都作为 2017 深双总策展人以“城市共生”（图 6）为展览主题，呼唤多样性的同时，也试图唤醒城中村从建筑空间到生活方式和本土文化，作为当代遗产的意识，正视城中村在改革开放 40 年间给深圳城市提供的养分和其所承载的城市记忆。

不拆真遗存，不建假古董[22]
——2017 年 9 月，住建部《关于加强历史建筑保护与利用工作的通知》（建规［2017］212 号）

试图通过大型文化事件驱动因历史遗留问题止步多年的南头改造，是需要极大的勇气和冒险精神的，总策展人孟岩表示，“双年展能在南头开幕就已经是成功了”。基于对南头古城场地的深入研究和分析，量身定制的城市设计空间改造思路与深双展场的空间叙事高度吻合。以原古城十字街格局为核心，疏通中山南北街，直接连通中山公园、古城和南门公园；选取一系列围绕这一主线的重要公共空间节点改造，并用作双年展展场；民居拆迁量近乎零。策展团队甚至提出双年展与租赁未到期的工厂同时运行的可能性，为此也在厂房展场设计中做出调整，试图尽可能地减少对城内居民生活的打扰，最大限度地提升古城公共空间品质，并为生活的延续、未来的发展留有足够的空间。至此，深港之源、千年古城遗址、城中村现状和双年展对未来的探索，在这样的契机下交织在了南头。

“城市共生”并未止步于空间改造层面的探索，而是一场没有先例、试图将多元生活方式交织在一起的社会实验，同时在观念的碰撞中探索未来的“大冒险”。深双作为一次国际性的学术型展览，带来全球各地的参展人和文化资源；他们短期驻扎在南头现场，像是投进南头的一粒石子，激起短暂的涟漪。奥地利参展人托恩·马

7　杨洪时任深圳市委常委、党组成员。

东的团队提到，在他们和村民的交流中常被问及一个问题：“你们为什么喜欢生活在我们南头古城？”对居住者而言，南头仅仅是迫于无奈的低成本的居住空间，而参展人的主动选择为他们提供了不同的视角，让他们开始重新思考“为什么喜欢生活在南头古城”。

图 4《特区一村》，姜珺主编，都市实践策划
Figure 4 *A Village by the SEZ*, edited by Jiang Jun, curated by URBANUS

4 如何看待城中村改造中的“士绅化”现象？

深双所促进的公共空间改造、文化资源入驻以及间接造成的原商户流失和置换，被戴上“士绅化”的帽子也是预料之中。但当“深双在南头”被放在城市尺度来讨论时，城市的人口结构中的主要份额正从务工者转变成年轻的初级白领，城中村居住环境的提升是基于市场规律的必然趋势。深双正是一次努力试图通过渐进式提升减缓士绅化的实验。

士绅化的讨论似乎是写在深圳城中村基因里的问题，但实际上士绅化的概念所指向的社会政治经济问题与城中村问题不尽相同。1964 年，社会学者、城市规划者露丝 · 格拉斯首次提出“士绅化”，是指“逆城市化”[8]后出现的“再城市化”现象。随后的几十年里，士绅化的定义不断扩展和演进，与驱赶低收入人群，特别是有色人种划上等号，并带有一定的美国特色的社会政治经济色彩，一直伴随着美国城市的城市复兴历史。近五年来，美国多个大城市逐步出台保障低收入居民居住权利的政策，以纽约为例，在 2016 年 3 月首次通过《强制包容住房条例》，要求开发项目中包含一定比例的可负担住宅。[23] 同时，非营利组织，如“城市的下一步”，也一直寻求针对士绅化带来的负面影响的城市策略。

在深圳的城市背景下讨论的核心更偏重“士绅化”带来的“人口置换”“居住权得不到保障”“单质化的大尺度规划”和“社会多样性（可负担性）的丧失”等问题。以“发展就是硬道理”为核心价值观的深圳，在建市四十周年之际未曾经历“逆城市化”就已经进入“二次城市化”的阶段，即产业结构和人口结构逐步转型，从“世界工厂”迈向“设计之都”[24]，从“人口红利”加速向“人才红利”发展[25]。换言之，高速发展的城市本身不可避免地成为全国甚至全世界最剧烈的、“士绅化”的现场。针对其引发的“失权”问题，近些年来政府也致力于推动一系列完善政策和制度建立的讨论和尝试，从而保障城市居民应有的权利。有着“落脚城市”意义的城中村，是在政策和制度还没有跟上的时候，保障城市居民居住权的最后一条、自我建立的、非正式的防线。因此，城中村的更新改造和“士绅化”问题在舆论上形成二元对峙的局面。

城中村随着深圳城市发展而迅猛生长，出于对“士绅化”的惧怕而拒绝城中村发展的可能性，因噎废食，是不可行的。城中村要生存的大前提，是必须跟上深圳城市发展的步伐，否则只会让城中村潜在的经济和文化价值不被认可，成为地产开发商眼中的“价值洼地”，继而被全新的商品房、写字楼、购物中心取代。地产思维的开发逻辑注重的是效率、标准化的生产，“一套图建一百栋楼”，简单高效地完成购地、设计、兴建和出售的流程；但在城市更新中，这种“简单”的代价是牺牲了城市的脉络、原有的生活方式、本土文化的自信和社会文化的多样性。只有适度地增长其自身的价值，避免因拆除重建式的城市更新而造成断崖式的“士绅化”，从而减缓士绅化的进程。相较有绝对优势的资本力量，深双更多依赖的是文化力量和部分行政力量。在各方力量相互博弈和制衡的城中村生态系统中，深双介入的力量足以驱动各方力量的对话和碰撞，在探讨城市未来的语境下共同寻找新的平衡点，避免过于强大的资本介入，形成吞并和一方独大的局面。

都市实践的展场改造是一次城中村改造范式的介入实验。深双期间自上而下和自下而上的改造，都是“试探”城市管理制度边缘的实验，开启了“南头模式”——另一种未来城市模式的探索。展览中“城中村实验场”的策划，以深双作为平台，促成房东和设计师交流的机会，让城市管理制度、空间设计和使用者需求因展览这个机会“被迫”放在实践这个平台上，寻求个体介入的力量、行政力量、居民利益的平衡点，同时也暴露了“一刀切”的城市管理制度在有机更新中未能精细化地针对个案执行的局限性。深双恰如其分地触碰单一模式的城市管理制度的限制，尝试以城中村方法（规则）量身定制新的规则，为有机更新的政策制定提供宝贵的实践经验，又通过文化事件赋予的力量，促成每一个人，每一个个体间的了解、尊重和自我价值的体现。

当前，我国城市更新已从传统的注重物质层面与拆

8 逆城市化是指 20 世纪 70 年代美国学者布莱恩 · 贝里提出的，与城市化相对应，大城市人口向小城市或小城镇分散式转移的现象。

图 5《湖贝老村》，都市实践
Figure 5 *Old Hubei Village*, URBANUS

旧建新过渡到以功能环境重塑、产业重构、历史文化传承、社会民生改善为重心的有机更新阶段。[26]
——何燕燕，王岩波 . 新华社新媒体中心，2018-7-27

作为改革开放的“试验田”，深圳一直肩负着探索城市未来的重任。活色生香的城中村似乎为当代无趣单一的规划城市之“死”提供了另一种重生的方案：如果未来城市发展的动机不再单纯地指向经济资本的发展，同时注重社会资本和文化资本的累积，是否有可能摒弃地产思维垄断的唯一模式，取而代之更多元、自下而上的个体开发模式？深双实验场是迈向“人人都是开发者”这一突破地产思维限制的“城市共生”的一次开端而非终点。策展团队意识到这是一个超越建筑学的，多视角、多维度、多方博弈的跨界问题。通过“共生实验室”的策划，策展团队试图推动南头古城“共建平台”的成立，保障后 2017 深双时代持续地触发和引导内城的动态更新，并呼吁城市管理者、建设者、学者、居民等利益相关方与关注者一起寻找未来城市的出路。

5　什么是未来城市的理想模式？

2018 年 3 月，深双在南头古城大家乐舞台举行闭幕式，标志着后 2017 深双时代的开启。2019 年 3 月，深圳市规划和自然资源局发布《深圳市城中村（旧村）综合整治总体规划（2019—2025）》提出全面推进城中村有机更新，规划期内保留一定比例的城中村，[9] 有序引导各区开展城中村综合整治，促进有机更新。如《南方都市报》所说，“对于深圳的城中村而言，起码短时间内，综合整治的大幕已经拉开。而未来，这些栖息着城市 2/3 人口的区域在新的城市化进程中将如何嬗变仍考验城市主官。”[2] 同年 10 月《关于推进城中村历史文化保护和特色风貌塑造综合整治试点的工作方案》中，把包括以万科为开发主体的南头古城在内的七个城中村列为“微更新”试点，旨在探索未来城中村治理的新模式。“深港同源”的南头古城保护开发更是在 2019 年纳入《粤港澳大湾区实施纲要》。同一时间，湖贝片区城市更新项目举行开工仪式，湖贝古村的有机更新也蓄势待发。

2005 年首次旧村改造政策出台后的 15 年间，城市对城中村当代价值的观念已经逐步转向，但在执行层面的手段和办法依然捉襟见肘。万科城市设计研究院（深圳）有限公司成立的城市星球研究所公众号文章提出了自发更新的局限性和难度，进而采用统一更新人口结构转型的介入策略。

“南头古城是一个以外来人口为主的城中村……当改造发生，房屋的所有者会享受最终价值增益，却大多不住在这里，住在这里的大多数人有改变现状的欲望，却不享受改造之后的经济收益，也无力负担这种支出，这导致单纯依靠社会力量的引导，现有的居民自发进行城市更新的方法很难进行下去……有效的存量更新应该考虑从“人口结构”入手，引入一批对城市和空间敏感的创意工作者，并让他们能够扎下根来。”[27]
——恒铭 . 城市星球研究所，2020-9-29

但是，深双期间不断在尝试的“社区营造”和“共同参与”等渐进式城市存量更新方式实现的可能性，不应仅是一句“很难进行下去”就被否决；更不应允许文化的累积在资本的裹挟下走捷径，以“尊重历史原真性”“延续街道生活记忆”“折射深圳改革开放 40 年建设史的光谱”[28] 为口号，利用与南头历史无关的民国风、岭南风和小资格调绑架人们对历史的认知和“对美好生活的想象”。这样一根筋地贩售所谓精致的生活方式容易使得深圳的精神文明建设，如同其空间建设，依然笼罩在挥之不去的“简单高效”之中。如此一来使得“历史层积”和“多元”成为一夜打造布景式网红地的噱头，也使得文化不断被移植却不允许它存活和生长。

2020 年，都市实践以建筑师的身份重返南头古城。在核心区建筑设计集群工作会上，都市实践提出，“在南头古城保护更新的同时，需要建立社区共建平台，最终达到新老建筑共生、新老居民共生、新老文化共生。[29] 深双期间通过社区共建和微改造等方式所挖掘出的“湛江牛杂店”“锦记商店”“九街糖水”“跌打医生”“包子西施”“黄记酒家”和“修鞋匠”等南头古城原有活跃的多元社群和特色的本土文化，应在古城保护更新中得以持续地赋能和发展。古城更新保护的不应仅仅是空间，更应该学会慢下来，保护原有的生活样态不被高速的城市化全盘摧毁。在一次性重金投入的主街更新后，有必要形成相关政策，引导和激发多层次的、持续的社会自发投入与内街自主更新。“文化沙漠”“没有历史”这些魔咒四十年来与深圳的发展如影随形，但急于一时以南头作为一剂打破魔咒的解药反而只会越陷越深。只有加以时间的滋养，社区平台机制的保障、引导和培育，本土文化自信的觉醒，才能使得深圳延续包容的“来了就是深圳人”的共建精神。

2021 年，在南街一次性更新后清新的建筑群中显得格格不入的“南头杂交楼”，可以被视为都市实践从城市尺度回归建筑尺度，对“城中村进化”模式的实验和

9　规划期内全市划定的综合整治分区用地规模为 55 平方公里，占比 56%。其中福田区、罗湖区和南山区综合整治分区划定比例不低于 75%，其余各区不低于 54%。

新旧共存的城市观念的践行。在这个“读图时代”，影像的主导地位和视觉霸权主义昭示了一个从语言主因型文化向图像主因型文化转变的时代。[30] 受困于消费主义下的城市和建筑成为依赖图像传达感知和想象的商品。这种符号化、单质化的视觉审美入侵至日常生活中，使得简单的“好看”与“难看”成为深入人心的评判标准。以杂交楼为代表的城中村自建房正是对这种全球消费主义下单一审美标准的反抗。矗立在中山南街与东街的转角处，五栋紧紧簇拥在一起的建筑高低错落、风格迥异，不能简单地用“好看”或“难看”去形容。它们的丰富多元代表的是一个五味杂陈时代特殊的文脉与生活气息，[31] 同时它们复杂、矛盾与冲突的建筑体量又彰显着南方生猛的张力。基于对其原生建造逻辑的理解和尊重的设计理念，通过内部空间和流线的梳理，“雕镂贯通”，[32] 使其更好地适应当下生活的需求，对外部进行有限度的重塑（图 7），“经年的加建改造、材料更替被记录，顶部还并置了一直一曲、一实一虚的新体量，隐藏了一个屋顶小园”（图 8）。[32] 改造后的“杂交楼”并未因建筑师的介入而变得圆滑世故，反而更彰显了它原生包容杂糅的个性。在建筑实践上，这是一次没有先例、向城中村学习的挑战，是对图像主导下越发趋同的精致的审美观“叛变”的宣言。而在城中村实践中，这是保留历史层积、叠加的渐进式更新的城中村进化样本。

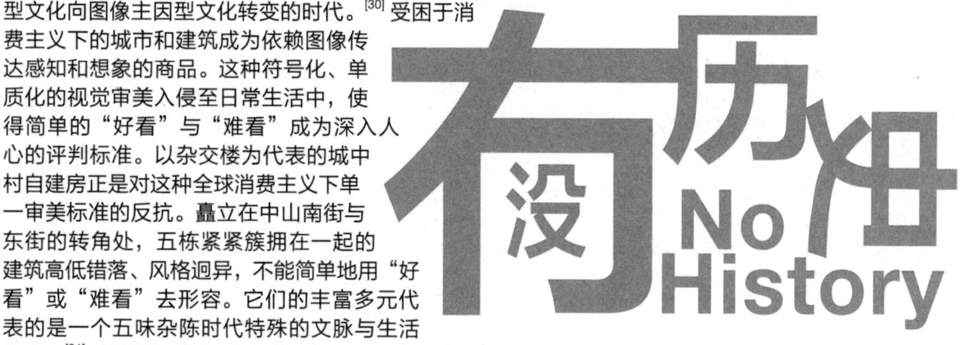

“南头杂交楼”并非通往“城中村进化”理想模式的唯一途径。现在比任何时代都更迫切地需要城市管理者、学者、建设者和相关利益方在各个层面的合作，以不同于以往地产思维主导的城市更新模式，探索个体价值得以体现，生活方式得以延续，人人都是开发者，从而导向多样化的空间和业态，多元价值共存的状态。通往理想从来不是一条通途，先行示范的深圳城市需要不放弃理想的实践者，在这个疯狂的都市中不断地实践！

Architectural Delirium: URBANUS, Their Practice, and the City of Shenzhen

Mo Sifei
Curatorial assistant of the 2017 Bi-City Biennale of Urbanism\Architecture (Shenzhen), URBANUS Researcher.

图 6《城市共生》展前册，黎莱恩摄
Figure 6 Catalogue of *Cities, Grow in Difference*, photograph by Li Laien

In 2002, URBANUS launched its urban research and practice for urban villages. Since then, the speed of urban development in Shenzhen has been described as "overtaking on a corner."[1] The city's unprecedented rate of urbanization is known as "Shenzhen Speed" and has posed a series of new conflicts in terms of urban village issues. In different time periods, tailored strategies have been created to cope with the distinct issues facing urban villages. Being an active urban practitioner, URBANUS proposes forward-looking and effective corresponding strategies through the acute observation of current urban issues. In general, URBANUS approaches intervention in urban villages from the perspectives of conservation and improvement. On one hand, it incorporates and mobilizes cultural resources, refusing to engage in the destruction of the urban village and its everyday life, while on the other, it explores local cultural resources and improves people's livelihood through public program interventions in public spaces, allowing the urbanized village to better integrate into the city (Figure 1).

1 In the maelstrom of rapid urbanization, how do we promote a shift away from the singular concept of urban construction?

In 1980, the Shenzhen Special Economic Zone (SEZ) was formally established with the launch of a long-term exploratory development within a new system. Shenzhen's predecessor was

1 A Common Phrase in Chinese refers to a crucial change at a key moment.

Bao'an County. The establishment of the SEZ initiated the growth of both Shenzhen as the original urban area was called and the original villages of Bao'an County. In terms of the city's geography, Shenzhen was divided into the SEZ and Bao'an, which later became known as the outer districts. Through the progress of reform and opening up, the two-tier land tenure and urban management systems have allowed both urban and rural land to expand at Shenzhen Speed, yet form completely different urban fabrics. Since Deng Xiaoping's 1992 Southern Tour, the advance of the reform and opening up has contributed to the rapid population growth in Shenzhen, which has led to a shortage in the formal housing market. For example, in 1979, the city's population was 310,000 in 1979, 2.68 million in 1992, and 8 million in 2005.[1] Conventional wisdom at the time was "fa bu ze zhong," meaning a law is left unenforced if it is violated blatantly by most people combined with a profitable rental market drove a boom of high-density informal tenements, which in turn came to house 60% of Shenzhen's population.[2] The initial lack of urban infrastructure and management in urban villages resulted in almost irreversible issues such as the "dirty, chaotic, and substandard"[2] condition and "illegal rushed buildings,"[3] which became a burden for Shenzhen to overcome, particularly in light of its ambitions of being an ultimate first-tier global city.

"Accordingly, in October 2005, the implementing entity for the Caiwuwei Financial Center Renewal project was confirmed to be Shenzhen Kingkey Real Estate Co. Ltd and Caiwuwei stock-holding corporation. On November 2, the companies acquired "Building Demolition Permission." As of October 23, 2006, more than 95% of the owners in Caiwuwei have signed demolition compensation and resettlement agreements with the demolishers. Yet, as of April 2007, there were still six households that had not reached an agreement with the demolisher on compensation for demolition. Among them, the couple, Cai Zhuxiang and Zhang Lianhao, express the strongest attitude. In September of this year, Cai Zhuxiang and Zhang Lianhao's compensation for demolition was settled through negotiation. Cai Zhuxiang received sky-high compensation, more than 10 million yuan, and agreed to demolition."[3]

In 2004, when Shenzhen had already been established for more than 20 years, all residual rural areas had been formally brought into the urban system, making the city's urbanization rate 100%. That year, the city issued an old village redevelopment policy, *Shenzhen Municipal Government Interim Provision of Shenzhen Urban Village (Old Village) Redevelopment*. The renewal model combined state-led development and market-oriented operations. In practice, renewal has been supervised and risk-controlled by the government and facilitated by premier development companies, improving the efficiency of renewals with a series of preferential policies such as land price reductions and exemptions. This has broken fresh ground for the developers, who have claimed markets by making precocious acquisitions.[4] In 2005, the demolition of Fishermen's Village *(Yunong Cun)*, Futian District was the pilot project for old village redevelopment, pushing urban villages inside the SEZ to the forefront of urban renewal and thereby encouraging villagers and developers to get in on the action. According to the rules of compensation in the *Interim Provision*, many of the 76 villagers of Fishermen's Village (with an average property of 1700 square meters per household) became multi-millionaires in the renewal.[4] Although developers paid villagers large compensation packages, nevertheless they also benefited from the preference of the land policy stated in the *Interim Provision*, which allowed high FAR development. In 2005, under the land supply restriction policy, there was no more developable land within the SEZ.[5] Developers faced with the problem of having zero land to develop (Figure 2). Given these circumstances, although the profit gained by the developers through urban village redevelopment is unknown by the public, nevertheless urban villages, which were previously considered low-value land within the center of the city, suddenly became potential game-changers for major developers to pursue.

2 Urban villages had been stigmatized by the popular medias using the phrases *Zang Luan Cha* meaning dirty, chaotic, substandard.
3 Due to the illegal nature of the self-built buildings and their extensions, the constructions usually happens rapidly within few days without being detected by the law enforcement.
4 New apartments for residents as compensation are not to exceed 480m² per household. The exceeding floor area is compensated with currency.
5 The land supply restriction policy started in 2002 with reducing in construction land supply and in 2005 stop offering new construction land inside the SEZ.

According to Shenzhen Economic Daily, data collected by real estate research institutions shows that Shenzhen's housing prices recorded the highest increase in history last year, with an average transaction price of 6,952 yuan per square meter, a 16% increase over 2004.[5]

In 2005, facing the prospect of urban villages possibly being razed in the near future, scholars and practitioners, including URBANUS, made urban villages part of the agenda of the first Shenzhen Biennale. In collaboration with Huang Weiwen,[6] URBANUS conducted research and curated the exhibition "Sections of City/Village," which was published the following year as the book *Village/City City/Village*[6] (Figure 3). It was a hypothetical deduction and prediction of the future renewal of urban villages, while proposing alternative approaches and policy-making directions for their future. With the biennale as a platform, practitioners, scholars and academic institutions, both aboard and domestic, initiated dialogue regarding the renewal of urban villages with decision-makers.

Village/City City/Village presents a hypothesis based on the real estate development model and a feasible demolition and rebuilding model: Even if villagers' growing expectations due to increasing compensation packages are overlooked, when the burden of compensation is placed entirely on the developer, at the very least a threefold increase of FAR is required to recoup such significant expenditure. Therefore, the renewal of urban villages seemingly overcomes the burden of the city while introducing a series of complications. Will current public space, facilities, infrastructure and transportation system be able to accommodate the sudden increase in population density? Who will pay the compensation? Is it worth sacrificing a time-honored historical context and hybrid urban life, replacing them with areas that meet global commercial standards and fit the image of an international city? How will social justice be secured if government advocates for huge profits by the villagers, when the value of illegal and low-quality assets is doubled? Where will alternative affordable housing come from after many urban villages have been eradicated?[6]

Meeting the challenges of renewal, URBANUS proposed tactical alternatives based on the case of Gangxia (in cooperation with Huang Weiwen), Fuxin, Xinzhou Village and Dafen Villages. Embarking on spatial interventions with design strategies such as partial demolition, patching, inserting and filling. They pushed for policy solutions to organic renewal and urban management by proposing the possibility of organic regeneration. Based on public recommendations by scholars as well as the emergence of complications and the scale of investment in the cases of Caiwuwei and Gangxia Villages, the old village redevelopment policy has become a diverse model for different types of villages instead of eradicating existing structures for redevelopment. Yet, the city continues to promote urban renewal of urban villages.[7]

The research and interventions of the Dafen case described in *Village/City City/Village* by URBANUS brought Dafen Village to the international stage. A traditional Hakka village on the edge of Shenzhen, Dafen developed into a typical urban village in parallel with the development of Shenzhen. It has also become an indispensable part of the global commercial oil painting industry chain, as 60% of all commercial oil paintings are produced by painters based in Dafen Village. This painting process involves individual painters participating in collective copying and standardized mass productions. This process has raised cultural discussions, although many recent members of the middle-to-high class consider this an example of low culture. URBANUS offers a different perspective based on their research, stating, "People, industry and urban regeneration are the focal points of the Dafen case, and there is no shame in meeting the market demand through reproduction and standardization, which are essential features of mass production."[7] Dafen Village is a paradigm for the integration of bottom-up and top-down urbanism, providing an antidote to the stigma of being dirty, chaotic, and substandard, which arose due to a lack of urban management. Indeed, it was the bottom-up development of a painting industry that prompted government infrastructure and sanitation improvements, where everyone is engaged in painting-related production and creation, from

6 At the time, Huang Weiwen was the Deputy Chief Planner of the Urban Planning, Land and Resources Commission of Shenzhen Municipality.

7 *Shenzhen Municipal Urban Renewal Measures*. (Shenzhen Municipal 2009 [No. 211]) On December 1, 2009, the Shenzhen Municipal Government issued the *Shenzhen Municipal Urban Renewal Measures*, which stated that the owner can be the main body facilitating renewal project, no "develope" is required, and the government encourages the owner to renovate by themselves; to break through the policy restriction that the land for renewal must be sold by tender, auction and listing, the projects facilitating by the owner can transfer the land by agreement.

small studios and factories to bookstores, coffee shops and design stores to attract members of the younger creative class.

Designed by URBANUS in 2007, the Dafen Art Museum became the first art institution built in an urban village. The museum has been widely praised by the international media while its site, Dafen Village, has gradually received international attention. In 2010, URBANUS partner, Meng Yan, was appointed as the chief curator of the Shenzhen Pavilion of the EXPO 2010 Shanghai. He joined with the project initiator Zhou Hongmei[8] and chief director of narrative Mou Sen to curate the exhibition (Figure 4). The inclusive and diversified development model of Dafen Village, which represented the city's urban villages, was presented at the EXPO as the "Urban Best Practice Area." In contrast to other participating cities, both domestic and international, that were eager to showcase their latest achievements in science, technology, and culture, then only 30 years old, Shenzhen displayed its history of healing urban traumas. This focus on the individuals who occupy the bottom rungs of an urbanizing society constituted a response to the fundamental question, who makes the city? Co-created by 500 painters, the "Dafen Lisa" itself did not represent Shenzhen. Instead, for the first time, on an international stage in contemporary China, the stories and dreams of the 500 ordinary people who created the painting represented the city, receiving widespread praise from all walks of life at home and abroad. The Shenzhen Pavilion showcased the creative, open-minded and avant-garde spirit of the city, including the determination of the participants of the Shenzhen Pavilion to lead the development of a Chinese city with new concepts, ideas and ideals, making Shenzhen a successful urban test ground.[8]

The Dafen case points to the establishment of a subversive concept of urban construction and development. An independent scholar, Qiu Feng's well-known interpretation of the 1982 Constitution "all urban land was declared state-owned" addressed the urban land of the time, rather than a provision for the future urbanized development of rural collective land. He believed that "Urbanization should not be restricted based on the land's ownership status."[9] Dafen's spontaneous urbanization suggests a future in which the elimination of urban villages is not the only way for them to urbanize. When public services and infrastructure are provided by the government within its jurisdiction, the inhabitants of urban villages can obtain rights to the city as citizens of the city, thereby achieving truly comprehensive urbanization.

2 How to find the balance between financial logic and cultural conservation after full urbanization?

In 2012, brownfield land exceeded vacant land for the first time in Shenzhen, which was facing the second wave of urbanization. Urban villages such as Hubei and Baishizhou were at the forefront of this new round of redevelopment. While Dafen Village had gained international attention, Old Hubei Village was included in the approved Old Hubei Village renewal master plan in 2011 without much notice. Located in the center of Shenzhen, Hubei maintained its original layout and was home to folk culture from the Chaozhou-Shantou region of Guangdong. The renewal plan was going to sweep away a 500-year-old village, its living history, and an example of hybrid everyday urban life and replace it with a globally standard commercial site. Led by Meng Yan, the research department of URBANUS immediately leapt into action, conducting research on Old Hubei Village, mobilizing their resources and generating dialogue. They invited stakeholders, architects, planners, social and humanities scholars, government representatives as well as media for a roundtable discussion on the possibilities of Old Hubei Village. Persistent research and communication about the village's value and its regeneration drew the attention of the Mayor of Luohu District, the senior management of the developer and the chairman of Hubei Stock-holding Corporation. The following year, URBANUS submitted a proposal, the first version of "Old Hubei Village Preservation Research,"(Figure 5) to the Luohu District Committee, outlining a strategy to preserve the fabric of three vertical and eight horizontal alleys of the old village by increasing the FAR of other areas of the redevelopment project.[10] Such an alternative allowed the government to acknowledge the possibilities of preserving the old village while fulfilling the required density of urban renewal. Later, in 2014, the developer commissioned URBANUS to submit a redevelopment proposal that explored a comprehensive and diversified model for the preservation and development of Hubei.

As the research on Old Hubei Village contin-

8 At the time, Zhou Hongmei was the Deputy Director of the Design Department of Urban Planning, Land and Resources Commission of Shenzhen Municipality.

ued, Baishizhou was also facing inevitable demolition, which was large in scale. In 2013, through the "Baishizhou Five Villages Urban Regeneration Research," URBANUS once again became the city's mediator, exploring an alternative solution to the contradiction between redevelopment and preservation, asking: Is there a way to meet the demanded FAR and promote social improvement in the post-urban village era, while stimulating the diversification of programs and inheriting the cultural context of an urban village? The urban design proposal, "Baishizhou Five Village Urban Renewal" weighed the financial requirements for redevelopment against the ideals of preservation, resulting in high-density megastructures with a FAR higher than the current policy allows. The goal was to find a way for the continuation of the urban village lifestyle and social fabric. High-density development would have allowed most of the area of the urban village to be preserved and to grow. The proposal seemed to be the opposite of organic regeneration, as it was another attempt to design a tool that would provide alternatives for the transformation of the urban policy and management system for Shenzhen as a city that owned its locality, urban culture and social system.[11]

> *"The Baishizhou project is located on Shennan Avenue, Nanshan District, Shenzhen, adjacent to the Science and Technology Park and the Overseas Chinese Town districts. It is a mega urban renewal project in Shenzhen. According to the approved urban renewal plan, the project has a total floor area of approximately 3.58 million square meters. The development will be carried out in three phases, and the overall development cycle is expected to be eight to ten years. At present, the Baishizhou project is actively advancing the preliminary work. The Baishizhou project obtained special planning approval at the end of 2018, and the procedures of compensation for the private property demolition contract were initiated in July 2019."*[12]

After rounds of negotiation, in 2016, China Resources announced the urban renewal plan for the Hubei district, officially launching the renewal project of Old Hubei Village. The proposal contradicted URBANUS' preservation and revitalization research, breaking the consensus of the previous four years, while exposing the contradictions between developer and public interests. The renewal plan completely razed the old village through the approaches of "relocation, demolition and rebuilding."[9] The company attempted to assuage the public's desire for preservation. Instead, the renovation proposal catalyzed a widely influential public discussion regarding the future of Old Hubei Village. Questioned by third party experts and receiving no answer through negotiations, the debate over the preservation and revitalization of Hubei Village moved to the public sphere, creating competition to control the discourse. According to a report from the *Southern Metropolis Daily*, the Culture Bureau questioned the justification of war initiated by the professionals, while the developers facilitating the redevelopment questioned the rationality and objectivity of the experts' appraisal of cultural value[13]. As urban renewal became a pressing issue, the conversation expanded to include more of the public, a cross-disciplinary alliance of people from the arts, social studies, planning and architecture launched the "The Hubei 120 Public Urban Plan," calling themselves Hubei 120, a reference the Shenzhen emergency number (equivalent to 911 in the United States). Their first design workshop was held at POSITION in Shenzhen. Through a series of public voices, exhibitions and mediations, the Urban Renewal Bureau of Luohu District made a renewal plan public in 2017, in which 10,350 square meters of Old Hubei Village would survive. Yet, that was far smaller than the recommended protected area in the plan proposed by Hubei 120, and many issues associated with the revitalization of the Old Village remained unresolved. Therefore, Hubei 120 called for the establishment of a platform for information sharing, the building of a regular and open communication channel as well as joining academic events hosted by any entity relating to the renewal project.[14]

Although the rapid pace of old village redevelopment continues, public exhibitions and discourses continue to drive the transformation of urban concepts. As early as 2013, Futian District Government supported the construction of municipal infrastructure and facilities in urban villages, with detailed implementation rules proposed for funding management. Since then, the district governments have progressively included urban villages in the urban management system. The improvement of the environment in urban villages

9 See details in "District Urban Renewal Planning of Hubei Dongmen Street Luohu District 2016" by Urban Planning & Design Institute of Shenzhen (UPDIS).

has gradually led to labels such as "dirty, chaotic, and substandard" and a "cancer of the city" being abandoned. This indicates that opportunities are ripe to push for the repositioning of the urban village within the city as well as a redefinition of its renewal. However, no alternative development model has been put forth, and no attempts have been made to resolve issues caused by real estate development, making the shortage of land for urban development a constraint for Shenzhen.

3 How can urban curation explore and serve as a paradigm for the possibility of urban diversity?

"The micro-research on the urban villages' renewal has been very thorough and comprehensive. The technical process of the planning and construction of old village redevelopment at the micro level, including demolition and relocation, compensation, land ownership transfer, construction of relocated housing and commodity housing, etc., are very mature and standardized. However, there is not enough understanding or research on the macro problems caused by the micro-renewal, such as the expulsion of non-registered permanent residents, the lack of public space, the unsustainability caused by reverse monopoly or violent demolition, etc."[15]

In 2016, Shenzhen's GDP was almost the same as that of Guangzhou, which at the time was the third largest in the country. In 2017, Shenzhen surpassed Guangzhou to become the third-largest city in China in terms of GDP. That year the Guangdong-Hong Kong-Macao Greater Bay Area was established. Shenzhen was transitioning from economic development to the accumulation of social capital and from a rapid spatial construction mindset in which "time is money, efficiency is life"[10] to an era of cultivating cultural and intellectual life. At the same time, redeveloping old villages through complete demolition and rebuilding was beginning to slow down. In 2016, the *13th Five-Year Plan for Urban Renewal in Shenzhen* proposed reducing 10 to 12 million square meters of the area covered by illegal buildings through urban renewal within the planning period (2016-2020). At the same time, it also advocated for organic renewal by relying mainly on comprehensive remediation supplemented by demolition and rebuilding.[28] The *Interim Measures on Strengthening and Improving the Implementation of Urban Renewal* stated that attention should be paid to the preservation of urban memory and the promotion of organic renewal in the urban renewal process. Also, the *Measures for Improving the Housing System for Talents* proposed to fully utilize existing housing resources and support rural collective economic entities to transform built tenements with incomplete legal status into rental housing for talents in compliance with the law. As part of these measures, the district government would provide an adequate amount of compensation for renovation and operation. At this point, urban village renewal related to the development of urban culture and everyday life was no longer a technical issue, but an urban issue with profound social significance that would deeply impact the future, changing Shenzhen's model of urban development.

Yang Hong [11] highly appreciated that the 2017 Bi-City Biennale of Urbanism\Architecture was planned to be held in an urban village. He believes that urban villages form a big family that houses the daily life of more than 7 million citizens. It can bring stories about Shenzhen's modern history, the history of reform and opening up, and the everyday life of ordinary citizens to a larger audience.[16]

In 2010, Nanshan District issued the *Nantou Old Town Preservation Plan 2010–2012*, which put forward the concept and principles of preservation planning, while integrating protection and development and highlighting historical and cultural themes. During the deliberative session for the plan, district leaders opined that Nanshan District would set up a leading group to advance the overall regeneration of Nantou Old Town, reinforce the protection of cultural heritage, and demolish illegal buildings.[17] However, nearly 90% of the buildings were classified as illegal, meaning that the proposed demolition area was quite large. Moreover, these buildings were home to more than 30,000 migrants. The plan to create a themed historical district required the relocation of 800 local residents, without taking into

10 "Time is money, efficiency is life" is a common phrase used as a slogan of Chinese economic reform quoted from Yuan Geng, an early proponent of the reform and opening up.

11 At the time, Yang Hong was a member of the Shenzhen Municipal Standing Committee.

account where the 30,000 migrant tenants would go.[17] Overnight, the "promise of an authentic ancient city"[18] became the point of view of the mainstream media. The 2012 *Research on the Implementation of the Nantou Old Town Preservation Plan* continued the concept of "onsite protection and localized remediation" to explore implementation strategies and paths, and it also put forward the proposition of "spatial replacement, cultural reconstruction." The recommended renovation proposal had significant limitations both conceptually and practically, calling for the demolition of 27,000 square meters of factories and 16,000 square meters[19] of residential buildings in the aesthetic control zone. According to average housing prices at the time, the compensation for relocation was estimated to be nearly 400 million yuan. In fact, in the past decade, due to complications and issues with the historical legacy, like the *Nantou Old Town Preservation Plan*, many research, planning and design proposals have remained unrealized. Many proposals have attempted to redevelop the cultural and tourism resources of the one-thousand-year-old town, focusing on its historical significance as the origin of Shenzhen and Hong Kong, while overlooking its contemporary value as the residence of more than 30,000 inhabitants with their own lifestyle and culture.

In 2016, embracing the dilemma of the development of an urban village and an ancient town, URBANUS was commissioned to conduct a new round of research and design on Nantou with an eye toward game-changing innovation. At the beginning of their research, the URBANUS team raised a question related to the future direction of the city: How can Nantou be accurately located while maintaining both its historical and contemporary heritage? As opposed to other urban villages, a millennia-old town and a lively urban village collided and integrated in Nantou Old Town. Thus, the value of urban villages as living contemporary heritage and Nantou's positioning as a millennia-old town were mutually constraining, limiting its development. Also, unlike Old Hubei Village, Nantou Old Town has experienced the ups and downs of history, leaving few historical remains. More than 80% of the buildings in Nantou Old Town were self-built buildings that had been constructed after the establishment of the SEZ. As an urban village, the hybrid urban life of Nantou was about to disappear, while as a heritage site, the millennia-old town had been reimagined to relive specific parts of history over and over again in the previous preservation plans.

By studying the historical context and urban fabric as well as analyzing cases and iterations of concepts of conservation projects at home and abroad, URBANUS positioned the regeneration of Nantou Old Town as follows: "The history of Nantou Old Town reflects a condensed and complete urban history of Shenzhen. Rather than a spatial renovation, the more needed efforts are to improve the quality of life of its inhabitants and to rejuvenate the local culture...only by respecting the authenticity of its history and cherishing the cultural layers and historical traces of each period of time can we shape a timelessly dynamic urban community rooted in local history and culture. We see today's Nantou not as an old historic town in the traditional sense, but as a historical heritage town, which carries on the history and culture of nearly two thousand years, and which preserves the spatial, social and cultural heritage of Shenzhen across every historical period. It is the only precious sample of Shenzhen's urban culture that displays both historical relics alongside a full record of China's rapid urbanization over the last three decades."[20] URBANUS suggested connecting history and the UABB in the proposal "Nantou Old Town Conservation and Regeneration." URBANUS partners, Meng Yan and Liu Xiaodu were appointed the chief curators of the 2017 UABB, choosing the theme "Cities, Grow in Difference." (Figure 6) They called for diversity, while acknowledging the value of urban villages being contemporary heritage sites in terms of space, lifestyles and local culture. As an intervention through cultural events, 2017 UABB's goals coincided with URBANUS' urban design to rejuvenate Nantou through incremental revitalization impacting larger clusters with micro-scale intervention, as well as organizing cultural events.[21] Indeed, the UABB had focused on Shenzhen's history and the contributions of the urban village to the city in the 40 years since the opening and reform.

No demolition of authentic remains, no construction of fake heritage.[22]

It took great courage and an adventurous spirit to experiment with large-scale cultural events as an incentive for the redevelopment of Nantou, which had been underdeveloped over the years due to issues with its historical legacy.

According to Meng Yan, the chief curator, "Just having the biennale held in Nantou should be considered a success." Through in-depth research on and analysis of Nantou Old Town, the tailored urban design for regeneration coincided with the spatial narrative of the UABB venue. With the traditional intersection as its spatial focus, Zhongshan Park, the Old Town and the Park outside the south gate were reconnected by opening up the north and south of Zhongshan Street. Also, a series of public space nodes along the street were selected to be renovated as venues for the biennale, while none of the residential buildings were demolished. The curatorial team proposed to have the biennale coexist with the operating factory, making adjustments for it in the design. The team tried to minimize interruptions to the everyday life of the inhabitants of Nantou Old Town, while maximizing the improvement of the spatial experience of the public space. The team also left space for the continuation of people's lifestyles and possible future development. At this point, the origin of Shenzhen and Hong Kong, the site of a millennia-old town, a living urban village and the exploration of the biennale have collided at Nantou Old Town.

Cities, Grow in Difference went beyond an exploration of spatial transformation; it was an unprecedented social experiment that attempted to mix diverse lifestyles and was a big adventure that celebrated the clash of ideas about the future of cities. As an international academic exhibition, the UABB brought together exhibitors and cultural resources from all over the world. Their short stay in Nantou was like a stone thrown into the water causing brief ripples. The Austrian exhibitor, Ton Matton and his team mentioned that while communicating with the local inhabitants, the most frequently asked question was "Why do you like living in our Nantou Old Town?" Migrant tenants had been pushed into Nantou for affordable living, whereas the exhibitors chose Nantou as a destination that provided a different perspective, which led them to rethink the things they liked about living in Nantou.

4 How to think about gentrification in urban village regeneration?

There is no doubt that the public space renovation, influx of cultural resources and displacement and indirect loss of local business in Nantou would be labeled as "gentrification." However, when the 2017 UABB is viewed from the perspective of Shenzhen, where the city's demographic was transforming from laborers to young white-collar workers, an improvement in the living environment of urban villages was inevitable. The biennale was an experiment in attempting to slow down gentrification through incremental improvements. Gentrification seems to be written in the genes of urban village issues, but in fact, the socio-political and economic issues behind the concept of gentrification are different from the issues that the urban village faces. In 1964, sociologist and urban planner Ruth Glass proposed the concept of "gentrification" for the first time, defining it as the urban phenomenon of re-urbanization after counter-urbanization.[12] In the following decades, due to certain U.S.-specific socio-political and economic issues, gentrification became a term associated with the displacement of low-income inhabitants, especially people of color. This tendency has accompanied the entire urban revival history of American cities. In the past five years, many major cities in the United States have introduced policies to protect the housing rights of low-income inhabitants. For instance, New York adopted a *Mandatory Inclusionary Housing Ordinance* in March 2016 that required developers to include affordable housing in development.[23] Also, non-profit organizations, such as "NEXT CITY," have sought urban strategies to counter the negative effects caused by gentrification.

In the urban context of Shenzhen, the discussion of gentrification has focused on issues such as population displacement, insecurity in the right to housing, uniform large-scale planning and the loss of social diversity (affordability), with the core idea that "developing is an unyielding principle."[13] A second phase of urbanization began at the fortieth anniversary of the city's establishment in the absence of a counter urbanization movement. A gradual shift in demographics and industrial structure resulted in Shenzhen's transformation from being the "world's factory" to the "design capital"[24] and from the "demographic dividend" to the "talent dividend."[25] In other words, rapidly developing cities have become scenes of intense gentrification both at home and abroad. Therefore, to protect the right to the city, the government has been committed to establishing and improving the system through discussions and experiments. In fact, the urban

12 Counter-urbanization was proposed by Brian Berry in the 70s, opposing to urbanization, the process of urban population relocated in rural area.
13 "Developing is an unyielding principle" a popular slogan of China's economic reform quoted Deng Xiaoping's speech in the 1992 Southern Tour.

village could be considered as the informal line of defense established by the people to protect their right to housing when policy and systemic action are absent. Therefore, the renewal of urban villages and questions about gentrification have become a dilemma in public discourse.

Urban villages have grown rapidly with the development of Shenzhen, and the development of urban villages should not be stopped out of a fear of gentrification. Keeping up with the overall development of Shenzhen is the key to the survival of urban villages. Otherwise, the undermining of the potential economic and cultural value of urban villages will result in replacement by brand-new real estate office towers and shopping centers due to their relatively low price for developers. Real estate development follows the principle of efficiency and standardized production, which involves the use of one set of construction drawings to build a hundred buildings. The fast and simple completion of the process of land purchase, design, construction, and sale in urban renewal tends to overlook the context, everyday life, local cultural confidence as well as social and cultural diversity. The process of gentrification can only be slowed by increasing urban villages' value incrementally and avoiding overnight gentrification due to demolition and reconstruction as part of urban renewal. Compared with powerful capital investment, the UABB relies on cultural forces and administrative power. In the system of an urban village in which various forces balance each other, the intervention of the UABB introduces new forces to mobilize stakeholders to generate conversations and search for a new equilibrium within the context of future cities. Each of the forces is mutually constraining to prevent the intervention from overpowering the conversation.

The regeneration of exhibition venues by URBANUS serves as a paradigm for an interventional experiment to transform urban villages. Both the top-down and bottom-up interventions during the 2017 UABB tested the flexibility of the system of urban management, starting with the Nantou Model as an exploration of an alternative future model. In the curation of the "Urban Village Laboratory," the biennale served as a platform to encourage communication between landlords and designers, allowing the system of urban management, spatial design and personal needs to interact. Together they sought a balance between the forces of personal and administrative intervention as well as inhabitants' interests. They also exposed the limitations of a system that has lacked the refinement needed to correspond to different practices. The biennale exposed the limitations of a single-mode urban management system, attempting to tailor new approaches according to the rules of urban villages. It provided valuable working experience for making policies with respect to organic renewal, while promoting the understanding and respect of individual value through empowerment by cultural events.

> *At present, urban renewal in China has transitioned from focusing on material aspects such as demolition and rebuilding to an organic renewal stage, which focuses on the reconfiguration of public space, industrial reconstruction, historical and cultural inheritance, as well as social and people's livelihood improvement.*[26]

As an "experimental field"[14] of reform and opening up, Shenzhen has shouldered the important task of exploring the future of cities. Vibrant urban villages provide an antidote to the monotony of contemporary planned cities, begging the question: What if the motivation for urban development in the future was no longer based on the desire for financial capital, but was also driven by the accumulation of social and cultural capital? Is it possible to abandon the real estate-based development model and replace it with a more diverse and bottom-up development model? The UABB is only a beginning rather than an end to an experiment with a development model in which everyone can be a developer that breaks old development patterns. Knowing this is a multi-disciplinary problem that goes beyond architecture and requires multiple perspectives from different entities, the curatorial team of the 2017 UABB tried to promote the establishment of the co-construction platform of Nantou Old Town through the "CGD Lab" to trigger and guide the continuous regeneration of the neighborhood in the post-biennale era, calling on city administrators, builders, scholars, residents and other stakeholders to work together for a way forward.

5 What is the ideal model for future cities?

In March 2018, the closing ceremony of the 2017 UABB was held at Dajiale Stage in Nantou Old Town, marking Nantou's post-2017 UABB era. In

14 A common expression of Shenzhen's role in China's economic reform.

March 2019, the Planning and Natural Resources Bureau of Shenzhen Municipality issued the *Shenzhen Urban Village (Old Village) Comprehensive Remediation Plan (2019–2025)*, which proposed to promote the organic renewal of urban villages by retaining a certain number of urban villages during the planning period and guide all districts to carry out comprehensive remediation of urban villages.[15] As the *Southern Metropolis Daily* reported, "For the urban villages in Shenzhen, at least within a short amount of time, comprehensive remediation begins. In the future, how to transform areas where 2/3 of the city's population live remains a test for leaders."[2] In October of the same year, the *Working Plan on Promoting the Comprehensive Remediation of the Historical and Culture Protection of Urban Villages and Creating a Distinctive Style* listed seven urban villages, including Nantou Old Town, that would be developed by Vanke as micro renewal pilot projects to explore a new model for urban village governance. In 2019, Nantou Old Town, being the origin of Shenzhen and Hong Kong, was included in the *Implementation Outline of the Guangdong-Hong Kong-Macao Greater Bay Area*.[16] At the same time, a groundbreaking ceremony was held for the Hubei district renewal project, while the organic renewal of Old Hubei Village was ready.

In the years since 2005, when the first old village redevelopment provision was promulgated, the contemporary value of urban villages has been recognized. However, the means and methods at the level of implementation remain insufficient. The offcial WeChat account of the Urban Planet Research Institution established by the Urban Research Institute of China Vanke has pointed out the limitations and challenges of bottom-up regeneration, adopting an intervention strategy that promotes unified renovation along with demographic transformation.

> *"Nantou Old Town, an urban village dominated by migrants...Most of the owners of the houses who benefit from the renovation do not live here, while the majority of inhabitants here who desire to improve the environment do not benefit from the increasing value of the properties, not to mention their ability to afford the cost of renovation. Therefore, relying solely on social forces is difficult to realize bottom-up spontaneous regeneration and mobilize current inhabitants...effective brownfield redevelopment should start from a "demographic dividend" by introducing a creative class that is aware of urban spaces and allow them to stay."*[27]

However, saying that it is "difficult to realize" should not be a reason to deny the possibility of the realization of incremental brownfield redevelopment that was explored during the 2017 UABB, including placemaking and community building. Furthermore, culture building should not take the market's approach to look for shortcuts. For example, using slogans such as "respect the authenticity of history," "inherit the memory of street life," and "reflect the spectrum of the 40-year architectural history of Shenzhen since reform and opening up,"[28] while using an away of unrelated architectural styles from Lingnan, the 1930s and the petite bourgeois style skews historic perception and how the good life is imagined. Selling a so-called exquisite lifestyle can easily lead to a simple and fast approach to the cultivation of Shenzhen's cultural and intellectual life of Shenzhen in ways that resemble rapid urbanization in previous eras. As a result, "layers of history" and "diversity" have become gimmicks for creating photogenic places for social media overnight and have not allowed transplanted cultures to survive and grow.

In 2020, as one of the invited architecture groups to design the core area of Nantou, URBANUS returned to Nantou Old Town. At the meeting of the Core Area Group Design, URBANUS proposed to establish a community co-construction platform during the preservation and renewal of Nantou Old Town for the co-existence of new and old architecture, inhabitants and culture.[29] They held that the vibrant community groups and local culture in Nantou Old Town, discovered during the 2017 UABB through small-scale interventions and community co-construction, should be empowered and developed as part of the preservation and renewal of Nantou Old Town, not only preserving the space, but also slowing down the pace and protecting the local lifestyle from rapid urbanization. After major investment into the renovation of the main streets, once and for all a

15 During the planning period, the size of the designated comprehesive remediation area is 55km^2, which is 56% of the renewal area. Futian, Luohu and Nanshan Districts should designated comprehensive remediation area no less than 75% of the renewal area while other districts no less than 54%.

16 Courtesy translation of *Outline Development Plan for the Guangdong-Hong Kong-Macao Greater Bay Area*. https://www.bayarea.gov.hk/filemanager/en/share/pdf/Outline_Development_Plan.pdf

图 7 都市实践改造后的杂交楼，张超摄于 2022
Figure 7 Renovated Hybrid Building by URBANUS, photograph by Zhang Chao, 2022

图 8 都市实践改造后的杂交楼鸟瞰图，张超摄于 2022
Figure 8 Bird's eyes photo of the renovated Hybrid Building by URBANUS, photograph by Zhang Chao, 2022

杂交楼

relevant policy must be developed to encourage sustained social reaction and spontaneous regeneration at multiple levels. The idea that Shenzhen is a "cultural desert" with "no history" has cursed over 40 years of urban development. Yet Nantou is no panacea. Only with time to grow as well as the assurance, guidance, the cultivation of a community platform, and awareness of local cultural confidence can Shenzhen continue the inclusive co-construction spirit of its slogan "*If you come, you are a Shenzhener*."

As of 2021, the Nantou Hybrid Building appears out of place with the renovated buildings of South Street. The Hybrid Building was one of URBANUS' experiments in urban village co-evolution and coexistence of the new and old on an architectural scale instead of on an urban scale. In an age dominated by a vision-centered paradigm and hegemony of vision, a transition from language-based culture to image-based culture is indicated.[30] Due to consumerism, the city and its architecture have become commodities that rely on visual perception and imagination. The invasion of the single-minded symbolization of aesthetics in everyday life makes the concepts of "good-looking" and "ugly" deeply rooted aesthetic standards. The Hybrid Building, like other self-built buildings in urban villages, pushes back against such single-minded aesthetic criteria under global consumerism. At the junction of Zhongshan South and East Street, five buildings are tightly clustered together in different heights and styles, and they cannot simply be described as "good-looking" or "ugly." The richness and diversity of the buildings represent an age filled with emotions, while the contradictory and conflicting architecture volumes showcase the tension of the vibrant south.[31] With acknowledgment and respect to the original building logic (Figure 7), the interior circulation is recreated through "repoussé and chasing"[32] to accommodate the needs of contemporary living. The maximum reuse of the exterior retains "the trace of additions and transformations as well as replacement of materials. Two new volumes, one rectangular and one curvilinear, one solid and one void, are juxtaposed on the rooftop (Figure 8), with a small roof garden hidden within."[32] The intervention by the architect does not make the hybrid building worldly. Instead, its vibrant character is enhanced. In terms of architectural practice, this is an unprecedented case of learning from urban villages and represents a manifesto of a rebellion against exquisite aesthetic criteria in an image-driven culture. Also, it is a paradigm of a gradual urban village regeneration that preserves the layers of history.

The Nantou Hybrid building does not represent the only way forward for the regeneration of urban villages. It is more pressing now than ever that cross-disciplinary cooperation occurs between the decision-makers of the city, scholars, practitioners and stakeholders. By establishing a mindset that is different from the real estate-based model of urban renewal, the exploration of individual value, inheriting lifestyles and the possibility of everyone being developers will lead to a vibrant space and a program in which diverse values coexist. There is never a shortcut to an ideal. For Shenzhen to remain a city of pioneers, it needs practitioners who will never give up their delirious urban practices.

[1] 深圳市统计局，国家统计局深圳调查队 . 深圳统计年鉴 2020[M]. 北京 : 中国统计出版社 , 2020.
[2] 孙雅茜 . 深圳城中村 : 从大拆大建到综合整治 - 栖息着城市 2/3 人口的城中村在新的城市化进程中将如何嬗变 [N], 南方都市报 , 2019-5-28.
[3] 梁永健 . 深圳蔡屋围钉子户得到 1700 万天价补偿 [N]. 南方都市报 , 2007-9-30.
[4] 南方都市报 . 圈地新思维 [M]// 南方都市报 . 未来没有城中村 . 北京 : 中国民主法制出版社 , 2011: 36-37.
[5] 董超文 . 深圳房价 2005 年涨幅创历年之最 [N]. 中新网 , 2006-1-30.
[6] 都市实践 . 村 · 城 城 · 村 [M]. 北京 : 中国电力出版社 , 2006.
[7] 孟岩 . 深圳 · 中国梦想实验场——2010 上海世博会深圳案例馆 [J]. 城市环境设计 , 2018(12): 314-319.
[8] 孟岩访谈 . 在深 21 年 : 都市实践 [J]. 建筑实践 , 2020(11): 90-101.
[9] 秋风 . 理解“城中村”现象 [M]// 南方都市报 . 未来没有城中村 . 北京 : 中国民主法制出版社 , 2011: 9-13.
[10] 都市实践 . 湖贝古村保护研究与“湖贝 120”公共计划 [J]. 城市环境设计 , 2018(12): 296-300.
[11] Travis Bunt. 白石洲五村城市更新研究 [J]. 城市环境设计 , 2018(12): 290-295.
[12] 绿景中国新闻中心 . 白石洲项目 80% 股权注入绿景 [EB/OL]. (2020-8-25) [2021-12-20]. http://www.lvgem-china.com/zh-hans/news-hans/2020/08/25/6229.html
[13] 晏婵婵，崔欣，陈博，郭锐川，谢湘南 . 湖贝旧村非文保单位 年底将启动一期清拆 [N]. 南方都市报 , 2016-7-5.
[14] 都市实践 . 湖贝大事记 [EB/OL]. [2020-11-10]. http://www.urbanus.com.cn/uabb/urban-village/hubei-timeline.
[15] 叶裕民 . 特大城市包容性城中村改造理论架构与机制创新——来自北京和广州的考察与思考 [J]. 城市规划 , 2015(8).
[16] 苏妮 . 第六届深港城市 \ 建筑双城双年展闭幕 [N]. 南方日报 , 2016-2-29.
[17] 郑恺 . 南头古城酝酿整体整治 [N]. 深圳商报 , 2013-9-4.
[18] 周昌和 . 许你一个真正的古城 [N]. 南方都市报 , 2011-12-12.
[19] 北京大学中国城市设计研究中心 , 中营都市设计研究院 .《南头古城保护规划》实施方案研究 [R]. 2014-2015.
[20] 孟岩 . 策展南头 : 一个城 / 村合体共生与重生的样本 [EB/OL]. (2017-11-4) [2020-11-15]. http://www.urbanus.com.cn/writings/nantou.
[21] 孟岩 , 林怡琳 , 饶恩辰 . 村 / 城重生：城市共生下的深圳南头实践 [J]. 时代建筑 , 2018(3): 58-64.
[22] 中华人民共和国住房城乡建设部 . 关于加强历史建筑保护与利用工作的通知 (建规〔2017〕212 号) [EB/OL]. (2017-9-20) [2020-11-15].https://www.mohurd.gov.cn/gongkai/fdzdgknr/zfhcxjsbwj/201709/20170922_233378.html.
[23] New York City Council. Mandatory Inclusionary Housing[EB/OL]. (2016) [2020-11-15]. https://council.nyc.gov/land-use/plans/mih-zqa/mih.
[24] 印朋 . 从“世界工厂”到“设计之都”——文化力赋能“深圳制造”[N]. 新华社 , 2020-9-13.
[25] 李凯 . 促进人口红利向人才红利转变 [N]. 人民日报 , 2021-12-31.
[26] 何燕燕，王岩波 . 我国城市更新项目将进入有机更新阶段 [N]. 新华社新媒体 , 2018-7-27.
[27] 恒铭 . 这些年，被埋没和误解的南头古城 [Z] . 城市星球研究所 , 2020-9-29.
[28] 丁侃 刘丹丹 . 深圳万科 追寻南头古城记忆 打造特区文化新名片 [N], 南方日报 , 2021-6-11.
[29] 都市实践 . 南头古城核心示范区集群设计工作坊共识与事项 [Z]. 2020-3-23.
[30] 王书东，徐荣伟 . 影像的视觉霸权 [J]. 青春岁月 , 2011(18): 33.
[31] 孟岩 . 城村共生中国城市的另一种图景 [EB/OL]. (2019-11-23) [2020-12-10].http://www.urbanus.com.cn/writings/selftalks-meng-yan.
[32] 都市实践 . 南头杂交楼 2022[EB/OL]. [2020-12-10]. http://www.urbanus.com.cn/projects/nantou-hybrid-building.

城 / 村计划进行时

罗祎倩
都市实践研究员
图片 © 都市实践

1 城 / 村计划源起

2020 年，深圳市成立四十周年之际，一项针对其两个重要组成部分——城市与城中村的“城 / 村计划”被正式提出。立足于公共性、学术性、实践性，总召集人、总策划人周红玫邀请核心的策展团队，包括总策展人孟岩，联合策展人刘珩、何健翔、赖希圣组成内环，以及学术委员会组成外环。

这一计划在“时不我待，只争朝夕”的重要时空节点中成形，实际上，它萌生于十数年的研究与实践中。2005 年首届深双即组织城中村专题，展出岗厦村、福新村、大新村改造提案。2010 世博以大芬城中村向全球介绍深圳，这种城市的异质存在引起国内外的大量关注。随后，城市中心尤其是主干道深南大道两侧的多个城中村，如蔡屋围、岗厦、白石洲、大冲等，在旧改中引发激烈的讨论。2017 深双，主展场南头进入大众视野，正面回应古城和城中村双重身份在当代城市中的角色，其后深圳城中村改造的相关政策发生转向。尽管其未有官方名称，“每一步实践都是城 / 村计划的一部分。”[1]

2 后 2017 深双时代的原关外时区

广及全国辽阔的疆域，小至深圳不足 2000 平方公里的市域，城市化进程存在着明显的“时差”。[2] 2010 年，深圳经济特区扩容至全市，方将位于原二线关外的宝安、龙岗两区纳入特区范围。而在（作为行政区划的）深圳市此前三十年的城市历史中，其城市化发展被二线关边界划分为关内和关外两个“时

图 1 “城 / 村计划：龙岗六村实践展”海报，烁设计
Figure 1 Poster of City/Village Initiative: Practice Exhibition of Six Urban Villages in Longgang District, SURE Design

区”，它们实行各自相应的政策和规定，不尽同步。在特区一体化后的十年间，“时区”的校准深受这段发展历史及其结果——空间结构和社会结构的影响。

在时间差异中，关于深圳城中村的研究和实践焦点有着多重的地域转移。从因城市用地紧张、功能划分等因素，原关内的数个村中村引起人们关注；到在世博会上构建深圳身份，原关外龙岗区的大芬村将城中村议题推向国际学术讨论的前沿；再到由于深圳进入存量发展、产业转移、形象塑造、房企逐利等多种动因，原关内毗邻城市中心的城中村不可避免地成为政治和资本的博弈场，亦成为社会热点。2019 年，已被旧改新闻环绕十余年的白石洲启动清租，这意味着原关内城中村命运的终结——村让渡于城，原本在此盘旋的各方逐渐转移阵地。与此同时，位于原关外的龙岗区分布着 499 个城中村，

1 周红玫，访谈，2021-3-30。
2 孟岩，访谈，2020—2021。

它们有着各自的起源和现状，因开发潜力较低而较少的强势介入，既面临着发展的机会，也面对着衰落的危机。在新的语境下，龙岗区重新成为城中村议题的地域焦点，如今我们讨论的不仅是大芬村所代表的城中村集合，而且是多种类型、多种属性的成百上千个城中村个案。

在后 2017 深双时代，城中村成为深圳城市发展研究和实践探索不再回避的议题。对于原关外的大量城中村，一方面，它们的概念仍有待厘清，其中有些未被通常意义上的城完全包围，紧邻山脉和水系；有些既有着通常意义上的城中村肌理和建筑，也有着城市中罕见的农田、围屋、未建设的宅基地；有些既是改革开放后外来移民的居所，也是历史上客家移民的聚落。另一方面，它们的价值被不断地定义，在中国特色社会主义先行示范区建设、大湾区发展、城市更新、保障性住房体系构建、房企寻求新的利润增长点等任务中扮演着不同的角色。尽管关于城中村的描述已从单一声调逐渐转向众声喧哗，但进一步地，何为症结，何为出路仍然错综复杂。

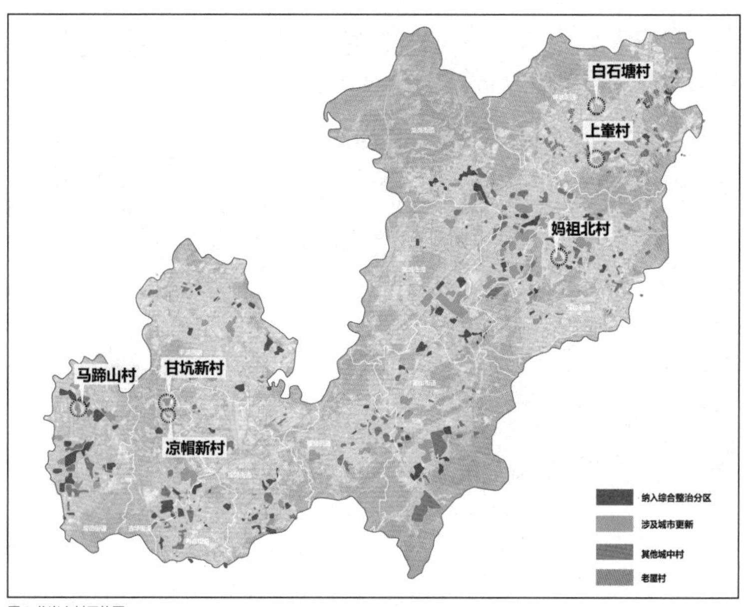

图 2 龙岗六村区位图
Figure 2 Location of the six selected urban villages in Longgang District

3 城 / 村计划机制

基于紧迫而关键的现实语境，城 / 村计划的首站落点龙岗区，就如周红玫所言，其在于“拥抱创新思维以挑战惯性思维、重新定义问题并提出创造性的解决方案”。城 / 村计划涵盖城市研究、城市设计与策展提案、试点实施、范例推广。如同深圳过往创新建立的公开竞赛、深双、新校园计划等新模式，城 / 村计划采用城市策展作为公共事件介入旧城更新，即在现行政策和规定下突破僵硬的管理边界，寻找合法的突破机会。通过展览媒介，呈现和探讨城市问题，公开展示解决方案，以充分考虑专业评论和公众意见，从而确立更为均衡的价值和评价体系、引进更为全面的专业资源和公众参与、形成更为强大的创新动力。如此，城中村更新从城市的公共性（社会与环境效益）及城市议题的学术性出发，避免被片面的追求政绩或商业效益误导甚至扭曲。同时，城 / 村计划试图对代建机制进行创新，将策展人 / 设计师、参展建筑师和艺术家的集群名单作为代建实施主体招标的重要条件，以此保证整体协调。

城 / 村计划内环的策展团队作为城中村改造的设计团队，需提出概念、整合思想、梳理内容、主持研究和设计、整合推动参与资源；还需指导、协调不同专业学科的研究和设计，组织、管理和监督实施最后的成果。与此同时，外环的学术委员会作为智囊团和评审团，确保项目的学术性定位；也作为一种常态化的公共参与机制，为专业人士基于公共立场介入城市规划提供路径，以维护公共利益。

基于此机制，“城 / 村计划：龙岗六村实践展”（图 1）成为首站的切入点，对试点城中村进行由点及面渐进式的激活提升。在形成更新机会的同时，通过艺术事件和公共活动，吸引全球文化资源注入城中村区域。由此，使文化与日常生活融合共生，创造新型城市人文景观。试点城中村的选址基于以下原则：① 2025 年前不涉及城市更新或拆迁计划的城中村；②交通便利、楼栋数量适中；③邻近城区品质较好；④具备一定的公共空间。最终选取白石塘村、甘坑新村、凉帽村、马蹄山村、上輋村、妈祖北区及周边（图 2）。实际上，它们是六类各有特点的城中村，每一个类别代表龙岗区内数十个情况类似的城中村。因此，在试点展开全方位的机制创新，可以在相对可控的范围内更具前瞻性、实验性、可行性，进而将试点转化为示范，推广其成功经验。

4 城 / 村计划：龙岗六村实践展

“城 / 村计划”作为方法围绕着“城 / 村共生”理念开展，是以另一种发展方式探索中国城市的未来图景。对于龙岗六村实践展，策展人以“城村栖居，多维生长”为主题，发出如下宣言：

城中村的迭新是以村的异质抵抗城的同质；
城中村为当代城市的“同我”；
城 / 村是未来中国城市栖居的别样类型。

在此共识下，总策展人孟岩，联合策展人刘珩、何健翔、赖希圣带领各自的团队都市实践、南沙原创、源计划、创始点开始试点六村的城市研究与设计工作，清华苑团队进行上位规划及现行政策梳理，共同从多个层面进行愿景重构和落地推进。最终通过介入实践展、主题文献展、小型展览和微介入计划三大展览呈现给社会各界。如此，龙岗六村本身即展场，展场即展品，观展体验与日常生活共构一种混合的现实。

图 3 城 / 村计划工作组在甘坑新村调研，璀璨年华摄于 2021 © 都市实践
Figure 3 Survey in New Gankeng Village, City/Village Initiative working group, photograph by Cuican Nianhua, 2021 ©URBANUS

4.1 城村研究：多维问诊

城 / 村计划以观察和发现切入六村研究，通过实地调研、非遗调查、乡贤对谈、企业访谈、工作坊等（图 3），梳理各村的基本情况。六村分别位于龙岗区的四个街道，各有特色，周边环境和上位规划各异。根据各村情况，城 / 村计划提出各自相应的六村发展定位和方向，试图针对性地解决问题，并为整体提升建立基础。

甘坑新村（图 4- 图 5）与凉帽村紧邻国家 5A 级旅游景区甘坑客家小镇，前者于原地重建，如今仍保留着客家舞麒麟的传统，可从文化符号消费转向创意生产力哺育；后者得名于手工凉帽生产，村既是厂，厂既是村，两面环山，可连接自然山水与文化生产，历史记忆与日常生活。马蹄山村紧邻“特区扩容示范区”坂雪岗科技城，被主干道包围，交通便捷，人流量大，可通过社会资本在此催生新型城村形态。白石塘村（图 6- 图 7）被劳动密集型制造工厂包围，老村位于新村北面，新村南面又被城市干道一分为二，“厂中村”可从落脚城市转向赋能村落。上輋村北临龙岗河，三面环山，整体规划为林地，内部有着多类建筑及遗存，应渐进发展，将设计作为一种助推力量。妈祖北区及周边有着高度混杂的多种“像素”，包括大田世居、农耕地、文体设施、工业园、城中村，可作为“万像”城，以拼合产生多元，并置创造机遇。由此可见，村村属性不同，面临的机遇和挑战不同，“一刀切”的更新模式无法扬长避短，反而会铲除有利要素，甚至制造新的问题。

实际上，六村均有数百年历史，而在近二十年因城市发展需求和居民对其的自发应对发生了剧烈的变化。以白石塘村为例，村庄始建于清朝，三面环山，南面为山间谷地，西面流过丁山河。初时，三堂屋布局的围屋建成，中轴线依次为月池、门厅和曾氏祠堂，因在挖水塘时挖出两块白色石头而取其吉利寓意为村名。而后随着人口增加，围屋扩建，并兴建萧氏祠堂；再随着部分村民搬离，围屋局部拆除，其南面新建房屋。直至 2003 年（全市城市化前夕），老屋南面的新村为低层楼房，仍有着南北贯穿的乡道、老屋前的月池、山脚的风水林。2008 年（深圳大运会举办前夕），新村楼房已进行加建，月池消失，风水林缩小，周边扩建工厂和修建盐龙大道。2021 年，原本的乡道被城市干道割裂，拆迁居民集体安置在村外西面的小区，成为一村三地。与此同时，在经历普遍的村落扩张和自然衰退后，白石塘村仍然保留着宗祠、土地庙、妈祖庙等历史线索；节庆时，麒麟队先在祠堂祭祀，再到各家拜访，并随着村民搬迁而修改路线，甚至应邀至 15 公里外的村民新居所。在某种程度上，白石塘村以文化活动维系着村的精神连接，以抵抗空间的割裂、区隔和衰落，突破了物理边界的限制——城中村已自发以自身的特质缓解城市问题。

通过六村采样，可见林林总总的问题相互交织，牵一发而动全身。抵抗“推倒重来”和“置之不理”两种极端做法，城 / 村计划试图建立一种“共生模式”，同时保障城村居民的生存权与发展权。期间，试点项目既要进行空间改造以对区域进行定位和提升，也要避免权属所有者借助更新的“东风”进行违章抢建。

梳理现有的实施路径，城 / 村计划对不同层面的政策和规定提出以下修正建议，以保障城村提升的有序推进。在宏观层面，以 2021 年颁布实施的《深圳经济特区城市更新条例》为前提，结合各相关职能部门，开展“综合整治 2.0 版”实施细则制定，作为政策引导。在中观层面，上位规划缺乏城村发展可能时，对城村给予针对性指引；上位规划与城村发展冲突时，重点标注备案城村情况，进行法定图则调整；并借助法定图则标准单元划定及国土空间规划，对现有规划展开检讨与修正。在微观层面具体操作实施时，梳理土地权属，对于新建建筑及增设的公共空间用地，可由政府与村集体或个人签订征转地协议后出让给股份公司；梳理物业权属，明确局部改造的建筑物的合法性，对非法建筑进行“新三规”处理，使之合法化。

城 / 村计划试图从公共利益切入空间设计，同时引入社会设计，由此提出三个面向的基点：①公共空间再定义：链接过去与未来的社区生活；②空间生产再创造：不仅是空间产品的创造过程，而且是社会关系的再创造过程；③空间发展再平衡：保护历史，传承记忆，并且寻求新机遇进行良性可持续的新发展。从跨学科的思路进行城村改造，从单体和局部的改造开始为整体区域的提升建立基础，避免“头痛医头，脚痛医脚”。城 / 村计划作为新的空间和文化战略，试图寻求当地自然生态、文化生态、社群关系、经济发展并行不悖的模式。

由此，城 / 村计划提出四类场所原型。第一类，城村新祠堂，思考各村的文化内核，从记录失落的城村故事到重塑各村的文化内核，旨在通过仪式性空间的构建、围绕新文化核心，展开关于日常生活和心灵归属的功能以重构当代城市精神空间。以甘坑新村为例，将麒麟馆和周边广场构建为文化内核和社区服务的复合体，包含麒麟剧场、博物馆及训练场、村史馆、图书馆、吹水台、室外展场等功能空间，也作为舞麒麟路线的节点（图 8- 图 9）。第二类，城村综合体，思考城村土地规划和使用，从平面式土地规划到立体精细化空间与运营规划。第三类，城村新市政，思考城村市政建设与管理权，从城村分离的市政建设管理到共建的动态化城村物业。第四类，城村新公社，思考城中村品质提升驱动力，从自上而下的资源导入到赋能社区自建能力实验室。新公社包括新社会关系的建立和赋能个人，产消结合的交换行为，交换行为的空间构建，空间组织及其资源链接，最后通过社会资本的累积提升文化资本。以白石塘村为例，在新老村交界打造能力实验室聚落，串联起创意沙龙、创意工坊、白石书屋、在地能力学堂等数个赋能工坊，并在其中分布创作间、图书亭等在地试验田和苗圃、花园、攒水、艺术装置等社区实验场（图 10）。城 / 村计划亦针对政策条例和商业模型分别提出城村发展共同体和城村新经济原型，从不同方面推动机制创新。

4.2　城村设计：共生模式

在研究工作的基础上，城 / 村计划对六村品质提升的设计工作由策展人带领的设计团队开展。各个团队以自身最擅长的方法，结合各村肌理和品质，提出各村的特定策略和空间改造设计——“精准打击”。各团队严谨充实的工作不能简而概之，本文以都市实践关于妈祖北区及周边的城市设计为例介绍规划研究成果。

妈祖北区及周边（图 11- 图 13）呈现一种混杂状态，包含龙岗区内许多城中村所面对的共同问题，但又因为这些问题的交织而出现当地特定的新问题。片区西、南

图 4 龙岗区甘坑新村，2021
Figure 4 New Gankeng Village, Longgang District, 2021

图 5 都市实践设计的“运动森谷”公园已落成，甘坑新村，2023
Figure 5 The "Sport Valley" park designed by URBANUS was built in New Gankeng Village, 2023

面均为郊野公园，隔山分别靠近龙岗中心城和宝龙科技城，东面临近深圳坪山高铁站，北面则临近深汕路和规划地铁线。在这种大尺度的自然生态、城市建设和基础设施之间，片区内部有着多种土地利用类型，并且特殊地包括大量的运动设施和大片的“空地”，呈现参差乃至无序的表象。

在片区中，始建于 1825 年、现今仍保存完整的大田世居无疑是现代城市中一种特殊的建筑类型，而正是它记录了该地的起源。历史上，陈氏一族几经迁徙，清代在龙岗墟后尾坜村（位于世居东北面数公里处）定居，建陈家祠，并创立源盛商号；而后定居大田村，建造大田世居，周边亦得名陈源盛村。

从建筑类型、土地权属、运营使用梳理区域现状，大田世居是典型的客家围屋，有三堂、两横、四角楼，前有月池，后有风水林；归属陈氏理事会，由深圳智慧文博运营，计划活化为文博综合体。源盛村拥有范围不小的用地，但仅保留紧邻世居西北角的六七层高的自建房自住和出租；风水林用作社区公园；其余用地出租给工厂，为成片多层板楼和铁皮棚房，设有公共活动空地。世居北面不远处有三块外卖地——妈祖北村、妈祖南村、源盛新村，均为六七层高的自建房，以出租为主（图 14）。这些居住区与原特区内的城中村不同，建筑间距一般可达三四米（在某种意义上，这种形态打破了关于深圳城中村“握手楼”的无差异印象），甚至有未建楼房的宅基地空地，部分用作篮球场和停车场。片区西侧有一块狭长的外卖地，为 6~14 层的独栋自建房，形成界面连续“大板楼”，商住混合；一块大面积的外卖地，用作农田和采摘园；一块集中的国家储备用地，暂时闲置。

若以现代城市规划的各项指标来评价，妈祖北区及周边无疑会表现为无序。然而，若从“前规划状态”来重新评价这个区域，可以发现其中的潜在结构。例如，虽然数个居住区均为自建房，但建筑形态和居住人群有所不同。源盛村为陈氏自留地，近似花园式小区，住宅每层居住面积约 120 平方米，以村民自住和长租家庭为主。源盛新村、妈祖北村、妈祖南村均为外卖地，住宅以年轻人租住为主，部分楼栋包含商业。大板楼亦为外卖地，靠近城市道路，建筑底层多用作商业，上部住宅以家庭长租为主，也作为厂工宿舍。可见其发展既遵循历史脉络（村民居住在最靠近大田世居的地方），也遵循经济逻辑（建筑功能考虑商业界面、消费客群等因素）。居民对各个时期的情况作出自发应对，因此提供了多样化的选择。

片区因一定程度的“共生”而形成的“高像素”便是此次城市设计的基础。基于对其发展逻辑的理解和价值的发掘，都市实践提出“跨过”当前的格式化规划，以城市设计对其历史、现状和未来提供衔接的可能性。根据片区内各个组团的具体情况，在现阶段提出四项计划。

针对以大板楼为主的村上城组团，社区营造计划从利益平衡的角度探索增量发展模式。大板楼密度高、体量大、权属复杂，设计通过加建具有建筑结构和公共服务设施双重功能的平台，为植入社区功能提供实体空间。设计通过政策引导底层界面升级，并局部纳入公共空间系统，例如架空部分空间作为城市厅堂，接驳轻轨；顶层加建平台，部分面积作为架空部分的返还，部分用作社会保障性住房的开发，部分提供公共服务配套。多个

图 6-7 龙岗区白石塘村，2021
Figure 6-7 Baishitang Village, Longgang District, 2021

主体可在此进行开发，包括政府建设基础设施提升公共空间、业主自主更新、政府与企业合作开发、政府与业主合作开发。这个过程也在探索政策许可、资本投入和空间出让的平衡点。

妈祖北组团位于公共交通枢纽、国家储备用地、妈祖庙和工厂中间，将其定位为多功能的复合社区，以回应现有的多种面向和未来的综合开发。通过社区营造计划，设计打开地面层，保持通透性，增强区域联系；在闲置的宅基地植入公共服务设施，打造口袋广场。通过微产业培育计划，设计改造工厂为创客中心和超级集市，并增加屋顶花园、能力实验室、共享宿舍等提供交流空间。通过多个空间改造的融合，将妈祖北发展成立体聚落，进而辐射周边（图 15）。

目前闲置的国家储备用地作为城乐园组团，城市服务核塑造计划将其作为区域发展的引擎。设计将组团作为服务组群的中心公园；在地下及半地下层设置停车场，以满足周边相对高密度的社区停车和人防需求；在地面层结合流线及自然景观资源，设置灵活多样的活动场地，包括水景、舞台、游廊、运动场等。

对于城中村、工厂、风水林、农田、运动场之间的边园组团，边界重塑计划使之串联周边的多元社区。设计在农田标高界面提供老人和儿童玩耍的空间；在工厂标高界面提供工人休憩驻足的空间，具有良好的视野，可远眺大田世居和大片田野；通过轻质的栈桥联系各个标高，缝合边界，并提供社区交流空间。

妈祖北区及周边集合众多城中村的共同问题，也就意味着需要汇聚多种解题思路，并且要求它们相互协调和补充（图 16）。当城 / 村计划在甘坑新村、凉帽村、马蹄山村、白石塘村、上雚村五村的研究和实践探索推导至极致状态时，将这些成果拼合在妈祖北区及周边，即成为一个理论模型——“极度差异化的城村”（图 17）。二十多年后，这与库哈斯在新世纪之初对珠三角城市所下的定义——“极度差异化的城市”形成互文。在这种意义上，城 / 村计划对妈祖北区及周边提出的发展模式即是“共生模式”的雏形。在如此模式下继续演变的城村将有可能成为中国城市土地利用的第三种模式，以破解当下村和城的困境。[3]

5　异质与共生

2021 年 5 月 20 日，城 / 村计划各参与方在水围村柠盟人才公寓 5 楼的青年之家参加联合工作坊，也作为第一次非正式的阶段成果汇报。此次工作坊亦发出明确的信号，其选择的发生地点即是一种新模式的探索案例——以人才保障房社区为导向的城中村改造试点项目。

在无法等待的（再）城市化情境下，城 / 村计划的行动者们需要一边厘清概念，一边改造现状。这意味着城村概念由于其载体和环境的不稳定而一直处于演变的状态，对其更新面临着历史文化原真性和未来城市愿景的双重诉求；同时，这也揭示了城村概念的本质，其始终不变的正是在直接、快速地应对现实时的持续演变，这也是此次计划的立足点之一。基于此，城村更新作为一种发展过程，而非最终的（消费）产品，无法定量描述。这无疑是城 / 村计划所面临的限制和挑战，然而，这也正是城 / 村计划发起的原因——以实际行动应对城市困境，寻求城市的另一种未来图景。

城村的异质性与它们在历史上的地理位置——南方和在近代的相对位置——特区密切相关，这导向了重重对照：新村及老村，村与城，人造环境与自然环境等，以及它们之间的多重边界。这些边界构成如今已被广泛使用的“城中村”这一名词的核心定义，直接或间接地导向城村的差异，甚至是表象上的对立，也在一定程度上导向当下关于都市性的理解和叙述。城 / 村计划提出的“共生模式”正是试图消解这种异质的非法性，使之成为未来城市中合规且常规的存在，以高像素的在地城市化样本回应无特质的规划城市。

或许当“共生”状态实现时，“城村”也将成为一个有着独立定义的名词，既生于村庄，又不同于村庄；既长于城市，又不同于城市；届时，城村也将孕育一种新的都市性。

3　孟岩，城 / 村计划联合工作坊，水围村，2021 年 5 月 20 日。

Ongoing City/Village Initiative

Luo Yiqian
URBANUS Researcher
Image ©URBANUS

1 Beginning of City/Village Initiative

At the fortieth anniversary of Shenzhen Municipality in 2020, the City/Village Initiative was formally proposed, concerning its two major constitutes, city and urban villages. With a standing point on public, academic, and practical aspects, the chief initiator and chief organizer Zhou Hongmei invited the core curatorial team, including chief curator Meng Yan and collaborated curators Liu Heng, He Jianxiang, and Chris Lai, to form an inner ring, and the academic committee to form an outer ring.

"In flying time, seize the days and nights." The Initiative was shaped in this critical temporal and spatial nexus. In fact, it had emerged within more than a decade of research and practice. In 2005, the first Shenzhen Biennale staged a themed exhibition on urban villages, including the redevelopment proposals for Gangxia, Fuxin, and Daxin Villages. The EXPO 2010 presented Shenzhen to global audiences via an urban village Dafen. The city's heterogeneous form attracted extensive domestic and global attention. Subsequently, a number of urban villages in the center of the city sparked heated discussions along with the reform projects, especially those urban villages located along Shennan Boulevard (Caiwuwei, Gangxia, Baishizhou, and Dachong). Via the 2017 UABB, society became aware of the Nantou main venue, responding positively to its dual identity as an old town and an urban village in the contemporary city. Subsequently, the relevant policies on urban village transformation changed. Although without an official name, "every step of the practice is part of the City/Village plan."[1]

2 The Outer Districts in the Post-2017 UABB Era

There have existed clear "time differences" in Chinese urbanization, whether we are speaking of the vast territory of the whole country or of Shenzhen's municipal area, which is less than 2,000 square kilometers.[2] Only in 2010 did the SEZ include Bao'an and Longgang Districts within its purview. During the first three decades of Shenzhen Municipality, its urbanization was divided into two "timezones"—inside-SEZ and outside-SEZ, which are referred to as the inner and outer districts, respectively. These zones were out of sync, following their respective policies and regulations. A decade after the expansion of the SEZ, the calibration of these "timezones" has been profoundly impacted by their developmental histories, which have informed their respective spatial and social structures.

Within and against these time differences, there have also been multiple shifts in the research on and practice focus in Shenzhen

1 Interview with Zhou Hongmei on March 30, 2021.
2 Interview with Meng Yan between 2020 and 2021.

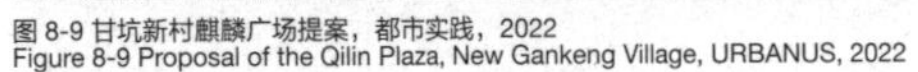

图 8-9 甘坑新村麒麟广场提案，都市实践，2022
Figure 8-9 Proposal of the Qilin Plaza, New Gankeng Village, URBANUS, 2022

urban villages. The trajectory expands from the attention to inner district urban villages due to limited land supply, functional division, and other factors, then the international academic frontier discussion when the Shenzhen identity was constructed at the EXPO via the outer district urban village Dafen, to the game frontier of the government and big capital and heated social topic of inner district urban villages adjacent to the center of the city due to several motives including the City's inventory development, industrial transfer, and image shaping, as well as real-estate enterprises' profit chasing. In 2019, after a decade of news reports on renewing Baishizhou, tenant evictions began. This suggested the end of the urban villages in the inner districts, as they made way for the city. And then the hovering parties shifted their battlefields. At the same time, there were 499 outer district urban villages in Longgang District. They have different origins and status quos, have experienced fewer powerful interventions because of their lower potential, and thus face both development opportunities and decline crises. In this new context, Longgang District has once again become the spatial focus of urban village issues. Importantly, we are no longer only debating a collective that was once represented by Dafen Village, but rather hundreds of individual urban villages of various types and specific attributes.

In the post-2017 UABB era, urban villages have become a necessary issue in the research and practical exploration of Shenzhen urban development. On the one hand, given the large number of urban villages in the outer districts, their definition needs to be further clarified because some are not surrounded by urban space, but rather adjacent to mountains and rivers; some not only comprise the typical fabric and architecture of urban villages, but also farmland, enclosed houses *(weiwu)*, and unbuilt homesteads, all of which are uncommon in the city proper, and; some have accommodated both recent reform and opening up migrants and historical Hakka migrants. On the other hand, their value is constantly being defined, as they play different roles in the mission to construct the pilot demonstration area of socialism with Chinese characteristics, to develop the Greater Bay Area, to build an affordable housing system, and to realize new growth by profit-seeking real-estate enterprises. Although the narrative of urban villages has shifted from a single to a heterogeneous perspective, nevertheless further debates on "what are problems and what are solutions" have become entangled.

3 The Mechanism of City/Village Initiative

Based on an urgent and critical context of reality, the first stop of the City/Village Initiative landed in Longgang District. Depicted by Zhou Hongmei, the chief initiator and chief organizer, the project aims to "embrace innovative thinking to challenge the conventional, to redefine problems, and to propose creative solutions."

The City/Village Initiative comprises urban research, urban design and curatorial proposal, pilot implementation, and paradigm promotion. Shenzhen promoted new models, such as open competition, UABB, and New School Initiative in order to break rigid management boundaries. Similarly, the City/Village Initiative adopts urban curation as a public event to intervene in old district regeneration, seeking legitimate opportunities under current policies and regulations. Through the agency of the exhibition, the initiative will present and discuss urban issues and overtly display resolving schemes in order to comprehensively consider professional critiques and public opinion. The initiative aims to establish more balanced value and evaluation systems as well as to introduce more holistic professional resources and public participation, therefore forming a strong motive for creation. In this way, urban village regeneration stands on the (social and environmental) publicness of the city and its relevant academic and/or urban topics, avoiding being misled or even distorted by the single-minded pursuits of political achievement and commercial benefit. At the same time, the City/Village Initiative attempts to innovate the agent-construction mechanism, listing the group of curators/designers, participating architects, and artists as an important condition to ensure overall coordination during project bidding.

The inner circle curatorial team is also responsible for the design of urban village transformations, proposing and integrating, studying materials, conducting research and design, and organizing and promoting participating resources, as well as guiding and coordinating the research and design of different disciplines, organizing, managing and supervising subsequent implementation. Parallelly, the outer circle academic committee plays the role of a think tank and jury, to guarantee the academic orientation of the project. It also constitutes a normalized mechanism for public participation, providing a path for professionals of public standing to intervene in urban planning and guard the public interest.

Based on the mechanism, the "City/Village Initiative: Practice Exhibition of Six Urban Villages in Longgang District" (Figure 1) became the entry point of the first stop with an eye toward incrementally activating and improving pilot urban villages. Along with regeneration opportunities, the initiative will introduce global cultural resources into urban villages through art events and public activities. Thus, it aims to create conditions of co-existence, integrating culture and daily and creating a new landscape of civic humanities. The project has selected six urban villages as the pilot areas for the initiative: Baishitang Village, Gankeng New Village, Liangmao Village, Matishan Village, Shangshe Village, and Mazu North and its surroundings (Figure 2). Villages were selected based on the following principles: first, they are not included in urban renewal or demolition and replacement plans that will be implemented before 2025; second, they have convenient traffic and moderate building numbers; third, they are located near good-quality urban areas, and; fourth, they have public spaces. In addition, these villages have different characteristics, representing dozens of urban villages with similar conditions throughout Longgang District. Therefore, the pilot areas can accommodate the holistic innovation mechanism in a more pioneering, experimental, and feasible way, leading to portable paradigms.

4 City/Village Initiative: Practice Exhibition of Six Urban Villages in Longgang District

As an approach, the City/Village Initiative elaborates on the concept of City/Village coexistence, exploring alternative urbanism in China through on development model. In terms of the Practice Exhibition of Six Urban Villages in Longgang District, the curators drew a theme "Dwelling in the City/Village, Growing in Multi-dimensions" and put forward a manifesto:

The evolution of urban villages resists the city's homogeneity with villages' heterogeneity;
Urban villages are the Alter Ego of the contemporary city;
The city/village is an alternative dwelling in China's future cities.

Under this consensus, chief curator Meng Yan and collaborated curators Liu Heng, He Jianxiang, and Chris Lai with their teams (URBANUS, NODE, O-Office, and DOFFICE, respectively) have started urban research

图 10 白石塘村“邻里造园”提案，都市实践，2022
Figure 10 Proposal of villagers' moving path, Baishitang Village, URBANUS, 2022

on and design of the six pilot villages and Tsinghua Yuan have studied urban planning and policy, working together to reconstruct a vision and push forward implementation at several levels. The goal is to present the public with three exhibitions, touching on intervening practice, themed documentary, and small display and intervention, respectively. In this sense, the six Longgang villages are venues, the venues are exhibits, and the exhibition experience and daily life co-construct a hybrid of realities.

4.1 City/Village Research: Diagnose in Multiple Dimensions

The City/Village Initiative approaches research on the six villages through observation and findings. The basic condition of each village will be determined through fieldwork, investigation into intangible cultural heritage, interviews with local worthies, interviews with enterprises, and community workshops (Figure 3). Located in four sub-districts, the six villages show distinct characteristics and are located in different environments, requiring different urban plans. Accordingly, the City/Village Initiative proposes individualized orientation and direction for the development of each village, attempting to resolve their problems with deliverable targets and to build a foundation for their overall improvement.

Gankeng New Village (Figure 4-5) and Liangmao Village are adjacent to Gankeng Hakka Town, which is listed as a national 5A-level tourist attraction. Gangkeng was rebuilt onsite, has kept alive traditional Hakka Qilin dance, and has the potential to transform from a cultural symbol into a creative incubator. Liangmao means "cool hats *(liangmiao)*, and has the dual roles of village and factory." It is enclosed by mountains on two sides, allowing for links between the natural environment and cultural production, as well as historical memory and daily life. Matishan Village abuts Ban Xue Gang High-Tech Zone (formerly Huawei High-Tech City), a "demonstration area of expanded SEZ." The village itself is enclosed within main roads, which provide convent access and allow for large traffic flows, predicating the incubation of a new form of city/village through social capital. Bashitang Village (Figure 6-7) is a "village amidst factories." It is surrounded by labor-intensive manufacturing factories, has a new village south to the old village and a city road that cuts across the south of the new village. It has the potential to transform from an arrival node into an empowerment village. Shenshe Village, adjacent to Longgang River to its north and surrounded by mountains on the other three sides, is planned as the woodland. Nevertheless, its extant buildings and remains mean that through proactive design, Shenshe can develop incrementally. Mazu North and its surroundings comprise highly hybrid pixels, including Datian Shiju, farmland, recreation facilities, industrial parks and urban villages. This layout allows for transformation into a "Mixed Pixel" city, generating diversity through re-combination and opportunities by juxtaposition. Each village has specific attributes and thus faces different opportunities and challenges. The indiscriminate renewal model ignores a village's specific advantages and liabilities, often eliminating what it contributes to the city and even creating new urban issues.

In fact, all six villages not only trace their roots back hundreds of years, but have also experienced huge changes over the recent twenty years due to the demands of urbanization and inhabitants' spontaneous responses to it. For instance, originally built during the Qing Dynasty, Baishitang Village is surrounded by mountains on three sides, with a valley to its south and Dingshan River to its west. Baishitang literally means "white stone pond," referring to the blessing bestowed by the two white stones that were found when the pond was dug. Historically, the village's moon pond *(yuechi)*, foyer, and the Zeng Clan Ancestral Hall comprised the central axis of the enclosed village, while village homes were laid out with three main rooms. As the population increased, the village enclosure was expanded through the construction of additional housing, and the Xiao Clan Ancestral Hall was built. When villagers moved away, some enclosed houses were demolished and new houses were built in the south. Before 2004, when Shenzhen achieved full urbanization, Baishitang comprised low-rise buildings standing in the new village to the south of the old village, a country road linking the north to the south, a moon pond in front of the old compound, and a fengshui woodland at the foot of the mountain. By 2008, as the city began preparations for the 2011 Shenzhen Universiade, Baishitang had taller houses in the new village, a reduced fengshui woodland, and the moon pond had been reclaimed. Nearby, factories had been built in the surrounding area and Yanlong Boulevard was constructed. As of 2021, a city road bisected the original country road and residents had been relocated to a neighborhood outside the village to the west, thereby disaggregating the village into three sections. Like other villages, Baishitang has weathered expansion and

图 11 龙岗区妈祖北区及周边，2021
Figure 11 Mazu North and its surroundings, Longgang District, 2021

图 12 龙岗区源盛村大田世居，2021
Figure 12 Datian Shiju, Yuansheng Village, Longgang District, 2021

decline. Clues to this history include its ancestral hall, Tudigong Shrine and Mazu Temple. During festivals and celebrations, the Qilin team first performs the ritual in the ancestral hall and then visits village homes. The procession includes relocated villagers and, when invited, the Qilin team will also call on new village houses, even housing that is located 15 kilometers away from the historic settlement. To some extent, cultural activities have enabled Baishitang Village to maintain spiritual connections within and despite spatial separation, isolation, and decline, breaking through the limitations of physical boundaries; the attributes of the urban village have actively ameliorated urban problems.

Intertwined issues were found in the six villages. In resistance to the extreme approaches of either "demolishing and reconstructing" or "ignoring" a village, the City/Village Initiative proposes constructing a "Co-existence Model" that protects inhabitants' rights to both survival and development in both the city and its villages. The pilot project has to transform the physical spaces in order to make orientation and improvement for the area, as well as to prevent stakeholders from illegal and rapid construction in the name of "regeneration."

Having considered current implementation paths, the City/Village Initiative put forward the following suggestions to amend policies and regulations at different levels, and ensure project progress. At the macro level, the initiative suggests taking into consideration the *Urban Renewal Regulations of Shenzhen Special Economic Zone*, which were promulgated and implemented in 2021 and collaborating with relevant bureaus to develop implementation details for "comprehensive remediation 2.0" as a guiding policy. At the middle level, the initiative has three recommendations: employing a targeted guide for the urban village when it lacks development chances in its upper-level planning; when upper-level planning conflicts with the village situation, adjusting statutory planning, and; examining and correcting current planning with the unit delineation of statutory planning and territory spatial planning. At the micro level, the initiative suggests determining land ownership and allowing the land for new-built architecture and public space to be transferred to the joint-stock company after the government has signed with the villager collective or individuals; determining property ownership and clarifying the legitimacy of partially-reconstructed buildings, and; legitimating illegal buildings with respect to the three new regulations.

The City/Village Initiative attempts to secure the public interest through intervening in spatial design, as well as to introduce social design, putting forward three basic points: first, to re-define public space, connecting community life past and future; second, to re-create spatial production, as the re-creation of both spatial products and social relationships, and; third, to re-balance spatial development, safeguarding history, inheriting memories, and seeking new opportunities for sustainable development. The initiative aims to regenerate urban villages via interdisciplinary ideas, transforming single and partial buildings to lay a foundation for overall improvement and avoiding stopgap measures. As a new spatial and cultural strategy, the City/Village Initiative attempts to find a model that facilitates the simultaneous unfolding of natural ecology, cultural ecology, social community, and economic development onsite.

To achieve these goals, the City/Village Initiative proposes four spatial prototypes. The first prototype is the new city/village ancestral hall, which takes into consideration the cultural core of each village, reshaping it in addition to documenting lost stories. In turn, this will develop daily life and urban belonging via a new cultural core with ritual space, reconstructing the spiritual space of a contemporary city. In Gankeng Village, for example, the Qilin pavilion and its plaza are reconstructed as a complex containing the cultural core and community services—a Qilin theatre, museum and practice court, village history hall, library, chatting room, and outdoor venue, as well as a performance node during the Qilin procession (Figure 8-9). The second prototype is the city/village complex, which approaches land planning and usage from a three-dimensional refined space and operations perspective, rather than from traditional two-dimensional land planning. The third spatial prototype is city/village infrastructure, transforming the historically separated infrastructure construction and management of the city and villages into collaboratively-built and dynamic properties. The fourth prototype is the novel city/village commune, which makes the top-down introduction of resources the driving force for improved quality of life and community empowerment. The new commune includes the construction of new social relationships, individual empowerment, exchange behaviors that connect production and consumption, spatial construction for these exchanges, and spatial organization and resource linking, thus ultimately increasing local cultural capital by accumulating social capital. In Baishitang Village, for example, the initiative is building a cluster of ability laboratories in the boundary area between the new and old villages, simultaneously connecting empowerment workshops, such as the idea salon, idea workshop, Baishi book house, and ability house, and distributing the experimental field of the workshops, book kiosks, farmland, gardens, water pond, and art installations (Figure 10). The City/Village Initiative also proposes the city/village development community as a prototype in terms of policy and regulation and the novel city/village economy in terms of commercial models, promoting mechanism innovation through different approaches.

图 13 龙岗区妈祖南天后古庙，2021
Figure 13 Tianhou Temple, Mazu South, Longgang District, 2021

4.2 City/Village Design: The Coexistence Model

Based on research results, curators and their design teams started the design work to improve the quality of each of the six villages in the City/Village Initiative. Each team approached the project through their strengths, considered the fabric and quality of each village, and proposed an individualized strategy and spatial design for each village, making precision strikes. Each team has conducted robust and rigorous work that is impossible to simplify. This essay will use URBANUS' urban design for Mazu North and its surroundings to illustrate this stage of the program.

Mazu North and its surroundings (Figure 11-13) manifest a hybrid state, containing many problems faced by the majority of Longgang urban villages as well as new problems that have arisen due to the interaction of these problems. The area is situated next to suburban parks to the west and south, and across the mountains, Longgang Central City and Baolong Technology City, respectively. Shenzhen Pingshan Railway Station is located to the east of Mazu North, while Shenshan Road and a planned metro line lie to its north. Within this large-scale natural ecology, urban construction and infrastructure, the area accommodates multiple types of land use, including sports facilities and large plots of "void space," presenting an irregular and even disordered appearance.

The most special landmark in the area is the Datian Generational Compound *(shiju)*, which was first built in 1825. Datian is a typical Hakka enclosed compound, with three main houses *(tang)*, two lines of row houses *(heng)*, and four towers *(jiaolou)*, a moon pond in the front, and a fengshui woodland in the back. Datian not only gives this urban space its special feeling, but it also records its origins. During the Qing Dynasty, the Chen family migrated several times, finally settling in Houweili in Longgang District, which is located several kilometers northeast of the generational compound. They built the Chen Ancestral Hall and founded the Yuansheng business house. Subsequently, they moved to Datian, establishing the eponymous village and building Datian Generational Compound. The surrounding area was named Chen Yuansheng Village, but over time the name was simplified to "Yuansheng Village."

In terms of the building types, property rights, and management and use, the status quo is as follows. The Datian Generational Compound belongs to the Chen Clan Council and is managed by Shenzhen Wisdom Wenbo, which plans to activate it as a culture and exhibition complex. Yuansheng Village owned a wide scope of land, but today only retains 6-7 story residential and rental properties that are located to the northwest of Datian. The Yuansheng fengshui woodland is now a community park, and the rest of the land has been rented to factories which have multi-story plants, shed houses and open spaces for public activities. North of Datian Compound are three plots of land—Mazu North Village, Mazu South Village, and New Yuansheng Village, which were sold to build 6-7 stories self-built rental properties (Figure 14). These residential areas are different from inner district urban villages. (To an extent, this typology breaks the stereotype of "handshake buildings" of Shenzhen urban villages.) They are situated 3-4 meters away from each other, and there are even unbuilt homesteads, some of which are used as basketball courts or parking lots. In the west of the area, there is a long and narrow stretch of land that was sold for commercial and residential mixed-use, forming a "big slab building" that comprises individual 6-14 stories self-built buildings. There is also a large area of land that was sold for farmland and a picking garden, and a reserved plot of land that is temporarily empty.

If Mazu North and its surroundings are evaluated by the indicators of modern urban planning, it is undoubtedly disordered. But when re-evaluated from a "pre-urban planning" perspective, its underlying structure is revealed. For example, although the residential areas within have self-built buildings, they have different architectural appearances and inhabitant groups. Yuansheng Village is retained by the Chen Clan, similar to a garden community, where the buildings have a living area of approximately 120 square meters of each floor, mostly occupied by villagers and long-term family renters. In contrast, New Yuansheng, North Mazu, and South Mazu Villages were sold off, and the buildings are mostly rented by young people and some provide commercial functions. The "big slab building" was also sold off. It is located close to the city road, and its ground floor is mostly commercial while its upper floors are rented by stable families or used as workers' dormitories. The development of the area manifests both its historical context (villagers living in the place closest to Datian Shiju) and economic logic (building functions taking into account commercial interfaces and consumer groups). In each historical period, the inhabitants

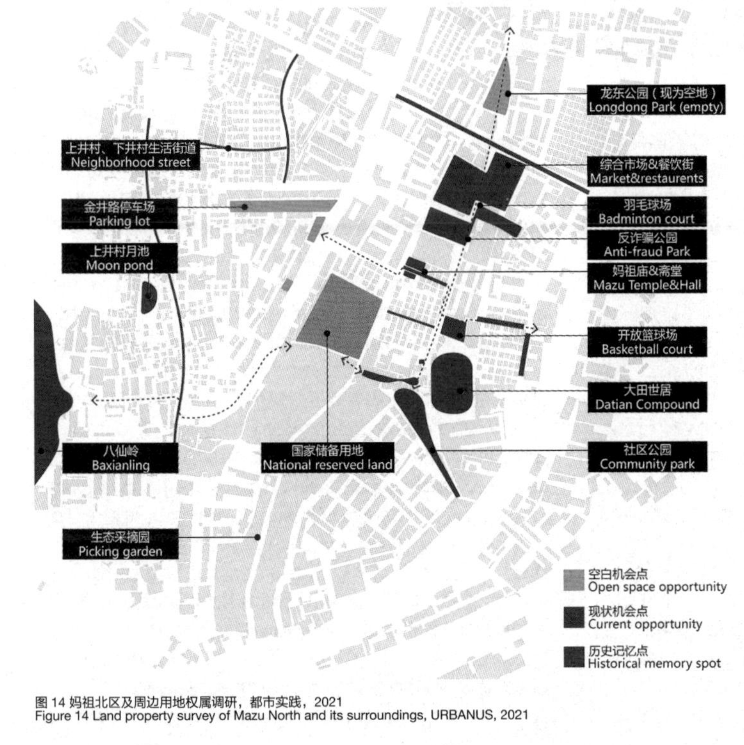

图 14 妈祖北区及周边用地权属调研，都市实践，2021
Figure 14 Land property survey of Mazu North and its surroundings, URBANUS, 2021

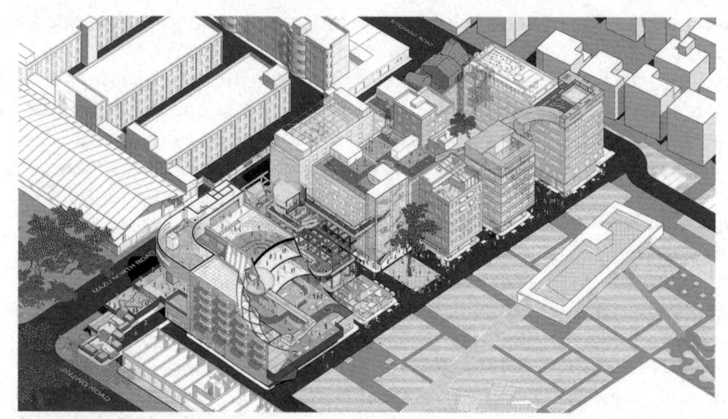

图 15 妈祖北立体街巷提案，都市实践，2022
Figure 15 Proposal of three-dimensional streets, Mazu North, URBANUS, 2022

图 16 妈祖北改造愿景，都市实践，2022
Figure 16 Vision of Mazu North, URBANUS, 2022

responded spontaneously to the actual situation, which in turn resulted in diverse choices.

A certain degree of "coexistence" informs the area's "high pixel" layout, which in turn becomes the foundation of the urban design project. Based on the understanding of its development logic and the exploration of its value, URBANUS proposes to "cross" the extant format, adopting an urban design to provide possible connections between the area's past, present, and future. URBANUS put forward four plans, which are based on the specific conditions of corresponding clusters.

A community empowerment plan is applied to the big slab building, which forms an urban cluster above the village. The goal is to explore inventory development from the perspective of balancing interests. The big slab building has high density, large volume, and complicated property rights. The design provides physical space for embedding community functions by adding a platform with dual functions of building structure and public service facilities. Policy guides interface upgrading on the ground floor and incorporates a part into the public space system (e.g. elevating some spaces for city hall and connections with the light rail). The design adds platforms on the top floor, which are used as compensation for space occupied by elevations, for the development of affordable housing, and in support of public services. Multiple entities participate in the design, which includes government infrastructure building and public space upgrading, owner's self-renewal, and collaborations between the government and enterprises or owners. This process also explores the balance between policy boundaries, capital investment, and space transfers.

The Mazu North cluster is located in the middle of a public transport hub, a plot of national reserved land, the Mazu Temple, and factories. It is positioned as a multi-functional composite community with the capacity to respond to existing conditions and comprehensive future development. Through the application of the community empowerment plan, the design opens up the ground floor of the building, keeping its porosity and enhancing connections. It inserts public service facilities on empty homesteads and builds pocket squares. Through the application of a micro-industry incubation plan, the design transforms the factories into a maker center and a special market and provides communication spaces, including a roof garden, capacity laboratory, and shared dormitory. Through the integration of these transformed spaces, Mazu North can develop into a three-dimensional settlement, and then radiate into its surroundings (Figure 15).

Through the urban service core shaping plan, the urban park cluster, consisting of the empty national reserved land, becomes an engine of regional development. The design is to make it a central park of the service group, placing parking lots on the underground and semi-underground levels; meeting neighborhood needs for parking and civil air defense for the neighborhood, and; in relation to people's flows and natural landscape resources on the ground level, setting up flexible venues for diverse activities, including a waterscape, stage, gallery and playground.

With respect to the border garden cluster comprising urban village housing, factories, fengshui woodland, farmland, and sports facilities, the design suggests applying a boundary reconstruction plan to connect with the nearby communities. The design provides recreational spaces for the elderly and children at the level of the farmland as well as resting spaces for workers at the level of factories, where there is a good view overlooking Datian Compound and vast fields. With lightweight bridges, the design links up different levels, sews up gaps, and provides communication spaces.

Mazu North and its surroundings contain common problems faced by the majority of urban villages, which indicates the need to integrate a variety of coordinating and complementary problem-solving ideas (Figure 16). If the research and practice of the City/Village Initiative achieve ideal conditions in Gankeng New Village, Liangmao Village, Matishan Village, Baishitang Village and Shangshe Village, then the results can be collaged in Mazu North and its surroundings. It is the formulation of a theoretical model—the city/village of exacerbated difference (Figure 17), raising intertextual concerns two decades after Rem Koolhaas's definition of the Pearl River Delta cities (CITY OF EXACERBATED DIFFERENCE©). In this sense, the development model of Mazu North and its surroundings proposed by the City/Village Initiative becomes the prototype of the "Coexistence Model." The evolution of the city/village under such a model has a chance to form a third model of land use in urban China, resolving the inherited either/or dilemma of village and city.[3]

5 Heterogeneity and Coexistence

On May 20, 2021, City/Village Initiative participants convened a workshop at the clubhouse on the fifth floor of a renovated youth community in Shuiwei Village, which was also the first informal presentation of their progress to date. By holding the workshop in a place that is part of a pilot project of urban village transformation oriented by an affordable housing community, the workshop also sent a message about its exploration of a new model.

In the context of unavoidable (re-)urbanization, the participants-in-action of the City/Village Initiative have to simultaneously clarify the concept and transform the situation. This means that the city/village has been evolving due to its unstable carrier and environs, and thus its regeneration must consider both the authenticity of historical culture and a vision of the city's future. This also reveals that the unchanging nature of the city/village is its constant change, its direct and rapid response to realities, which in turn become a basis for ongoing regeneration. Given this situation, the regeneration of the city/village is regarded as a developmental process, instead of an end product (for consumption); it is unable to settle in a quantitative narrative. This is the challenge and limit the City/Village Initiative faces, and also the reason it came into being—we respond to urban conundrums by our actions, seeking an alternative urban future.

The heterogeneity of the city/village is closely linked to its geographical location in history—the South—and its relative position in the modern era in the Special Economic Zone. This positioning has generated reference pairs, including new and old villages, city and village, and artificial and natural environments, as well as the borderlines between them. The borderlines have constituted the core definition of the widespread term "urban village (village amidst the city)," and directly or indirectly have led to differences, or even contrasts, between cities and villages, as well as to a certain extent, resulted in our current understanding of and narrative about urbanity. The "Coexistence Model" put forward in the City/Village Initiative attempts to resolve the putative illegality of this heterogeneity, making it the legitimate and normal existence of the future city, where the response to homogenous planned cities is in situ urbanization.

When the status of "coexistence" is realized, the "city/village" will become a term with an independent definition that originates from villages and grows in cities, but diverges from neither. At that moment, the city/village will engender a new concept of urbanity.

3 Meng Yan, City/Village Initiative Workshop, held in Shuiwei Village on May 20, 2021.

图 17 极度差异化的城村，都市实践
Figure 17 City/Village of Exacerbated Difference, URBANUS

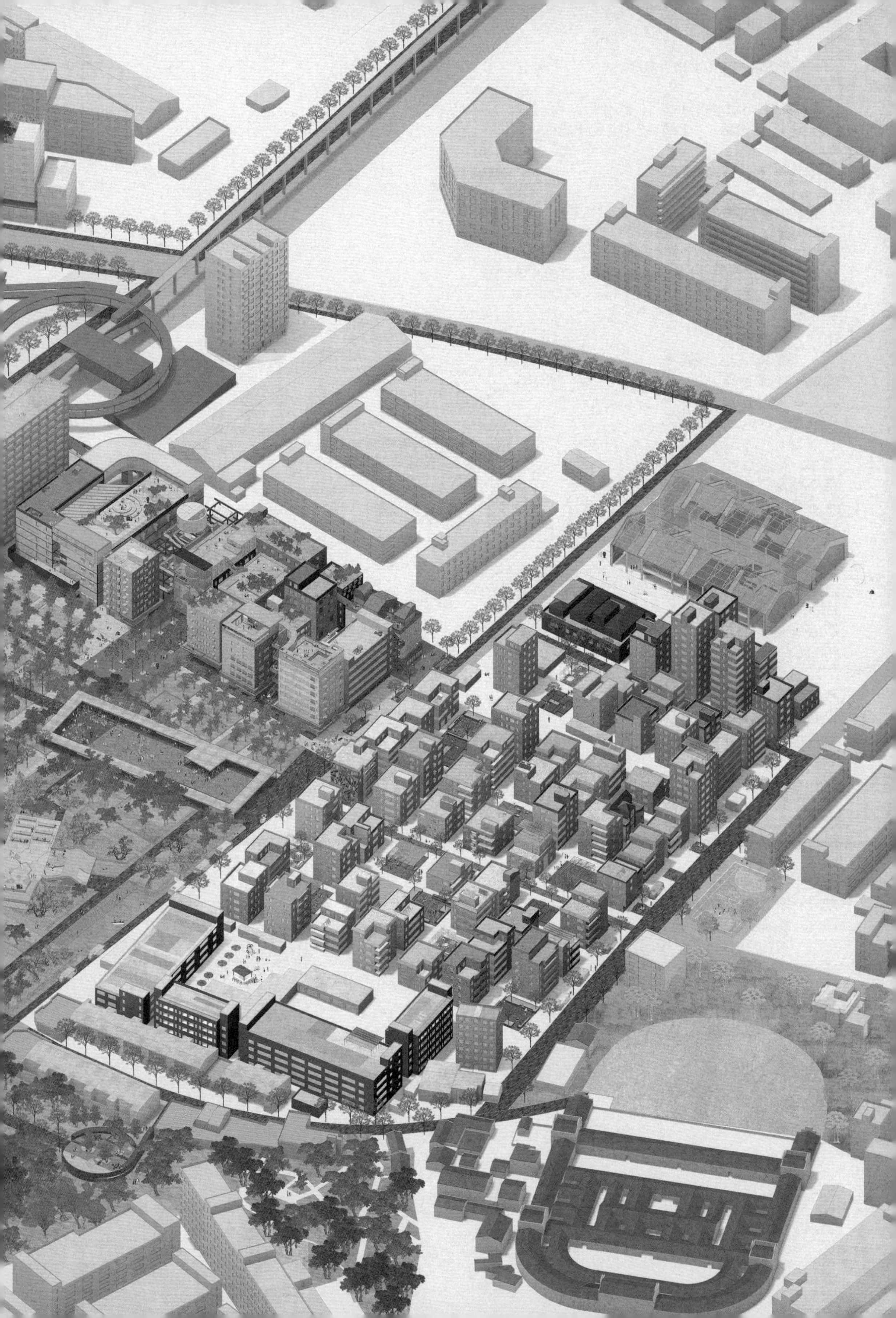

图书在版编目（CIP）数据

城村共生 = Urban Village Coexistence : 汉英对照 / 都市实践编著 . -- 上海 : 上海文化出版社，2024.3
ISBN 978-7-5535-2865-6

Ⅰ . ①城… Ⅱ . ①都… Ⅲ . ①城市规划 – 研究 – 深圳 – 汉、英 Ⅳ . ① TU982.265.3

中国国家版本馆 CIP 数据核字 (2023) 第 220509 号

主　　编 孟岩
编　　辑 莫思飞、罗祎倩
编辑顾问 董超媚
英文审校 马立安
出版联络 张云、陈文赟

出 版 人 姜逸青
责任编辑 江岱
平面设计 huangyangdesign（黄扬、张永、吴诗怡）

Editor-in-chief: Meng Yan
Managing editor: Mo Sifei, Luo Yiqian
Editorial consultant: Dong Chaomei
Copy editor (English): Mary Ann O'Donnell
Public relation: Zhang Yun, Chen Wenyun

Publisher: Jiang Yiqing
Publishing editor: Jiang Dai
Graphic design: huangyangdesign (Huang Yang, Zhang Yong, Wu Shiyi)

书　　名 城村共生
作　　者 都市实践
出　　版 上海世纪出版集团 上海文化出版社
地　　址 上海市闵行区号景路 159 弄 A 座 3 楼　201101
发　　行 上海文艺出版社发行中心
地　　址 上海市闵行区号景路 159 弄 2 楼　201101
印　　刷 雅昌文化（集团）有限公司
开　　本 787×1092　1/16
印　　张 25　彩插 72
版　　次 2024 年 3 月第 1 版　2024 年 3 月第 1 次印刷
书　　号 ISBN 978-7-5535-2865-6/TU.023
定　　价 218.00 元

告读者 如发现本书有质量问题请与印刷厂质量科联系。
联系电话：0755-86083235